IoT and Sensor Networks in Industry and Society

IoT and Sensor Networks in Industry and Society

Editors

Daniele D. Giusto
Virginia Pilloni

MDPI • Basel • Beijing • Wuhan • Barcelona • Belgrade • Manchester • Tokyo • Cluj • Tianjin

Editors
Daniele D. Giusto
University of Cagliari
Italy

Virginia Pilloni
University of Cagliari
Italy

Editorial Office
MDPI
St. Alban-Anlage 66
4052 Basel, Switzerland

This is a reprint of articles from the Special Issue published online in the open access journal *Energies* (ISSN 1996-1073) (available at: https://www.mdpi.com/journal/energies/special_issues/IoT_and_Sensor_Networks_in_Industry_and_Society).

For citation purposes, cite each article independently as indicated on the article page online and as indicated below:

LastName, A.A.; LastName, B.B.; LastName, C.C. Article Title. *Journal Name* **Year**, *Volume Number*, Page Range.

ISBN 978-3-0365-0216-8 (Hbk)
ISBN 978-3-0365-0217-5 (PDF)

Contents

About the Editors

Daniele D. Giusto Giusto received his Laurea (M.S.) degree in electronic engineering and his Dottorato di Ricerca (Ph.D.) degree in telecommunications from the University of Genoa, Genoa, Italy, in 1986 and 1990, respectively. Since 1994, he has been a permanent faculty member of the Department of Electrical and Electronic Engineering, University of Cagliari, Italy, where he became a full professor of Telecommunications in 2002. His research interests are in the area of smart cities, sensor networks and IoT, mobile and professional networks, and digital media. Dr. Giusto is a senior member of IEEE, was a member of the IEEE Standard committee (2007–2010), was the Italian delegate in the ISO-JPEG, and is the Italian delegate in the ISO-Smart Cities committees.

Virginia Pilloni received her PhD degree in Electronic and Computer Engineering with Doctor Europaeus mention from the University of Cagliari, Italy, in 2013. She currently is an Assistant Professor at the University of Cagliari, Italy. Her main research interests are focused on Internet of Things and Wireless ad-hoc networks, with particular attention to the improvement of their performance through task allocation. She has been an expert evaluator of EU H2020 and BELSPO proposals. She has been involved in several national and international research projects. She is an Editor of Elsevier Computer Networks and Frontiers in Communications and Networks. She is a member of the ITU-T SG20 on Internet of Things (IoT) and smart cities and communities (SC&C). She has served as organization chair and program committee member of several international conferences. She is a member of the IEEE ComSoc Member Services Board, and the IEEE ComSoc Young Professionals Standing Committee, where she currently serves as the awards chair for the IEEE ComSoc Best YP Awards.

Preface to "IoT and Sensor Networks in Industry and Society"

The exponential progress of Information and Communication Technology (ICT) is one of the main elements that fueled the acceleration of globalization. Internet of Things (IoT), Artificial Intelligence (AI), and big data analytics are some of the key players of the digital transformation that is affecting every aspect of daily life, from environmental monitoring to healthcare systems, from production processes to social interactions. In less than 20 years, everyday life has been revolutionized and concepts such as smart homes, smart grids, and smart cities have become familiar even to non-technical users.

The integration of embedded systems, ubiquitous Internet access, and machine-to-machine (M2M) communications have paved the way for paradigms such as IoT and cyber-physical systems (CPS) to be introduced in high-requirement environments such as those related to industrial processes, under the forms of the industrial Internet of Things (IIoT or I2oT) and cyber-physical production systems (CPPS). As a consequence, in 2011, the German High-Tech Strategy 2020 Action Plan for Germany first envisioned the concept of Industry 4.0, which is rapidly reshaping traditional industrial processes. The term refers to the promise to be the fourth industrial revolution. Indeed, the first industrial revolution was triggered by water and steam power. Electricity and assembly lines enabled mass production in the second industrial revolution. In the third industrial revolution, the introduction of control automation and programmable logic controllers (PLCs) gave a boost to factory production. As opposed to the previous revolutions, Industry 4.0 takes advantage of Internet access, M2M communications, and deep learning not only to improve production efficiency but also to enable the so-called mass customization, i.e., the mass production of personalized products by means of modularized product design and flexible processes.

Less than five years later, in January 2016, the Japanese 5th Science and Technology Basic Plan took a further step by introducing the concept of Super-Smart Society or Society 5.0. According to this vision, in the upcoming future, scientific and technological innovation will guide our society into the next social revolution after the hunter-gatherer, agrarian, industrial, and information eras, which respectively represented the previous social revolutions. Society 5.0 is a human-centered society that fosters the simultaneous achievement of economic, environmental, and social objectives, to ensure a high quality of life for all citizens. This information-enabled revolution aims to tackle today's major challenges such as an aging population, social inequalities, depopulation, and constraints related to energy and the environment. Accordingly, citizens will be experiencing impressive transformations in every aspect of their daily lives.

This book offers insights into the key technologies that are going to shape the future of industry and society. It is subdivided into five parts: Part I presents a horizontal view of the main enabling technologies, whereas Parts II–V offer a vertical perspective on four different environments.

Part I, dedicated to IoT and sensor network architectures, encompasses three chapters. In Chapter 1, Peruzzi and Pozzebon analyze the literature on the subject of energy harvesting solutions for IoT monitoring systems and architectures based on low-power wireless area networks (LPWAN). The chapter does not limit the discussion to long-range wide-area network (LoRaWAN), SigFox, and Narrowband-IoT (NB-IoT) communication protocols, but it also includes other relevant solutions, such as DASH7 and long-term evolution machine type communication (LTE-M). In Chapter 2, Hussein et al. discuss the development of an Internet of Things message protocol that supports multi-topic messaging. The chapter further presents the implementation of a platform, which integrates the proposed communication protocol, based on a real-time operating system. In Chapter 3, Li et al. investigate the heterogeneous task scheduling problem for data-intensive scenarios, to reduce the global task execution time, and, consequently, reducing data centers' energy consumption. The proposed approach aims to maximize efficiency by comparing the cost of remote task execution and data migration.

Part II is dedicated to Industry 4.0 and includes two chapters. In Chapter 4, Grecuccio et al. propose a solution to integrate IoT devices by leveraging a blockchain-enabled gateway based on Ethereum, so that they do not need to rely on centralized intermediaries and third-party services.

As it is better explained in the paper, where the performance is evaluated in a food-chain traceability application, this solution is particularly beneficial in Industry 4.0 domains. Chapter 5, by De Fazio et al., addresses the issue of safety in workplaces by presenting a smart garment that integrates several low-power sensors to monitor environmental and biophysical parameters. This enables the detection of dangerous situations, to prevent or at least reduce the consequences of worker accidents.

Part III consists of two chapters based on the topic of smart buildings. In Chapter 6, Petroșanu et al. review the literature about recent developments in the smart building sector, related to the use of supervised and unsupervised machine learning models of sensory data. The chapter focuses particular attention on enhanced sensing, energy efficiency, and optimal building management. In Chapter 7, Oh examines how much the education of prosumers about their energy consumption habits affects power consumption reduction and encourages energy conservation, sustainable living, and behavioral change, in residential environments. In this chapter, energy consumption monitoring is made possible thanks to the use of smart plugs.

Smart transport is the subject of Part IV, including three chapters. In Chapter 8, Roveri et al. propose an approach that leverages the small world theory to control swarms of vehicles connected through vehicle-to-vehicle (V2V) communication protocols. Indeed, considering a queue dominated by short-range car-following dynamics, the chapter demonstrates that safety and security are increased by the introduction of a few selected random long-range communications. In Chapter 9, Nitti et al. present a real-time system to observe and analyze public transport passenger mobility by tracking them throughout their journey on public transport vehicles. The system is based on the detection of the active Wi-Fi interfaces, through the analysis of Wi-Fi probe requests. In Chapter 10, Miler et al. discuss the development of a tool for the analysis and comparison of efficiency indicated by the integrated IT systems in the operational activities undertaken by road transport enterprises (RTEs). The authors of this chapter further provide a holistic evaluation of the efficiency of telematics systems in RTE operational management.

The book ends with the two Chapters of the Part on smart environmental monitoring V. In Chapter 11, He et al. propose a sea surface temperature prediction (SSTP) model based on time-series similarity measurement, multiple pattern learning, and parameter optimization. In this strategy, the optimal parameters are determined by means of an improved particle swarm optimization method. In Chapter 12, Tsipis et al. present a low-cost, WSN-based IoT system that seamlessly embeds a three-layered cloud/fog computing architecture, suitable for facilitating smart agricultural applications, especially those related to wildfire monitoring.

We wish to thank all the authors that contributed to this book for their efforts. We express our gratitude to all reviewers for the volunteering support and precious feedback during the review process. We hope that this book provides valuable information and spurs meaningful discussion among researchers, engineers, business people, and other experts about the role of new technologies in industry and society.

Daniele D. Giusto, Virginia Pilloni
Editors

energies

MDPI

Review

A Review of Energy Harvesting Techniques for Low Power Wide Area Networks (LPWANs)

Giacomo Peruzzi and Alessandro Pozzebon *

Department of Information Engineering and Mathematics, University of Siena, 53100 Siena, Italy; peruzzi@diism.unisi.it

* Correspondence: alessandro.pozzebon@unisi.it; Tel.: +39-0577-233-702

Received: 27 May 2020; Accepted: 30 June 2020; Published: 3 July 2020

Abstract: The emergence of Internet of Things (IoT) architectures and applications has been the driver for a rapid growth in wireless technologies for the Machine-to-Machine domain. In this context, a crucial role is being played by the so-called Low Power Wide Area Networks (LPWANs), a bunch of transmission technologies developed to satisfy three main system requirements: low cost, wide transmission range, and low power consumption. This last requirement is especially crucial as IoT infrastructures should operate for long periods on limited quantities of energy: to cope with this limitation, energy harvesting is being applied every day more frequently, and several different techniques are being tested for LPWAN systems. The aim of this survey paper is to provide a detailed overview of the the existing LPWAN systems relying on energy harvesting for their powering. In this context, the different LPWAN technologies and protocols will be discussed and, for each technology, the applied energy harvesting techniques will be described as well as the architecture of the power management units when present.

Keywords: energy harvesting; LPWAN; IoT; LoRa; LoRaWAN; Sigfox; DASH7; NB-IoT

1. Introduction

The rapid growth of Internet of Things (IoT) applications and markets has been the main driver for the emergence of a wide range of innovative data transmission technologies, whose main objective is to satisfy a different set of requirements with respect to the ones typically targeted by human-centered personal communication devices. In this sense, while the key features for traditional Internet-based systems are low latency, large bandwidth, and high bit rates, for IoT-based architectures the crucial requirements turn to be low power, low cost, and large transmission ranges.

In order to fulfill these requirements, in the last years several novel technologies have been developed, leading to the definition of a novel category of data transmission technologies, the so-called Low Power Wide Area Networks (LPWANs). As the name suggests, all the technologies belonging to this family aim at targeting two of the features listed before, i.e., the low power consumption and the large transmission range; nevertheless, while the third requirement, the low cost, it is not explicitly cited in the denomination, it is still fulfilled by every LPWAN technology as, within the IoT vision, which foresees the presence of billions of interconnected objects all around the world, the low cost of the devices as well as of the connections is intrinsically mandatory to make these technologies actually employable.

LPWANs currently include proprietary and open platforms, cellular and non-cellular technologies, systems operating in the unlicensed Sub-GHz bands as well as in LTE licensed bands [1]. While the most widespread LPWAN technologies currently are Long Range (LoRa), SigFox, and Narrowband-IoT (NB-IoT), a plethora of other solutions can be found on the market, and it is foreseeable that a large number of novel technologies will emerge in the next years with the pervasive adoption of IoT systems as well as with the emergence of novel cellular standards, not only within the upcoming 5G framework, but also with future sixth-generation technologies.

In general, as anticipated all LPWAN technologies fulfill the three requirements listed before, i.e., long range, low cost, and low power. Concerning long range, all LPWAN technologies are usually able to wirelessly transmit data from distances up to some kms in urban areas and some tens of kms in rural areas: these performances allow to set up large scale data acquisition infrastructures which are crucial in several application scenarios, as for example the Smart City and the Smart Industry ones. The term Smart City encompasses all those technological infrastructures that, thanks to the information and communication technologies, allow to optimize and make more efficient the daily processes that take place in urban environments, like, for example, traffic, waste management or public transport [2]. Data transmission in these contexts is usually required at a city scale, that means coverages that are in the order of the tens of square kms. As for an example, the Historic Centre of Rome has an approximate extension of 20 km^2, making thus impossible to cover such a large area with more traditional transmission technologies like WiFi, Bluetooth or ZigBee. Conversely, tests proved that with LPWAN technology it is possible to cover the centre of a medium sized city with even a single access point [3].

Regarding the second requirement, i.e., low cost, LPWANs are seen as an alternative to traditional cellular networks due to the lack or to the limited presence of subscription costs: in any case, even when a fee is required, the cost is proportional to the small amount of data to be transferred. This means that it is by far lower than the rates applied by the mobile operators and becomes sustainable even when a large quantity of interconnected devices has to be deployed. Moreover, the cost of the radio modules is also very low, usually in the order of few euros for the transmitting devices and of few hundreds of euros for the gateways which are in charge of receiving and managing the transmitted messages for that technologies which are not provided with a proprietary infrastructure.

The third requirement, i.e., the low power, is probably the most significant one. Indeed, any kind of IoT infrastructure is based on the integration of a large number of interconnected devices that are in charge of acquiring data, often in remote places where the connection to an electrical grid is technically unfeasible. Moreover, in those cases where connections to the grid are available, as, for example, in Smart Home scenarios, the number of devices to be powered makes a wired connection practically unachievable. Only two alternatives are then viable: the use of batteries or the exploitation of any kind of renewable energy by means of harvesting solutions. Batteries can be a convenient option due to their low cost; nevertheless, they require to be replaced and, in case of non-rechargeable ones, they must be disposed after being used, with a not negligible environmental impact. Instead, energy harvesting represents an eco-friendly alternative, and as many sources of energy are constantly available, it allows to continuously power interconnected devices for long periods without requiring any human intervention. For these reasons, energy harvesting systems have always been applied to interconnected devices, in particular within Wireless Sensor Network (WSN) architectures, that can be seen as the forerunners of the more complex distributed IoT infrastructures based on LPWAN technologies. Regarding LPWANs, in the last years several energy harvesting based solutions have been proposed in literature, exploiting a wide range of different power sources. The aim of this paper is then to provide a structured survey of the existing solutions, focusing on the architectures of the proposed systems, as well as on the application scenarios, in order to point out the features of the harvesting techniques developed to power these kinds of interconnected devices. To this end, the literature was surveyed by resorting to appropriate keywords, as it will be explained later on, and by excluding works published more than five years ago. The outcome of this search were works describing energy harvesting systems along with their integration within the chosen network architecture. In so doing, several papers were taken into account and 44 of them will be analyzed in the core of this review paper. Besides these contributions, some others will be treated as well, as they deal with energy harvesting techniques by tackling the theme from a more general point of view.

Therefore, a comprehensive review of the existing energy harvesting solutions for the powering of LPWAN-based monitoring systems and architectures was drawn up.

The rest of the paper is structured as follows. Section 2 reports the research methodology that was followed in order to draw up this review. Section 3 provides a detailed overview of the most common LPWAN technologies, with a particular focus on the LoRa technology, and the associated LoRaWAN protocol, which currently represents the most widespread solution, and the one to which the most part of the works found in literature are related to. Section 4 provides an overview of the energy harvesting techniques developed for WSN architectures, and Section 5 is the core of the paper and surveys the existing harvesting solutions for LPWANs. Section 6 discusses the main topics that arose during the review, and Section 7 provides some conclusive remarks.

2. Research Methodology

The review of the state-of-the-art concerning energy harvesting techniques exploited in LPWAN-based applications was conducted consulting the most common tools for the research of scientific papers and contributions: *Elsevier Scopus*, *Google Scholar*, and *IEEExplore*. In order to identify the suitable contributions, the research was conducted on the three tools using the following keywords; *LPWAN Energy Harvesting*, *LoRa Energy Harvesting*, *LoRaWAN Energy Harvesting*, *SigFox Energy Harvesting*, *Narrow Band IoT Energy Harvesting*, *NB-IoT Energy Harvesting*, *DASH7 Energy Harvesting*, *Weightless Energy Harvesting*, *IoT Energy Harvesting* and *Internet of Things Energy Harvesting*. For each of these keywords, at least 100 papers were examined. For the most significant ones (i.e., *LPWAN Energy Harvesting*, *LoRa Energy Harvesting*, *LoRaWAN Energy Harvesting*, *IoT Energy Harvesting* and *Internet of Things Energy Harvesting*) at least 200 papers were analyzed.

In order to be discussed in this work, only works describing the energy harvesting subsystem and its actual integration in a LPWA network architecture were chosen. Following this selection phase, a total number of 44 papers were discussed in Section 5: these contributions are to the best of our knowledge the only ones that satisfy the inclusion requirement. Apart from these papers, a few other contributions were found, integrating off-the-shelf solar panels in LoRa sensor nodes, without describing however the harvesting system architecture. We decided not to include them in the survey as solar-powered LoRa systems are the only ones that are relatively common and, at the level of description of those works, were reputed of few interest for what concerns the target of this survey.

3. Low Power Wide Area Network Technologies

A dichotomous listing of LPWAN-enabling technologies could be the one amid cellular and non-cellular standards: the former ones are license-based while the latter ones are license-free. As a direct consequence, adopting cellular technologies entails pretty high running costs due to subscription to plans provided by telecom operators. Therefore, adopting these solutions may be hardly feasible whenever connectivity has to be provided to a large quantity of objects. However, the most favorable benefit that such technologies offer is that very wide coverages are ensured along with very little data losses and low latency, such that lack of connectivity does not turn out to be an issue any longer. On the other hand, license-free technologies are cheaper than cellular ones, at least most of the time. This is due to the fact that they exploit the spectrum belonging to Industrial, Scientific, and Medical (ISM) bands. Such frequencies are unlicensed though; therefore, they are intrinsically subject to interference that can hinder link quality. For the same reason, coverage of unlicensed technologies is relatively restricted and data losses could be only limited but not avoided. Another peculiarity characterizing license-free technologies is that their carrier frequencies are region depending. This means that hardware devices need to be carefully designed bearing in mind the place in which the network will be deployed. Conversely, cellular technologies are usually not affected by this issue. Hereinafter, some of the most important LPWAN enabling technologies are listed and briefly described while at the end of the Section they will be recapped in Table 1. A similar approach was followed in a review paper of few years ago [4] that performed a comparable contribution with respect to the one reported in this

Section and in this review in general. Therein, LPWAN technologies (either cellular and non-cellular, although the latter ones were treated more in detail) to be exploited within IoT contexts were reviewed by focusing on each specification peculiarity and suitability for different IoT paradigms so as to meet the relative requirements. In addition, design specifications of LPWANs were surveyed along with the fact that challenges, future insight and research directions were pointed out too.

3.1. LoRa

In 2012, The American company Semtech developed the Long Range (LoRa) modulation and since that moment it held the relative patent. LoRa is a robust technology allowing for quite wide coverage. Such features are allowed by the modulation fundamentals: indeed, LoRa is grounded on the Chirp Spreading Spectrum (CSS) modulation and adopts the Additive Links On-Line Hawaii Area (ALOHA) technique to access the communication channel (i.e., transmission may occur at any time). It is currently managed and controlled by the LoRa Alliance that is a consortium composed by more than 500 companies coping with hardware as well as software. This association periodically issues successive releases of the Long Range Wide Area Network (LoRaWAN) standard: a MAC layer communication protocol having LoRa modulation as physical layer. In contrast to the fact that LoRa is a proprietary modulation by which Semtech collects royalties from chip vendors embedding LoRa modules in their own boards, LoRaWAN standard is openly accessible. On the other hand, LoRaWAN is not the only standard built on top of LoRa technology, indeed Link Labs developed an alternative LoRa-based LPWAN. As it was previously mentioned, LoRa is a long range wireless technology allowing coverage extending from few kilometers in urban ares, up to tens kilometers in rural environments. Moreover, like other LPWAN standards, LoRa features single-hop communications within star topologies networks.

3.2. Sigfox

Sigfox is a French company having more than 10 years of activity during which Sigfox networks have been set up and run all around the world by operating on different sub-GHz ISM bands according to the region. Users exploiting the network are required to pay a subscription which varies depending on different plans. Sigfox is designed following the paradigm of star topology allowing for either uplinks and downlinks during which small size payloads (i.e., at most 12 B) can be transmitted. This turns out to be a major drawback along with a limited number of maximum transmitted packets per day in relation to the subscribed plan. On the other hand, a big benefit Sigfox provides is that it offers a full network infrastructure and a cloud service to collect, retrieve and analyze all the incoming data. Therefore, the only concern users have to deal with is the design and the realization of sensor nodes because all of the heavy lifting is accomplished by Sigfox itself. For what concerns bandwidth and data-rates, Sigfox requires extremely narrow band (i.e., 100 Hz) and it only allows for pretty slow data-rate (i.e., 100 bps). Finally, Sigfox makes use of the Binary Phase Shift Keying (BPSK) modulation and of the Random Frequency Time Division Multiple Access (RFTDMA) technique to access the channel.

3.3. DASH7

DASH7 is an open source protocol for LPWANs and, in general, for WSNs. It exploits three sub-GHz ISM bands (i.e., 433 MHz, 868 MHz, and 915 MHz) and its main features are extended battery lifetime due to limited consumption, quite large coverage range, low latency especially for moving nodes, and fairly high data-rate (i.e., up to 167 kbps). Moreover, DASH7 allows for bidirectional communications. DASH7 inherits its default parameters from the ISO 18000-7 standard and boosts the latter by defining a full communication stack starting from the physical layer up to the application one. DASH7 is also known as BLAST network technology due to the acronym of its principal attributes: Bursty (i.e., short and sporadic sequences of data are transmitted), Light (i.e., packet size is limited to 256 B), Asynchronous (i.e., there is no periodic synchronization within the network),

Stealth (i.e., the nodes only communicate towards pre-paired gateways) and Transitive (i.e., nodes may freely move within the environment and, at the same time, do not lose connectivity). Such a standard is managed and distributed by the DASH7 Alliance that is a nonprofit consortium made of either companies and universities. DASH7 exploits Gaussian Frequency Shift Keying (GFSK) modulation and provides for either a star, tree or node-to-node network topology. Concerning the MAC layer, DASH7 makes use of Carrier Sense Multiple Access with Collision Avoidance (CSMA/CA) protocol.

3.4. Ingenu

More than a decade ago, Ingenu was released and it was an innovative technology within the LPWAN framework as it exploits ISM bands having higher carrier frequencies than, for instance, Sigfox: Ingenu, indeed, runs at 2.4 GHz, thus sharing the same frequency of Bluetooth or WiFi. Such a feature entails pros and cons: this is an ISM band which is globally unlicensed thus developers do not have to think about the region their products will be employed; in addition, such a band allows for broader bandwidth in comparison with other sub-GHz ISM bands. On the other hand, though, within such a band link quality is worse than different sub-GHz frequencies due to the physics of the problem. Together with Differential Binary Phase Shift Keying (D-BPSK) modulation, the essence of Ingenu LPWAN is its proprietary and patented technology: the Random Phase Multiple Access (RPMA) that is either a physic and a Medium Access Control (MAC) layer. RPMA was specifically designed so as to meet the requirements of an LPWAN: long battery lifetime thanks to limited consumption, robustness towards interference, and wide coverage and high capacity of a single RPMA access point. Additionally, RPMA technology is also suitable to deal with bidirectional communication and broadcast transmission. Concerning the network infrastructure, Ingenu offers a full architecture likewise Sigfox does. Eventually, Ingenu also set up the first wireless machine network that is the largest IoT network worldwide which is especially dedicated to provide connectivity for machine. Sadly, though, such a facility was only established in few dozens of cities, the bulk of which are located in the US.

3.5. Weightless

Weightless has been managed by the English Weightless Special Interest Group since 2012 and it encloses a set of wireless enabling technologies for LPWANs because, at its beginning, three standards were issued: Weightless-P, Weightless-N, and Weightless-W. However, as time goes by, two of those standards (i.e., Weightless-N and Weightless-W) were deprecated: Weightless-N only allowed for uplinks thus it was a monodirectional communication technology, while Weightless-W was devised to work within the unexploited frequencies belonging to the TV bands (i.e., 470 ÷ 790 MHz). Therefore, Weightless-P (or simply Weightless) is the only outlasting standard: it is a bidirectional standard requiring narrow band to run and occupying all of the frequencies belonging to the unlicensed ISM sub-GHz bands (i.e., 163 MHz, 433 MHz, 470 MHz, 780 MHz, 868 MHz, 915 MHz, and 923 MHz). Moreover, Weightless is also an open standard that adopts both the Gaussian Minimum Shift Keying (GMSK) and the Offset Quadrature Phase Shift Keying (OQPSK) modulations and the Time-Division Multiple Access (TDMA) scheme to access the communication channel.

3.6. LTE-M

Long Term Evolution (LTE) Machine Type Communication (LTE-M) operates, focusing on Machine-to-Machine (M2M) communication and IoT, by using the cellular LTE standard. Therefore, it is fully compatible with actual cellular networks. Indeed, telecom companies do not necessitate any additional device to be installed as they just have to update the firmware running on their base stations to the newer versions. Because of this, there exist some telecom providers that have already started activating such services making some experiments. In comparison with standard LTE, it offers optimized sleep modes from the point of view of consumption resulting in higher power budgets. Amid the cellular LPWAN technologies, LTE-M has the highest data-rate. However, it requires the broader channel band. It exploits the 16 Quadrature Amplitude Modulation (16-QAM) and, for what concerns the channel access, it makes use of two methods stemming from the Frequency Division Multiple Access (FDMA): Single Carrier FDMA (SC-FDMA) for uplinks and Orthogonal FDMA (OFDMA) for downlinks.

3.7. NB-IoT

Notwithstanding that NB-IoT and LTE-M come from the same organization (i.e., the 3GPP), they differ for several aspects. First, NB-IoT has smaller data rate and narrower bandwidth than LTE-M. Second, NB-IoT does not necessarily work within LTE bands: it is designed to operate in a subset of LTE bands or even in the unexploited Global System for Mobile (GSM) bands. Even though it has lagged behind LTE-M, more and more telecom firms are investing on it at the moment. It has been notably used especially in Europe where it is finding its IoT alcove (e.g., asset tracking). NB-IoT standard specifications aver that it requires less power than LTE- M. This is a double-edged sword though: being thrifty from the point of view of consumption is always a good feature wherever sensor nodes relying on batteries are employed, but it also means having a worse penetration capability through obstacles. Regarding the physical layer, it may exploit both the Quadrature Phase-Shift Keying (QPSK) modulation and the Binary Phase-Shift Keying (BPSK) modulation. For what matters to the access to the channel methods, it shares all of them with LTE-M.

Table 1. LPWAN main enabling technology comparison.

	Type	Frequency Band [MHz]	Data-rate [kbps]	Channel Bandwidth [kHz]	Message Payload [B]	Coverage Range [km]
LoRa	Non-cellular	433 (CN) 868 (EU) 915 (US, AU)	$0.3 \div 50$	125 250 500	$\leq$255	$\simeq$5 (urban) $\simeq$15 (rural)
Sigfox	Non-cellular	868 (EU) 902 (US) 920 (AU)	100	100	$\leq$12	$\simeq$40
DASH7	Non-cellular	433 (CN) 868 (EU) 915 (US)	9.6, 55.5 or 166.6	25 or 200	$\leq$256	$\simeq$5
Ingenu	Non-cellular	2400	624 (uplink) 156 (downlink)	1000	$\leq$10,000	$\simeq$45
Weightless	Non-cellular	Any sub-GHz ISM band	$200 \div 100$	12.5	$\leq$48	$\simeq$2
LTE-M	Cellular	LTE bands	1024	1400	$\leq$256	$\simeq$5
NB-IoT	Cellular	LTE subset bands	250	180	$\leq$1600	$\simeq$1 (urban) $\simeq$10 (rural)

4. Energy Harvesting Techniques for Wireless Sensor Networks

Sensor nodes belonging to LPWANs, or more generally to WSNs, typically rely on a limited source of power (e.g., batteries), which can act either as the only powering method or as a backup source. Nodes falling in the first case necessarily have to put into practice energy saving strategies (e.g., effectuating duty cycling or embedding low power components) in order to extend their lifetime. On the other hand, for nodes that are part of the second case it is fundamental to contrive energy harvesting techniques. Regardless of this distinction, equipping nodes with an energy harvesting system, whenever the deployment environment consents it, is always a serviceable method to extend their lifetime [5–7] as well as to reduce both human intervention for maintenance and environmental impact by decreasing the amount of batteries requiring disposal. However, the optimum would be to design self-powered sensor nodes [8] but this is extremely tough and, most of the time, it turns out to be challenging or even unfeasible.

In the case of outdoor operative scenarios, the most immediate renewable source of power is the solar one and a way to exploit it could be to provide nodes with photovoltaic cells. Such systems are grounded on a common structure: a rechargeable battery and a solar panel are simultaneously controlled by a Battery Management System (BMS) with the aim of correctly powering a device acting as load. In other words, the solar panel is exploited as a primary source of energy for either recharge the battery and power the load; whenever irradiation decreases (e.g., during nights or rainy and cloudy weather) the BMS disables solar panels and enables the battery (which, in the best cases should be fully charged) to power the load until favorable weather conditions occur once again. For instance, BMSs may be realized by employing circuits for Maximum Power Point Tracking (MPPT), that is, an algorithm guaranteeing that the load can utilize the almost peak power produced by the solar panel, thus obtaining an intelligent BMS carrying out its tasks thanks to hardware components rather than software instructions [9]; however, similar outcomes could be achieved by resorting to Pulse Width Modulation (PWM) in place of MPPT [10]. As a direct consequence, photovoltaic and solar panels should be thoroughly designed and identified. Indeed, sizes are not always directly proportional to performances as manufacturing processes and constructing materials play a key role [11]. Moreover, photovoltaic panels performances could be enhanced by supplying panels with cooling systems [12]. Without losing consistency, photovoltaic energy harvesting may be also employed within indoor environments provided that ad hoc systems are adopted [13,14]. Concluding, the literature has plenty of examples resorting to solar energy harvesting techniques to power WSN nodes due to the fact that it is straightforward. Instances range from simple and cheap solutions [15] to energy efficient and optimized alternatives [16] or systems aiming at maximizing sensor nodes lifetime [17].

One of the most instantaneous effects of sunlight is the occurring of heat and such a phenomenon may be exploited for energy harvesting too. Indeed, whenever a thermal gradient (i.e., a temperature difference) between two ends of a thermoelectric material is experienced, charges moving through the material due to heat generate a thermoelectric effect. Charges migrate from the hot junction towards the cold one causing a potential difference: this phenomenon is also known as Seebeck effect [18]. The interesting aspect of this technique is that it can be put into effect not only in situations in which heat is caused by sun rays, but whenever a thermal gradient arises due to the most disparate sources. For instance, in automotive applications, energy autonomous sensor nodes may be designed [19] exploiting the heat produced by the engine. Another idea could be that also heat dissipating by electronic devices could be harvested [20], or even exploiting thermal gradient within scenarios in which temperature monitoring is accomplished [21].

Another renewable source of energy available outdoor that can be exploited for energy harvesting purposes is wind. Aeolian harvesting systems commonly work exploiting windmills acting on DC motors: indeed, by resorting to reversibility, the latter ones behave as generators whenever the shaft is rotated by external torques. Ideally, due to such a property, a voltage at motor leads that is linearly dependent on the number of revolutions of the shaft is generated. However, such a phenomenon may also end up in a drawback: if wind speed would not be constant, as it could likely happen, voltage would not be constant too. Therefore, it would not be suitable for directly powering a DC sensor node. To this purpose, an optimized AC/DC buck-boost converter for aeolian energy harvesting was investigated [22]. Another shortcoming that is implicitly entailed by aeolian harvesting is that performances of such a technique is strongly depended on wind forecast, therefore a system that automatically copes with this issue may be useful [23] or, alternatively, setting up a method to predict power provided by wind could be a remedy as well [24].

Hydroelectric and aeolian harvesting share a similar scheme. Indeed, the main difference is that water flowing within pipes or conducts causes the production of energy rather than wind. Indeed, water pushing on mill blades causes mill rotation which is exploited to generate electricity by resorting to DC motor reversibility property as it happens in aeolian harvesting systems. The use cases may be various; however, this technique provides the best performances whenever it is exploited in scenarios in which water is forced to flow inside pipelines as it causes a pressure increase which directly implies a greater amount of harvested energy. For instance, this strategy may be suitable for sensor nodes accomplishing the measurement of flow rates within pipes in domestic applications [25] or even for environmental monitoring issues inside underground Medieval aqueducts [26]. Concerning energy harvesting from sea waves, several studies proved its feasibility. Such a technique has a wide application within sensor nodes that are offshore deployed, either floating or underwater, as it is commonly inconvenient to reach such devices to perform maintenance procedures. Harvesting energy from sea waves is feasible as oscillations stemming form such a motion may be converted in electrical energy [27]. Moreover, that phenomenon may be enhanced by resorting to duck-shaped structures [28].

Energy harvesting from sea waves inherits similar concepts that are adopted in systems harvesting energy from mechanical vibrations: indeed, both the sources of energy cause oscillations that are converted in electric power for running sensor nodes. Such a task is achieved by making use of piezoelectric crystals: they are materials characterized by the property that whenever they are subject to a mechanical deformation, a voltage is generated and vice versa. Therefore, this sort of harvesting could be exploited in a various number of scenarios presenting vibrating objects: vehicles [29], aircrafts [30], vibrating machineries [31], bridges [32], mainframe computers stacked in racks within data centers [33], or even in underwater environments [34]. Moreover, piezoelectric harvesting does not necessarily require either objects strongly vibrating [35] or bulky devices [36] thus making it widely suitable. In addition, piezoelectric materials stand at the base of acoustic energy harvesting [37,38] in which mechanical deformation is provoked by air waves generated by acoustic signals. However, acoustic harvesting may be also performed by falling back on Helmholtz resonators [39].

An alcove of WSN is the one related to wireless network whose nodes are attached to human body becoming wearable sensor nodes: the so called Wireless Body Area Networks (WBANs). Even in this context there exist harvesting techniques [40]. In particular, they exploit human body activities (e.g., movement [41,42], pace [43], and so on) in order to power sensor nodes. In other words, any of the methods that were developed for harvesting energy in a broad sense may be transplanted within the framework of WBANs.

Electromagnetic fields may be also harnessed for energy harvesting purposes. Such an idea is the basis on which wireless power transfer is grounded: nowadays this technique is massively adopted in docking stations that are capable of simultaneously charging several sorts of devices (e.g., smartphones, tablets, smartwatches, and so on). From a theoretical point of view, electromagnetic energy harvesting is ruled by Faraday law of induction stating that the variation of the flux of a magnetic field that is concatenated with an electrical circuit causes an induced electromotive force in the circuit itself generating a current opposing the flux variation gave rise to it. Therefore, multiple sources of dynamic electromagnetic field may be considered for energy harvesting: high voltage cables [44], power cables for smart grids [45], human body motions [46], antenna panels of satellites [47], and many more.

5. Energy Harvesting Techniques for LPWANs

In this section, we will discuss the existing literature focusing on energy harvesting solutions for LPWAN-based systems. While the research methodology was already presented in Section 2, before discussing the survey we would like to point out that the works were selected according to the principle that the presented results should be focused on the description of a working network architecture whose sensor nodes are powered by means of energy harvesting. For this reason, several works focusing only on the harvesting technique without contextualizing it within a specific LPWAN infrastructure were not taken into account. Similarly, we do not take into account works dealing with transmission technologies that fall within the IoT context but cannot be classified as LPWAN (e.g., Bluetooth Low Energy or WiFi).

This choice is justified by the fact that some surveys focusing on energy harvesting techniques for IoT in a broader sense have been already proposed. In particular, a preliminary analysis of energy harvesting techniques to be employed to power up wirelessly connected devices within the IoT framework is proposed in [48], while a deeper analysis of the state-of-the art is faced in [49]. In this review paper, all the possible harvesting techniques for IoT devices are analyzed in detail, focusing on a wide range of works that partially includes some LPWAN-based architectures.

As the focus of this review is not only on the harvesting techniques, but also on the transmission technologies, we decided to structure this section according to the different LPWAN standards. While a large part of the reviewed papers exploits the LoRa technology and the LoRaWAN protocol, the related subsection has been furthermore structured according to the harvesting principle. The other subsections focus on SigFox, DASH7, and cellular technologies. Regarding other LPWAN solutions, like Ingenu or Weightless, to the best of our knowledge the literature lacks of works dealing with energy harvesting solutions to power sensor nodes exploiting such communication technologies.

All of the energy harvesting sources that are reviewed in this section and the relative LPWAN enabling technologies for which instances were retrieved within the literature are schematized in Figure 1.

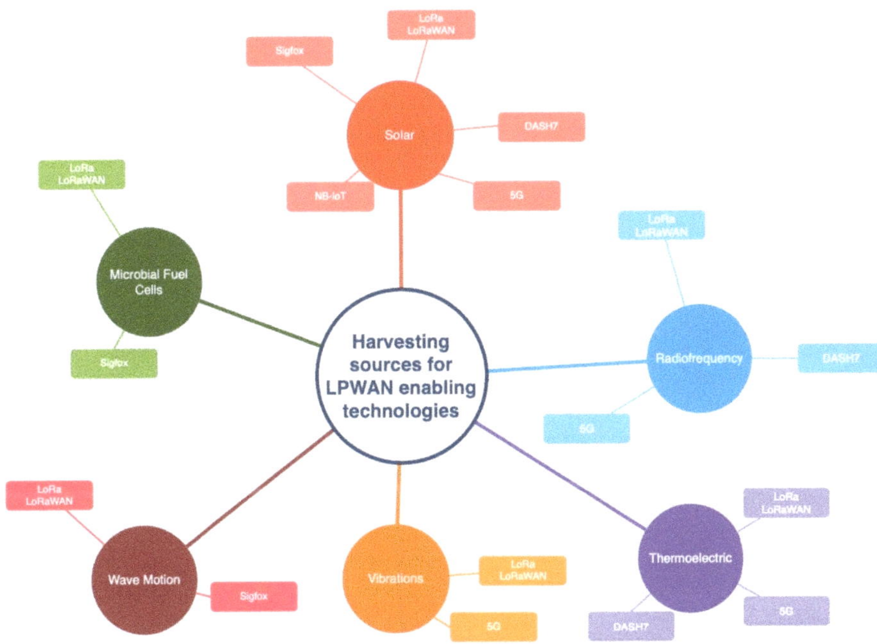

Figure 1. Harvesting sources for LPWAN enabling technologies.

5.1. LoRa and LoRaWAN

LoRa is currently the most employed LPWAN transmission technology for the realization of IoT architectures in a wide range of application scenarios. For this reason, together with several papers focusing on the description of systems exploiting a specific harvesting technique, some contributions were found in the literature dealing with specific architectures, not focused on a particular energy source. These contributions are precious for the design of the overall LoRaWAN systems and are then listed below. Papers dealing with systems powered by well-defined energy sources will be discussed then in the Sections 5.1.1–5.1.7, according to the specific harvesting principle.

Gleonec et al. [50] proposed one of the first works discussing the adoption of energy harvesting systems to power up LoRaWAN sensor nodes. While they do not adopt a specific harvesting technique, they focus on the management of multiple power sources, that are simulated by means of voltage generators. The proposed solution is validated by means of an experimental setup whose aim is to demonstrate the feasibility of a powering system multiplexing multiple energy sources. Such a multisource approach was also proposed in other works which will be discussed in Section 5.1.7.

A general architecture focusing on the functioning of a battery-less LoRaWAN sensor node, powered by an undefined source of energy, is presented in [51]. The model proposed in this work is interesting for any system powered by any kind of renewable energy: indeed, the architecture is based on a capacitor for energy storage and operates according to functioning model that allows to guarantee the continuous operation of the system. While the model is effective, it is based on the assumption that a constant energy is provided to the harvesting system: nevertheless, this requirement cannot be satisfied by some of the most common energy sources such as the solar one. The same battery-less approach is presented in [52]: in this case, instead of focusing on the architecture of the LoRaWAN device, the paper tackles protocol issues related to packet collisions and minimum throughput maximization.

Finally, one last paper was identified focusing on the performances of LoRaWAN devices powered by an energy harvesting system that may fit with different power sources. Indeed, in [53], the impact deriving from the adoption of a renewable source of energy for LoRaWAN networks in industrial monitoring applications on the overall systems costs is analyzed: a comparison of the maintenance costs due to replacement of batteries in battery-powered devices with or without the use of harvesting solutions is provided, pointing out the benefits deriving from the adoption of these techniques in the proposed application context.

5.1.1. Solar

As already underlined in Section 4, solar energy is the most common environmental power source for the realization of self-powered wireless systems. For what concerns the LoRa technology, a relatively large number of papers have been identified, dealing at different levels with the adoption of photovoltaic (PV) panels for the powering of either LoRa nodes [8,54–61] or LoRa Gateways [62,63]. In general, all those papers share the same configuration for what concerns the energy harvesting system: this includes the use a small scale PV panel, a BMS, and an energy storage system that may be whichever amid a battery or a supercapacitor.

Starting from the works describing End Nodes architectures, the majority of them propose an energy harvesting system to be employed for a specific application scenario; indeed, only one contribution was found dealing with a general purpose sensor node [54]. This paper provides an effective theoretical way to dimension the energy harvesting components of the node as well as the energy storage devices (i.e., batteries and supercapacitors) and the PV panels. The structure of a solar-powered sensor node is then described in detail: in this system, the charge process is managed by means of an SPV1050 low-power harvester by STMicroelectronics, implementing the MPPGT function. An INA226 power monitor by Texas Instruments is also introduced in the system to check the battery status. The energy consumption of the sensor node is carefully evaluated and field test results are provided for a fortnight operation period. This paper is especially interesting since the proposed architecture can be also used with other harvesting techniques (e.g., the adoption of Peltier modules instead of PV panels is suggested in the text), by simply replacing the harvesting component.

A first group of papers mainly focuses on the overall system architecture, without discussing in detail the energy harvesting solution: in these works, a basic structure employing commercial PV panels and supercapacitors or batteries for energy storage is presented, shifting the main focus on the application scenario. Polonelli et al. [55] present the architecture of an energy self-sufficient LoRaWAN sensor node to be employed for the measurement of the displacement of cracks in buildings within the frameworks of Structural Health Monitoring (SHM). The energy harvesting system presented in this work is based on the use of a solar panel connected to an SPV1040 step-up converter by STMicroelectronics. An L6924D battery charger (by STMicroelectronics too) is used to manage the charge of a 2 F supercapacitor. A similar, but less detailed architecture for what concerns the solar harvester, is presented in [56]; in this case, the proposed system is expected to be floating on the sea surface, while collecting data concerning water temperature and transmitting them ashore by means of LoRaWAN protocol. The harvesting unit encompasses a solar panel connected to two parallel

7.5 F supercapacitors, while no charge management electronic component is used. Another paper [57] focuses on earthquake detection and discusses the architecture of a LoRaWAN sensor node embedding accelerometers and other environmental sensors. In this case, the node is powered by a 2000 mAh LiPo battery whose charge is managed by a commercial Arduino-type Solar Charger Shield by Seeed Studio, connected to a 1.5 W off-the-shelf solar panel. Despite the simplicity of the technical solution, the paper also proposes a detailed analysis of the node power consumption which is useful to evaluate the sizing of the energy harvesting system. Matthews at al. [58] present an interesting platform integrating LoRa data transmission and UHF RFID contactless identification for vehicle recognition: the described platform is powered by means of a PV system which is not described in detail. Finally, a last paper [59] presents a simple architecture for environmental monitoring based on the off-the-shelf STMicroelectronics I-Nucleo-LRWAN1 multi-sensor shield: the system is powered by means of a 4 F supercapacitor charged by means of a PV panel, but no details are provided concerning the charge management system.

A second group of papers, while being related to a specific application scenario, discusses more in detail the implemented solar harvesting system. Rossi and Tosato [60] present a LoRa sensor node to be employed for environmental pollution monitoring, providing a detailed description of the energy harvesting solution. In this system, the charge management unit is composed of a BQ25570 harvester power management circuit from Texas Instruments, which is used to charge a lithium battery. As the sensor node has a large operating current absorption, an additional stage was added to supply it, which is composed of a TPS63000 Buck-Boost Converter from Texas Instruments, regulated by means of a Texas Instruments TPL5110 timer, periodically enabling the current drainage. Regarding the PV panel, in this paper three different solutions are tested, two custom-made and one off-the-shelf: the first two are, respectively, based on 50 Ixys cells arranged on an x-shaped matrix, and 12 Sanyo micro-PVs in parallel, while the commercial one is a 1 W panel from Seeed Studio. Following a set of system tests, the authors demonstrate that the best choice is represented by the Seeed Studio commercial PV panel which provides in general better performances with respect to the other solutions.

In [8], another interesting application scenario is faced: indeed, the paper presents a wearable LoRa sensor node, to be used for personal environmental data collection. One of the most challenging requirements that such a system has to fulfill concerns the energy harvesting system dimensions, as the monitoring device has to be worn by users. To overcome this limitation, the proposed solution is centered on a 3 cm radius circular PV panel that is arranged into a watch-shaped device integrating the whole sensor node components. The charge process is managed by an ADP5090 controller from Analog Devices, which regulates the charge of a 12.5 F supercapacitor. The overall power consumption of the node is evaluated in detail and field tested in order to identify a duty-cycling able to guarantee a continuous operation of system is assessed too. Finally, a last more recent contribution was identified providing a detailed description of a solar energy harvesting system for environmental sensor nodes [61]: the structure is similar to other contributions, being based on a LTC3105 DC/DC converter from Analog Devices on an AM-5412CAR amorphous silicon solar cell from Panasonic Eco Solutions, charging a Li-ion battery.

Regarding solar energy harvesting systems used to power LoRaWAN Gateways, only two contributions were identified. In this case, the power requirement is by far higher than for End Nodes, as Gateways are required to always listen to incoming packets from End Nodes, and then no duty-cycling policy can be applied. For this reason, in [62] the sizing of the PV plant to be used to power the Gateway is discussed in detail, proposing an algorithm to estimate the system power consumption. As the Gateway also has to forward the packets received by the End Nodes to the LoRaWAN Network Server, the energy requirement must include the powering of another data transmission technology too. As the system is expected to be cable-less, the consumption of 4 different technologies (i.e., LTE, WiMax, Satellite and Long-Range WIFi) is discussed as well as the analysis of carbon footprint. Then, the achievable CO_2 saving with the four different typologies of data transmission is investigated too. In [63], an edge computing paradigm within LPWAN framework is tackled. The proposed architecture includes Gateways, that rely on harvested energy, receiving data from End Nodes. In contrast with traditional harvesting systems, the authors propose a stochastic model grounded on a Markov decision process to manage the scavenged energy. Finally, simulations proved the feasibility and the effectiveness of the system. In particular, Gateway cooperation is stochastically controlled, as it has just been mentioned, so to look for an optimum state allowing computation and data-forwarding tasks with the aim of cleverly managing the harvested energy. In so doing, a Gateway may offload a task to another one if such an action is believed to be the one minimizing the required energy. Concerning the harvesters, they are supposed to be systems scavenging energy from ambient light. In addition, harvesting is modeled as a stochastic process accounting for harvesting rate rather than harvesting volume of energy to be assessed within variable temporal slots.

5.1.2. Radio Frequency

Wireless power transfer by means of magnetic or inductive coupling is widely used in several contexts, from passive Radio Frequency Identification (RFID) systems to wireless chargers for mobile phones. While such a powering technique requires a limited distance between the Radio Frequency (RF) source and the device to be powered, thus contradicting the wide area operational principle of LPWANs, some LoRa-based systems powered by means of RF harvesting were found in literature. A first solution exploiting the wireless RF channel to power a LoRa node is presented in [64]. In this paper, the architecture of the harvesting circuit is described in detail, focusing on the antenna design as well as on the RF-DC circuit that is required to convert the Alternate Current (AC) signal generated by the receiver antenna to a DC one that can be used to power a sensor node. Following the description of the system, the realization of a wirelessly powered LoRa node is presented and its functioning is validated by means of field tests. While the wireless power transfer is achieved, the systems suffers from the short transmission range limitation pointed out before since the tested operational range is shorter than 2 m, and it is then in contrast with the wide area requirement.

While the short range is clearly a limitation for the adoption of this type of harvesting for LoRa systems, Peng et at. [65] propose the so-called PLoRa system, that aims at providing long range transmission from passive sensor nodes by means of backscatter transmission. Regarding the node powering, this exploits environmental RF signals from any possible source in the 900 MHz frequency band as well as solar energy when the former one is not available: the nodes are provided with a 0.33 F supercapacitor for energy storage. While the proposed system is able to achieve data transmission at distances up to some hundreds of meters, it cannot be fully considered as RF-powered as the presence of the solar panel is crucial to guarantee the long range operation.

Another interesting contribution regarding wireless power transfer for a LoRaWAN sensor node is presented in [66]. In this paper, a whole experimental setup is described in detail comparing the performances of the system according to the use of different typologies of electronic components. In particular, the performances of two different RF-DC converters are investigated, correlating a commercial P2110B converter from Powercast with a device realized ad-hoc. Similarly, the performances of different typologies of antennas were analyzed and compared. As in previous cases, while the system proved to be effective to power a LoRaWAN node, the actual operating distance was in general very short, below 3.5 m.

A final contribution is provided in [67]: this paper presents a theoretical discussion concerning wireless power transfer in general, and comparing different techniques: inductive coupling, ultra-dense millimetre-wave small cells and Magnetically Coupled Resonance (MCR). A solution based on this last technique (i.e., MCR), exploiting ferrite structures, is then presented and discussed, albeit a practical implementation of the system is not provided. Nevertheless, while the proposed system may be interesting, the lack of its actual implementation prevents from discussing its potential performances.

5.1.3. Thermoelectric

Thermoelectric energy is harvested by means of thermoelectric generators which exploit temperature gradients to supply electrical power, as it was previously mentioned in Section 4. In so doing, self-powered LoRa End Nodes to be employed for industrial monitoring issues may be designed and implemented as the work in [68] illustrates. Indeed, the authors propose a flexible thermoelectric generator to be wrapped around heat pipes reaching the temperature of 70 °C. Additionally, the sensor node is in charge of monitoring sundry parameters (e.g., pipe temperature, ambient temperature, humidity, CO_2 concentration, and organic compound concentration). Moreover, the authors point out that within indoor environment like an industrial one, thermoelectric generators are by far suitable for energy harvesting than, for instance, photovoltaic devices, as the former devices are able to constantly harvest energy regardless of the available amount of light as only a temperature gradient is needed. The developed sensor node has a common architecture which comprehends the harvester, a BMS, a rechargeable battery, a DC–DC converter, dedicated sensors, a microcontroller, and a transceiver enabled by LoRa modulation.

Continuing in the same vein of industrial monitoring, the authors of [69] propose a battery-less and maintenance-free LoRaWAN End Node harvesting energy from industrial cooling pipes at 80 °C. Such a device is exploited to monitor vibrations of machineries. Due to the fact that this study follows the footsteps of the previous one, the sensor node architectures of both the nodes resemble each other. However, the main dissimilarity amid the two prototypes concerns energy storage since [69] makes use of a 5 F supercapacitor rather than a rechargeable battery.

Thermoelectric generators are also able to fully operate even when they are employed for harvesting energy form trees [70]. Such a study is therefore interesting because the relative test campaigns, that were set up throughout several months across seasons, demonstrate the feasibility and the effectiveness of the system. In this setting, the thermoelectric generator makes use of the temperature gradient arising between the tree trunk and the ambient air: indeed, the inner temperature slowly varies while the outer one experiences more sudden changes due to sun rising and setting. Therefore, such an event takes place on a daily basis thus allowing a theoretically perpetual power supply for a LoRaWAN End Node.

Finally, the authors of [71] propose an autonomous LoRa End Node powered by a thermoelectric generator harvesting unused or wasted heat coming from disparate sources like hot water pipes or factory machineries. The device is employed for security and environmental monitoring issues. To achieve such tasks by only relying on the harvester, it embeds the same classes of components that were listed so far for comparable devices supplied by similar systems: besides the harvester there are a DC–DC step-up converter, a BMS, a storage element (i.e., in this case a supercapacitor), a voltage regulator, and the sensor node itself.

5.1.4. Vibrations

Mechanical systems as well as structures that are subject to stress, strain, or any other sort of external forcing term experience vibrations which can be promptly exploited for powering up sensors nodes by means of suitable harvesters. Such an idea is also exploited for supplying LoRa End Nodes too. Orfei et al. [72] developed a battery-less LoRa node for the monitoring of the asphalt of a bridge. Energy sufficiency is ensured by an electromagnetic energy harvester scavenging energy from bridge vibrations (which, for instance, are engaged by vehicles crossing it) that exploits Halbach array for permanent magnets arrangement so to increase the magnetic fields resulting in a miniaturization of the harvester still preserving its performances. The collected energy is firstly rectified and regulated and then stored within a supercapacitor. Overall, the node is capable of fulfilling its tasks (i.e., data sensing and data transmission) by leveraging on the energy harvested from bridge vibration occurring any time a vehicle passes through.

Harvested vibrations through electromagnetic devices are also the source of energy that is employed in [73] within industrial contexts. Indeed, it shows two sensors nodes that are supplied by the vibrations generated by a standard industrial compressor: the former is a LoRaWAN End Node, while the latter exploits Bluetooth connectivity, and therefore it is neglected in this review as it falls outside the scope of LPWAN communication protocols. Similarly as before, the harvester requires a conditioning electronics achieving rectification and regulation, in a first stage, and a supercapacitor so to store the collected energy. Test results point out that the LoRaWAN node is able to reach the minimum sampling rate of 30 seconds which in most of the cases suffices monitoring requisites of slowly varying physical phenomena.

5.1.5. Wave Motion

Energy harvesting lays its own foundation on energy conversion principles. Such a phenomenon gains a notable momentum whenever renewable sources of energy are exploited. Water waves intrinsically carry kinetic energy which is potentially boundless, therefore its harvesting could be a bold move. Chandrasekhar et al. [74] put into effect this idea so to power a LoRa position tracker hosted in a smart buoy for marine scopes. The harvesting system is enclosed within the buoy and it consists of a triboelectric nanogenerator and an electromagnetic generator for recovering kinetic energy of water waves. In addition to them, a solar cell is placed on top of the buoy just as a backup harvester whenever calm wave conditions are experienced. Despite the fact that heterogeneous energy sources are exploited, the work in [74] is not included within Section 5.1.7 because solar energy only plays a secondary role. The aforementioned generators are capable to efficiently convert kinetic energy into electrical energy that is stored either in a capacitor and in a Li-ion battery via a BMS. The triboelectric and the electromagnetic generators are simoultaneosly combined for pursuing the harvesting scope. Indeed, the former operates under contact and separation of triboelectric units that are especially manufactured. They are actuated by means of a cylindrical tube on which a coil is wounded along its outer surface while in its inner side a moving magnet is present thus forming the electromagnetic generator. This coupling allows to independently generate electrical power during the same mechanical motion due to waves.

5.1.6. Microbial Fuel Cells

Aquatic environments that are characterized by favorable conditions (e.g., enough dissolved oxygen) are well suited as a developing habitat for floating microbial fuel cells. In spite of this, such bio-electrochemical systems proved to be able to operate even in anoxic conditions [75] and to power up LoRa End Nodes. The study finds out that especially designed microbial fuel cells are able to supply enough power for the LoRa node provided that a DC-DC converter is employed since a microbial fuel cell in itself would only scavenge an insufficient amount of energy. However, just very infrequent transmissions (i.e., twice a day) may be tolerated by such harvesting system though.

Microbial fuel cells may take place even in conjunction with plants, and the related harvested energy could suffice for enabling battery-less LoRa sensor node performing environmental monitoring within a smart cities framework [76]. Like comparable systems, such device is capable to generate a stable output voltage permitting the correct supplying of the node that embeds sensors, microcontroller, transceiver, and miscellaneous electronics (e.g., a DC–DC converter and a supercapacitor) employed for the harvester.

5.1.7. Hybrid Techniques

A simultaneous exploitation of multiple sources of energy for harvesting purposes is advisable whenever the deployment scenario consents it. Indeed, premises for achieving zero impact sensor nodes from an energy point of view could be easily met. On the other hand, there exist studies comparing different sorts of harvesting techniques with the aim of finding out potential pros and cons. For instance, in the work in [77], solar, thermal, and piezoelectric harvesting techniques are investigated and correlated in order to design autonomous LoRa End Nodes. The latter has a standard architecture including the harvester, a BMS, a rechargeable battery, and the proper sensor node which is in turn composed of sensors, microcontroller, transceiver, and so on. The principal results point out that solar harvester ensures enough chances to provide the node a self-sustaining state while the remainder two drastically decrease such probability. Unfortunately, though, all of those techniques have not been integrated yet within the aforesaid research work.

Solar and thermal energy harvesting are exploited in LoRa End Nodes so to obtain an energy-efficient device allowing for either short- or long-range communication [78]: indeed, besides standard LoRa transmissions enabling long range broadcast, also an energy aware wake-up radio is embedded in the system permitting to wirelessly trigger the node forcing data sampling and data sending tasks. In so doing, a twofold scope may be achieved: self-sustainability, due to harvesting and ultra-low-power building components selection, and transmission control, by exploiting the possibility to ping the sensor node via the wake-up radio.

The combination of solar and thermoelectric energy harvesting also drives the study put forth in [79]. Therein, an innovative floating device scavenging energy from both sun rays and thermal gradients is investigated. It makes use of LoRa modulation to covey data (i.e., samples of environmental parameters like temperature, humidity, and water pH) to the nearest Gateway. The mixture of the two energy harvesters allows for slightly less than 10 days of fully operation without sun light exposure. Therefore, such a floating LoRa End Node is able to harvest enough power to be completely autonomous.

At this stage it is crystal clear that solar energy harvesters are massively adopted as sunlight is the most immediate source of renewable energy, as it was stated earlier on in Section 4. Moreover, such a strategy usually plays a pivotal role in multiple sources harvesters. An additional instance is the one in [80] where it is in tandem with a radio frequency harvesting module to power up a LoRaWAN End Device for environmental monitoring scopes. Such a device is designed accomplishing energy self-sufficiency owing to the fact that an energy saving policy (i.e., duty-cycling between sleep and active mode) is actuated along with a backup rechargeable battery which does not need to be replaced and disposed thanks to the scavenging capabilities of the node.

5.2. *Sigfox*

Over the years, Sigfox has constantly gained approval amid the framework of IoT and LPWAN. Indeed, despite the fact that users have to subscribe and pay fees to exploit it, as the whole network infrastructure is in charge of Sigfox itself, such a technology proved to be quite plug-and-play because only sensor nodes need to be designed and implemented thus absolving users from concerns related to gateways and server side. Sigfox nodes share a common feature with others belonging to networks which are enabled by different communication standards: they are commonly designed to operate only relying on a limited source of power. Therefore, they still require to be thoroughly designed so to extend their lifetime by minimizing energy consumption. To this end, the authors of [81] put forth a study whose outcome is the derivation of a model, which is based on measurements on Sigfox hardware modules, so to assess hardware performances stemming from realistic usages. In particular, the authors claim that by exploiting a 2400 mAh battery an ideal lifetime spanning from 1.5 to 2.5 years while hourly transmitting six messages with a data rate ranging from 100 bit/s to 600 bit/s may be achieved. Moreover, an asymptotic lifetime of 14.6 years can be reached provided that message broadcasting rate is diminished. Some of the benchmarks that are accounted in the model are uplink physical layer data rate, payload size, unidirectional or bidirectional communication, and message losses. The overall result looks very promising and even satisfactory as it is. However, employing harvesting techniques could permit either to make use of smaller batteries so to reduce both node size and cost or to consent a bigger amount of transmitted data to fulfill application scenario requirements.

Likewise, different enabling technologies, solar energy harvesting is widely adopted in Sigfox sensor nodes. For instance, within water monitoring systems, self-sufficiency from the energy point of view could be achieved by making use of solar energy harvesting [82]. In particular, by keeping sensor node duty cycle below 0.4%, self-sufficiency is reached by resorting to a 720 mAh rechargeable battery, an off-the-shelf solar shield acting as BMS and a 2 W solar panel delivering a maximum of 330 mA during favorable light states so to completely charge the battery in less than 3 h. Energy autonomy for Sigfox nodes employed within environmental monitoring was also investigated in [83]. Such a result stems from the fact that the node is powered by a solar cell, a 90 mAh coin cell battery both managed by an optimized BMS. In so doing, transmissions may take place every half an hour so to ensure 8 h of operation in full darkness that can be compared to the duration of nights during summer. Alternatively, such a duty cycle could be tuned thus resulting in a transmission every 5 min under overcast condition and still the autonomy is ensured.

The literature additionally presents works dealing with floating microbial fuel cells that are exploited to scavenge energy so to power up Sigfox nodes [84]. Floating microbial fuel cells may operate for more than a year as a sort of floating gardens which live within aquatic environments characterized by a large amount of dissolved oxygen. The energy harvesting system needs to be especially designed so to collect energy from a various number of the aforesaid cells that reside in the same water medium. In addition, since a maximum of 800 mV can be usually harvested form a single cell, DC–DC step-up converters are required to be embedded within the harvesting system. In particular, each microbial fuel cell is connected to its DC-DC converter, while the output storage element (i.e., a single capacitor) is in common with the aim of powering the load. In so doing, a maximum overall amount of 5 V can be reached which is largely suitable to power up sensor nodes. However, solely a very slow data transmission (i.e., once a day) can be ensured.

Kinetic energy of sea waves may be also harvested so to power Sigfox sensor nodes by exploiting oceanic undrogued drifters [85]. The harvester consists of a gymbal system, a gearing transmission capable of converting oscillations into rotations, a test mass, a flying wheel, and a micro-generator. Due to motion caused by sea waves, the harvester (that is enclosed within the drifter) rotates according to certain angles of roll, pitch, and yaw. Next, the test mass attempts to oppose such a motion in order to maintain a vertical position; therefore, it properly rotates owing to the gymbal structure. This phenomenon causes a torque that is transmitted through the gears to the flying wheel and, in turn, to the micro generator hosting a DC motor which transforms this rotation into DC current by exploiting the reversibility principle. Finally, such power is processed and managed by especially conceived devices so to provide up to 0.22 mW. Therefore, the whole system is not already suitable for being the main power supply and hence it is used as a backup power of a sensor node that is mounted on the drifter itself which communicates, by leveraging on Sigfox, for coastal communications.

5.3. DASH7

DASH7 networks are well suited for including either sensors and actuators owing to the intrinsic bidirectional capabilities of the protocol itself. Therefore, both sensors and actuators could be equipped with energy harvesters with the vision of being autonomous nodes. D'Elia et al. [86] put into practice this idea within the context of heat distribution systems in domestic buildings. Sensors and actuators communicate via DASH7: the former ones sample temperature and the latter ones drive motors for the tuning of radiators valves. Each class of nodes is provided with a dedicated harvester: the actuators scavenge energy from heat thus embedding a thermal energy generator, while the sensors are powered via photovoltaic cells. Likewise other systems previously described, the harvesters work along with DC–DC converters, BMSs, and rechargeable batteries.

Autonomous DASH7 nodes could be also realized by resorting to radio frequency harvesting, and specifically battery-less self-sustained nodes are shown in [87]. This task is solved by employing a tailor-made rectenna (i.e., the bulk of radio frequency harvesters) connected to a DC-DC converter along with a buffer capacitors and control circuits so to correctly power the sensor node. Of course, all of the building blocks are ultra-low-power components so to overcome to the little amount of harvestable energy coming from environmental radio frequency signals. Despite it, the prototype is capable to operate by intercepting signals within a radius of 17 m.

Energy harvesting could be the key for achieving completely autonomous sensors especially when such devices are designed to operate along with a wake-up radio triggering their functioning only when a wake-up packet is received as it is presented in [88]. Therein, the same prototype that was beforehand illustrated in [86] is equipped with wake-up radios and the relative impact on the nodes lifetime extension is discussed. In so doing, autonomy is elongated as the required energy to be harvested form the environment to correctly power the nodes drastically decreases. Of course, the wake-up radio definitely needs to be meticulously designed so to be as low power as possible in order to not significantly undermine the energy balance resulting form the harvesters. For instance, the wake-up radio that is adopted in [88] only draws up to 1 μA.

5.4. Cellular Technologies

Energy harvesting techniques are still scarcely adopted in systems based on cellular technologies. Indeed, the literature is not plenty of works dealing with this theme. A reasonable motivation could be that such communication standards generally require a vaster amount of energy in comparison with non-cellular technologies due to the fact that the former ones customarily offer higher performances with respect to the latter ones. However, some instances of harvesters for nodes communicating embracing cellular technologies are presented in [89], concerning NB-IoT, and in [90], regarding 5G.

Challenges and opportunities for energy harvesting to be used in NB-IoT sensor nodes with the aim of extending battery lifetime are listed within [89]. Precisely, energy harvesting for NB-IoT nodes is investigated in a smart home scenario where ambient light, either indoor and outdoor, can be scavenged. For what concerns indoor settings, window sills and books shelves are taken into account. The results from this study point out a predictable outcome: outdoor settings by far outperform indoor ones but still all of them provide a significant lifetime lengthening. Specifically, and on a yearly basis, indoor harvested ambient light can provide slightly less than 3000 additional messages of 50 B and 200 B length that only need a good signal coverage to be broadcast, while those figures decrease to few more than 250 whenever a deep coverage is necessary. On the other hand, additional messages that can be sent by exploiting outdoor ambient light roughly double up, respectively, for each of the aforementioned conditions.

Energy harvesting for the brand new 5G technology are summarized in [90]. Therein it is claimed that radio frequency energy harvesting is a favorable method for an alternative energy supply in order to be exploited within 5G communication systems. Along with radio frequency harvesting, also other renewable sources of energy (e.g., thermal energy, sun, light, and mechanical energy) are surveyed. Moreover, the paper additionally points out constraints related to energy harvesting such as causality: when some energy is harvested at the moment, it would be only available in the future. Due to this issue and the remainder that are listed, radio frequency harvesting seems to be a reliable and predictable energy source 5G networks may rely on.

6. Discussion

The just surveyed papers remark potentialities, efficiency, as well as effectiveness of energy harvesting techniques to be exploited within LPWAN contexts. Therefore, such methods will definitely have a booming and flourishing future as it is highly expected that the need of designing sensor nodes able to run relying on the aforesaid strategies will be more and more increasing for multiple reasons. First, if the number of fully autonomous sensor nodes (i.e., the ones which do not require batteries due to the fact that the energy they scavenge is sufficient) would blow up, then no more batteries will have to be substituted and disposed thus contributing to the reduction of pollution embracing a greener perspective. Second, for such nodes that do not reach an utter self-sufficiency, a notable reduction of human interventions for maintenance will be experienced resulting in an overall simplification of upkeep procedures. Additionally, even though harvesters may be more expensive than batteries, the overall node costs will be amortized throughout the whole lifetime since it experiences a remarkable lengthening. Therefore, there exist all the conditions for considering them as full-fledged forms of investment.

As it was stated earlier on (see Section 4), the most immediate renewable source of power is the solar one. As a consequence, such a form of energy is also massively exploited for harvesting; indeed, this fact is confirmed by the conspicuous number of the reviewed studies dealing with it. Moreover, thermoelectric harvesters are also indirectly affected by solar energy since heat is the outcome of sun rays impacting on objects. However, albeit other sources of energy are not as mainstream as the solar one, some solutions exploiting them were cited as well as they proved to be as reliable and efficient as solar energy.

Regardless of the source of energy that is scavenged, sensor nodes relying on harvesters are pooled by a common architecture: the harvester itself is controlled by a BMS deciding whether to directly exploit the scavenged energy to power the node or to store it within rechargeable batteries or supercapacitors. In addition, whenever the source of energy is not constant, further electronics (e.g., AC–DC converters) is needed so to correctly supply the node: aeolian harvesters or systems exploiting mechanical vibrations usually suffer from such a matter. On the other hand, hydroelectric generators working within pipes do not ordinarily necessitate of those additional systems especially whenever the inner water flux is constant. Another discriminant dictating the energy harvesting technique to adopt is the application scenario. While outdoor solar is massively employed, indoor other strategies should be considered. For instance, within industrial environments thermal harvesting (e.g., from cooling systems) or mechanical vibrations harvesting (e.g., from vibrating machineries) are widely put into effect. For what concerns aquatic environments, microbial fuel cells are often taken into account provided that the aquatic context is suitable for cells living, while sensor nodes deployed overboard may take advantage of sea waves motion although the harvester needs of a thorougher design.

Apart from the exploited energy harvesting technique, the bulk of the reviewed papers deals with LPWANs to be employed for environmental or industrial monitoring. However, such facilities may be of precious aid in order to face and manage delicate issues like the one lashed out the entire world during 2020: pandemics. Indeed, the globe was stroke by what was named as Corona Virus Disease 2019 (COVID-19). Albeit a recovery phase was gradually started after some months from the outbreak, subsequent waves (the first of which presumably could occur for late 2020) were predicted. Therefore, as it is also suggested in [91], IoT infrastructures could play the role of crucial supporters to such a fight: for instance, a city-scaled pervasive monitoring network for the measurement of body temperature by leveraging on thermal cameras to be installed at the entrance of public places (e.g., schools, bars, shops, and so on) whose data could be retrieved and analysed by a central core (e.g., hospitals) so to cope with the spread of contagion could be a solution. However, the same concept could be applied to other disease monitoring scenarios too. Another subtle point arising whenever data flows through networks like LPWANs and IoT systems is the one of cyber risk that has to be intended in a broad sense: from personal information leakage to data transfer especially when sensible information is involved like in the case of the aforesaid example coping with COVID-19. On the concerns associated with the communication technologies introduced earlier on and with IoT in general, [92] gives a deep outlook in standardisation of IoT cyber risks as well as future perspectives by identifying IoT cyber risk vectors (i.e., IoT attack vectors from particular approach used so to mine into big data vulnerabilities) and by integrating them in models aiming at determining cyber risks impact. Therefore, we deem that besides the future development of energy harvesting techniques for LPWANs, also countermeasures for risks (which naturally emerge in this context) will experience a notable spike.

7. Conclusions

The aim of this paper was to provide a comprehensive review of the existing energy harvesting solutions for the powering of LPWAN-based monitoring systems and architectures. The literature was carefully reviewed and, at the best of our knowledge, all the most significant contributions dealing with the proposed topic have been included in this survey. While the total number of reviewed papers may appear to be low (i.e., a total number of 44 contributions was discussed) we would like to point out that all the works where published in the last five years, since LPWAN technologies are relatively new in the context of IoT.

As LPWAN-based IoT systems are spreading very fast, the emergence of energy harvesting techniques is expected in the next few years. In this sense, we would like to point out that some renewable power sources, that are widely exploited in several technological context, like wind or water flow, were never applied to LPWAN systems. We believe that, at least for these two energy sources and hopefully for others, the appearance of prototypes and working systems over the next few years will be seen.

Author Contributions: The authors equally contributed to the paper. All authors have read and agreed to the published version of the manuscript.

Funding: This research received no external funding.

Conflicts of Interest: The authors declare no conflicts of interest.

References

1. Mekki, K.; Bajic, E.; Chaxel, F.; Meyer, F. A comparative study of LPWAN technologies for large-scale IoT deployment. *ICT Express* **2019**, *5*, 1–7. [CrossRef]
2. Zanella, A.; Bui, N.; Castellani, A.; Vangelista, L.; Zorzi, M. Internet of things for smart cities. *IEEE Internet Things* **2014**, *1*, 22–32. [CrossRef]
3. Addabbo, T.; Fort, A.; Mugnaini, M.; Panzardi, E.; Pozzebon, A.; Vignoli, V. A city-scale IoT architecture for monumental structures monitoring. *Measurement* **2019**, *131*, 349–357. [CrossRef]
4. Boulogeorgos, A.A.A.; Diamantoulakis, P.D.; Karagiannidis, G.K. Low power wide area networks (lpwans) for internet of things (iot) applications: Research challenges and future trends. *arXiv* **2016**, arXiv:1611.07449.
5. Jain, P.C. Recent trends in energy harvesting for green wireless sensor networks. In Proceedings of the 2015 International Conference on Signal Processing and Communication (ICSC), Noida, India, 16–18 March 2015; pp. 40–45.
6. Shaikh, F.K.; Zeadally, S. Energy harvesting in wireless sensor networks: A comprehensive review. *Renew. Sust. Energy Rev.* **2016**, *55*, 1041–1054. [CrossRef]
7. Bui, N.; Rossi, M. Staying alive: System design for self-sufficient sensor networks. *ACM Trans. Sens. Netw.* **2015**, *11*, 1–42. [CrossRef]
8. Wu, F.; Redouté, J.M.; Yuce, M.R. We-safe: A self-powered wearable IoT sensor network for safety applications based on LoRa. *IEEE Access* **2018**, *6*, 40846–40853. [CrossRef]
9. Li, Y.; Shi, R. An intelligent solar energy-harvesting system for wireless sensor networks. *EURASIP J. Wirel. Commun.* **2015**, *1*, 179. [CrossRef]
10. Mhetre, C.; Narkhede, P.; Pashte, Y.; Patankar, N.; Nishan, P. Solar Powered Wireless Sensor Network Using Zigbee Module. In Proceedings of the 2019 International Conference on Intelligent Computing and Control Systems (ICCS), Madurai, India, 15–17 May 2019; pp. 468–473.
11. Savanth, A.; Weddell, A.; Myers, J.; Flynn, D.; Al-Hashimi, B. Photovoltaic cells for micro-scale wireless sensor nodes: measurement and modeling to assist system design. In Proceedings of the 3rd International Workshop on Energy Harvesting & Energy Neutral Sensing Systems, Seoul, Korea, 1 November 2015; pp. 15–20.
12. Wang, J.C.; Liao, M.S.; Lee, Y.C.; Liu, C.Y.; Kuo, K.C.; Chou, C.Y.; Huang, C.K.; Jiang, J.A. On enhancing energy harvesting performance of the photovoltaic modules using an automatic cooling system and assessing its economic benefits of mitigating greenhouse effects on the environment. *J. Power Sources* **2018**, *376*, 55–65. [CrossRef]
13. Tsai, D.L.; Wu, H.H.; Wei, C.L. A low-power-consumption boost converter with maximum power tracking algorithm for indoor photovoltaic energy harvesting. In Proceedings of the 2017 IEEE Wireless Power Transfer Conference (WPTC), Taipei, Taiwan, 10–12 May 2017; pp. 1–3.

14. Pubill, D.; Serra, J.; Verikoukis, C. Harvesting artificial light indoors to power perpetually a Wireless Sensor Network node. In Proceedings of the 2018 IEEE 23rd International Workshop on Computer Aided Modeling and Design of Communication Links and Networks (CAMAD), Barcelona, Spain, 17–19 September 2018; pp. 1–6.
15. Chien, L.J.; Drieberg, M.; Sebastian, P.; Hiung, L.H. A simple solar energy harvester for wireless sensor networks. In Proceedings of the 2016 6th International Conference on Intelligent and Advanced Systems (ICIAS), Kuala Lumpur, Malaysia, 15–17 August 2016; pp. 1–6.
16. Sharma, H.; Haque, A.; Jaffery, Z.A. Modeling and optimisation of a solar energy harvesting system for wireless sensor network nodes. *J. Sens. Actuator Netw.* **2018**, *7*, 40. [CrossRef]
17. Sharma, H.; Haque, A.; Jaffery, Z.A. Maximization of wireless sensor network lifetime using solar energy harvesting for smart agriculture monitoring. *Ad Hoc Netw.* **2019**, *94*, 101966. [CrossRef]
18. Haug, B. Wireless sensor nodes can be powered by temperature gradients; no batteries needed: Harvesting energy from thermoelectric generators. *IEEE Power Electron. Mag.* **2017**, *4*, 24–32. [CrossRef]
19. Mehne, P.; Lickert, F.; Bäumker, E.; Kroener, M.; Woias, P. Energy-autonomous wireless sensor nodes for automotive applications, powered by thermoelectric energy harvesting. *J. Phys. Conf. Ser.* **2016**, *773*, 012041. [CrossRef]
20. Abdal-Kadhim, A.M.; Leong, K.S. Application of thermal energy harvesting from low-level heat sources in powering up WSN node. In Proceedings of the 2017 2nd International Conference on Frontiers of Sensors Technologies (ICFST), Shenzhen, China, 14–16 April 2017; pp. 131–135.
21. Hou, L.; Tan, S.; Zhang, Z.; Bergmann, N.W. Thermal energy harvesting WSNs node for temperature monitoring in IIoT. *IEEE Access* **2018**, *6*, 35243–35249. [CrossRef]
22. Haidar, M.; Chible, H.; Di Zitti, E.; Caviglia, D.D. An Optimized AC/DC Buck-Boost Converter for Wind Energy Harvesting Application. In Proceedings of the 2019 IEEE International Conference on Environment and Electrical Engineering and 2019 IEEE Industrial and Commercial Power Systems Europe (EEEIC/I&CPS Europe), Genova, Italy, 11–14 June 2019; pp. 1–4.
23. Jushi, A.; Pegatoquet, A.; Le, T.N. Wind energy harvesting for autonomous wireless sensor networks. In Proceedings of the 2016 Euromicro Conference on Digital System Design (DSD), Limassol, Cyprus, 31 August–2 September 2016; pp. 301–308.
24. Wu, Y.; Li, B.; Zhang, F. Predictive power management for wind powered wireless sensor node. *Future Internet* **2018**, *10*, 85. [CrossRef]
25. Kroener, M.; Allinger, K.; Berger, M.; Grether, E.; Wieland, F.; Heller, S.; Woias, P. A water-powered Energy Harvesting system with Bluetooth Low Energy interface. *J. Phys. Conf. Ser.* **2016**, *773*, 012040. [CrossRef]
26. Mecocci, A.; Peruzzi, G.; Pozzebon, A.; Vaccarella, P. Architecture of a hydroelectrically powered wireless sensor node for underground environmental monitoring. *IET Wirel. Sens. Syst.* **2017**, *7*, 123–129. [CrossRef]
27. Mirab, H.; Fathi, R.; Jahangiri, V.; Ettefagh, M.M.; Hassannejad, R. Energy harvesting from sea waves with consideration of airy and JONSWAP theory and optimization of energy harvester parameters. *J. Mar. Sci. Appl.* **2015**, *14*, 440–449. [CrossRef]
28. Ahmed, A.; Saadatnia, Z.; Hassan, I.; Zi, Y.; Xi, Y.; He, X.; Zu, J.; Wang, Z.L. Self-powered wireless sensor node enabled by a duck-shaped triboelectric nanogenerator for harvesting water wave energy. *Adv. Energy Mater.* **2017**, *7*, 1601705. [CrossRef]
29. Mouapi, A.; Hakem, N.; Delisle, G.Y.; Kandil, N. A novel piezoelectric micro-generator to power Wireless Sensors Networks in vehicles. In Proceedings of the 2015 IEEE 15th International Conference on Environment and Electrical Engineering (EEEIC), Rome, Italy, 10–13 June 2015; pp. 1089–1092.
30. Somov, A.; Chew, Z.J.; Ruan, T.; Li, Q.; Zhu, M. Piezoelectric energy harvesting powered WSN for aircraft structural health monitoring. In Proceedings of the 2016 15th ACM/IEEE International Conference on Information Processing in Sensor Networks (IPSN), Vienna, Austria, 11–14 April 2016; pp. 1–2.
31. Panthongsy, P.; Isarakorn, D.; Sudhawiyangkul, T.; Nundrakwang, S. Piezoelectric energy harvesting from machine vibrations for wireless sensor system. In Proceedings of the 2015 12th International Conference on Electrical Engineering/Electronics, Computer, Telecommunications and Information Technology (ECTI-CON), Hua Hin, Thailand, 24–27 June 2015; pp. 1–6.
32. Khan, F.U.; Ahmad, I. Review of energy harvesters utilizing bridge vibrations. *Shock Vib.* **2016**, *2016*, 1340402. [CrossRef]

33. Garcia, R.; Combette, P.; Poulin, Y.; Foucaran, A.; Podlecki, J.; Hassen, S.B.; Grilli, M.A.; Hess, O.; Briant, F. Piezoelectric energy harvesting: application to data center monitoring. *Sens. Rev.* **2015**, *35*, 401–408. [CrossRef]
34. Li, H.; Tian, C.; Lu, J.; Myjak, M.J.; Martinez, J.J.; Brown, R.S.; Deng, Z.D. An energy harvesting underwater acoustic transmitter for aquatic animals. *Sci. Rep.* **2016**, *6* 33804. [CrossRef] [PubMed]
35. Han, Y.; Feng, Y.; Yu, Z.; Lou, W.; Liu, H. A study on piezoelectric energy-harvesting wireless sensor networks deployed in a weak vibration environment. *IEEE Sens. J.* **2017**, *17*, 6770–6777. [CrossRef]
36. Du, S.; Jia, Y.; Zhao, C.; Amaratunga, G.A.; Seshia, A.A. A nail-size piezoelectric energy harvesting system integrating a mems transducer and a cmos sshi circuit. *IEEE Sens. J.* **2019**, *20*, 277–285. [CrossRef]
37. Salim, N.; Idros, M.F.M.; Al-Junid, S.A.M.; Razak, A.H.A. Study on the capability of Acoustic Energy Harvesting for low power device application. In Proceedings of the 2016 IEEE Student Conference on Research and Development (SCOReD), Kuala Lumpur, Malaysia, 13–14 December 2016; pp. 1–4.
38. Khan, F.U. Electromagnetic energy harvester for harvesting acoustic energy. *Sadhana* **2016**, 41, 397–405. [CrossRef]
39. Khan, F.U. Three degree of freedom acoustic energy harvester using improved Helmholtz resonator. *Int. J. Pract. Eng. Man.* **2018**, *19*, 143–154.
40. Boumaiz, M.; El Ghazi, M.; Mazer, S.; Fattah, M.; Bouyad, A.; El Bekkali, M.; Balboul, Y. Energy harvesting based WBANs: EH optimization methods. In Proceedings of the International Workshop on Microwave Engineering, Communications Systems and Technologies (MECST'2019), Leuven, Belgium, 29 April–2 May 2019; pp. 1040–1045.
41. Magno, M.; Kneubühler, D.; Mayer, P.; Benini, L. Micro kinetic energy harvesting for autonomous wearable devices. In Proceedings of the 2018 International Symposium on Power Electronics, Electrical Drives, Automation and Motion (SPEEDAM), Amalfi, Italy, 20–22 June 2018; pp. 105–110.
42. Ju, Q.; Li, H.; Zhang, Y. Power management for kinetic energy harvesting IoT. *IEEE Sens. J.* **2018**, *18*, 4336–4345. [CrossRef]
43. Kuang, Y.; Ruan, T.; Chew, Z.J.; Zhu, M. Energy harvesting during human walking to power a wireless sensor node. *Sens. Actuators A Phys.* **2017**, *254*, 69–77. [CrossRef]
44. Zeng, X.; Li, B.; Li, H.; Chen, S.; Chen, Y. Non-invasive energy harvesting for wireless sensors from electromagnetic fields around 10 kV three-core power cables. In Proceedings of the 2017 1st International Conference on Electrical Materials and Power Equipment (ICEMPE), Xi'an, China, 14–17 May 2017; pp. 536–539.
45. Khan, F.U. Energy harvesting from the stray electromagnetic field around the electrical power cable for smart grid applications. *Sci. World J.* **2016**, *2016*, 3934289. [CrossRef]
46. Chamanian, S.; Ulusan, H.; Zorlu, O.; Baghaee, S.; Uysal-Biyikoglu, E.; Kulah, H. Wearable battery-less wireless sensor network with electromagnetic energy harvesting system. *Sens. Actuators A Phys.* **2016**, *249*, 77–84. [CrossRef]
47. Takacs, A.; Okba, A.; Aubert, H.; Charlot, S.; Calmon, P.F. Recent advances in electromagnetic energy harvesting and Wireless Power Transfer for IoT and SHM applications. In Proceedings of the 2017 IEEE International Workshop of Electronics, Control, Measurement, Signals and their Application to Mechatronics (ECMSM), Donostia-San Sebastian, Spain, 24–26 May 2017; pp. 1–4.
48. Adila, A.S.; Husam, A.; Husi, G. Towards the self-powered Internet of Things (IoT) by energy harvesting: Trends and technologies for green IoT. In Proceedings of the 2018 2nd International Symposium on Small-scale Intelligent Manufacturing Systems (SIMS), Cavan, Ireland, 16–18 April 2018; pp. 1–5.
49. Curry, J.; Harris, N. Powering the Environmental Internet of Things. *Sensors* **2019**, *19*, 1940. [CrossRef] [PubMed]
50. Gleonec, P.; Ardouin, J.; Gautier, M.; Berder, O. Architecture exploration of multi-source energy harvester for IoT nodes. In Proceedings of the 2016 IEEE Online Conference on Green Communications (OnlineGreenComm), Piscataway, NJ, USA, 14 November–17 December 2016; pp. 27–32.
51. Delgado, C.; Sanz, J.M.; Famaey, J. On the Feasibility of Battery-Less LoRaWAN Communications Using Energy Harvesting. In Proceedings of the 2019 IEEE Global Communications Conference (GLOBECOM), Waikoloa, HI, USA, 9–13 December 2019; pp. 1–6.

52. Benkhelifa, F.; Qin, Z.; McCann, J. Minimum Throughput Maximization in LoRa Networks Powered by Ambient Energy Harvesting. In Proceedings of the ICC 2019 IEEE International Conference on Communications (ICC), Shanghai, China, 20–24 May 2019; pp. 1–7.
53. Sherazi, H.H.R.; Imran, M.A.; Boggia, G.; Grieco, L.A. Energy Harvesting in LoRaWAN: A Cost Analysis for the Industry 4.0. *IEEE Commun. Lett.* **2018**, *22*, 2358–2361. [CrossRef]
54. Mabon, M.; Gautier, M.; Vrigneau, B.; Le Gentil, M.; Berder, O. The Smaller the Better: Designing Solar Energy Harvesting Sensor Nodes for Long-Range Monitoring. *Wirel. Commun. Mob. Comput.* **2019**, 2878545. [CrossRef]
55. Polonelli, T.; Brunelli, D.; Guermandi, M.; Benini, L. An accurate low-cost Crackmeter with LoRaWAN communication and energy harvesting capability. In Proceedings of the 2018 IEEE 23rd International Conference on Emerging Technologies and Factory Automation (ETFA), Turin, Italy, 4–7 September 2018; pp. 671–676.
56. Kobayashi, M.; Matsumoto, T.; Beering, J. Development of Water Temperature Measuring Application Based on LoRa/LoRWAN. In Proceedings of the 2019 4th International Conference on Information Technology (InCIT), Bangkok, Thailand, 24–25 October 2019, pp. 254–258.
57. Boccadoro, P.; Montaruli, B.; Grieco, L.A. QuakeSense, a LoRa-compliant Earthquake Monitoring Open System. In Proceedings of the 2019 IEEE/ACM 23rd International Symposium on Distributed Simulation and Real Time Applications (DS-RT), Cosenza, Italy, 7–9 October 2019; pp. 1–8.
58. Matthews, V.O.; Ajala, A.O.; Atayero, A.A.; Popoola, S.I. Solar photovoltaic automobile recognition system for smart-green access control using RFID and LoRa LPWAN technologies. *J. Eng. Appl. Sci.* **2017**, *12*, 913–919.
59. Petrariu, A.I.; Lavric, A.; Coca, E. Renewable Energy Powered LoRa-based IoT Multi Sensor Node. In Proceedings of the 2019 IEEE 25th International Symposium for Design and Technology in Electronic Packaging (SIITME), Cluj-Napoca, Romania, 23–26 October 2019; pp. 94–97.
60. Rossi, M.; Tosato, P. Energy neutral design of an IoT system for pollution monitoring. In Proceedings of the 2017 IEEE Workshop on Environmental, Energy, and Structural Monitoring Systems (EESMS), Milan, Italy, 24–25 July 2017; pp. 1–6.
61. Nikolov, D.N.; Ganev, B.T.; Rusev, R.P. Energy harvesting power supply for an autonomous environmental sensor node. In Proceedings of the 2019 IEEE XXVIII International Scientific Conference Electronics (ET), Sozopol, Bulgaria, 12–14 September 2019; pp. 1–4.
62. Sherazi, H.H.R.; Piro, G.; Grieco, L.A.; Boggia, G. When renewable energy meets LoRa: A feasibility analysis on cable-less deployments. *IEEE Internet Things J.* **2018**, *5*, 5097–5108. [CrossRef]
63. Lin, H.; Chen, Z.; Wang, L. Offloading for Edge Computing in Low Power Wide Area Networks With Energy Harvesting. *IEEE Access* **2019**, *7*, 78919–78929. [CrossRef]
64. Janhunen, J.; Mikhaylov, K.; Petajajarvi, J. Experimental RF-signal based wireless energy transmission. In Proceedings of the 2017 European Conference on Networks and Communications (EuCNC), Oulu, Finland, 12–15 June 2017; pp. 1–6.
65. Peng, Y.; Shangguan, L.; Hu, Y.; Qian, Y.; Lin, X.; Chen, X.; Fang, D.; Jamieson, K. PLoRa: A passive long-range data network from ambient LoRa transmissions. In Proceedings of the 2018 Conference of the ACM Special Interest Group on Data Communication, Budapest, Hungary, 20–25 August 2018; pp. 147–160.
66. Tjukovs, S.; Eidaks, J.; Pikulins, D. Experimental Verification of Wireless Power Transfer Ability to Sustain the Operation of LoRaWAN Based Wireless Sensor Node. In Proceedings of the 2018 Advances in Wireless and Optical Communications (RTUWO), Riga, Latvia, 15–16 November 2018; pp. 83–88.
67. Molefi, M.; Markus, E.D.; Abu-Mahfouz, A. Wireless power transfer for LoRa low-power wide-area networks (LPWANs). In Proceedings of the 2019 Southern African Universities Power Engineering Conference/Robotics and Mechatronics/Pattern Recognition Association of South Africa (SAUPEC/RobMech/PRASA), Bloemfontein, South Africa, 28–30 January 2019; pp. 105–110.
68. Kim, Y.J.; Gu, H.M.; Kim, C.S.; Choi, H.; Lee, G.; Kim, S.; Yi, K.K.; Lee, S.G.; Cho, B.J. High-performance self-powered wireless sensor node driven by a flexible thermoelectric generator. *Energy* **2018**, *162*, 526–533. [CrossRef]
69. Alegret, R.N.; Aragones, R.; Oliver, J.; Ferrer, C. Exploring IIoT and Energy Harvesting Boundaries. In Proceedings of the IECON 2019-45th Annual Conference of the IEEE Industrial Electronics Society, Lisbon, Portugal, 14–17 October 2019; pp. 6732–6736.

70. Meli, M.; Hegetschweiler, L. Harvesting energy from trees in order to power LPWAN IoT nodes. In Proceedings of the Wireless Congress 2018, Munich, Germany, 21–22 March 2018; pp. 1–11.
71. Wang, W.; Chen, X.; Liu, Y.; Wang, X.; Liu, Z. Thermo-electric Energy Harvesting Powered IoT System Design and Energy Model Analysis. In Proceedings of the 2019 IEEE 13th International Conference on Anti-counterfeiting, Security, and Identification (ASID), Xiamen, China, 25–27 October 2019; pp. 303–308.
72. Orfei, F.; Mezzetti, C.B.; Cottone, F. Vibrations powered LoRa sensor: An electromechanical energy harvester working on a real bridge. In Proceedings of the 2016 IEEE SENSORS, Orlando, FL, USA, 30 October–3 November 2016; pp. 1–3.
73. Rodriguez, J.C.; Nico, V.; Punch, J. A vibration energy harvester and power management solution for battery-free operation of wireless sensor nodes. *Sensors* **2019**, *19*, 3776. [CrossRef]
74. Chandrasekhar, A.; Vivekananthan, V.; Kim, S.J. A fully packed spheroidal hybrid generator for water wave energy harvesting and self-powered position tracking. *Nano Energy* **2020**, *69*, 104439. [CrossRef]
75. Gajda, I.; Cristiani, P.; Greenman, J.; Pizza, F.; Bonelli, P.; Ieropoulos, I. Long Term Feasibility Study of In-field Floating Microbial Fuel Cells for Monitoring Anoxic Wastewater and Energy Harvesting. *Front. Energy Res.* **2019**, *7*. [CrossRef]
76. Ayala-Ruiz, D.; Castillo Atoche, A.; Ruiz-Ibarra, E.; Osorio de la Rosa, E.; Vázquez Castillo, J. A Self-Powered PMFC-Based Wireless Sensor Node for Smart City Applications. *Wirel. Commun. Mob. Comput.* **2019**, *2019*, 8986302. [CrossRef]
77. Ferrero, F.; Le-Quoc, H. Multi-harvesting solution for autonomous sensing node based on LoRa technology. In Proceedings of the 2017 International Conference on Advanced Technologies for Communications (ATC), Quy Nhon, Vietnam, 18–20 October 2017; pp. 250–257.
78. Magno, M.; Aoudia, F.A.; Gautier, M.; Berder, O.; Benini, L. WULoRa: An energy efficient IoT end-node for energy harvesting and heterogeneous communication. In Proceedings of the Design, Automation & Test in Europe Conference & Exhibition (DATE), Lausanne, Switzerland, 27–31 March 2017; pp. 1528–1533.
79. Lee, W.K.; Schubert, M.J.; Ooi, B.Y.; Ho, S.J.Q. Multi-source energy harvesting and storage for floating wireless sensor network nodes with long range communication capability. *IEEE Trans. Ind. Appl.* **2018**, *54*, 2606–2615. [CrossRef]
80. Wang, Y.; Huang, Y.; Song, C. A New Smart Sensing System Using LoRaWAN for Environmental Monitoring. In Proceedings of the 2019 Computing, Communications and IoT Applications (ComComAp), Shenzhen, China, 26–28 October 2019; pp. 347–351.
81. Gomez, C.; Veras, J.C.; Vidal, R.; Casals, L.; Paradells, J. A Sigfox energy consumption model. *Sensors* **2019**, *19*, 681. [CrossRef] [PubMed]
82. Di Gennaro, P.; Lofu, D.; Vitanio, D.; Tedeschi, P.; Boccadoro, P. WaterS: A Sigfox-compliant prototype for water monitoring. *Internet Technol. Lett.* **2019**, *2*, e74. [CrossRef]
83. Joris, L.; Dupont, F.; Laurent, P.; Bellier, P.; Stoukatch, S.; Redouta, J.M. An Autonomous Sigfox Wireless Sensor Node for Environmental Monitoring. *IEEE Sens. Lett.* **2019**, *3*, 01–04. [CrossRef]
84. Schievano, A.; Colombo, A.; Grattieri, M.; Trasatti, S.P.; Liberale, A.; Tremolada, P.; Pino, C.; Cristiani, P. Floating microbial fuel cells as energy harvesters for signal transmission from natural water bodies. *J. Power Sources* **2017**, *340*, 80–88. [CrossRef]
85. Carandell, M.; Toma, D.M.; Carbonell, M.; Gasulla, M.; del Río, J. Design and development of a kinetic energy harvester device for oceanic drifter applications. In Proceedings of the 2019 IEEE International Instrumentation and Measurement Technology Conference (I2MTC), Auckland, New Zealand, 20–23 May 2019; pp. 1–6.
86. D'Elia, A.; Perilli, L.; Viola, F.; Roffia, L.; Antoniazzi, F.; Canegallo, R.; Cinotti, T.S. A self-powered WSAN for energy efficient heat distribution. In Proceedings of the 2016 IEEE Sensors Applications Symposium (SAS), Catania, Italy, 20–22 April 2016; pp. 1–6.
87. Pizzotti, M.; Perilli, L.; Del Prete, M.; Fabbri, D.; Canegallo, R.; Dini, M.; Masotti, D.; Costanzo, A.; Franchi Scarselli, E.; Romani, A. A long-distance RF-powered sensor node with adaptive power management for IoT applications. *Sensors* **2017**, *17*, 1732. [CrossRef]
88. Perilli, L.; Scarselli, E.F.; La Rosa, R.; Canegallo, R. Wake-Up Radio Impact in Self-Sustainability of Sensor and Actuator Wireless Nodes in Smart Home Applications. In Proceedings of the 2018 Ninth International Green and Sustainable Computing Conference (IGSC), Pittsburgh, PA, USA, 22–24 October 2018; pp. 1–7.

89. Haridas, A.; Rao, V.S.; Prasad, R.V.; Sarkar, C. Opportunities and challenges in using energy-harvesting for NB-IoT. *ACM SIGBED Rev.* **2018**, *15*, 7–13. [CrossRef]
90. El Hassani, S.; El Hassani, H.; Boutammachte, N. Overview on 5G Radio Frequency Energy Harvesting. *ASTESJ* **2019**, *4*, 328–346. [CrossRef]
91. Radanliev, P.; De Roure, D.; Van Kleek, M. Digitalization of COVID-19 pandemic management and cyber risk from connected systems. *arXiv* **2020**, arXiv:2005.12409.
92. Radanliev, P.; De Roure, D.C.; Nurse, J.R.; Montalvo, R.M.; Cannady, S.; Santos, O.; Maddox, L.; Burnap, P.; Maple, C. Future developments in standardisation of cyber risk in the Internet of Things (IoT). *SN Appl. Sci.* **2020**, *2*, 169. [CrossRef]

energies

Article

Internet of Things (IoT) Platform for Multi-Topic Messaging

Mahmoud Hussein [1,2,†,‡], Ahmed I. Galal [2,‡], Emad Abd-Elrahman [1,*,†,‡] and Mohamed Zorkany [1,†,‡]

1 National Telecommunication Institute (NTI), Cairo 11768, Egypt; mah.hussein@nti.sci.eg (M.H.); m_zorkany@nti.sci.eg (M.Z.)
2 Faculty of Engineering, Minia University, Minia 61519, Egypt; galal@mu.edu.eg
* Correspondence: emad.abdelrahman@nti.sci.eg; Tel.: +20-1000-720-268
† Current address: 5 Mahmoud El Miligui Street, 6th district-Nasr City, Cairo 11768, Egypt.
‡ These authors contributed equally to this work.

Received: 1 May 2020; Accepted: 26 June 2020; Published: 30 June 2020

Abstract: IoT-based applications operate in a client–server architecture, which requires a specific communication protocol. This protocol is used to establish the client–server communication model, allowing all clients of the system to perform specific tasks through internet communications. Many data communication protocols for the Internet of Things are used by IoT platforms, including message queuing telemetry transport (MQTT), advanced message queuing protocol (AMQP), MQTT for sensor networks (MQTT-SN), data distribution service (DDS), constrained application protocol (CoAP), and simple object access protocol (SOAP). These protocols only support single-topic messaging. Thus, in this work, an IoT message protocol that supports multi-topic messaging is proposed. This protocol will add a simple "brain" for IoT platforms in order to realize an intelligent IoT architecture. Moreover, it will enhance the traffic throughput by reducing the overheads of messages and the delay of multi-topic messaging. Most current IoT applications depend on real-time systems. Therefore, an RTOS (real-time operating system) as a famous OS (operating system) is used for the embedded systems to provide the constraints of real-time features, as required by these real-time systems. Using RTOS for IoT applications adds important features to the system, including reliability. Many of the undertaken research works into IoT platforms have only focused on specific applications; they did not deal with the real-time constraints under a real-time system umbrella. In this work, the design of the multi-topic IoT protocol and platform is implemented for real-time systems and also for general-purpose applications; this platform depends on the proposed multi-topic communication protocol, which is implemented here to show its functionality and effectiveness over similar protocols.

Keywords: internet of things (IoT); real-time system (RTS); real-time operating systems (RTOS); IoT protocols

1. Introduction

The proliferation of wireless connectivity in Internet of Things (IoT) devices is rapidly expanding. This is leading to the launch and integration of many IoT services and applications. The IoT is a technology that is widely used for interconnecting devices ("Things") through the Internet. It is used in many applications and fields, such as security, E-health, home automation, emergencies, logistics, smart metering, industrial control and smart cities [1]. Recently, a concept has been applied to IoT platforms that involves the perception of the conditions of the network, the analysis of the gathered knowledge, the making of smart decisions and the performance of actions adaptively [2]. This targets the maximization of the performance of the entire network.

Moreover, an IoT system can integrate cooperative algorithms and mechanisms that can ameliorate problems and promote performance by achieving intelligent actions [3]. This system can detect the network conditions, analyze the gathered knowledge, make intelligent decisions and perform automatic and adaptive actions that maximize the network performance. In this process, multi-domain integration can increase network capacity. Despite the limited research into the intelligent IoT field [4], there are many applications of the technology in different directions such as smart homes and cities [5], drone applications [6], agriculture and farming. Another IoT domain—called the cognitive domain—adds computing algorithms and mechanisms to IoT platforms so that the system devices can make decisions and actions [7]. These devices should have an IoT communication protocol that is responsible for establishing the connection between clients (devices) and the system's main broker server. The server (broker) executes the protocol algorithm with the connected devices, where the algorithm describes the sequences required for a successful communication process.

The common network architecture for IoT systems is centralized networks, as shown in Figure 1; the system may have a large number of nodes which require transmissions between multipoints (clients). In most IoT platforms, to achieve IoT systems requirements, a centralized network between the server and clients is established. In IoT systems, to connect between two devices (nodes, clients), devices should establish a connection with the server, and then the node can send and receive various messages which are described by IoT protocols.

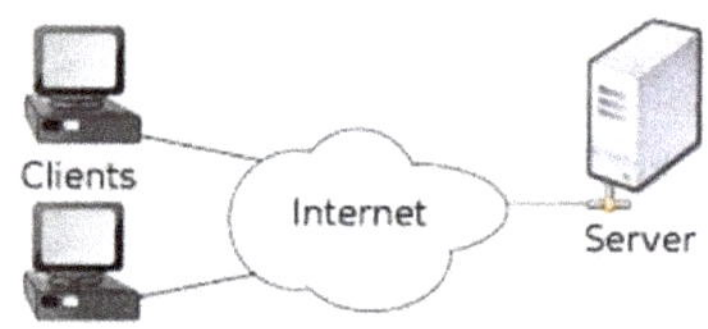

Figure 1. Architecture of the Internet of Things (IoT) network.

MQTT and CoAP message protocols are the most common IoT data protocols. MQTT messages have less of a delay than the CoAP protocol and a smaller message size compared with CoAP, and it is based on a transmission control protocol (TCP) connection, while CoAP uses user datagram protocol (UDP) connections; thus, MQTT has higher reliability than CoAP. MQTT is more suitable than CoAP for real-time systems as fewer overhead bytes are added to the messages being transferred [8].

This paper can be considered as an extension of our conference paper [9] which initiated the main framework. In this research, a new message protocol is proposed for IoT applications. The proposed protocol is designed to overcome an issue which has appeared with MQTT: multi-topic non-support messages, which require extra bytes and cause increased delays from the overhead side. On the other hand, the proposed data protocol handles the multi-topic messages which have become a feature of the system; this will be discussed and detailed in the following sub-sections.

In this work, the MQTT protocol is selected as the most famous IoT standard protocol for comparison with our proposed protocol. There are two ways to simulate the proposed protocol compared with standard MQTT: using ready-made solutions that depend on using open source programs (e.g., Mosquitto (https://mosquitto.org/)— Eclipse Mosquitto is an open-source message MQTT broker (Eclipse public license (EPL)/Eclipse distribution License (EDL)) that can be used to implement the MQTT protocol and support different versions 3.1, 3.1.1, and 5.0. Mosquitto supports Windows, Mac, Linux, Debian, Ubuntu, and Raspberry Pi that are commonly used, or building a new IoT platform from scratch. In this research, we decided to design an IoT platform from scratch to support real-time applications; thus, the simulation and real implementation results are discussed after implementing the IoT protocol based on the proposed architecture. Then, the proposed protocol is compared with the standard MQTT protocol in terms of message overheads and transmission delays.

Real-time systems (RTSs) are systems in which the output occurs in real time and must be correct; thus, most IoT systems should be RTSs. It is important to tailor to critical systems that have deadlines at a critical time, and these can be classified into two types: soft and hard RTSs. If a delayed time requirement is accepted, the RTS system is called a soft RTS; if not, then it is called a hard RTS. To meet the critical deadlines in an RTS, real-time operating systems (RTOSs) will be used [10]. An RTOS is an operating system (OS) that is used for embedded RTSs. This OS is used as it guarantees capabilities such as compactness, high performance, predictability, reliability and modularity [11]. RTOS supports many services such as time, memory and task management, as well as providing multi-tasking. Taking advantage of all of these features of RTOS, the proposed platform will be implemented based on this system. Research perspectives regarding RTOS for IoT investigating such themes as adjusting RTOS to work with IoT systems, the implementations of IoT platforms, IoT frameworks and IoT performance evaluation were discussed in [1].

IoT platforms make IoT development simpler, as all IoT clients (devices with Internet access) are connected to the broker (server). Using an IoT platform, we can publish sensor data from clients to other interested clients (subscribers) through the IoT and take further actions through actuator nodes. Much research has been undertaken into implementing IoT platforms [12–14], but most of them support neither the nature of RTSs, where time is critical, nor specific applications. Furthermore, most research has used ready-made protocols such as MQTT [15]. Thus, in our research, an IoT platform based on RTOS will be implemented using the proposed multi-topic communication protocol.

In the rest of this work is structured as follows: Section 2 presents an overview of the relevant IoT protocols. Section 3 introduces the proposed multi-topic IoT protocol. The proposed IoT platform based on RTOS is detailed in Section 4. The experimentation setup phases are highlighted in Section 5; then, experimental and simulation results are shown in Section 6. In Section 7, the main characteristics of the multi-topic protocol are discussed. Finally, the work is concluded with some prospective research directions in Section 8.

2. State of the Art

2.1. IoT Protocols

IoT systems are characterized by remote monitoring and control aspects. These aspects allow IoT components to communicate together through a remote service that is powered by Internet communications. System nodes are connected through a predefined communication protocol. The data protocol of IoT systems provides different numbers of message frames which enable remote messaging between IoT system nodes. There are many IoT protocols for IoT systems, including MQTT, MQTT-SN, AMQP, DDS, CoAP and SOAP, but MQTT and CoAP are the most common protocols [16].

MQTT is the most dominant IoT communication protocol; it is a pub/sub message system for limited-resources device and unreliable networks and was developed by IBM [17] and standardized by the Organization for the advancement of structured information standards (OASIS).

Another common IoT data protocol is CoAP; this is a recently developed protocol which must be used for communication by constrained devices [18]. It depends on the "representational state transfer" (REST) mechanism, which supports "request–response" models such as HyperText Transfer Protocol (HTTP).

Many IoT protocols support single-topic messaging. This type of messaging implies that one topic only per message can be sent through the network (a single topic such as publishing the temperature, pressure, humidity, etc.). The message topic is important information that is required to be published to subscribers. A subscriber node is a node that explicitly requests any published messages for a specific topic, as shown in Figure 2.

For example, if an MQTT client (publisher) has three sensors (such as temperature, pressure and humidity) and we need to send each sensor reading to a specific application instance client, (for example, sensor_1 sends to application instance_1, sensor_2 sends to application instance_2 and sensor_3 sends to application instance_3), we must send three different messages from the publisher, and each message has a unique topic.

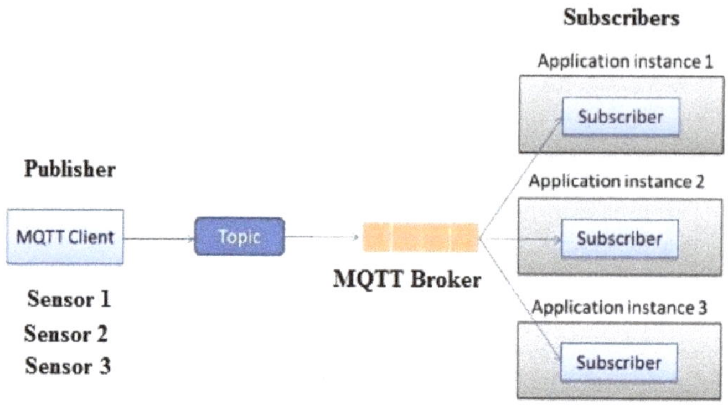

Figure 2. Publish/subscribe architecture based on single-topic messaging.

An IoT device should be able to send multiple messages for different topics. Therefore, the IoT protocols enable nodes to send many messages, but every message contains one topic only (i.e., the status of one sensor only). In our research, we propose the multi-topic feature in which a message can contain many topics for different subscribers without any waiting delays such as incurred by the message batching technique.

In recent years, the technique of batching multiple messages has been introduced in some Cloud system applications such as the Google Cloud Pub/Sub system which supports multiple message batching (https://cloud.google.com/pubsub/docs/publisher). However, batching multiple messages does not mean multi-topic messaging; batching messages puts messages into a queue until completion, which implies latency, and so batching is not suitable for real-time system applications. On the other hand, the proposed multi-topic messaging technique automatically sends any ready number of topics without waiting or latency making, it more suitable for supporting real-time system applications.

IoT systems could be adapted with promising technologies such as edge computing. This type of computing does not replace the MQTT protocol. Companies such as Cisco benefit from edge computing in the IoT field by adopting MQTT; this was highlighted by a senior manager at Cisco (https://blogs.cisco.com/internet-of-things/setting-a-simple-standard-using-mqtt-at-the-edge#comments).

Although MQTT is suitable for embedded real-time systems, the feature of multi-topic messaging is not supported by the protocol. Thus, in this work, a modified multi-topic messaging protocol that supports the multi-topic feature is implemented, as discussed in the next sections. This feature reduces network traffic and reduces the delay required to publish multi-topic messages.

2.2. *Multi-Message versus Multi-Topic Techniques*

In this part, we differentiate between two concepts: the multi-messages technique and our proposed multi-topics messaging technique.

2.2.1. Multi-Messages Technique (Batching)

The multi-messages techniques (i.e., batching), as shown in Figure 3, involves batching and buffering on senders to group multiple messages and send them as one batch to increase the throughput and cut down simple queue service (SQS) costs; however, this method has an impact on latency. In this method used, for example, in (https://codeahoy.com/2017/08/03/message-batching-to-increase-throughput-and-reduce-costs/), the customization of the batching algorithm can depend either on a specific number of messages (for example, 15 messages) or wait until threshold time value (for example, 50 ms). Thus, the batch will be formed either for a certain number or value and then sent out. This could cause applications to crash due to the queuing and processing sequence of messages. Therefore, we can lose some or all messages, and this method is thus not suitable for critical and real-time applications. Examples of these methods include message-batching, cloud.google (https://cloud.google.com/pubsub/docs/publisher), pulsar.apache (https://pulsar.apache.org/docs/ja/concepts-messaging/) and cloudkarafka (https://www.cloudkarafka.com/blog/2019-09-11-a-dive-into-multi-topic-subscriptions-with-apache-kafka.html).

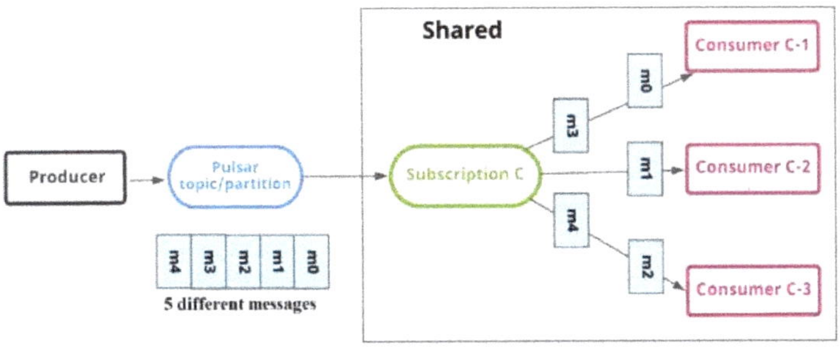

Figure 3. Multi-messages technique using batching; for example, pulsar.apache.

2.2.2. Multi-Topic Messaging Technique

The approach shown in Figure 3 is the multi-messages technique; however, in our proposal, we consider multi-topic messaging rather than multi-messaging. In our proposed multi-topic messaging protocol, we can send a single message with many topics; for example, as shown in Figure 4, three topics (i.e., temperature, humidity and pressure) are all sent in one message by the publisher to the broker. Then, the broker (proposed server) can detect and split this message into the number of new messages (three messages according to the three inherent topics in the received one) and resend each topic in a message to its specific client (each subscriber) without delay and loss in the data. The system acts intelligently as if it had received three separate messages from the publisher although it received one.

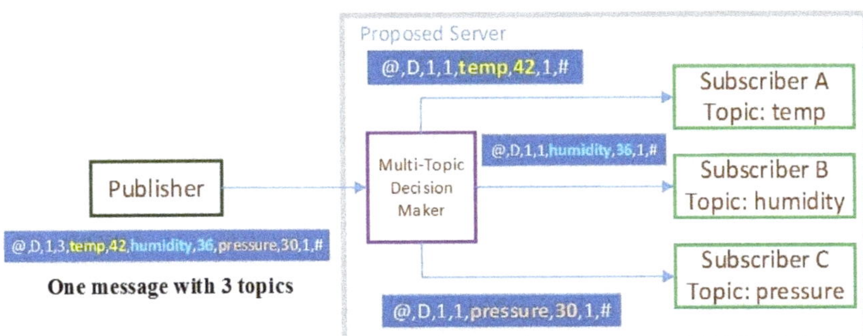

Figure 4. Proposed message with the three-topics technique.

Table 1 summarizes the main differences between the proposed protocol and four of the most common approaches in the IoT field; MQTT, Google Pub/Sub, Pulsar, and Karafka. This summary shows three different aspects: multi-topic subscriptions, multi-topic publications and multiple-messages queuing. The multi-topic subscription message means that a client can send one message that subscribes to many different topics. The clients at the beginning of the communication send this message only once to tell the broker that this client is interested in these topics. One publishing message contains many topics, but multiple-messages queuing is a batching-based technique that aggregates multiple different messages into a queue.

Table 1. Comparison against relevant protocols.

	MQTT [17]	Google Pub/Sub	Pulsar	Karafka	Proposed
Multi-topic subscriptions	Yes	No	Yes	Yes	Yes
Multi-topic publications	No	No	No	No	Yes
Multiple-messages queuing (batching)	No	Yes	Yes	Yes	No

2.2.3. Studied Use-Case Scenario

Let the IoT client (a patient with medical sensors: an electroencephalogram (EEG) signal and electrocardiography (ECG) signal and temperature, pressure and glucose level monitors) have five sensors and our aim be to send each sensor's status to a specific doctor (physician). The current solution in MQTT and batching or multi-messages methods, such as in the famous apache "pulsar.apache.org" method, is to send five different messages, with each message having frame headers; quality of service (QoS); for example, a QoS equal to 2 requires four acknowledgment messages between the client and server for one message) means more delays and more overhead bytes. The second solution is to let the client (a patient with five medical sensors) concatenate these five messages into one message (with a single topic), but in this case, the server will send this message to all destinations (five physicians). In this case, it would not be possible to separate the message to send each topic to a specific physician, as shown in Figure 3. On the other hand, in our proposal, we can send a single message with five topics (five patient sensor statuses) in a single message to the broker. Then, the broker can make the decision whether to separate this message into five new messages and resend each message to a specific physician, as shown in Figure 4 for the three-topic example.

3. Proposed IoT Multi-Topic Messaging Protocol

The proposed multi-topic data protocol is simply highlighted in this section. It is introduced either to establish the connection or to start the communication between IoT nodes. Moreover, it is designed to solve the single-topic messaging problem by supporting multi-topic messaging. This multi-topic feature acts as a brain for the messaging IoT broker. Acting as an intelligent IoT system, it converts normal IoT nodes to smart nodes that obtain the published messages and analyze the obtained messages to select the message destination. The resulting system can make the forwarding decisions implemented by the broker. While a client can send many types of data with different content in the same message without sending the data using many messages, the multi-topic feature provided by the proposed protocol can reduce the required traffic for the nodes, meaning that the protocol could be used in low-bandwidth networks and with limited hardware requirements.

Moreover, the IoT broker is modified from simply acting as a connecting point between clients to processing the received messages as it can separate one multi-topic message into many messages and display each message as a new message from a new client. Thus, it appears that the IoT broker has a "brain" and can make a simple decision when decomposing a multi-topic message.

We therefore propose new software (i.e., the broker IoT server) that can mimic human brain function (as it reads/inspects messages to check the number of topics and re-create a new number of messages according to the number of topics in the received message). Moreover, it makes decisions by sending each message from these newly generated messages to different destination clients. Thus, the single incoming message to the proposed broker will be split into many messages and each message re-transmitted to different destinations. Furthermore, for the content aspect of intelligent IoT platforms, our proposal adjusts the content of the message and re-creates new messages for a particular type of audience, as shown in the studied use case in Section 2.2.3.

3.1. Protocol Architecture

As shown in Figure 5, the proposed protocol architecture is similar to MQTT except for the added value of the multi-topic messaging feature in our proposal. The proposed data protocol depends on TCP/IP and consists of a centralized server (broker), IoT nodes (clients) and a multi-topics communication protocol:

> Protocol Application Range: All applications that the IoT supports can use the proposed protocol as a communication IoT protocol. The application range includes enterprise, utilities, mobiles and home applications.
>
> Communication Nodes: The server (broker), which is topic-based, is responsible for connecting nodes (client/devices). The different IoT nodes communicate with others through the server to accomplish the functionality of the system. A sequence diagram that describes each node type is shown in the next sections.

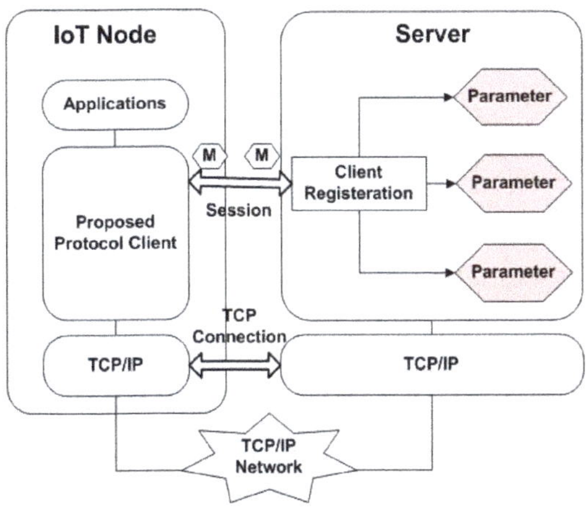

Figure 5. Architecture of the proposed multi-topic IoT protocol.

Nodes can be classified into four categories; sensor nodes, actuator nodes, normal or hybrid Nodes and finally, monitor nodes.

3.1.1. Sensor Nodes

The sensor node opens a TCP connection by sending a "connect" message to the broker server, as shown in Figure 6. This node must also send an identification number (ID) to the broker server to identify this node on the system. Then, it waits for an acknowledgment message (ACK) from the server; after that, it will be ready to publish the sensing data (either periodic or at interrupt times) or the information to the broker.

An IoT node can publish a message periodically or when an event occurs. Events may be generated from hardware interruptions (e.g., a push-button is pressed) or software triggers (e.g., a temperature exceeds a specific value).

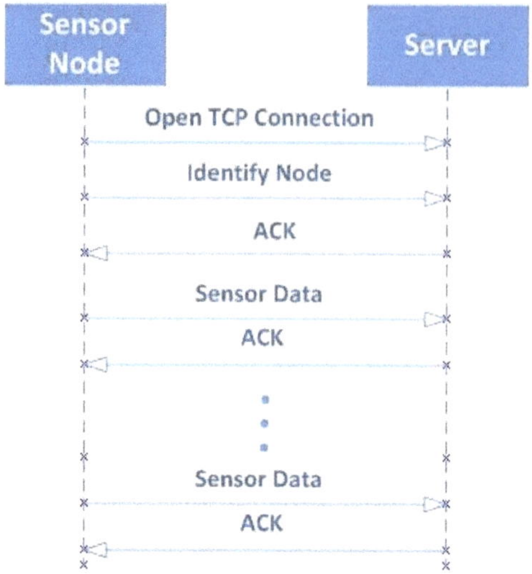

Figure 6. Sequence diagram for a sensor node. ACK: acknowledgement message.

3.1.2. Actuator Nodes

The actuator node and the main server first open a TCP connection. Then, the identification process is performed. Afterwards, this node will receive commands or messages which are transmitted by the monitor nodes (i.e., monitor clients) through the main server, as shown in Figure 7.

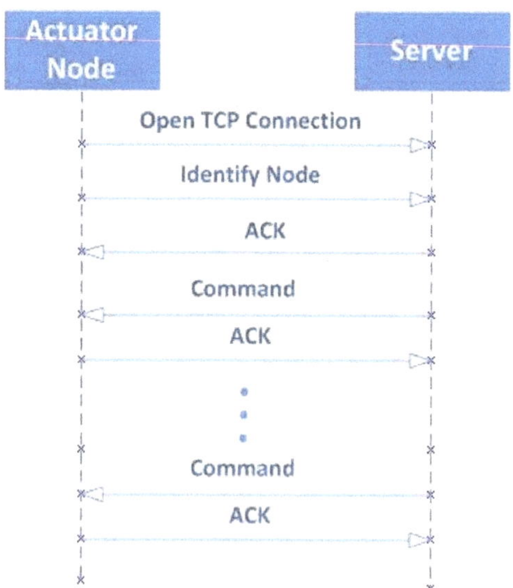

Figure 7. Sequence diagram for an actuator node.

3.1.3. Normal Nodes

A normal node performs the functionality of actuator and sensor nodes: it opens the TCP connection, then it identifies itself to the server with an identification number. Afterwards, it sends the sensed data to the server, and any actions or commands received from the server by the monitoring notes will be also executed through this node, as indicated in Figure 8.

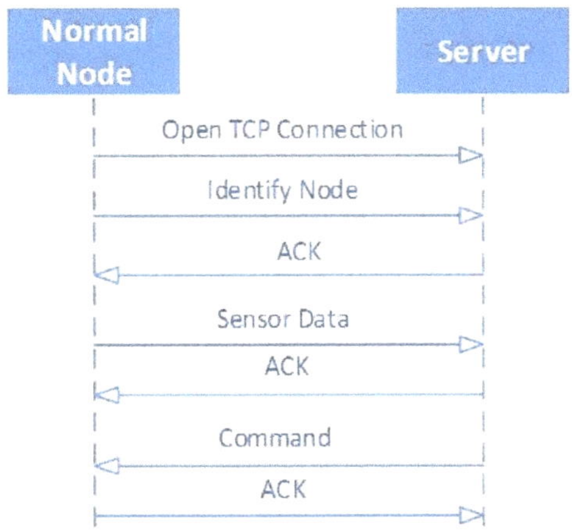

Figure 8. Sequence diagram of a normal node.

3.1.4. Monitor Nodes

The monitor node opens a TCP connection with the server after the identification process is performed; it registers to receive data from certain sensor nodes and then sends its messages to a specific actuator via the broker server, as shown in Figure 9.

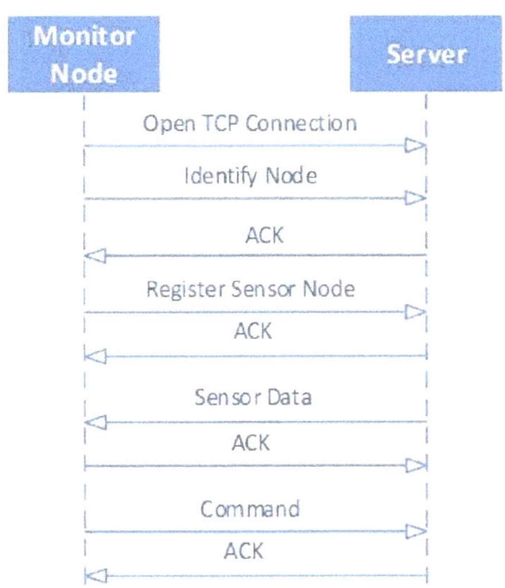

Figure 9. Sequence diagram of a monitor node.

3.2. Node to Node Communication

All node to node (client to client) communications must be done through the broker. Each client/node will try to establish a TCP connection and identify itself with the server. Then, it waits for an ACK message back from the server. After receiving the ACK message, successful communications between clients or nodes through the broker server can be done by sending and receiving different communication packets, as shown in Figure 10.

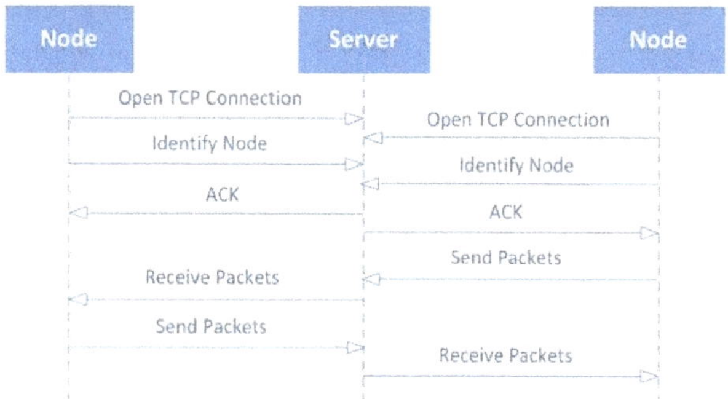

Figure 10. Sequence diagram for node to node communication.

3.3. Protocol Frame Format

Any kind of Internet connection, such as Wi-Fi, Ethernet, General packet radio services (GPRS), and 3G or 4G data connections, can be used to communicate between IoT nodes. Wi-Fi technology is selected in our simulation to communicate nodes with the server. The Wi-Fi hardware implementation module has a built-in TCP/IP stack as the proposed protocol depends on a TCP/IP protocol such as the MQTT protocol. The proposed multi-topics protocol has four type frames: identification, ACK, registration and finally, data frames. Table 2 indicates the common fields used in the proposed frame structure; all frames must have the same start (@) and end (#) of frame indicator. The frame types are shown in Table 3.

Table 2. Description of common frame fields.

Frame Field	Field Description
@	Start of Frame
ft	Frame Type
nid	Node ID
#	End of Frame
,	Frame delimiter
sn	Sequence Number
pc	Parameter Count
as	ACK State
pid	Parameter ID
pv	Parameter Value

Table 3. Frame Types.

Frame Field	Field Description
I	Identification frame
R	Registration frame
A	Acknowledgment frame
D	Data frame

3.3.1. Identification Frame

The identification frame is first transmitted by a node to declare and identify itself to the server. This phase is done by using a specific identification number (ID), and the system should wait until receiving the ACK message from the broker server. After that, each node can transmit (publish) or receive (by re-publishing from the server) frames. The frame structure needed for the identification node is shown in Figure 11. This frame is used in sensor nodes, actuator nodes, monitor nodes and hybrid or normal nodes. The function of this frame is similar to the "connect" message in the MQTT protocol.

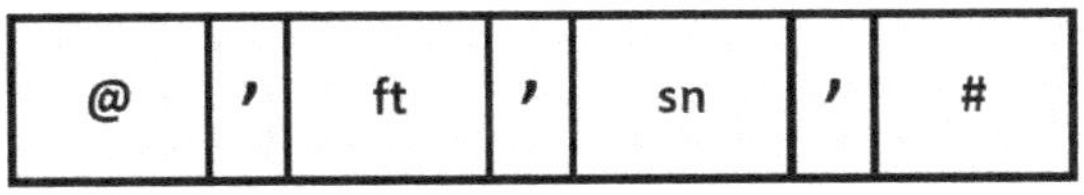

Figure 11. Identification frame structure.

3.3.2. Acknowledgment Frame

The acknowledgment (ACK) frame message is sent back from the broker server to the sender client to acknowledge the transmission process. This frame is transmitted from the server "broker" to the client "node". If the requested acknowledgement field is set, then an ID of the frame must be followed in the ACK frame. Figure 12 shows an ACK frame. This frame is used by the server as an acknowledgment to the sender node for all frame types. The function of this frame is similar to the "Connect ACK" message in the MQTT protocol.

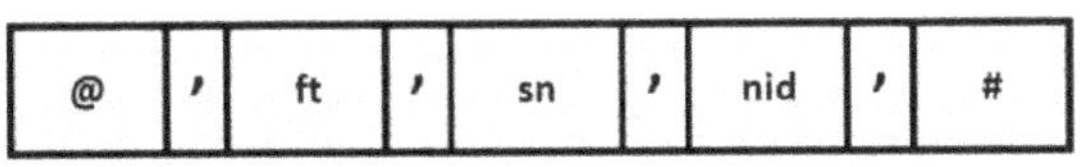

Figure 12. Acknowledgment frame structure.

3.3.3. Registration Frame

When the node listens to (receives a message) a specific parameter (topic), it must first send the registration frame with PIDs (parameter IDs) and PCs (parameter counts) as indicated in Figure 13. After the registration process, any message sent from other nodes with those parameters (topics), will be published to the node through the server; the ID frame should be sent first. This frame is used in actuator nodes, monitor nodes and normal or hybrid nodes. The function of this frame is equal to the "subscribe" message in the MQTT protocol.

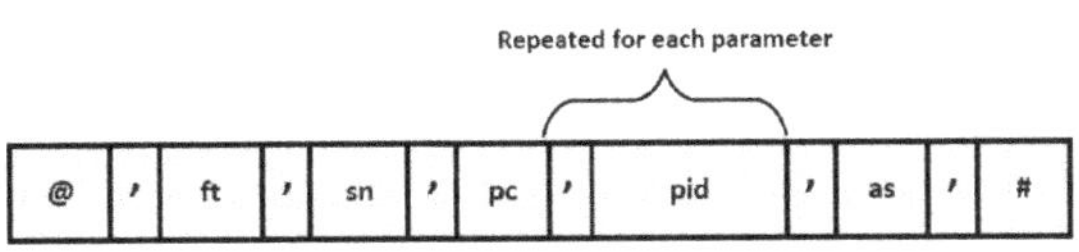

Figure 13. Registration frame structure.

3.3.4. Data Frame

With the data frame, the node can share and publish its data to other clients (nodes), as shown in Figure 14. The data frame has multiple parameters, such as a count field (PC), an ID (PID) and a value (PV). These parameters must be registered to use PID. Furthermore, the ID should be transmitted at the beginning of the process. This frame is used in sensor nodes, monitor nodes and normal or hybrid nodes. The function of this frame is equivalent to the "publish" message in the MQTT protocol, but it supports multi-topic messages.

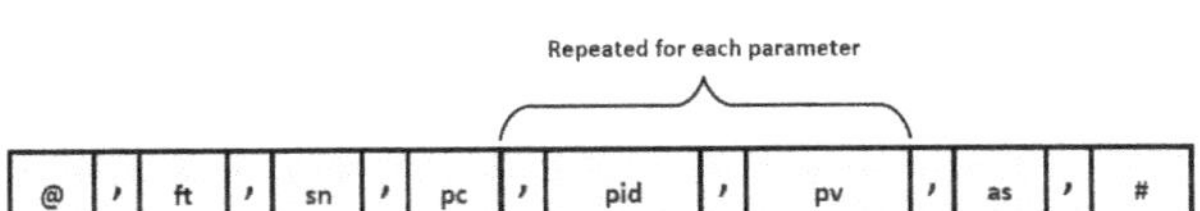

Figure 14. Data frame structure.

4. Proposed IoT Platform Based on RTOS

In this section, the design and implementation of the proposed IoT platform will be highlighted for critical and general applications and also for normal and real-time systems. The system uses FreeRTOS [19] as an RTOS to meet critical deadlines on time. Our platform depends on the proposed multi-topic protocol to support the "multi-topic" messaging feature. This will enhance the traffic throughput and decrease the delay required for multi-topic messages. The platform consists of subsystems such as IoT nodes and a broker server which communicates between nodes. Moreover, the proposed multi-topics protocol manages messages between IoT nodes and the broker server. This platform implementation uses the frame structure, which indicates how to exchange the data. The proposed platform can be easily used in a wide range of applications as it provides IoT connectivity and reliability and it uses the Wi-Fi module (we can use any type of Internet access) for the TCP/IP stack to decrease the hardware cost. The advanced RISC machine (ARM) Cortex-M4 is used to implement the proposed platform, which is powerful and has good power consumption. Furthermore, the IoT node can be a smartphone (i.e., a smartphone with the proposed IoT client application). The communication between system nodes through the Internet is based on the use of the broker as a backbone. This is proposed to be a topic-based broker, as per the MQTT message protocol, and it is designed based on RTOS.

4.1. *Design of Real-Time System (RTS)*

An RTS is a system that has a correct output that must be executed at the right time. Moreover, any RTS has many time requirements and critical deadlines that must be satisfied. These requirements must be handled by the RTOS used in system development. Therefore, the RTOS must have specific features that consider IoT challenges simultaneously, including scalability, connectivity, modularity, safety and security.

- Scalability refers to the IoT system's ability for future extension (i.e., system expandability).
- Connectivity refers to whether the RTOS is compatible with and supports many communication protocol standards.
- Modularity refers to the support of the implementation of the modules of the system. This will simplify the addition and integration of new features to smart devices.
- Safety refers to the prevention of any malfunction behavior that can lead to undesirable action.
- Security refers to the importance of countermeasures against either threats or malicious attacks.

Figure 15 shows the RTOS architecture in the proposed system. The RTOS has a great significance for the proposed IoT systems. This is because the use of the RTOS in development will add new advantages to the system, such as increased reliability, efficiency and predictability. Besides, it can simplify the management of the system [20]. As the platform design is done using RTOS, the system design could be listed in four system tasks, as shown in Figure 16.

Task "T_Connect" is the responsible task for connecting with the broker through the Wi-Fi communication module. Task "T_Comm" is responsible for receiving and transmitting the data from and to the remote main IoT server. Task "T_Sensor" gathers the information from the sensors in a special form and then sends the data to the main server. Finally, task "T_Actuator" executes the received commands by the "T_Comm" which are sent from the remote monitor node.

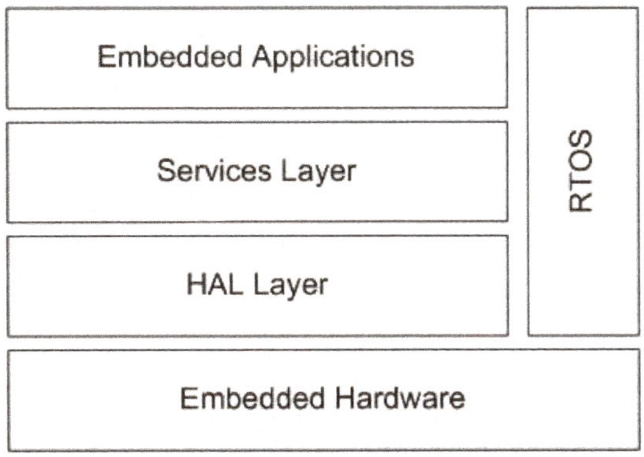

Figure 15. Real-time operating system (RTOS) architecture.

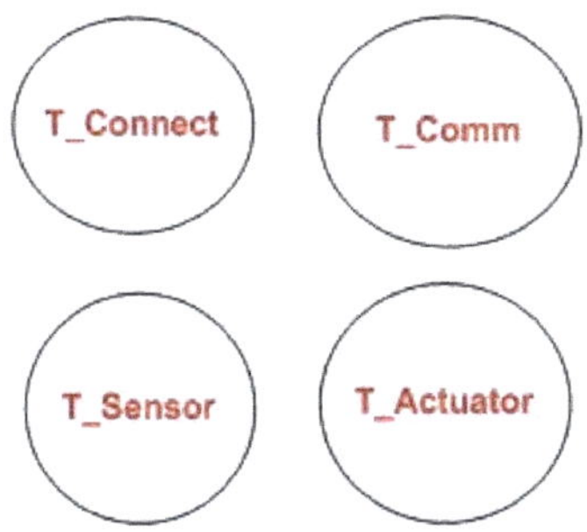

Figure 16. System tasks.

The system implementation scenario is as follows: the system peripheral initialization is done first, and then the operating system services used in the real-time design are created and initialized with the default state parameters. The system tasks should be created to allocate the required memory to become ready to operate. Finally, the operating system should be started to enable the scheduling of system tasks by the operating system scheduler.

The proposed IoT platform is implemented based on FreeRTOS (V10.3.1, Real Time Engineers Ltd., MIT License). It is possible for any task to block on a specific synchronization event with a specific time; it will exit from the blocked state even if the waiting event does not occur due to a time-out. The ready state tasks are able and ready to run but not in the running state currently as they have a lower priority than the already running task; all system tasks and the transitions between them are shown in Figure 17.

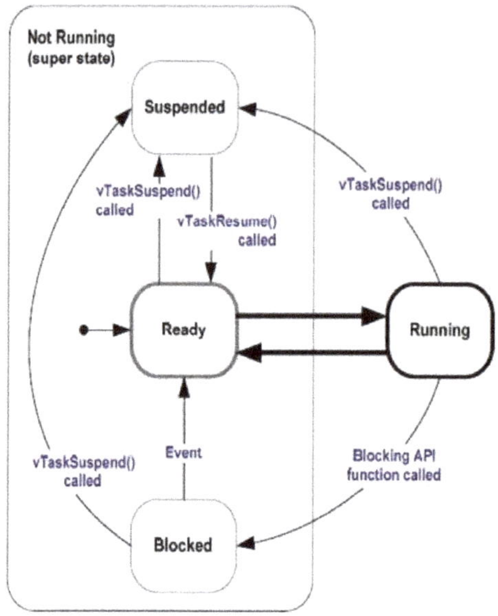

Figure 17. Task state diagram.

4.2. Proposed IoT Nodes

The proposed platform is an embedded system that could be implemented and designed for any type of micro-controller to meet system requirements. The proposed IoT node was designed and implemented in our laboratory at the National Telecommunication Institute (NTI) (http://www.nti.sci.eg/pcblab/) (as a printed circuit board (PCB) prototype) and passed all tests (PCB, unit, simulation and field tests) successfully. The proposed prototype of the IoT node was designed and implemented based on STMicroelectronics (STM) Nucleo Board. This board uses the ARM Cortex M4 processor. The ARM processor in the IoT node is developed for high-performance and to decrease the cost of devices as well as decreasing the power consumption.

Each node consists of different units, such as a micro-controller to manage node tasks and a Wi-Fi hardware stack to connect the client to the wireless network as an Internet access method. Furthermore, the proposed node has different sensor interfaces to sense a process or an environment and output interfaces (actuators) to control the environmental effects automatically.

All system nodes are connected together through the main broker server for specific designed tasks. For example, the proposed generic node that contains three sensors (i.e., a patient IoT unit with three medical sensors: an ECG signal and temperature and glucose level sensors), and we need to send each individual sensor status to the relevant physician. In addition, the proposed client has two actuators (i.e., turn-on alarm sound and automatic insulin dose) as shown in Figure 18.

In detail, we designed the IoT client unit to monitor the status of the patient and publish a message periodically (i.e., every 30 min) if the sensor readings correspond to normal conditions. However, if the status of any sensor is changed to an abnormal state (i.e., a patient temperature exceeds a threshold value, or the glucose level or ECG signal exceeds or drops below a certain value), the proposed IoT client understands the changes and makes the following automatic decisions;

- Publish this sensor status immediately to the physician.
- Increase the sending rate so that it is more suitable for the new changing rate (i.e., every 30 s).
- Run actuators, such as turning on the alarm sound or automatically injecting an insulin dose to save a patient's life.

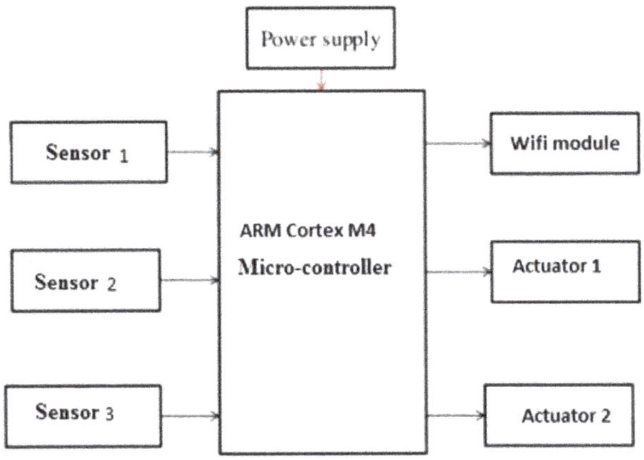

Figure 18. Block diagram of the proposed embedded IoT client.

This process is performed in an adaptive way for one sensor or many sensors, as clarified in our proposal for the IoT client.

- Understand the sensor status.
- Take an automatic decision to publish a message periodically (in normal status) or when an event occurs (patient abnormal).
- Send all sensor statuses in one message with multi-topics to the IoT server (to minimize delay), then send each sensor reading to its related subscriber (physician).
- Make the decision to run a specific actuator if required.

Based on the IoT node tasks, we can categorize nodes into four main types: sensor nodes, actuator nodes, normal nodes and monitor nodes.

The sensor nodes sense the environment and send the sensed data to the server periodically with a certain configuration period. These nodes include one sensor or more and do not include output interfaces (actuators). Figure 19 shows the sensor node flowchart.

Actuator nodes can affect or control the environment through messages (commands) from other monitor nodes via the broker server. These nodes include one or more actuators and do not include sensors. Figure 20 shows the actuator node flowchart.

Normal nodes have the functionality of the sensor and the actuator nodes; these nodes include actuators and sensors, as shown in Figure 21. A node's behavior is to communicate first with the IoT server and then send its sensor data to the server. It receives and executes the commands that come from monitor nodes through the broker server.

Monitor nodes could be proposed as hardware nodes or smartphones which monitor and control system nodes; they are the nodes which do not include sensors or actuators. These nodes can monitor sensors or control the actuators through publishing commands (sending messages), also receiving and processing the sensor data. The node communicates with the IoT broker server, and if the connection is established correctly, then it identifies itself to the server. It will register for topics and the parameters required from other nodes in the IoT system. Whenever it has new commands for the other nodes in the system, it will send them to the server directly for the command-specific topics or parameters.

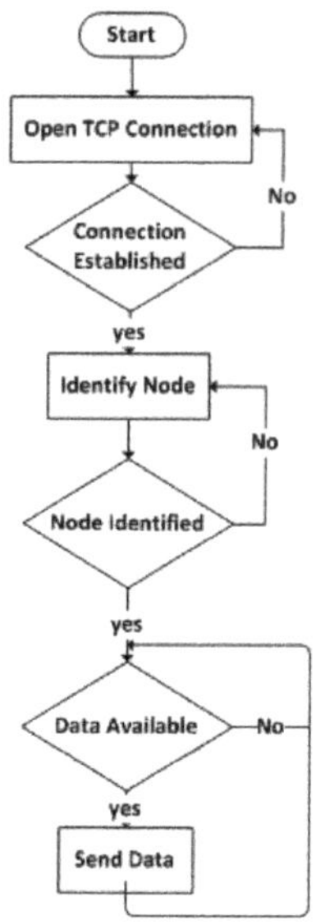

Figure 19. Sensor node flowchart.

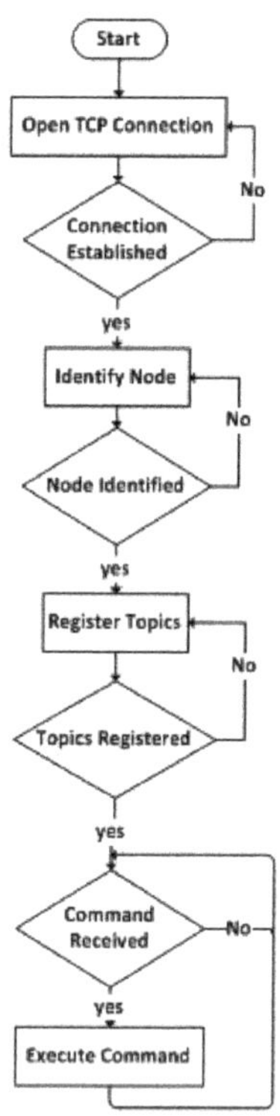

Figure 20. Actuator node flowchart.

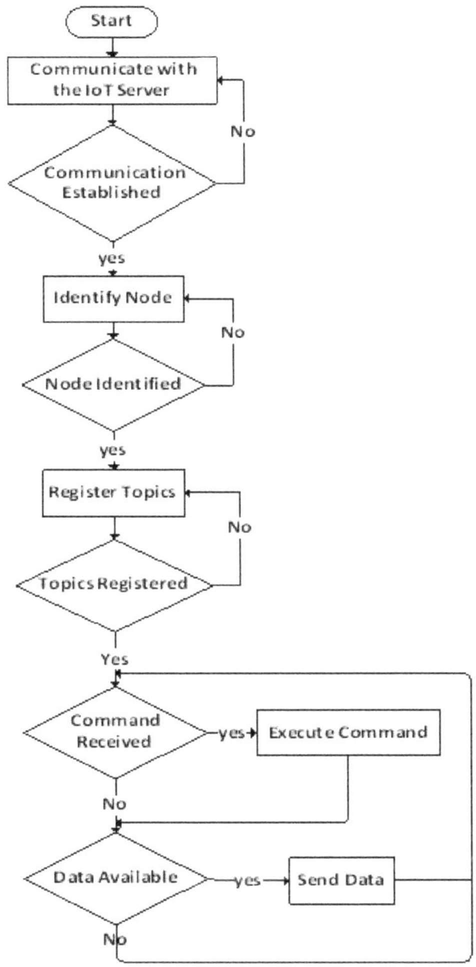

Figure 21. Normal node flowchart.

4.3. Android Device as a Client Node

As the monitor node could be an Android device; the Android application layouts are designed. The main layout is responsible for the communication with the IoT server, and the communication layout is responsible for the data exchange with the other nodes. This node application is implemented based on Java programming for Android development by the Eclipse IDE (release: 4.15, Eclipse Foundation Inc.); the main activity is responsible for handling the main application layout components, which are used to connect the broker with the IP address and the port number of the server.

4.4. IoT Server (Broker)

The IoT server is designed and implemented from scratch, independent of any ready-made server solutions; it is implemented based on the Java programming language. The basic block in the system is the IoT broker application, which communicates and connects all IoT Nodes. Thus, an IoT server is implemented for the proposed system, and the server is able to communicate with all types of nodes and is responsible for enabling monitor nodes to visualize the data of different sensors. The server's behavior, as shown in Figure 22, is to communicate the IoT Nodes.

If the sensor node is connected to the server, it forwards messages to the monitor node (registered node); if the connected node is an actuator node, the server will send the monitor message commands to it; if it is a normal node, the server will do the same thing for the actuator and sensor nodes; and if it is a monitoring node, it sends commands (messages) to the server and then forwards them to the actuator node.

The proposed broker server thread is implemented based on Java programming. It starts by listening to a specific server port to become ready for any node connection. If a new node tries to connect, then a new thread is created; the created thread is called the IoT device thread and is responsible for handling the IoT client node connection for the data exchange between this node and the other nodes in the system.

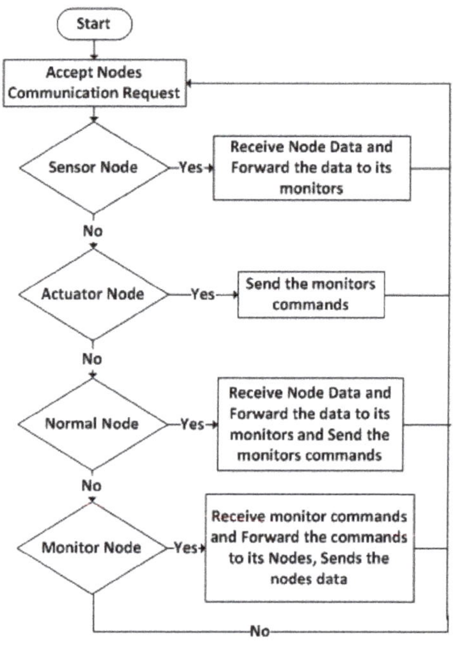

Figure 22. Server (broker) flowchart.

5. Experiment Setup

To study the performance of our proposed protocol, we carried out experiments. Additionally, the proposed protocol was compared with the standard MQTT protocol. This was done by setting different parameters of the network which affect the protocol's performance. The experimentation phase was executed inside our NTI premises on the real network infrastructure of the institute. We chose two performance metrics in the experiments: the delay time and the total transmitted bytes. The first metric concerns the delay of the transmitted message, defined by the interval time between sending frames (i.e., the publishing of a message) and the returned ACK message answered by the server, while the second metric concerns the total transmitted bytes per successfully sent message.

5.1. Hardware Setup

The hardware setup phase, as shown in Figure 1, consisted of three PCs/Laptops. The first one is used for the WANem (version 3.0) software (http://wanem.sourceforge.net/) to simulate the effects of channel losses to simulate transmission losses. Additionally, using the WANem server, we can simulate communication delays. The server (the computer which acts as a broker) for node communication and another laptop (client node) act as a node that publishes the message and waits for it to be acknowledged. Wireshark Software (https://www.wireshark.org/) is installed on the node (client) to monitor the traffic that is used later on for the analysis phase. The proposed multi-topic protocol and the MQTT protocol run on the same server. Each message is published from the IoT node to the broker server, going through the WANem node (the machine that simulates the network delay and packet drops). Finally, the server's ACK message will be returned to the transmitted node through the WANem server machine.

5.2. Software Setup

The Mosquitto (MQTT broker) as an open source MQTT broker is implemented based on the last updated version of the MQTT standard, which is v3.1, and the proposed multi-topic protocol broker-sides are implemented using Java coding. The proposed software of the broker is designed, implemented and tested with the same scenarios of the MQTT protocol. The software tools used in the practical experiment are the Wireshark software for packet analysis and the WANem software for network emulation.

The Wireshark tool can be used to measure and calculate metrics such as packet length and delay time; the software can monitor the network packets, the time elapsed for each packet and also the total traffic over a network. It can also generate reports after the monitoring of the network at a certain time; these reports are helpful for accurately studying the network traffic.

The WANem software is used for network delay and packet loss insertions. Network conditions do not have a fixed behavior, and so a network-related protocol evaluation and comparison cannot be fair because of the variant network conditions. For fixed network conditions, the WANem software is used; it can insert different network conditions for each test, including packet delays and packet loss percentage. These insertions help us to test under specific network conditions for a practical network test rather than simulation tests.

6. Simulation and Experimental Results

To test the proposed platform, we built an IoT platform consisting of one broker server, the WANem machine (PC), and a client, which were connected as a simple setup for the experiment. Both protocols (the proposed IoT communication protocol & the MQTT protocol) can achieve message delivery without applying a packet loss rate percentage. Successfully achieving this scenario would indicate that the two protocols have good communication and message delivery rates under different packet loss ratios. Thus, we can study the performance either in terms of the message delay or the successful amount of transmitted data, as indicated in the next sub-sections.

6.1. Message Delay

The message delay can be defined as the time interval between sending a message (i.e., publishing one) from the IoT node and receiving the ACK (returned message reply) from the broker node/server. This is calculated as shown in Equation (1). The message delay metric is a significant parameter for real-time systems where time is critical or sensitive. Different packet loss rates are applied to have ameliorate the message delay as a result of message re-transmission. Moreover, the MQTT protocol, which has a quality of service (QoS) equal to 1, is compared against our proposed multi-topic protocol with An ACK data state (QoS) equal to 1 for the same messages.

$$\overline{D} = \frac{1}{n}\sum_{i=1}^{n}(T_{ri} - T_{si}) \tag{1}$$

where $\overline{D}$ is the average message delay at a certain loss percentage, n is the number of messages, T_{ri} is the reception time of the message ACK and T_{si} is the sending time of the message.

MQTT has a small frame size compared to the multi-topic proposed frame, and so the protocol message delay in the case of a single-topic is less than our proposed one, as shown in Table 4.

In Figure 23, when the network losses are less than or equal to 15%, the average message delay of MQTT and the proposed protocol can be seen to be approximately the same. After 15% of network losses, the introduced message delay by the proposed protocol is greater than MQTT. The difference between the protocols increases with increasing network losses. Here, one single-topic message has a lower frame overhead than one multi-topic message. This is a normal case due to the new frame headers added to achieve the multi-topic functionality. However, the difference is not significant.

Table 4. MQTT versus proposed protocol average end-to-end delay in seconds for different rates of losses (%).

	0% Loss	10% Loss	20% Loss	30% Loss
MQTT	0.001081	0.065235	0.695212	2.357077
Proposed	0.001095	0.071645	0.721798	2.505362

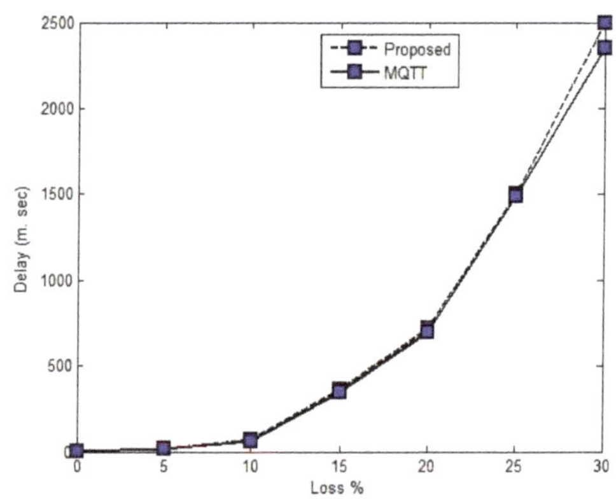

Figure 23. The average message delay (in seconds).

6.2. Message Data Transfer

We found that the message data transferred per message should be considered as an important metric; thus, the traffic generated by the network should be as small as possible. The message data transferred can be calculated as the ratio of the total generated bytes to the number of successful messages delivered, as shown in Equations (2) and (3). We depend on the Wireshark tool output to calculate this metric at different percentages of packet loss rates.

$$\overline{L} = \frac{1}{n}\sum_{i=1}^{n}(\overline{L_i}) \tag{2}$$

where $\overline{L}$ is the number of bytes on average per successfully transmitted message, $\overline{L_i}$ is the number of bytes on average per successfully transmitted message for each trial and n is the number of trials.

$$\overline{L_i} = \frac{L_i}{M} \tag{3}$$

where $\overline{L_i}$ is the number of bytes on average per successfully transmitted message for each trial, *Li* is the total traffic or number of bytes per trial and M is the number of successful messages that replied with an ACK.

In Figure 24, the average message bytes of MQTT and the proposed protocol are close when the network losses are less than or equal to 10%. After that, the message bytes of the proposed protocol are greater than MQTT. When the network losses increase, the difference between the protocols increases. Here, a lower protocol overhead is introduced by a single-topic message, but a small increase is found for the multi-topic message. The new frame headers that are added to achieve the multi-topic functionality are the reason for this behavior.

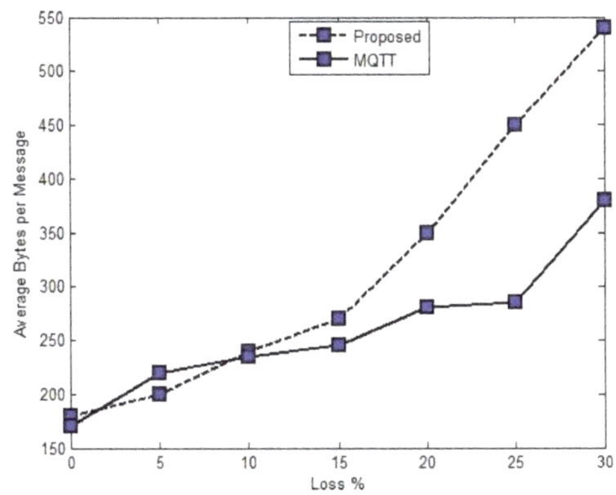

Figure 24. Average traffic per message (bytes).

6.3. Multi-Topic Messages

In many cases, we need to publish multi-topics to many clients or nodes at the same time. For this use case, the MQTT protocol cannot support these features. Thus, in MQTT, we need to publish each topic in a separate message; however, in our multi-topic protocol, which supports multi-topic features, we can publish these different topics in a single message. In this case, the proposed protocol delay is smaller than for MQTT, as shown in Figure 25, and the overhead bytes for each multi-topic message are shown in Figure 26 for no losses.

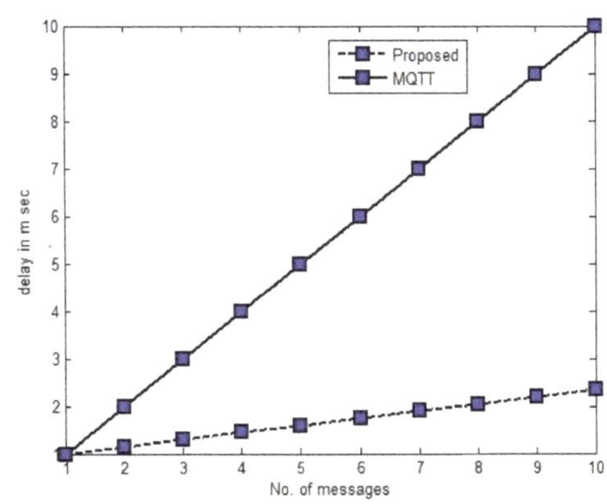

Figure 25. The multi-topic measured message delay (in milliseconds).

In Figures 25 and 26, when the number of the transmitted messages is equal to one, MQTT and the proposed protocol have approximately the same message delay and message size. However, when the number of transmitted messages is equal to two, the introduced message delay and message size from the MQTT are larger than the proposed approach. This difference increases as the number of transmitted messages increases. These observations and results arise because sending multiple single-topic messages adds more overhead than sending one multi-topic message, as MQTT requires a number of messages to be sent that equal the number of topics (one message per topic). On the other hand, the proposed protocol sends only one multi-topic message that has a lower overhead and achieves more throughput than the single-topic messages.

To conclude, our proposed multi-topic protocol has a lower delay and traffic compared to the standard MQTT protocol. This fact resulted from the addition of the multi-topic feature to the proposed protocol, adding some bytes for more topics as opposed to the approach in the MQTT protocol.

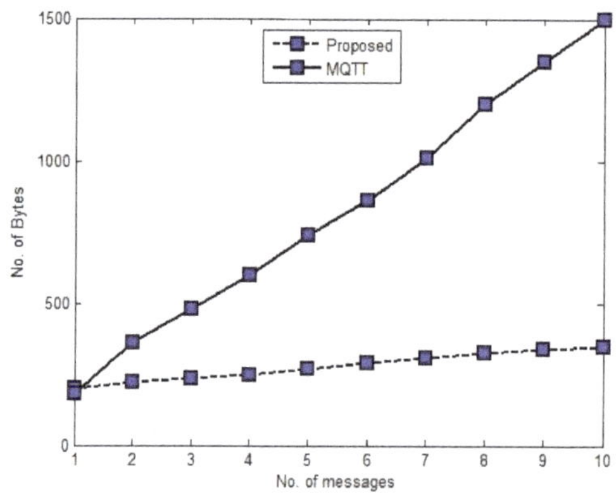

Figure 26. The total traffic measured per multi-topic message in bytes.

6.4. Worst-Case Scenario

Many current IoT applications depend on real-time systems. Real-time implies the enhancement of two parts of IoT communication systems: the first part describes the communication between the IoT node and the connected sensors or actuators (the proposed IoT client based on ARM Cortex-M4), and the second part involves the communication between the IoT client and the IoT broker. In our research, we addressed the two parts as follows.

- The proposed embedded IoT client is handled by using FreeRTOS (a common real-time operating system on the market). It provides multi-tasking to guarantee that many tasks work correctly in a semi-concurrent way and handle deadlines. Thus, the proposed IoT client is a real-time embedded system. Moreover, using RTOS in the proposed IoT embedded unit can decrease the processing time and minimize the delay.
- The second part (which involves the communication between the client and the server) is handled in our research by using the multi-topic feature to enhance the delay required for publishing many messages. Overall, the proposed system has a lower delay than similar systems due to the use of FreeRTOS and the multi-topic feature, as shown in Table 5.

Table 5 shows that the multi-topic messaging in the proposed protocol is better than MQTT for more than one topic (two topics or more), and the single-topic messaging in MQTT is near to the proposed protocol.

Table 5. Comparison for the worst-case scenario.

Worst-Case Scenario / Network Losses	0%	5%	10%	15%	20%	25%	30%
One message vs single-topic (MQTT)	1.14	16	85	375	1253	3153	5894
One message vs single-topic (Proposed)	1.21	22	108	627	1342	3245	6361
Two messages vs two-topic message (MQTT)	2.24	31	165	725	2452	6253	10,194
Two messages vs two-topic message (Proposed)	1.28	25	118	687	1388	3275	6450
Five messages vs five-topic message (MQTT)	5.4	75	405	1850	6253	12,153	21,894
Five messages vs five-topic message (Proposed)	1.31	26	122	714	1391	3315	6550

From this result, and according to the condition of meeting a 5 second deadline, we can state the following:

- In the case of a single topic, the proposed protocol, as well as standard MQTT, can transmit the data even if the loss percentage in the communication network reaches 25%; however, if loss reaches 30%, neither method can transmit the data within 5 s.
- In the case of two messages vs a single message with two topics, the proposed protocol can transmit the data even if the loss percentage in the communication network reaches 25%, while standard MQTT can only tolerate 20%.
- In the case of five messages vs a single message with five-topics, the proposed protocol still can transmit the data even if the loss percentage in the communication network reaches 25%, while standard MQTT can only tolerate 15%.

Thus, the proposed protocol is better and more able to work in worst-case conditions than standard MQTT.

7. Protocol Characteristics

This section will highlight the main implementation characteristics of our proposed protocol compared to the MQTT standard.

7.1. Main Characteristics

The proposed protocol implementation characteristics can be summarized as follows;

- Simplicity: This is an important feature for any protocol or system and can be summarized as follows: it must be easy to learn—and easy to use—the protocol or the system.
- Low protocol overhead: The protocol overhead is the number of bytes required for each designed frame. The frame structure shows that our protocol has a small overhead: i.e., the start of the frame byte and the frame type byte, where the frame elements are just comma-separated elements. The protocol has four main types of frames which have low overheads (few bytes): the identification frame has 6 bytes, the acknowledgement frame has 5 bytes, the registration frame has 8 bytes and the data frame has 9 bytes.
- Constrained network capability: The network challenges increase in the IoT systems, as billions of nodes require a network connection that maintains its data from corruption or being lost. In such networks, the network bandwidth should be limited for each node, as the network capacity might be extended; our protocol requires a small frame size for all the frame types.
- Constrained devices availability: The huge number of devices is a reason for minimizing the total cost for each device; choosing controllers with limited resources reduces the total cost of the IoT device. This has limited the use of such systems to date, although more powerful processors have been developed. The proposed protocol consists of four simple frames with a small number of bytes for each frame. The design simplicity and the low protocol overhead make the protocol able to work with the constrained devices which have controllers with an 8-bit processor architecture and limited resources.

- Asynchronous communication model: An IoT node receives its required data once the data are available; this system is called an event-oriented system, which is not polled in a periodic way to detect the data. The protocol enables the IoT nodes to use this feature, meaning that they are always up-to-date with data from the source.
- TCP-Oriented protocol: The TCP is the industry standard; it is considered to be the language of the Internet. TCP is a reliable protocol because the protocol itself checks everything that was transmitted if it was delivered at the receiving end correctly or not; if not it will re-transmit the data. Our protocol depends on a TCP connection.
- Publish/Subscribe model: The pub/sub messaging protocols allow messaging with a topic name, which is an array of characters; the node which sends the message is called a publisher and the node which receives the message is called a subscriber. This model or architecture has many features such as message decoupling and one-to-many messaging, which is discussed below. The model of the proposed protocol is similar to this model.
- Data decoupling and one-to-many messaging: The data decoupling feature is one of the features of pub/sub systems shared by the proposed protocol. This means that more than one node can publish or send messages on the same topic or parameter name. The senders will send data to the broker server with a certain topic or parameter name; the server will send these messages to the nodes which register the same parameter name, as in Figure 27.

  One-to-many messaging is the second feature of the pub/sub systems shared by the proposed protocol, which means that a node can send the same message to many nodes at once. The sender node will send a message with a specific parameter name to the main server, and the server will send the same message to all the registered nodes with the same parameter name, as shown in Figure 28.

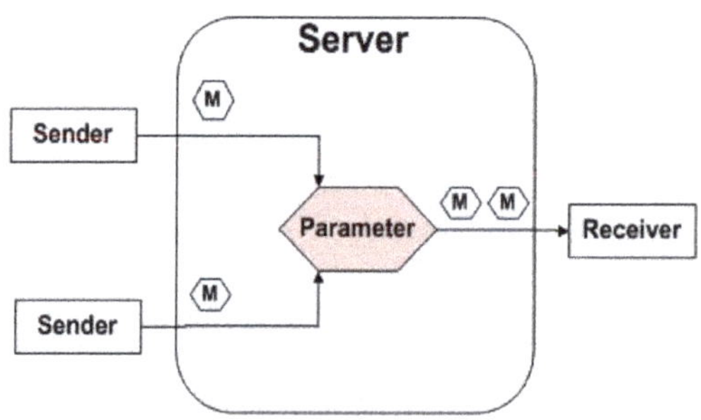

Figure 27. Data decoupling.

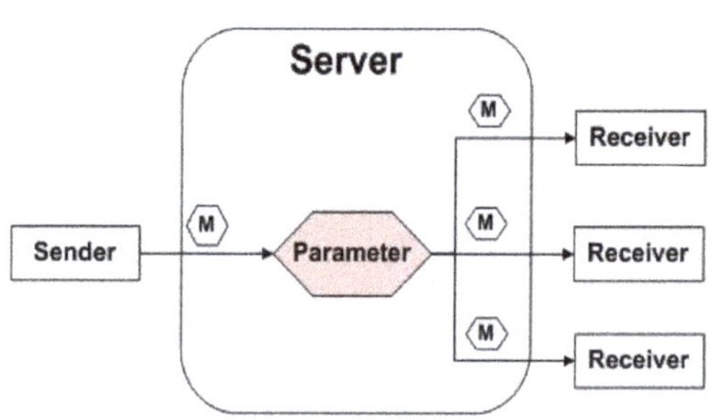

Figure 28. One-to-many messaging.

- Multi-functional communication: Any IoT node can perform both message decoupling and one-to-many messaging. It can send a message with any parameter to the server, and the server will automatically flood the message to all the registered nodes to that parameter. A node is capable of registering to any different parameters, and when the server receives any messages for these parameters, it will forward them to the node directly; both operations are shown in Figure 29.

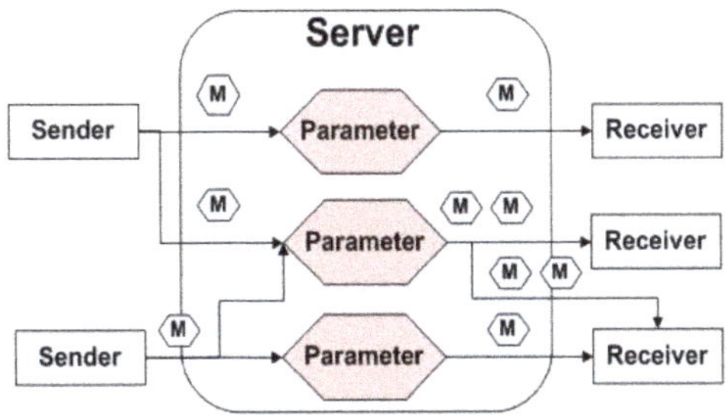

Figure 29. Multi-functional communication.

- Reliability: The proposed protocol is based on the TCP protocol, which adds a sort of physical layer of reliability. One of the protocol frames is called the acknowledgment frame; this frame is required by the receiving node to acknowledge the reception if it succeeded—the server also sends this frame to acknowledge the message received from any IoT node. Moreover, the identification frame should be replied with an ACK; the registration and data frames have the choice of enabling the ACK to be expected from the receiver or not.
- Scalability: The simplicity of the proposed multi-topic communication protocol tends to lead to a simpler server algorithm. The server main thread starts the main server algorithm and listens for the new IoT node connection when a new node tries to connect to the server. Thus, the scalability is not limited by the protocol itself as it consumes a few bytes and a small number of frames.

7.2. Comparison between Standard MQTT and Our Proposed Protocol

Our protocol has many features supported by MQTT protocol, such as a TCP-oriented protocol, message delivery mechanism and a publish/subscribe architecture. The protocols are different in some features, such as the multi-topic feature, which is supported only by the proposed protocol and not the MQTT, and the frame field-determining mechanism. A brief comparison between the standard MQTT and the proposed IoT multi-topics protocol is shown in Table 6.

Table 6. The proposed protocol and MQTT comparison. QoS: quality of service.

	The Proposed Protocol	MQTT Protocol
Transport layer	TCP-oriented	TCP-oriented
Message delivery mechanism	Simple ACK mechanism	QoS mechanism
Architecture	Publish–subscribe architecture	Publish–subscribe architecture
Multi-topic support	Yes	No
Frame field-determining mechanism	Field-delimiters mechanism	Field-size mechanism

In our evaluation, the average case was calculated for the end-to-end messaging for a fair comparison between the two messaging protocols (the MQTT protocol and the proposed one) with different network losses. It is worthy of note that an embedded system is not simply a real-time system that must take care of the worst-case deadlines within the device itself, but the end-to-end messaging is a network evaluation test that must be tested with many message transmissions. This evaluation is done for many trials that must be calculated on average for a successful comparison between the two messaging protocols.

8. Conclusions

In this research, the definition and architecture of the proposed IoT multi-topic protocol are presented. Additionally, the main protocol feature, which is called the multi-topic feature, and the application range that could be applicable for that protocol are discussed. The multi-topic IoT communication sequence diagrams are shown, and the protocol frame formats are designed in this research. The characteristics of the proposed protocol are also listed. Furthermore, the general IoT architecture based on RTOS is presented, and a new multi-topic IoT platform is proposed to implement the IoT nodes. The node–server communication is introduced and the experiment setup is shown; finally, the experimental results are discussed. Experiments were carried out to study and compare the performance of our IoT multi-topic protocol versus the standard MQTT using different parameters of the real network which affect the performance of protocols. Practical experiments were executed in a real environment based on our NTI network infrastructure. The obtained results showed that our protocol had less delay and lower traffic for multi-topics compared to the MQTT. Therefore, the proposed protocol outperformed the MQTT protocol in the case of multi-topic messaging. Moreover, our protocol was better than the batching of multiple messages for real-time system applications.

Author Contributions: All authors have equally participated in contributions to this research article, either in terms of the concept, methodology or validation phases of the proposed protocol. All authors have read and agreed to the submitted version of the manuscript.

Funding: This research received no external funding.

Conflicts of Interest: The authors declare no conflict of interest.

References

1. Hussein, M.; Zorkany, M.; Abdelkader, N. Real Time Operating Systems for the Internet of Things, Vision, Architecture and Research Directions. In Proceedings of the IEEE Conference Publications, WSCAR, Cairo, Egypt, 12–14 March 2016; pp. 72–77.
2. Zhang, M.; Zhao, H.; Zheng, R.; Wu, Q.; Wei, W. Cognitive internet of things: Concepts and application example. *Int. J. Comput. Sci. Issues (IJCSI)* **2012**, *9*, 151.
3. Zhang, Y.; Ma, X.; Zhang, J.; Hossain, M.; Muhammad, G.; Amin, S. Edge Intelligence in the Cognitive Internet of Things: Improving Sensitivity and Interactivity. *IEEE Netw.* **2019**, *33*, 58–64. [CrossRef]
4. Abdurahman, M. A Survey on the Concepts, Trends, Enabling Technologies, Architectures, Challenges and Open Issues in Cognitive IoT Based Smart Environments. *Int. J. Sci. Res. Sci. Eng. Technol. (IJSRSET)* **2018**, *4*, 512–522. [CrossRef]
5. Park, J.; Salim, M.; Jo, J.; Sicato, J.; Rathore, S.; Park, J. CIoT-Net: A scalable cognitive IoT based smart city network architecture. *Hum. Cent. Comput. Inf. Sci.* **2019**, *9*, 1–20. [CrossRef]
6. Ding, G.; Wu, Q.; Zhang, L.; Lin, Y.; Tsiftsis, T.; Yao, Y. An Amateur Drone Surveillance System Based on Cognitive Internet of Things. *IEEE Commun. Mag.* **2018**, *56*, 29–35. [CrossRef]
7. Gad, R.; Talha, M.; Abd El-Latif, A.; Zorkany, M.; El-Sayed, A.; EL-Fishawy, N.; Muhammad, G. Iris Recognition Using Multi-Algorithmic Approaches for Cognitive Internet of things (CIoT) Framework. *Future Gener. Comput. Syst. (FGCS)* **2018**, *89*, 178–191. [CrossRef]
8. Thangavel, D.; Ma, X.; Valera, A.; Tan, H.; Tan, C. Performance Evaluation of MQTT and CoAP via a Common Middleware. In Proceedings of the 2014 IEEE Ninth International Conference on Intelligent Sensors, Sensor Networks and Information Processing (ISSNIP), Singapore, 21–24 April 2014; pp. 1–6.
9. Hussein, M.; Zorkany, M.; Abdel Kader, N. Design and Implementation of IoT Platform for Real Time Systems. In *International Conference on Advanced Machine Learning Technologies and Applications;* Springer: Cham, Switzerland, 2018; pp. 171–180.
10. Hahm, O.; Baccelli, E.; Petersen, H.; Tsiftes, N. Operating Systems for Low-End Devices in the Internet of Things: A Survey. *IEEE Internet Things J.* **2016**, 3, 720–734 [CrossRef]
11. Labrosse, J. MicoC/OS-II:The Real-Time Kernel, 2nd ed.; pp. 1–648. Available online: https://www.amazon.com/MicroC-OS-II-Real-Time-Kernel-ebook/dp/B07CSRL7WY (accessed on 10 February 2020).

12. Zamfi, M.; Florian, V.; Stanciu, A. Towards a Platform for Prototyping IoT Health Monitoring Services. In *International Conference on Exploring Services Science*; Springer International Publishing: Cham, Switzerland, 2016; pp. 522–533.
13. Bellagente, P.; Ferrari, P.; Flammini, A.; Rinaldi, S. Adopting IoT Framework for Energy Management of Smart Building: A Real Test-Case. In Proceedings of the 2015 IEEE 1st International Forum on Research and Technologies for Society and Industry Leveraging a Better Tomorrow (RTSI), Turin, Italy, 16–18 September 2015; pp. 138–143.
14. Haghi, A.; Burney, K.; Kidd, F.; Valiente, L.; Peng, Y. Fast-paced development of a smart campus IoT platform. In Proceedings of the Global Internet of Things Summit (GIoTS), Geneva, Switzerland, 6–9 June 2017; pp. 1–6.
15. Happ, D.; Karowski, N.; Menzel, T.; Handziski, V.; Wolisz, A. Meeting IoT Platform Requirements with Open Pub/Sub Solutions. *Ann. Telecommun.* **2017**, 72, 41–52. [CrossRef]
16. Karagiannis, V.; Chatzimisios, P.; Vazquez, F.; Alonso, J. A Survey on Application Layer Protocols for the Internet of Things. *Trans. IoT Cloud Comput.* **2015**, 3, 11–17.
17. IBM. MQTT V3.1 Protocol Specification, Protocol Standard. pp. 1–81. Available online: http://docs.oasis-open.org/mqtt/mqtt/v3.1.1/os/mqtt-v3.1.1-os.pdf (accessed on 21 June 2020).
18. Shelby, Z.; Hartke, K.; Bormann, C. RFC 7252: The Constrained Application Protocol (CoAP), Internet Engineering Task Force, (IETF) PROPOSED STANDARD. 2014. pp. 1–112. Available online: https://tools.ietf.org/html/rfc7252 (accessed on 21 June 2020).
19. FreeRTOS. Available online: https://www.freertos.org/ (accessed on 21 June 2020).
20. Gaur, P.; Tahiliani, M. Operating Systems for IoT Devices: A Critical Survey. In Proceedings of the 2015 IEEE Region 10 Symposium, Ahmedabad, India, 13–15 May 2015; pp. 33–36.

energies

MDPI

Article

Data-Intensive Task Scheduling for Heterogeneous Big Data Analytics in IoT System

Xin Li [1], Liangyuan Wang [1], Jemal H. Abawajy [2], Xiaolin Qin [1], Giovanni Pau [3] and Ilsun You [4,*]

1 College of Computer Science and Technology, Nanjing University of Aeronautics and Astronautics, Nanjing 210023, China; lics@nuaa.edu.cn (X.L.); myteng@nuaa.edu.cn (L.W.); qinxcs@nuaa.edu.cn (X.Q.)
2 School of Information Technology, Deakin University, Locked Bag 20000, Geelong, VIC 3220, Australia; jemal.abawajy@deakin.edu.au
3 The Faculty of Engineering and Architecture, Kore University of Enna, 94100 Enna, Italy; giovanni.pau@unikore.it
4 Department of Information Security Engineering, Soonchunhyang University, Asan-si 31538, Korea
* Correspondence: isyou@sch.ac.kr

Received: 13 July 2020; Accepted: 17 August 2020; Published: 1 September 2020

Abstract: Efficient big data analysis is critical to support applications or services in Internet of Things (IoT) system, especially for the time-intensive services. Hence, the data center may host heterogeneous big data analysis tasks for multiple IoT systems. It is a challenging problem since the data centers usually need to schedule a large number of periodic or online tasks in a short time. In this paper, we investigate the heterogeneous task scheduling problem to reduce the global task execution time, which is also an efficient method to reduce energy consumption for data centers. We establish the task execution for heterogeneous tasks respectively based on the data locality feature, which also indicate the relationship among the tasks, data blocks and servers. We propose a heterogeneous task scheduling algorithm with data migration. The core idea of the algorithm is to maximize the efficiency by comparing the cost between remote task execution and data migration, which could improve the data locality and reduce task execution time. We conduct extensive simulations and the experimental results show that our algorithm has better performance than the traditional methods, and data migration actually works to reduce th overall task execution time. The algorithm also shows acceptable fairness for the heterogeneous tasks.

Keywords: big data analysis; heterogeneous data-intensive task; IoT system; service response delay; task scheduling

1. Introduction

Big data analysis plays an important role in many IoT application scenarios, it supports various services for users. The cloud data center is the promising way to deal with the big data analysis tasks and support data-intensive [1] services. However, it is becoming hard due to the big data analysis tasks are becoming heterogeneous, it may be periodic data analysis tasks for enterprises to analyze user behaviors, or online tasks from edge environment [2] to support various services [3]. For these tasks, it has different expectation on task execution time. While scheduling heterogeneous tasks, the system not only needs to reduce the execution time for periodic tasks, but also response online tasks as quickly as possible, usually within a few minutes, even a few seconds [4]. Hence, it is an important issue to schedule the heterogeneous tasks in data center to support various services, which is also important for energy reduction since longer task execution also consumes more energy.

Researchers have proposed many approaches to meet the demand for periodic tasks or online tasks [5]. Since the scenarios of applying big data are getting more and more complicated, only scheduling periodic tasks or online tasks cannot satisfy the needs of users. However, the computing resources in data centers are not unlimited, periodic tasks and online tasks have to compete for resources. While scheduling the heterogeneous tasks consist of periodic tasks and online tasks simultaneously, it is difficult to reduce the execution time of periodic tasks and respond online tasks timely. In this paper, we propose an algorithm to solve this problem of scheduling heterogeneous tasks and compare it with some classical task scheduling methods. To reduce the task execution time, we prefer the task is executed in locality mode, which means the task can access the required data from the server which hosts the task. It is easy-understanding that remote data access is time consuming for network delay. Otherwise, the task will be executed in remote mode, which indicates the task will access the data remotely. To reduce task execution time, it is preferred to schedule a task from the queue to occupy the idle resource with locality mode. However, it is impossible to achieve locality mode for all tasks due to the limited resource. The delay scheduling [6] is proposed to achieve more data locality by delaying the task execution until the desired server has idle resource. It is an efficient method to improve the data locality, and this also motivate us move the data from one server to another actively, but not passively like the delay scheduling.

In this paper, we investigate the task scheduling problem, and propose the task scheduling algorithm for heterogeneous tasks with data migration. We take the data block as an important part since it affects the data locality directly. It is acceptable that we cannot schedule all tasks to be executed in locality mode since the storage space is limited in data centers. The competition for computing resources between periodic tasks and online tasks is also a problem. However, some traditional task scheduling methods are no longer applicable to heterogeneous tasks like FIFO. To satisfy the demands of time reduction, in our algorithm, we first established a time model for time calculation of tasks executed in different modes based on some of our previous work [7]. In order to execute tasks in locality mode as much as possible, we migrate the required data block to the server when scheduling a task. We also compare the heterogeneous task scheduling algorithm with other classical algorithms and conduct some simulation experiments. The results show that our algorithm can guarantee better data locality and acceptable fairness for tasks so that improve the efficiency of whole system.

In summary, we make the following contributions in this paper.

- We formulate the scheduling problem for heterogeneous tasks in data centers, and propose a data-migration based algorithm to achieve better data locality for task scheduling. To the best of our knowledge, this is the first attempt to introduce the active manner to schedule both periodic tasks and online tasks simultaneously.
- The proposed algorithm take both delay scheduling and data migration into account to improve the data locality. We establish the task execution model to predict the task execution time, and make measurable comparison for various execution cost to achieve better decision.
- We conduct extensive simulation experiments to evaluate the efficiency of the algorithm. The results show that the algorithm has better performance on task execution time reduction. In addition, the fairness of the tasks is also acceptable.

The rest of this paper is organized as follows. Section 2 describes the related work on task scheduling and data deployment. Section 3 introduces how to establish the model of server selection and how to calculate different time consumptions. Then, we introduce the heterogeneous task scheduling strategy in Section 4. In Section 5, we conduct some experiments and evaluate the efficiency of the algorithm. Finally, we conclude this paper in Section 6.

2. Related Work

The research community have conducted the task scheduling from various aspects, like makespan reduction, resource sharing, energy saving, fault tolerance and so on. In addition to the scheduling in cloud data centers, it is also a hot topic in Internet of Things (IoT) system [8–10], and Mobile Edge Computing (MEC) [11,12]. In this paper, we focus on the big data analysis task scheduling issue in cloud data centers. For the data-intensive tasks, except the resources, data locality is the core issue to improve the task scheduling efficiency, it greatly affects the execution time of tasks [13].

MapReduce [14] is the typical data processing paradigm for big data analysis. For the Hadoop [15], the opensource implementation for MapReduce, the default task scheduling strategy is First In First Out (FIFO), which is simple and has low overhead. However, it is unacceptable for the heterogeneous tasks with different Quality of Service (QoS) requirements, and the performance cannot be guaranteed since the data distribution could be not suitable for task execution due to frequent data transmission. Hence, Fair Scheduler is proposed based on multi-user environment [16]. This scheduling method sets a pool of resources for each user in cluster which can ensure that the users can occupy the same amount of resource with high probability. The Fair Scheduler support some kind fairness among different tasks, and also brings resource waste or resource shortage since the tasks have heterogeneous resource requirements, which limits the performance improvement for Fair Scheduler. To improve the resource utilization, Capacity Scheduler [16] is proposed by providing resource-aware task scheduling. The core idea is to adjust the tasks in the queue by sensing the resource usage case.

Most of the former task scheduler, like ShuffleWatcher [17], BAR [18], LATE [19], and SAMR [20], mainly focus on the resources, especially for the computing resources, since the computation with CPU is the main work for the task. However, there will be frequent I/O operation for the big data analysis tasks, which make the data locality more important since the local data I/O could reduce the task execution time efficiently than remote data access. Delay scheduling [6] is the representative scheduler to schedule the online tasks dynamically by achieving better data locality. The core idea is to let the task waits for a few moment if the server with the desired data has no sufficient computing resource until some task release resources or reach the threshold time. The method works since it can achieve data locality with more opportunities. We can also regard it makes a tradeoff between the waiting time and the data transmission time.

The delay scheduling could be a passive method to achieve data locality for online task scheduling. Since the data placement is important for data locality, there are also some work focus on the data placement issue to provide more data locality opportunity. CoHadoop [21], Scarlett [22], and PACMan [23] are presented to place the data reasonably for achieving better data locality. The joint task and data placement is another approach to achieve data locality for offline tasks [24], which analyzes the relevancy between task and data to decide how to arrange the data placement and task assignment comprehensively. In addition, with the different idea, the migration approach [7] is proposed to achieve data locality actively. The main idea is to compare the cost between waiting cost and migration cost. The data will be moved from some server to the desired server if the data migration can achieve data locality and with less cost.

As more and more date is generated in the mobile edge devices, the task scheduling issue will be discussed in edge computing environment. The cache-aware task scheduling method in edge computing is proposed to achieve better data locality [25], which obeys the similar idea is this paper. This trend also prompt the task scheduling issue in System on a Chip (SoC), an energy-aware task scheduling is proposed for heterogeneous real-time MPSoCs in IoT [26].

For the most of the approaches discussed above, they work for scheduling only periodic tasks or only online tasks, and work in passive manner. In this paper, we propose the scheduling strategy to schedule heterogeneous tasks consist of periodic and online tasks with active manner, which is a new attempt for task scheduling. The comparison of the features is shown in Table 1.

For the most scheduling strategies discussed above are applied to schedule only periodic tasks or only online tasks. They cannot satisfy the demand for scheduling both two kinds of tasks simultaneously. Hence, we propose our scheduling strategy to schedule heterogeneous tasks consist of periodic and online tasks and take the input data into consideration.

Table 1. The feature comparison for representative approaches.

Approach	Periodic Task	Online Task	Passive Manner	Active Manner
FIFO	√	√	√	
ShuffleWatcher		√	√	
DealyScheduling		√	√	
CoHadoop	√			√
Scarlett		√		√
This Paper	√	√		√

3. Problem of Server Selection

3.1. Task Execution with Data Migration

For the data center with homogeneous servers, all the periodic tasks and online tasks share the resources and data and resources in the cloud data center. Because the tasks we discussed are data-intensive, for each of them, the execution is associated with two important factors: resources and input data. For each server in the cluster, it will be split into uniform resource units like Containers or Virtual Machines (VMs). Therefore, an idle server can allow several tasks to be executed on it at the same time. However, in this paper, to simplify the problem, it is acceptable that we assume each server as one resource slot and each time it could only execute one task. It is easy to expand one server with one resource slot to one server with many resource slots. In addition, the input data is also necessary. The location of data blocks affects the execution of tasks. For each task, there could be two execution modes: locality mode or remote mode. The locality mode indicates the task is executed on the server who hosts the input data for the task. Otherwise, the task executes in remote mode by accessing the input data from another server. It is easy to understand that we always want tasks to be executed in locality mode as much as possible since better data locality means less task execution time. However, the resources is limited and it is impossible to execute all tasks in locality mode.

To evaluate the efficiency of our algorithm, we need to calculate the execution time of all tasks in system. In this paper, the heterogeneous tasks consist of periodic tasks and online tasks. We use P_i to indicate the ith periodic task in the task queue, and O_i represents the ith online task. For the task execution time, we use T_r as the task execution time indicator in remote mode while T_l as the indicator in locality mode.

For the data center, it contains N servers and K different data blocks for task execution. For each data block D_i, it has 3 data replicas deployed on different servers. We use f_i to indicate the required input data of task P_i or O_i. According to the location of f_i when scheduling tasks, we will use the following indicators to classify the task execution mode.

$$\phi(P_i, f_i) = \begin{cases} 0, & \text{locality mode;} \\ 1, & \text{remote mode.} \end{cases} \tag{1}$$

where locality mode means the periodic task P_i is executed locally.

$$\phi(O_i, f_i) = \begin{cases} 0, & \text{locality mode;} \\ 1, & \text{remote mode.} \end{cases} \tag{2}$$

where locality mode means the online task O_i is executed locally.

In our algorithm, we combine delay scheduling with data migration to improve data locality. For the task scheduling decision when some idle resource is available, we need to make a decision by a comparison between the costs of delay scheduling and data migration. With the delay scheduling manner, the selected task will wait until the desired server has idle resource to achieve data locality.

Another method to achieve data locality is to migration the input data from some server to the server with idle resource which will host the selected task. This method is known as data migration in this paper. However, the data migration also bring cost by data transmission. Hence, it is necessary to make a comparison between data migration and delay scheduling. We use T_m to indicate the data migration time. It means the time of transferring a data block from a server to another server is T_m.

3.2. Server Selection

In this paper, we have assumed that each server has one resource slot and can execute one task at each time. Hence, the task scheduling problem can be treated as assign the task to some server or select some server to host the task. Therefore, we can transfer the problem to select a proper server for each task. In a data center, the heterogeneous tasks consist of periodic tasks and online tasks. Periodic tasks are in a queue and wait for scheduling while online tasks arrive in another queue randomly. For a periodic task, it has much longer response time than online task. The goal of scheduling periodic tasks is to shorten the execution time so that improve the efficiency of whole system. While for online tasks, the response time is short. An online task cannot wait for a long time because the users need the system to respond as quickly as possible. To meet this demand, the goal of scheduling online tasks is to scheduling them before the deadline of response. This is more urgent than time reduction. A heterogeneous task scheduling strategy which can satisfy the requirements of periodic tasks and online tasks is necessary.

We use S_j to represent the idle server in data center. As mentioned above, D_i indicates the ith data block, and f_i means the replica of D_i. Therefore, we can define the indicators to represent the assignment decisions on idle server S_j as follows.

$$\pi(D_i, S_j) = \begin{cases} 0, & \text{there is no replica of } D_i \text{ on } S_j; \\ 1, & \text{otherwise.} \end{cases} \tag{3}$$

$$\pi(P_i, S_j) = \begin{cases} 1, & \text{Periodic task } P_i \text{ is assigned to server } S_j; \\ 0, & \text{otherwise.} \end{cases} \tag{4}$$

$$\pi(O_i, S_j) = \begin{cases} 1, & \text{Online task } O_i \text{ is assigned to server } S_j; \\ 0, & \text{otherwise.} \end{cases} \tag{5}$$

We want to reduce the task execution time to improve the efficiency in cluster. The running time of whole system is determined by the running time of each server. In addition to the calculation of task execution time, we also need to calculate the running time of servers. We assume that the time is split into time-slots and the workload of each server is used to represent the required running time. Hence, for the workload of some server, say S_j, it contains two parts, the initial workload the tasks assigned the server. The initial workload can be represented by the symbol $L^i(S_j)$. Then, the total workload for server S_j, which also reflects the whole task execution time, could be represented like the following equation:

$$L(S_j) = T(P_i) \cdot \pi(P_i, S_j) + L^i(S_j)$$

Generally, we summarize the related notations in the Table 2, so as to describe the problem properly.

Table 2. Notations.

Symbol	Description
N	number of servers in the data center
K	number of data blocks
$response\_time$	response time of online tasks
$\pi(D_i, S_j)$	data assignment indicator
$\pi(P_i/O_i, S_j)$	task assignment indicator
$\phi(P_i/O_i, f_i)$	task execution mode indicator
$T(P_i/O_i)$	task execution time for task P_i or O_i
$L(S_j)$	workload of server S_j
T_m	data migration time

4. Heterogeneous Task Scheduling

4.1. Scheduling Heterogeneous Tasks with Data Migration

For a data center consists of multiple uniform servers with a default data (blocks) placement, which ensures that there are 3 data replicas for each data. The problem is to assign the heterogeneous tasks on the servers such that the task execution time is minimized. It could be formulated as:

$$min.maxL(S_j), 1 \leq j \leq N$$
$$s.t. \forall t, \sum_{1}^{K} \pi(D_i, S_j) \leq M, 1 \leq j \leq N$$
$$\forall J \in (\cup O) \cup (\cup P), J.start\_time \leq J.response\_time$$

where $L(S_j)$ is the workload, which indicates the task execution time as mentioned above. The number of servers in the data center is N, and the number of data (blocks) is K.

The first constraint means that the hosted data (blocks) for each server should not exceed its storage capacity, represented by M. The second constraint means that the task response time, indicated by $J.response\_time$, should be greater than the task start time, $J.start\_time$, where $\cup O$ means the set of online tasks while $\cup P$ represents the set of periodic tasks, and we have $|\cup O| = m$, $|\cup P| = n$.

Since there are N servers working at the same time, each server has a workload $L(S)$. If we find that server S_j has the maximum workload and we minimize the maximum load of S_j, the whole job execution time of system will be reduced. To achieve this goal, each time we assign a task to the server, we select the server with the least workload in the system, and execute the task in the mode with the least execution time. Obviously, it could be better to achieve data locality for the tasks to reduce the task execution time. Hence, the problem is to pick one task from the queue to occupy the available computing resources. If there is no proper task, which means none of the tasks could be executed locally, we need to make a decision between waiting and remote access to achieve better benefit. Therefore, we propose the heterogeneous task scheduling as shown in Algorithm 1.

Algorithm 1 Heterogeneous task scheduling.

Require: $\cup O$: the set of online tasks which have arrived the system;
$\cup P$: the set of periodic tasks which have arrived the system;
S_v: the server with minimum workload;
S_i: the ith server;
t: the final completion time;

$t \leftarrow 0$;
for $i \leftarrow 1$ to N **do**
  if $S_i.slot > 0$ **then**
    if $\exists O_j \in \cup O$ **and** $\pi(D_j, S_i) = 1$ **then**
      assign(O_j, S_i);
    else if $\forall O_j \in \cup O, \pi(D_j, S_i) = 0$ **then**
      if $\pi(D_j, S_v) = 1$ **then**
        if $\Delta T = L^i(S_v) - t + T_l(O_j)$ **and** $L^i(S_v) - t < O_j.response\_time$ **then**
          assign(O_j, S_v);
        else
          assign(O_j, S_i);
      else if $\pi(D_j, S_v) = 0$ **then**
        ScheduleTaskwithDataMigration(O_j);
    else if $\cup O = \emptyset$ **and** $\exists P_j \in \cup P, \pi(D_j, S_i) = 1$ **then**
      assign(P_j, S_i);
    else if $\cup O = \emptyset$ **and** $\forall P_j \in \cup P, \pi(D_j, S_i) = 0$ **then**
      if $\pi(D_j, S_v) = 1$ **then**
        $\Delta T = min(L^i(S_v) - t + T_l(P_j), T_r(P_j))$;
        if $\Delta T = L^i(S_v) - t + T_l(O_j)$ **then**
          assign(P_j, S_v);
        else
          assign(P_j, S_i);
      else if $\pi(D_j, S_v) = 0$ **then**
        ScheduleTaskwithDataMigration(P_j);
$t \leftarrow t + 1$;
update();
return t;

In Algorithm 1, S_i is an idle server with available resource slots and we need to assign a task to it. Because the heterogeneous tasks consist of periodic tasks and online tasks, we first check whether there is an online task in the queue. If the queue of online tasks is not empty, we will select one task O_j and judge if it can be executed locally. We assign the task O_j to S_i if it can access its input data from S_i. Otherwise, we find the server S_v with minimum workload.if S_v has the required data of task O_j in the data center, we make a comparison between two kinds of time: (1) the waiting time until S_v is idle, and the execution time locally, represented by by $L^i(S_v) - t + T_l(O_j)$; (2) the time of executing task O_j on server S_i in remote mode, the time is represented by $T_r(O_j)$. We compare them and find the less one and indicate it with ΔT. Here, with the result of ΔT, we have two choices as follows:

- $\Delta T = L^i(S_v) - t + T_l(O_j)$. It means we can schedule task O_j to be executed on server S_v so that the remote task can be a local task and reduce the execution time. However, this a an online task scheduling problem. As mentioned before, an online task has a constraint that it has to be executed before its deadline of response time. It means the waiting time until the server S_v to be idle cannot exceed the response time of O_j. Therefore, before we schedule O_j, we should guarantee that the waiting time $L^i(S_v) - t$ is less than $O_j.response\_time$. This manner of scheduling refers to delay scheduling.

- $\Delta T = T_r(O_j)$. If the time of executing task O_j in remote mode is minimum, we can assign task O_j to server S_i and execute it in remote mode immediately.

If task O_j cannot be executed locally on server S_i and there is no local data blocks of task O_j on server S_v, we could consider whether to introduce data migration to execute O_j. The strategy of scheduling tasks with data migration will be interpreted in next subsection.

If the queue of online tasks is empty, we will schedule the periodic tasks. We first check if there is a local periodic task P_j and assign it to the idle server S_i. If all periodic tasks cannot be executed locally on S_i, we need to find the task (say P_j) with the least execution time, indicated as ΔT. Then, we still need to check the execution mode for P_j. If $\Delta T = L^i(S_v) - t + T_l(P_j)$, the task P_j will be assigned to server S_v with locality mode by data migration. Or, the task P_j will be assigned to server S_i with remote execution mode, as shown in the algorithm. Since periodic tasks have much longer response time than online tasks, we do not need consider the waiting time and response time.

4.2. Data Migration

In Section 3 of this paper, we have discussed that data migration will bring extra time consumption. Because data transferring costs network, it must influence the execution of remote tasks. Therefore, we must consider the impacts on remote tasks before migrating data blocks. We explain the process of scheduling a task with data migration in Algorithm 2 as follows.

Algorithm 2 ScheduleTaskwithDataMigration(J).

Require: S_i: the i_th server; J:the selected task; S_v:the server with minimum workload in cluster without D_J.

if There is a remote task on S_v **and** $(L^i(S_v) - t \leq T_m)$ **then**
  $\Delta T = minL^i(S_v) - t + T_m + T_l(J), T_r(J)$;
else if There is a remote task on S_v **and** $(L^i(S_v) - t > T_m)$ **then**
  $\Delta T = min2(L^i(S_v) - t) + T_m + T_l(J), T_r(J)$;
else if There is a local task on S_v **and** $(L^i(S_v) - t \leq T_m)$ **then**
  $\Delta T = minT_m + T_l(J), T_r(J)$;
else if There is a local task on S_v **and** $(L^i(S_v) - t > T_m)$ **then**
  $\Delta T = minL^i(S_v) - t + T_l(J), T_r(J)$;
if $\Delta T = T_r(J)$ **then**
  assign(J, S_i);
else
  migration(f_J, S_v);
  assign(J, S_v);

In Algorithm 2, task J can be executed on S_i in remote mode or executed on S_v after data migration. Both of the two execution modes need data transmission, which brings transmission cost. We need to compare two time costs: $L^i(S_v) - t$ and T_m. The first $L^i(S_v) - t$ means the task on server S_v still needs to take $L^i(S_v) - t$ to execute. T_m is the time of transferring required data block by selected task J to server S_v. We consider data migration while the S_v is executing a previous task. If T_m is larger than $L^i(S_v) - t$, which means the data migration will not finish when S_v is idle. Hence, task J needs to wait for the end of data transferring before it can be executed. Otherwise, data migration will finish before S_v being idle, and task J can be executed when S_v completes its previous task. Based on this time comparison and decision on if there is a remote task on server S_v, we can list four scenarios as follows:

- There is a remote task on S_v and $(L^i(S_v) - t \leq T_m)$: In this condition, the time of migrating data is over the rest execution time of previous task on S_v, task J needs to wait the end of data migration before it can be executed. Because the previous task on S_v is remote task, $L^i(S_v) - t$ and T_m will double. The work load of S_v will increase $(L^i(S_v) - t)$, so that the new workload of S_v is :

$$L^i(S_v) + (L^i(S_v) - t)$$

Because the previous task on S_v will finish before data migration, the time of migrating data will not be twice when S_v is idle. The rest time for task J to wait and be executed when S_v is idle is:

$$\frac{2T_m - 2(L^i(S_v) - t)}{2} + T_l = T_m - (L^i(S_v) - t) + T_l$$

Therefore, the task execution time with data migration could be:

$$L^i(S_v) + (L^i(S_v) - t) + T_m - (L^i(S_v) - t) + T_l - t = L^i(S_v) - t + T_m + T_l$$

The time of executing task J on S_i in remote mode is $T_r(J)$, we compare these two kinds of time and select the mode with less time.

- There is a remote task on $S_v \&\& (L^i(S_v) - t) > T_m$: In this condition, data migration has finished before S_v is idle. The task J could be executed directly on S_v. $L^i(S_v) - t$ and T_m will also double when executing a remote task and migrating data at the same time on S_v. If data migration is finished, S_v will still take some time to execute its remote task, and the time is:

$$\frac{2(L^i(S_v) - t) - 2T_m}{2} = (L^i(S_v) - t) - T_m$$

The new workload of server S_v is:

$$L^i(S_v) + 2T_m + (L^i(S_v) - t) - T_m$$

Therefore, the whole time for task J to wait and be executed on S_v is:

$$L^i(s_V) + 2T_m + (L^i(S_v) - t) - T_m + T_l - t = 2(L^i(S_v) - t) + T_m + T_l$$

We also compare the time above with $T_r(J)$ and select the less one.

- There is a local task on $S_v \&\& (L^i(S_v) - t) \leq T_m$: Because local tasks do not need network, data migration will not affect local tasks. Similarly, we compare T_m and $L^i(S_v) - t$, and calculate the whole time for task J to be executed on S_v, the time is T_m. We choose the less one between $T_m + T_l(J)$ and $T_r(J)$ and execute J in corresponding mode.
- There is a local task on $S_v \&\& (L^i(S_v) - t) > T_m$: Similar to the situation above, we calculate the whole time for task J to be executed on S_v, the time is $L^i(S_v) - t + T_l(J)$. We compare it with $T_r(J)$ and choose the less one.

From the case analysis, we find that we will execute J on S_i in remote mode with few cases. Otherwise, we migrate the required data of J to from other server to server S_v and execute it in locality mode. Here, we will introduce more about data migration. The algorithm on how to migrate a data block is described in Algorithm 3.

For the migration procedure, we first judge if the destination server has enough storage capacity ($S_v.space > 0$), we can migrate the data block to it and the number of data replica ($f_J.replica$) increases by one. If target server has no space to store the data block, we will find the data block D_d with least degree, which means there is few task needs the data block. If the replica number of D_d is more than one, we can replace it with the migrated data block. This is because we have to guarantee that each data block in the data center must has at least one replica. If the replica is the only one, we give up the operation of data migration.

Algorithm 3 migration(f_J, $\bar{S_v}$).

Require: J: the selected task; S_v: the destination server; *num*: the number of replicas of each data; *degree*: the number of tasks which require data f_J as input;S_v: the target server for data migration.

if $S_v.space > 0$ **then**
  assign f_J to S_v;
  $f_J.replica += 1$;
else
  for $i \leftarrow 1$ to M **do**
    $D_d \leftarrow minimalDegree(S_v)$;
    if $D_d.num > 1$ **then**
      replace f_d with f_J;

5. Experimental Results

To evaluate the proposed approach, we first establish a time model by some real experiments of task execution and data migration. Then, we evaluate our heterogeneous task scheduling algorithm by comparing it with other scheduling approaches.

5.1. Time Model

To validate and establish the time model, we transfer data blocks with different size and research the effect of network competition on time. We test the data transmission time on five groups with various data sizes. For each group, first, we transfer one data block each time from one server to another. Then we transfer two data blocks with different size at the same time. The results are shown in Table 3, where the time unit is millisecond (ms). Learn from results, we can conclude that the network competition will almost doubles the time of transferring time.

Table 3. Time of transferring different data blocks (ms).

Data Size	1 Data Block	2 Data Block
64 MB	947,829	1,519,609
128 MB	1,495,165	2,669,139
256 MB	2,648,865	4,950,684
512 MB	4,939,071	9,511,114
1024 MB	9,535,483	20,151,628

Because HDFS has a default data block size of 64 MB, we ran a sorting program on a server and record the results with and without data migration, which means that the program reads files from local server and from another remote server respectively. We run the program for multiple times and calculate the average value. The final results are shown in Table 4. We analyze the results and find the conclusion as:

$$T_r = \alpha \cdot T_m, \alpha \approx 1.3$$

which means remote tasks spend more time than local tasks, and the remote access time is about 1.3 times of local data access.

Second, data migration will prolong the execution time of remote tasks, but it will not influence local tasks. Our algorithm needs to schedule online tasks and local tasks at the same time. This conclusion tells us that before migrating a data block, we must consider the effects on remote tasks.

Table 4. Tasks executed in two modes (ms).

	With Data Migration	Without Data Migration
Locality mode	1,578,563	1,593,488
Remote mode	3,584,282	2,120,580

5.2. Simulation Analysis

We conduct extensive simulations with various settings and introduce the results on our algorithms. We compare our heterogeneous task scheduling strategy with traditional FIFO strategy and delay scheduling. We also compare it with HRTPS [27], a dynamic scheduling algorithm with precedence constraints for hybrid real-time tasks. The HRTPS classifies the process of task scheduling into static situation and dynamic situation. In static situation, it assigns some of servers to periodic tasks to guarantee that all periodic tasks could be scheduled in a period of time. Then it considers the dynamic situation with real-time tasks arriving randomly. It reserves some servers for these real-time tasks so that they can be scheduled in time. As a result, the HRTPS can schedule heterogeneous tasks simultaneously.

In our experiments, we make the simulation settings as follows. (1) Set $N = 50$ for the data centers, and $M = 20$ for each server; (2) There are $K = 300$ data blocks and 3 replicas for each data block. The replicas are placed in the servers randomly while any server will not host the same replicas; (3) The input data for each task is randomly assigned; (4) The task execution time with data locality is a random value; (5) There are $m = 500$ periodic tasks and $n = 500$ online tasks; (6) The online tasks arrive randomly within 100 time slots; (6) The online tasks must be finished before its deadline, which is a random value within 20 time slots.

We conduct the simulation with different scheduling approaches, the basic results are shown in Figure 1. In the figure, the y-coordinate is the CDF value, which increases by time increasing in x-coordinate. For each point, it means the percentage of completed tasks under the fixed time-slot. The performance of FIFO is not good, and the proposed heterogeneous task scheduling method improves the performance significantly. Meanwhile, the heterogeneous task scheduling method has better performance than HRTPS and delay scheduling.

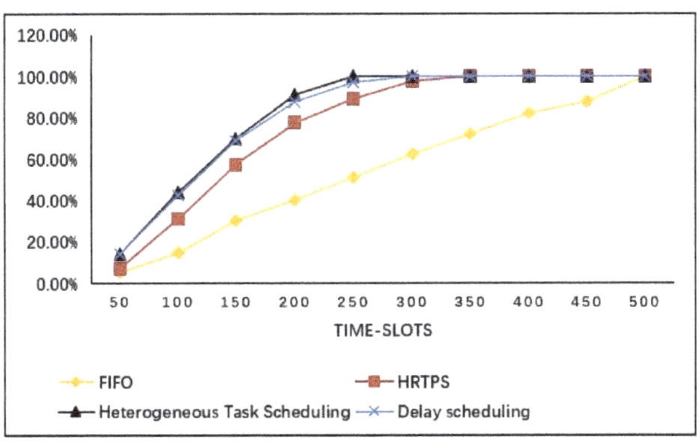

Figure 1. Experimental results of scheduling heterogeneous tasks.

Figure 2 describes the server utilization of each scheduling approach for given time slots. We use $N1$ indicates the number of servers which are occupied. The server utilization can be calculated by $N1/N$ and N is the number of all servers in data center. From the result, we can know that our algorithm has the best server utilization among all three methods at the beginning and decreases from the intermediate period of the experiment, this is because most of tasks have been completed.

The result shows that our heterogeneous task scheduling algorithm has nice performance on server utilization and can reduce the execution time of tasks.

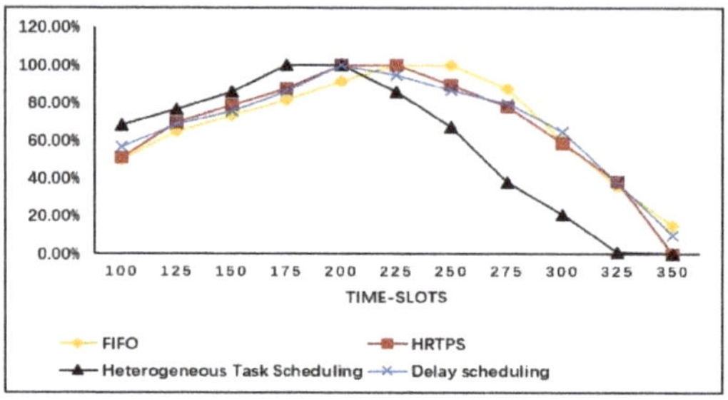

Figure 2. Experimental results of server utilization.

We also conduct experiments to study the differences under different ratios of online tasks to periodic tasks. The number of all tasks are 1000. The results is described in Figure 3. The x-coordinate represents the ratios of online tasks to periodic tasks. The y-coordinate is the time slots which indicate the finish time of all tasks in system. From the results we can conclude that the heterogeneous task scheduling strategy has better performance when the amount of online tasks is large. The result also shows that when the ratio is bigger, the whole execution time is larger. This is because the online tasks will be scheduled to occupy the resources with high priority. However, the limited resource cannot meet the locality task execution for all tasks. Hence, some online tasks are executed in remote mode or locality mode with data migration, both of which brings extra data transmission time. At the same time, the larger ratio means more online tasks are executed without locality mode and the execution time increase.

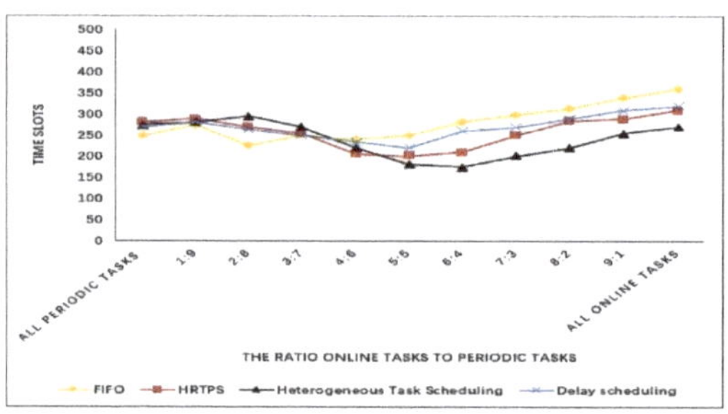

Figure 3. Experimental results of different ratios.

For the above simulations, we also record more data for different approaches. (1) The average server utilization: The server utilization is shown as Figure 2, and we calculate the average server utilization based on the value of each time slot. (2) Overtime online tasks: We record the completion time for each online task and check if it is larger than its deadline. (3) Waiting time: The time duration between the arrival time and scheduled time for the online tasks. We compare these factors with the FIFO, delay scheduling, and HRTPS. The results are shown in Table 5. Learn from the results, we find that our heterogeneous task scheduling algorithm has the least average waiting time, which is the major focus in this paper. For the average server utilization, the value of delay scheduling and HRTPS is close to our algorithm though our algorithm has less task execution time as shown in Figures 1 and 2. This is because the tasks occupy the server with longer time due to remote data access for the delay scheduling and HRTPS algorithm. For the number of overtime online tasks, our algorithm and the delay scheduling algorithm will lead to some overtime cases since the waiting mechanism for tasks. Actually, this could be fixed by adjusting the default waiting time or weight for the online tasks.

Generally, the proposed heterogeneous task scheduling algorithm can improve the performance on task execution time reduction by reasonable data migration. It is a valuable attempt for task scheduling with active manner.

Table 5. Comparison of three methods in three aspects (time slots).

	Averag Server Utilization	Number of over Time Online Tasks	Average Waiting Time
FIFO	42.3%	5	197
Delay scheduling	64.0%	3	68
HRTPS	68.3%	0	72
Heterogeneous	64.4%	2	64

6. Conclusions

In this paper, we proposed a heterogeneous task scheduling method to schedule periodic tasks and online tasks simultaneously for IoT system. The method take both the delay scheduling and data migration into account to improve data locality. The core idea is to compare the cost of different decisions and choose the reasonable one to schedule the task. We conduct extensive simulations, and the experimental results show that our algorithm has better performance on task execution time reduction compared the FIFO, delay scheduling, and HRTPS. This work indicates that the data migration is an efficient way to improve data locality, which actively occupy resource by moving data from one server to anther server. This is total different from the classical delay scheduling, which waits for idle resource passively. However, it is still a challenging problem to improve the performance of data migration since the system is complex. In addition, it is valuable to discuss how to control the number of data replicas dynamically such that there would be more opportunities for data migration.

Author Contributions: Conceptualization, X.L., L.W. and X.Q.; formal analysis, X.L. and L.W.; funding acquisition, X.L. and I.Y.; investigation, X.L., L.W., J.H.A., X.Q., G.P. and I.Y.; methodology, X.L. and I.Y.; project administration, X.L. and I.Y.; software, X.L. and W.L.; supervision, J.H.A., X.Q. and I.Y.; writing—original draft, X.L., L.W. and J.H.A.; writing—review & editing, X.L., X.Q., G.P. and I.Y. All authors have read and agreed to the published version of the manuscript.

Funding: This research was funded by the National Natural Science Foundation of China under Grant 61802182 as well as in part by the Soonchunhyang University Research Fund.

Conflicts of Interest: The authors declare no conflict of interest.

References

1. Yu, B.; Pan, J. Location-aware associated data placement for geo-distributed data-intensive applications. In Proceedings of the 2015 IEEE Conference on Computer Communications (INFOCOM), Kowloon, Hong Kong, China, 26 April–1 May 2015; pp. 603–611.
2. Shi, W.; Gao, J.; Zhang, Q.; Li, Y.; Xu, L. Edge computing: Vision and challenges. *IEEE Internet Things J.* **2016**, *3*, 637–646. [CrossRef]
3. Zhu, C.; Zhou, H.; Leung, V.C.M.; Wang, K.; Zhang, Y.; Yang, L.T. Toward big data in green city. *IEEE Commun. Mag.* **2017**, *55*, 14–18. [CrossRef]
4. Li, X.; Tatebe, O. Data-Aware Task Dispatching for Batch Queuing System. *IEEE Syst. J.* **2017**, *11*, 889–897. [CrossRef]
5. Li, D.; Wu, J.; Chang, W. Efficient Cloudlet Deployment: Local Cooperation and Regional Proxy. In Proceedings of the International Conference on Computing, Networking and Communications, Maui, HI, USA, 5–8 March 2018; pp. 757–761.
6. Zaharia, M.; Borthakur, D.; Sarma, J.S.; Elmeleegy, K.; Shenker, S.; Stoica, I. Delay scheduling: A simple technique for achieving locality and fairness in cluster scheduling. In Proceedings of the 5th European Conference on Computer Systems, Paris, France, 13–16 April 2010; ACM: New York, NY, USA, 2010; pp. 265–278.
7. Li, X.; Wang, L.; Lian, Z.; Qin, X. Migration-Based Online CPSCN Big Data Analysis in Data Centers. *IEEE Access* **2018**, *6*, 19270–19277. [CrossRef]

8. Sharma, V.; You, I.; Kumar, R. Energy Efficient Data Dissemination in Multi-UAV Coordinated Wireless Sensor Networks. *Mob. Inf. Syst.* **2016**, *2016*, 8475820:1–8475820:13. [CrossRef]
9. Sun, R.; Yuan, J.; You, I.; Shan, X.; Ren, Y. Energy-aware weighted graph based dynamic topology control algorithm. *Simul. Model. Pract. Theory* **2011**, *19*, 1773–1781. doi:10.1016/j.simpat.2010.09.002. [CrossRef]
10. Deng, Y.; Chen, Z.; Zhang, D.; Zhao, M. Workload scheduling toward worst-case delay and optimal utility for single-hop Fog-IoT architecture. *IET Commun.* **2018**, *12*, 2164–2173. [CrossRef]
11. Chen, M.; Hao, Y. Task Offloading for Mobile Edge Computing in Software Defined Ultra-Dense Network. *IEEE J. Sel. Areas Commun.* **2018**, *36*, 587–597. [CrossRef]
12. Zhang, W.; Zhang, Z.; Zeadally, S.; Chao, H. Efficient Task Scheduling with Stochastic Delay Cost in Mobile Edge Computing. *IEEE Commun. Lett.* **2019**, *23*, 4–7. [CrossRef]
13. Wang, W.; Zhu, K.; Ying, L.; Tan, J.; Zhang, L. Map task scheduling in MapReduce with data locality: Throughput and heavy-traffic optimality. *IEEE/ACM Trans. Netw.* **2016**, *24*, 190–203. [CrossRef]
14. Dean, J.; Ghemawat, S. MapReduce: A flexible data processing tool. *Commun. ACM* **2010**, *53*, 72–77. [CrossRef]
15. Apache Hadoop. Available online: http://hadoop.apache.org/ (accessed on 15 June 2020) .
16. Thomas, L.; Syama, R. Survey on MapReduce scheduling algorithms. *Int. J. Comput. Appl.* **2014**, *95*, 9–13. [CrossRef]
17. Ahmad, F.; Chakradhar, S.T.; Raghunathan, A.; Vijaykumar, T.N. Shufflewatcher: Shuffle-aware Scheduling in Multi-tenant Mapreduce Clusters. In Proceedings of the 2014 USENIX Annual Technical Conference (USENIXATC 14), Philadelphia, PA, USA, 19–20 June 2014; pp. 1–12.
18. Jin, J.; Luo, J.; Song, A.; Dong, F.; Xiong, R. BAR: An Efficient Data Locality Driven Task Scheduling Algorithm for Cloud Computing. In Proceedings of the 2011 11th IEEE/ACM International Symposium on Cluster, Cloud and Grid Computing, Newport Beach, CA, USA, 23–26 May 2011; pp. 295–304.
19. Lee, Y.C.; Zomaya, A.Y. Energy conscious scheduling for distributed computing systems under different operating conditions. *IEEE Trans. Parallel Distrib. Syst.* **2011**, *22*, 1374–1381. [CrossRef]
20. Chen, Q.; Zhang, D.; Guo, M.; Deng, Q.; Guo, S. SAMR: A self-adaptive MapReduce scheduling algorithm in heterogeneous environment. In Proceedings of the IEEE International Conference on Computer and Information Technology, Bradford, UK, 29 June–1 July 2010; pp. 2736–2743.
21. Eltabakh, M.Y.; Tian, Y.; Ozcan, F.; Gemulla, R.; Krettek, A.; McPherson, J. CoHadoop: Flexible Data Placement and Its Exploitation in Hadoop. *Proce. VLDB Endow* **2011**, *4*, 575–585.
22. Ananthanarayanan, G.; Agarwal, S.; Kandula, S.; Greenberg, A.G.; Stoica, I.; Harlan, D.; Harris, E. Scarlett: Coping with Skewed Content Popularity in Mapreduce Clusters. In Proceedings of the European Conference on Computer Systems, Proceedings of the Sixth European Conference on Computer systems (EuroSys 2011), Alzburg, Austria, 10–13 April 2011; pp. 287–300.
23. Ananthanarayanan, G.; Ghodsi, A.; Warfield, A.; Borthakur, D.; Kandula, S.; Shenker, S.; Stoica, I. PACMan: Coordinated Memory Caching for Parallel Jobs. In Proceedings of the 9th USENIX conference on Networked Systems Design and Implementation, San Jose, CA, USA, 25–27 April 2012; pp. 267–280.
24. Li, X.; Wu, J.; Qian, Z.; Tang, S.; Lu, S. Towards location-aware joint job and data assignment in cloud data centers with NVM. In Proceedings of the 36th IEEE International Performance Computing and Communications Conference (IPCCC 2017), San Diego, CA, USA, 10–12 December 2017; IEEE Computer Society: Washington, WA, USA, 2017; pp. 1–8.
25. Li, C.; Tang, J.; Tang, H.; Luo, Y. Collaborative cache allocation and task scheduling for data-intensive applications in edge computing environment. *Future Gener. Comput. Syst.* **2019**, *95*, 249–264. [CrossRef]
26. Zhou, J.; Sun, J.; Cong, P.; Liu, Z.; Zhou, X.; Wang, T.; Wei, T.; Hu, S. Security-Critical Energy-Aware Task Scheduling for Heterogeneous Real-Time MPSoCs in IoT. *IEEE Trans. Serv. Comput.* **2020**, *13*, 745–758. [CrossRef]
27. Yin, J.; Gu, G.; Zhao, J. Dynamic scheduling algorithm for hybrid real-time tasks with precedence constraints. *Comput. Integr. Ated. Manuf. Syst.* **2010**, *16*, 411–416.

energies

MDPI

Article

Combining Blockchain and IoT: Food-Chain Traceability and Beyond

Jacopo Grecuccio [1], Edoardo Giusto [1,*], Fabio Fiori [2] and Maurizio Rebaudengo [1]

1 DAUIN, Politecnico di Torino, Corso Duca degli Abruzzi, 24, 10129 Torino, Italy; jacopogrecuccio1994@gmail.com (J.G.); maurizio.rebaudengo@polito.it (M.R.)
2 Foodchain S.p.A.,Via Cavour, 2, 22074 Lomazzo, Italy; fabio.fiori@food-chain.it
* Correspondence: edoardo.giusto@polito.it; Tel.: +39-011-0907068

Received: 20 June 2020; Accepted: 21 July 2020; Published: 25 July 2020

Abstract: Recently, the interest around the Blockchain concept has grown faster and, as a consequence, several studies about the possibility of exploiting such technology in different application domains have been conducted. Most of these studies highlighted the benefits that the use of the blockchain could bring in those contexts where integrity and authenticity of the data are important, e.g., for reasons linked to regulations about consumers' healthcare. In such cases, it would be important to collect data, coming in real-time through sensors, and then store them in the blockchain, so that they can become immutable and tamper-proof. In this paper, the design and development of a software framework that allows Internet-of-Things (IoT) devices to interact directly with an Ethereum-based blockchain are reported. The proposed solution represents an alternative way for integrating a wide category of IoT devices without relying on a centralized intermediary and third-party services. The main application scenario for which the project has been conceived regards food-chain traceability in the Industry 4.0 domain. Indeed, the designed system has been integrated into the depiction of a use case for monitoring the temperature of fish products within a warehouse and during the delivery process.

Keywords: internet of things; blockchain; distributed ledger technologies; smart societies; industry 4.0; supply chain; food traceability

1. Introduction

During the last ten years, the industrial processes commonly applied in any business field, from manufacturing to agriculture and many others, have been subject to a radical evolution led by the arising of new technologies interconnecting the digital and the physical world and making up the Cyber-Physical-Systems (CPSs). CPS consists of a system of collaborating computational elements controlling physical entities, in which mechanical and electronic systems are embedded and networked using software components. They use a shared knowledge and information of processes to cooperate for accomplishing a specific task [1,2]. The trend of embedding interconnected electronic devices, capable of interacting with the environment in which they live in and enabled to communicate over the Internet, has spread not only in the industrial and manufacturing fields but also in other business sectors such as agriculture, breeding, and food transformation processes [3]. The resulting scenario is known under the context of Internet-of-Things (IoT), which can be defined as a digital overlay of information over the physical world. Each object (referred to as a *"thing"*) connected in such a network is uniquely identifiable and able to sense and react with the environment, as well as with other users or objects [4,5]. The IoT devices market nowadays represents a huge portion of the whole Information-Technology (IT) and electronic devices market with more than $235 billion spent in 2017, and it is expected to double in 2021, growing up to $520 billion [6].

Besides IoT, another technology arose and has grown even faster: the Blockchain. A Blockchain (BC) is a distributed ledger based on a peer-to-peer (P2P) network, in which participants, called *nodes*, agree on a unique version of the distributed data storage through a shared consensus mechanism. All information stored inside the ledger is digitally signed employing cryptographic primitives and data-authenticity is guaranteed by the use of asymmetric key-pairs. The first definition of BC was given in November 2008, in a white-paper titled "Bitcoin: a Peer-to-Peer Electronic Cash System" sent to the subscribers of a mailing list about cryptography by a mysterious author known under the pseudonym of Satoshi Nakamoto [7]. After the advent of Bitcoin, several different implementations of Blockchain have been proposed, targeting not only crypto-currency and finance application but also business processes automation and certification [8]. Several studies nowadays are exploring the possibility of applying the BC technology for product certification processes, especially in the food industry, and most of them believe that this technology would represent a powerful partner for solving problems related to consumer's trust in food products and to implement a traceability system able to find possible sources of contamination in-time, thus reducing the risks for citizens' health [9].

In this paper, a possible integration strategy that combines both IoT and BC for improving business processes is presented. The proposed architecture is able to fully exploit the concept of a public Blockchain. In fact, IoT devices are able to directly sign transactions and store them in the BC in such a way that there is no third party involved in the process, which could lead to a series of undesirable situations. The final result of this study is a *blockchain-enabled gateway* for IoT devices to be integrated in an Industry 4.0 domain and in many other scenarios in which the combination of both technologies brings advantages to both business processes and customer satisfaction. In particular, the selected use-case in which the architecture has been applied and tested is related to the agri-food supply chain, also known as *food-chain*.

The remainder of this paper is organized as follows. Section 2 gives a high-level overview of the Blockchain's basic elements with a specific focus on Ethereum in Section 2.3. Section 3 describes the current landscape of IoT-BC integration, highlighting its main weaknesses. In Section 4, the proposed integration strategy is presented, along with implementation details in Section 5. Section 6 depicts a simulation of a use-case application related to the cold-chain monitoring, also describing the main issues related to the agri-food supply chain. Section 7 displays and comments the results obtained with the proposed architecture. In Section 8, conclusions are drawn, also proposing possible extensions.

2. Background

The decentralized network proposed by Nakamoto [7] is based on a Peer-to-Peer architecture, where all the nodes have the same functionalities and provide the same services without the distinction of roles as it happens in a Client–Server architecture. Each peer in the network, known as a **node**, stores locally a copy of all the history of the **transactions** published on the network. A Blockchain can be formally defined as a structured database of **blocks**, each one containing several transactions, linked together by including in each block the cryptographic hash of the prior block in the chain. This allows us to check the integrity of the whole chain going backward until the genesis block, which is the very first block of the chain. Sometimes separate blocks can be produced concurrently leading to the creation of a temporary fork of the chain. To solve this problem, any BC defines an algorithm for scoring different versions of the history so that only the one with the highest score is considered valid and selected over the others. The mechanism used for verifying new transactions to be added to the chain is called **mining**, and not only makes the users agree on a unique version of the chain but at the same time allows for create new coins that can be spent in the network, exploiting a **reward mechanism**.

2.1. Blockchain Principles

As it appears from the previous section, the fundamental elements upon which a Blockchain is built are blocks and transactions. The need for each node of the network of storing locally the entire chain implies the usage of an efficient data structure that has to occupy a small quantity of memory and at the same time to guarantee that the data stored in the chain are immutable and tamper-proof. In order to be scalable, in Bitcoin BC, each block's hash is the hash of its **block-header** (a data structure containing a timestamp, a nonce and the hash of the previous block) and the root hash of a multi-level data structure known as Merkle-Tree. A **Merkle-Tree** is a binary tree composed of a set of nodes with a large number of leaf nodes containing the underlying data at the bottom. Starting from the leaves each intermediate node stores the hash of its two children up to the root node that is at the top of the tree. The main purpose of this data structure is to allow for retrieve the required data efficiently and eventually piecemeal from different nodes in the network. Moreover, the hash mechanism behind this structure, combined with the shared consensus, ensures that the entire chain can not be changed or altered. However, nowadays the effectiveness of this protocol is argued not to be sustainable for the long-term. Indeed, as of June 2020, the total space occupation of the Bitcoin Blockchain is roughly 270 GB and it keeps growing faster. This factor makes unaffordable the implementation of a node on devices with limited storing and computing capabilities like mobile and embedded devices [10].

2.2. Digital-Signature and Elliptic-Curve Cryptography

Asymmetric key cryptography, also known as **public-key cryptography**, is a cryptographic scheme that makes use of a pair of large numbers (*keys*) that are somehow related together but they are not identical. One key, referred to as *private* because it has to be known only by the owner and not shared with any other parties, is used for decrypting a message over an insecure channel. Instead, the other key of the pair referred to as *public* because the owner shares it with other parties, is used by the message's sender for encrypting the message to be sent to the other party. It appears clearly that such mechanism works under three fundamental assumptions:

- Private- and public-key should be related, but at the same time it should be unfeasible to obtain the private key knowing only the public one,
- The decryption function gives the correct result if and only if the correct private key (i.e., the one related with the public key used for the encryption) is given and for no other different value,
- Private keys should be kept secret and not shared with any other parties.

Therefore, asymmetric cryptography protocol implementations differ mostly in the algorithm used for the key pairs generation and the encryption/decryption functions. For example, the RSA algorithm is based on the *large integer factorization problem* and, simply speaking, uses the product of two large prime numbers as part of the public key, and derives the private key from the same number as well. The encryption strength relies mostly on the key-size and its robustness increases exponentially as such parameter grows. Actually, RSA [11] keys can be typically 1024 or 2048 bits long, but experts believe that RSA 1024 is near to be cracked and it should not be considered secure anymore, especially with the advent of quantum computing [12]. Another implementation of the public key cryptography's protocol relies on the algebraic structure of elliptic curves over finite fields and is known as Elliptic Curve Cryptography (ECC). It is based on the *elliptic curve discrete logarithm problem* that consists in finding the discrete logarithm of a random elliptic curve element concerning a publicly known base point. The approach of using elliptic curves in cryptography was proposed for the first time in 1985 by Neal Kolbiz [13] and Victor Miller [14], but it became widely used starting from 2004. It is used in Blockchain implementations such as Bitcoin and Ethereum as a Digital Signature scheme [15,16]. The security of ECC stands in the fact that, roughly speaking, given a point $k * G$ on the curve it is difficult to go back to the original value of k, this is known as the **Discrete Logarithm Problem**, which is said to be one of the NP-hard problems, and, compared to RSA, ECDSA offers the same level of security but with smaller keys.

2.3. Ethereum

Ethereum is an Open-Source project aiming at the creation of a public BC for distributed computing. It was born in 2013 thanks to Vitalik Buterin, a Canadian developer and researcher in the field of cryptocurrencies. Through a crowdfunding campaign, he was able, in 2014, to implement his idea, published and available online the following year [17].

The project intended to create an alternative protocol for building decentralized applications (DApp), allowing a different set of trade-offs to be used in those applications where rapid development time, security and the interaction between different distributed applications or systems is crucial. Following this idea, Ethereum provides a fundamental layer composed of a Blockchain and a Turing-complete programming language so that anyone can write its own DApp, implementing its arbitrary rules for ownership, transaction formats, and state-transition functions. The most interesting and explored functionality of this new architecture is one of the so-called *Smart-Contracts*. *Smart-Contracts* can be seen as cryptographic entities containing a value that can be represented by a certain amount of currency or information or both, unlocked only if an application-defined set of conditions are met. To achieve the goal of creating a simple framework for building powerful decentralized-application, the basic philosophy pursued in Ethereum development is based on a few principles:

- **Simplicity**: the protocol has to be as simple as possible, even at the cost of some inefficiency in terms of execution time or data storage. The idea is that an average programmer should be able to follow and implement the entire specification.
- **Universality**: Ethereum should not provide features. Instead, it has to provide a complete fundamental layer upon which any kind of application can be easily built.
- **Modularity**: the framework has to be structured in modules, as separate as possible.
- **Agility**: the protocol's architecture should not be extremely fixed, and any meaningful further proposal bringing significant benefits or improving scalability and sustainability is accepted.
- **Non-discrimination** and **non-censorship**: the protocol should not attempt to actively restrict or prevent specific categories of usage.

Ethereum's fundamental currency is *Ether* (*ETH*) and is used to pay for *gas*, which represents the unit of computation used in a transaction and other state transitions, such as Smart-Contracts execution.

2.4. Smart-Contracts

The term Smart-Contract (SC) was introduced for the first time in the 1990s by Nick Szabo, a computer scientist and cryptographer, who realized that the Distributed Ledger Technology can be exploited for securing relationships over a network [18]. Bitcoin was the first Blockchain implementation to support a basic form of Smart Contracts through its scripting language, which was however somewhat limited. The Ethereum project, on the other hand, was born with Smart-Contracts in mind and aiming at realizing a framework for DApps running over a BC ledger in a peer-to-peer network. Just as the term suggests, Smart-Contracts are somehow related to "normal" contracts, with the difference that the latter ones are signed by partaking entities and enforced by the law, whereas the former ones are digitally signed and define relationships among parties exploiting cryptographic mechanisms. Basically, a Smart-Contract is a self-executing application that runs on top of the BC and is therefore immutable. It can be seen as a complex *if-then* statement that is executed if and only if a set of conditions is met. Smart-Contract can either produce an output or even trigger the execution of another Smart-Contract for performing a sophisticated task. Because of their decentralized nature and the possibility to exploit BC's features for handling trust-less contexts, SCs are becoming an attractive topic for various kinds of business fields spreading from finance, insurance, manufacturing, and many others.

To better understand how SCs work and how big their potential can be if exploited in some applications domain, it is worth describing an example use case (Figure 1). Let's assume that farmer

A sells tomatoes to company B which then uses them for producing tomato sauce. The transfer of tomatoes from A's farm to B's warehouse is in charge of a third delivery company C. The actual most common scenario would be the one in which A, B, and C have their own digital business layer, generally made of a centralized data storage (in a local Database or using some cloud services) and their own software products that deal with such data structure for registering shipping, invoices, inventory and so on. Moreover, this software may differ between one company from another, making it difficult to automatize their cooperation. Let us assume now that the same business scenario was managed through the use of Smart-Contracts and a Blockchain in a Peer-to-Peer network at which all of the three companies participate. A smart DApp developer would implement three Smart-Contracts able to interact together:

- A first Smart-Contract that can handle the transfer of ownership of the batch of tomatoes from A to C, ensuring that the latter would be responsible for any loss or damage during shipping.
- Another Smart-Contract that can handle the transfer of ownership from the delivery company to recipient B.
- A third Smart-Contract, triggered by the previous one that handles payments between the parties in something similar to the following statement: *If tomatoes are delivered without any damage a certain amount of funds will be transferred to both the selling and shipping companies. If tomatoes are damaged then company C has to refund the selling company, and so the proper amount of crypto-money will be transferred from C's account to A's one.*

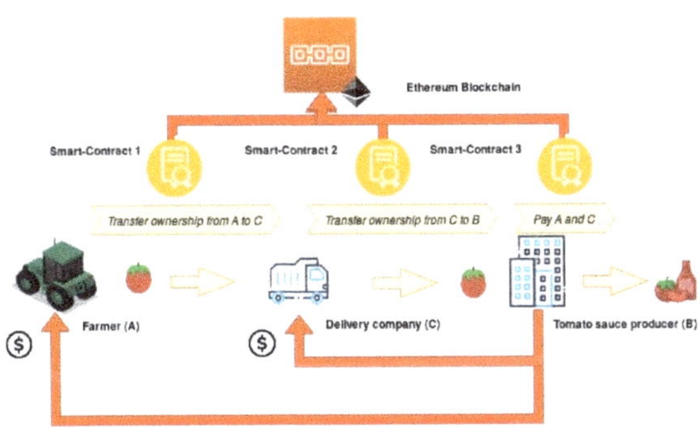

Figure 1. Application of Smart Contracts to traceability in the food-chain domain (image created using https://draw.io).

Thanks to the BC, all operations occurred between the parties will be registered in the ledger forever, and so it is straightforward to think to more complex mechanisms to be built upon the simple scenario described above (insurance for the delivery process, supply-chain traceability, etc.).

2.5. *Quadrans: The Industrial Blockchain*

The Quadrans [19] platform is an open-source and public Blockchain network that runs Smart-Contracts and decentralized applications. The first implementation of this platform, born in 2012 as a fork of Ethereum, has been conducted under the Foodchain S.p.A. brand, and its initial goal was to enable traceability, transparency, and authenticity of the information within agri-food supply chains. In August 2018 Quadrans Foundation was officially formed and acknowledged by Swiss local authorities. The goal of the foundation is to guarantee continuous advancement in technology in an ethical and publicly open way. Quadrans' infrastructure is public, and its open-source blockchain has been designed to overcome scalability issues and to reduce the instability of operational costs affecting other existing blockchains. The current protocol, used by Quadrans, ensures a high throughput (average block-time is about 5 s) that can maintain at the same time both high security and decentralization grades and supports all the core Ethereum's features, such as EVM (Ethereum Virtual

Machine) and Smart-Contracts. Having reached these goals, Quadrans represents a valuable solution for effectively applying blockchain technology in many business scenarios.

3. Literature Review

Having given enough details on the basic principles of BC's technology, it is possible to introduce the real goal of this article. As stated in the introduction, the project aims to find a new method of interaction between the BC infrastructure and the world of internet-connected devices.

The most common solution applied for this scenario consists of a Client–Server communication (as in Figure 2) between a central server and the IoT devices. The server is in charge of (*a*) gathering the data streams collected from the field and (*b*) storing these data, or part of them, in the Blockchain. Even if this solution is starting to be applied in different proof-of-concept and business applications, this has some point of weakness:

- The central server may become a single-point-of-failure that can inhibit some data to be stored in the BC when it is in a fault condition,
- Assuming that redundancy is applied, , in order to minimize the probability of failure, there is still the presence of some kind of centralization that makes difficult the interaction between cooperative business parties,
- The data which is then sent to the BC is not signed-in-place, a way in which its authenticity and integrity can be guaranteed to start from the source, but it is signed only when it is received by the central server before posting it into the Blockchain,
- Nevertheless, such kinds of systems are now mostly realized using cloud solutions for the IoT field. Experts believe that the cost of such services is going to increase rapidly and to become a significant component of business costs for most companies.

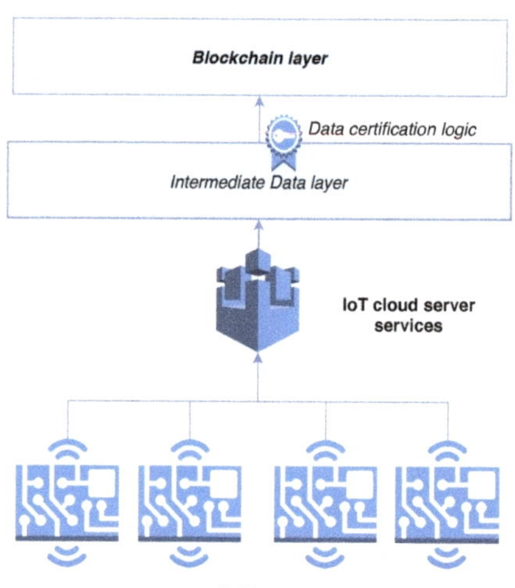

Figure 2. The common architecture used for storing data collected from IoT devices into the blockchain (image created using https://draw.io).

The surveys in [10,20] present extensive study of these and other issues. The authors of [20] describe the issue of storage size in relation to an application in the IoT field. The solution we present solves this issue by implementing light-nodes that act as a gateway. These light-nodes only perform the tasks of signing and sending their own transactions. They do not participate in the consensus mechanism (there is no mining), and thus they do not have to locally store the whole blockchain.

A relevant paper on the quest for decentralization, secure data storage, and IoT compatibility can be found in [21]. This paper describes a model to handle data-streams coming from IoT devices with a decentralized access mechanism control. This solution has some undesired aspects though. It uses an intermediate level between the cloud and the BC, meaning that the data are not set directly to the BC by the device. Moreover, it uses the Bitcoin BC, even if in its VirtualChain version, supposedly lighter and faster.

The solution described in [22] seems focused mainly on performing some kind of action based on the conditions of Smart-Contracts on the Ethereum Blockchain. It has interesting propositions regarding the use of smartphones to set/update conditions for Smart-Contracts, and the possibility of IoT devices to act according to these conditions. In addition, some pieces of Smart-Contracts are shown in the paper. Despite these characteristics, the whole architecture is not clearly described, especially for what concerns the communication with Ethereum blockchain, where the blockchain itself resides, and the signing functionality of the IoT device.

In [23], BPIIoT is presented. The authors refer to it as a key-enabler for Cloud-Based Manufacturing, enabling peers of a peer-to-peer decentralized network to interact among them without a trusted intermediary. They also implement Smart-Contract handling on a BeagleBone single-board computer (SBC) which seems to be connected to an Arduino board for sensing capabilities. However, this paper also does not give many details about the implementation of such system, especially regarding the connection with the Ethereum blockchain and the signing capabilities.

Authors in [24] present AgriBlockIoT, a decentralized traceability system for the Agri-Food supply chain management. AgriBlockIoT is nicely organized in layers and can support either Ethereum or Hyperledger Sawtooth. The paper also presents an interesting *from-farm-to-fork* use case, depicting all actors, materials, and tasks involved in such a process. Our approach can be considered an evolution of this, since we remove the need of running a full node of the blockchain in the stakeholder's local network. Moreover, the use of IoT devices is only simulated and not practically implemented.

A blockchain-IoT-based food traceability system (BIFTS) is proposed in [25]. Not only it does integrate BC and IoT, but it also proposes an appealing quality decay evaluation of perishable goods based on fuzzy logic. A critical point in this architecture is the fact that, to get to the BC, the data have to pass through a cloud server. This represents in any case some sort of centralization, which we would like to avoid. In fact, it is not possible, inside the cloud, to authenticate the data with respect to a certain device.

In [26], authors present a blockchain-based fair nonrepudiation service provisioning scheme for industrial IoT scenarios. It is a complex and interesting framework in which the BC acts as an evidence recorder and a service publication proxy. A limitation of this proposal is that their devices communicate through a BC node, like the first intermediate solution proposed later in Section 4.

The perspective presented in [27] is really compelling. Their solution features the direct signing capabilities of devices, which is one of our aims, and the offloading of communication with the BC to a proxy service. The framework is built in such a way that the proxy server itself cannot forge the devices' signatures, thus maintaining the authenticity of the data. In addition, the applicability to the cold-chain use case is the same as our approach. However, their solution relies on the use of Hyperledger Fabric as a permissioned BC, which is not public by definition. A permissioned BC implies the presence of a certain authority issuing credentials and certificates, which is indeed a form of centralization. Moreover, the SDK proposed does not seem ready yet to target very constrained IoT devices built upon a microcontroller.

The advantages of the proposed solution are the following:

- Implemented devices do not perform mining and do not need to store the whole BC, making this solution suitable for low-end hardware (low-power and low-cost),
- The use of a BC (Quadrans) directly derived from Ethereum and natively supporting smart-contracts, with average block times of around 5 s,
- IoT and BC levels communicate directly, without a cloud intermediary,

- Every device has an associated *wallet*, meaning that it is possible to guarantee principles of data-authenticity and non-repudiation since the device itself signs the data before sending it to the BC.

4. Proposed Architecture

The proposed architecture is intended for solving the weaknesses presented above, and possibly to introduce some improvements to the described context. It consists of enabling commercial or ad hoc Internet-of-Things devices to communicate directly with the BC infrastructure through a *gateway* device to which they are connected using wired or wireless protocols. Such a gateway device should be able to sign-in-place the data sent from the IoT data-logging device and then to directly push on the BC. The most straightforward method of doing so would be to let the gateway be a full-node of the chain (i.e., a node in the peer-to-peer network that can store and verify the whole chain and keep it updated too). However, this is unfeasible as the target application context lies in the embedded-systems domain, which commonly consists of limited computing capability devices, for sure equipped with not enough storage to maintain an entire copy of the chain.

A first intermediate solution (Figure 3) that has been explored is based on an architecture where the IoT layer communicates directly with an Ethereum full-node placed within the same Local-Area-Network in which the devices are connected. Such a local node is then in charge of signing and broadcasting the transactions requested by the devices, as well as keeping their private key storage and mappings (between IoT device ID and private key). The implementation of this solution resulted in being fast and straightforward because the only part on which one has to deal with requires only the communication mechanism between the local BC node and the IoT devices, defining the data-model of the message to be exchanged and eventually some encryption mechanism to secure the communication channel. However, this solution suffers from the same problems related to centralization and security as the common case previously described. Nevertheless, it would only be worse if no mechanism for device authentication was implemented for the communication with the local full node.

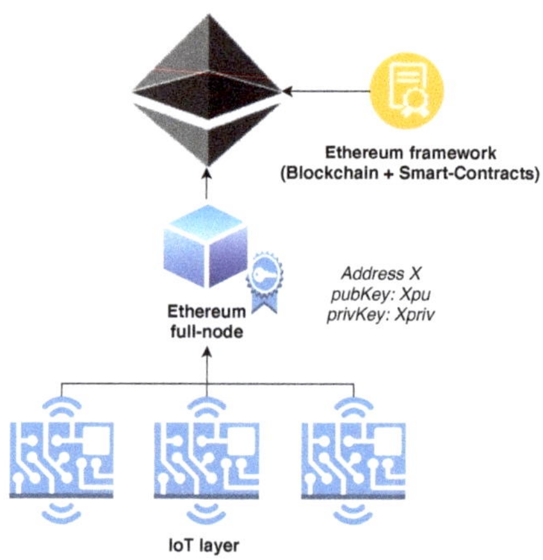

Figure 3. A representation of the first explored solution. IoT devices communicate with the local full node which is in charge of signing and broadcasting transactions for them. However, the IoT layer remains agnostic of the existence of the Blockchain. It simply sends data to a local centralized entity (image created using https://draw.io).

Therefore, the path followed in the project exploits the Remote-Procedure-Call RPC protocol [28] for enabling an embedded device (acting as a gateway) to trigger the execution of some procedure on a remote server exposing such service. Such a device has to be physically placed in the very near proximity of the IoT device (e.g., as expansion hardware for the device) or even be the IoT device itself. The RPC server exposes to calling clients (i.e., IoT gateways) the Web3 Ethereum API (Application Programming Interface that allows for interacting with Ethereum's BC, so that any embedded device can access the information stored in the chain, exchange coins (e.g., for implementing machine-to-machine payments) and call the execution of Smart-Contracts for implementing complex logic on top of the blockchain layer. Moreover, exploiting the events mechanism of Smart-Contract [29], and the possibility to search for those events in the chain by selectively download block headers through RPC calls, it could even be possible to trigger IoT devices directly within distributed applications and keep the history of such requests. To apply this solution, however, there should be one strict requirement above the others: *the gateway must be identifiable on the networks through its address, and should also be able to sign the transaction locally and offline, using its private-key, before sending it to the RPC server*. This functionality is essential to avoid the possibility that the transaction is maliciously manipulated when it is sent to the server.

The final overall proposed architecture is shown in Figure 4.

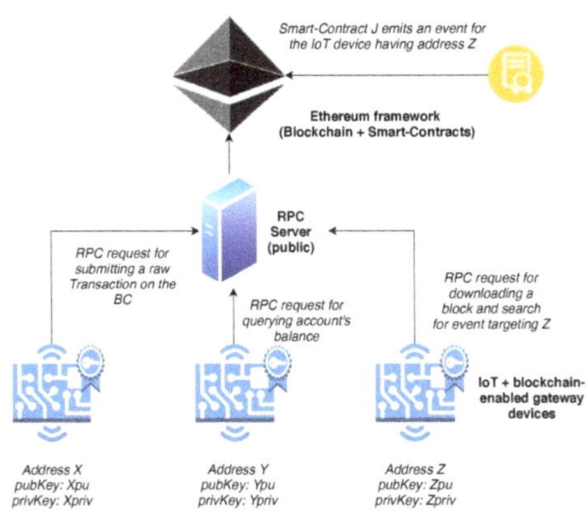

Figure 4. A representation of the final proposed architecture. Each IoT device has its own gateway and can sign transactions locally and offline. Each of them is also identified within the blockchain through its address and can be thus a target for possible Smart-Contract events (image created using https://draw.io).

5. Software Architecture

Most of the project complexity has been demanded from the software layer, meaning that the vast majority of the operations are performed by the CPU without any additional special-purpose peripheral or hardware accelerator.

The developed software framework is intended to be as easy and user-friendly as possible. In this way, it can be used by the majority of IoT developers, with minimal knowledge of the details behind the Blockchain architecture. Following this philosophy, the resulting framework can be represented as a layered architecture (Figure 5) where the top-most software components are those providing the interaction services to the common Internet-of-Things application.

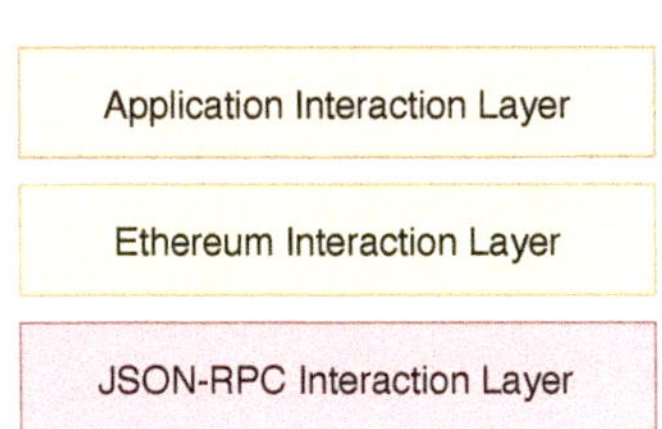

Figure 5. Representation of the Software Architecture (image created using https://draw.io).

5.1. The JSON-RPC Layer

The Remote-Procedure-Call is based on the Client–Server model and is used in those contexts where one program wants to request a service from another one, executed in a remote host. An RPC call is a synchronous event requiring the caller program to be suspended until the remote procedure returns its result. An example of this behavior can be found in Figure 6.

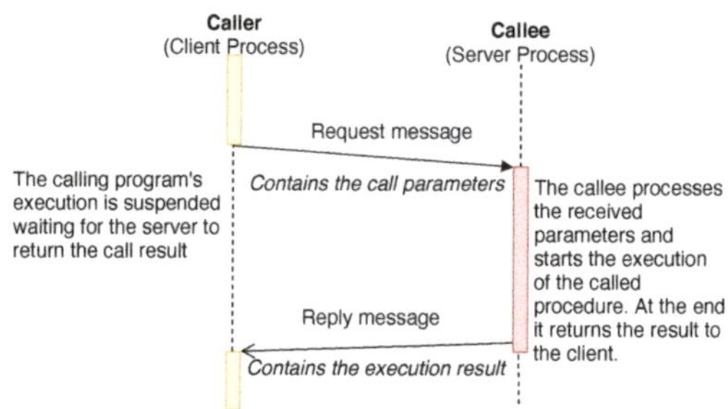

Figure 6. A sequence diagram showing the flow of execution of a remote-procedure-call (image created using https://draw.io).

JSON-RPC represents a state-less and light-weight implementation of an RPC. It is an extension of the original RPC protocol, which uses the JSON data-format for remote-procedures, parameters, and results encoding [30].

In the developed framework, JSON-RPC Client is the software component in charge of performing JSON-RPC calls over the TCP socket on which the RPC server is listening. Its implementation consists of a queue, where other tasks push a data-structure for each RPC-Call that they request. This data-structure is used to model the standard fields of a Remote-Procedure-Call and, in addition to those fields, another one is provided to store the current state of the call. Each call, in this implementation, can assume the following logical states:

- `QUEUED`: it is the first state assumed by a call-object. It describes a call that has been pushed into the queue by a caller task and it is waiting to be executed,
- `SENT`: the call has been sent to the RPC server successfully,
- `WAIT_RESPONSE`: the client is waiting for the server response for this call,
- `SUCCESS`: the call has been executed successfully and its result is now available,
- `ERROR`: an error occurred during the call processing,
- `DELETED`: the call has been removed from the queue.

A graphical representation of the state-transition diagram for a call can be seen in Figure 7. The calls queue is accessed in a circular-buffer manner and its dimension, in terms of the number of calls, is static and does not rely on dynamic memory allocation. The module's user can then set the dimension, according to the application requirements, through a C-language `define` pre-compiler instruction and therefore select the optimal trade-off between memory occupation and performance. Summarizing, the `JSONRPC_Client` interface provides functions for the upper layers , allowing the caller to:

1. Push a new RPC call in the queue,
2. Retrieve the status of a call previously submitted,
3. Retrieve the result of an executed call,
4. Delete a call,
5. Check the queue status (empty/full).

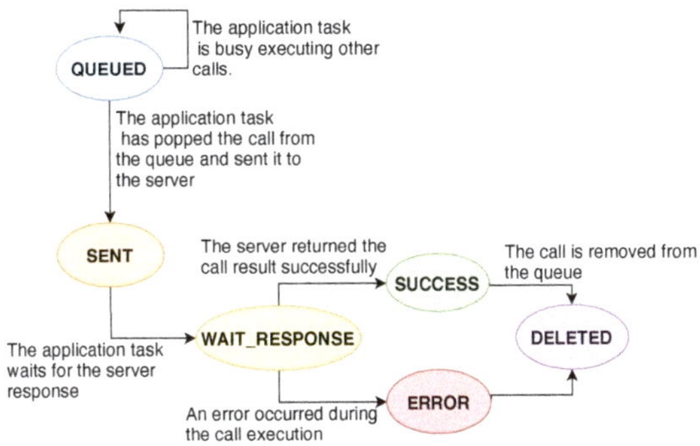

Figure 7. The state-transition diagram for a call-object (image created using https://draw.io).

In addition, the module implements the application-task function (`JSONRPC_Task`) in charge of maintaining the queue, converting calls-object into JSON, sending calls over the TCP socket, retrieving the response, and parsing it from JSON. The sequence diagram in Figure 8 illustrates an example of an RPC call execution.

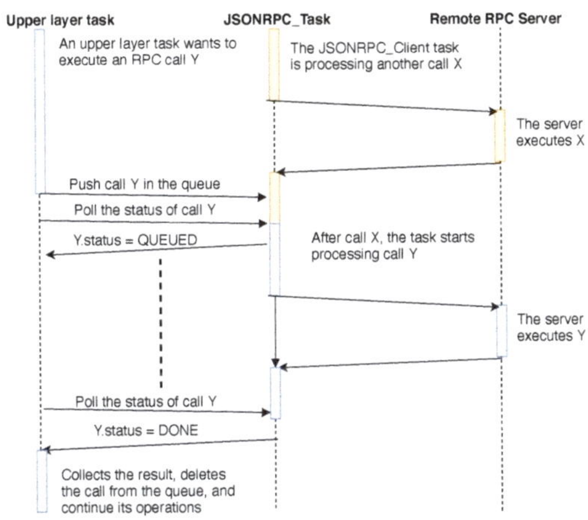

Figure 8. A sequence diagram illustrating how the developed JSONRPC client interacts with upper layers requesting a remote call to be executed (image created using https://draw.io).

5.2. The Ethereum Interaction Layer

The Ethereum interaction layer is one of the frameworks that implements a subset of the functions provided by the Web3 API. Each of the provided functions is in charge of encoding the necessary parameters for the correspondent RPC call to be executed, as well as adapting the returned result into user-friendly C-language types.

The provided functions allow the caller to check the status of an account (i.e., nonce and balance), perform transaction and message calls, and request a specific set of blocks from the chain. The fundamental data-structure of this layer is the `WEB3_ETH_OBJ` that holds the information about the RPC call object as well as the state of the Web3 call. The implementation of these functions follows the state-machine model,with a given web3-call evolving through the following states:

- `CALL`: Parameters are encoded into JSON format and the proper RPC call is pushed into the queue of the JSONRPC client module,

- WAIT: Polling on the call-object status to check when the result is ready or an error is returned,
- DONE: The result is ready to be processed,
- ERROR: An error occurred.

A graphical representation of the state-transition diagram for a Web3 call can be seen in Figure 9.

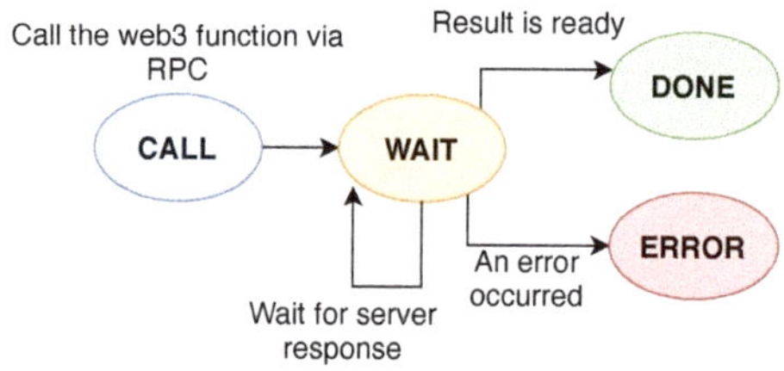

Figure 9. The state-transition diagram for a web3 call in the developed framework (image created using https://draw.io).

5.3. The Application Interaction Layer

The Application interaction layer represents a wrapper of the Ethereum interaction layer in the sense that it hides the details for performing specific operations on the blockchain (e.g., signing a transaction before sending it as raw). It wraps all the functions of the layer below and performs the pre-processing and post-processing operations needed for some of them.

Again, they are internally implemented as state-machines, cycling through the following possible states:

- WEB3TASK_INIT: In this state, all the pre-processing is performed before the web3 function is called,
- WEB3TASK_WAIT_RESPONSE: The state machine waits for the web3 call to be executed and return its result to perform the post-processing operations,
- WEB3TASK_IDLE: Nothing to be done,
- WEB3TASK_ERROR: An error occurred.

6. Use Case Application

This chapter presents a use case covering the supply chain of fish and seafood products. The use case has been simulated in collaboration with Foodchain S.p.A., a company whose mission is to enhance the traceability process of food's supply chain through the Blockchain technology.

To simulate the use case scenario, an IoT device has been built and programmed in such a way that it can autonomously communicate with the Blockchain by directly signing its own transactions. Moreover, to corroborate the usefulness of the system, an actual use case scenario has been depicted later in this section.

For what concerns the Blockchain side, Foodchain S.p.A. has developed a platform, based on Quadrans' infrastructure that provides services for both companies and final buyers. Companies can use the platform as a common Enterprise Resource Planning (ERP) software, registering products, batches, and processesand any information related to the product's lifecycle (documents, certificates, media, etc.). Such information is immutably registered into the BC, on top of which the platform lies, using Smart-Contracts. At the end of the product's transformation process, the platform generates some kind of Smart-Label (like a QR code or an RFID tag) that uniquely identifies the product and can therefore be scanned by the product's buyer for viewing the entire traceability process. Another important feature provided by the platform is that it enables many companies and producers to cooperate together within the network, allowing them to follow the transformation cycle of a product even when it passes through the production and distribution chains of different actors.

6.1. Food-Chain Issues

Recent studies have demonstrated that customers' loyalty in food companies and in all the products they sell has drastically decreased year after year. In 2018, the Center for Food Integrity published its annual report about consumers' habits and feelings when buying food. The investigation showed that an increasing number of customers are concerned about what they eat, thus they are asking for more information about the food they buy, to choose the healthy one [31]. However, most of them do not trust the information written behind the product's package, especially when it does not directly come from farmers, but it went through some transformation processes instead. In addition, the awareness of consumers about the damages caused to the environment by some farming and breeding processes is encouraging them to not buy entirely certain categories of products [32].

The lack of trust in food producers is the consequence of a minimal customers' knowledge about the whole supply chain and also the fear of a globalized market that may allow fraudulent producers to sell dangerous or counterfeited products. In particular, this last aspect is critical both for consumers, who are subjected to risks related to their health, and "honest" producers, who not only suffer the economical damages but are also affected by the side-effects of the distrust of buyers. In the year 2018, the Italian Central Inspectorate of Quality Protection and Fraud prevention of food products (ICQRF) has registered an increment of 58%, compared to the previous year, of crimes related to food frauds, seizing about 17.6 tons of goods, for an equivalent value of over 37 million USD. The wine sector resulted to be the most affected by this phenomenon [33].

6.2. The Concept of Food Trust

From the previous analysis, it is clear that consumers' confidence in the agri-food industry is the result of their concerns or certainties on various aspects related to the product they intend to buy. One of the most important factors to restore a good level of reputation of the food industry is to guarantee **food safety**, which is every day undermined by the numerous cases of food hazards. The lack of information about transformations and events under which the final product has gone through during the supply chain poses the basis for a serious alarm about the health of citizens and, despite the efforts of government in control and prevention, the current infrastructure turns out to be unsuitable and inefficient in the fight against this phenomenon. Another important aspect of food trust is related to the concepts of **food-integrity** and **food-authenticity**, which have a significant impact both on the healthcare and the economy of countries. Food-Integrity defines the state of being undiminished and unaltered with respect to its original nature, which means that no unapproved and undeclared transformations and alterations have been made on such products. The Food-authenticity definition, instead, is more related to economical aspects, like frauds that generally do not represent a risk for the health of consumers. However, food frauds and counterfeiting are hurting the markets of many countries, causing losses for billions of euros per year. The most subjected products to counterfeiting threats are those that earn a great economical value because of their territory of origin or brand, like oil, wine, and cheese. In July 2016, the European Union Intellectual Property Office (EUIPO) has published research about the economical costs of intellectual property infringement in spirits and wine. The results showed enormous damage to the legitimate industries, with estimated losses for approximately 1.3 billion euros of revenue annually [34]. The last but not least component of the concept of food trust is linked to customers' attention, concerning the sustainability of processes and the agri-food chain. Consumers are becoming ever more aware of the role they play in the game for building a sustainable food industry. Most of them, especially the young ones, base their decision-making when buying a product not only on the "classical" factors (price, nutritional values, brand, etc.) but also on the impact that the product they buy has on the environment. The expansion of this new category of customers, known as *Ethical Consumers*, is leading this new market sector to grow very fast, registering in 2015 a growth of 5.3% and 9.7% in 2017 (UK Market) [35,36].

6.3. Combining IoT and Blockchain to Empower the Traceability Process

There is a strong necessity for an enhancement in the monitoring and certification processes to overcome these issues. In the actual situation, most of the compliance data and information are audited by trusted third parties and stored either on paper or in centralized databases. However, these approaches suffer from many informational problems, such as the high cost and inefficiency of paper-based processes, resulting in fraud, corruption, and error happening both on paper and using IT systems.

Thus, there is a need for a new trustworthy approach for registering information about the food chain. The availability of this information to customers is finding a potential powerful partner in the arising Blockchain technology. Indeed, the application of the BC technology in the agri-food sector could lead not only a better and more trustworthy traceability process, but it also could represent a powerful partner in the assessment of the product's value for the buyer. For these reasons, there are actually several experiments and proof-of-concepts aiming to find a way to apply the BC technology into this business field. This potential can be truly unlocked by enabling the blockchain platform to collect, and register on the ledger, data gathered directly from the environment (such as farm fields, food transformation chains, logistic processes, etc.) employing IoT devices [24].

6.4. System Architecture

The scenario in which the use case is applied is about the cold-chain monitoring during logistics operation. In such a context, the IoT device is assumed to be installed into a refrigerated truck that transports seafood from a harbor's warehouse to the final fish-shop. The device can connect to the Internet through a mobile network connection and is in charge of monitoring the fridge's temperature and truck's position employing a temperature sensor and a GPS module, respectively. As soon as the data are collected, the device then sends them to a proper Smart-Contract, in charge of storing them in the BC, using the IoT-blockchain-gateway presented in this paper.

The gateway's hardware configuration used for this specific simulation consists of:

- A PIC32MX 32-bit microcontroller
- A 256-byte EEPROM (used for storing device keys in the developed Proof-of-Concept, keeping in mind that in a real-life application this should become a secure-storage element to avoid the stealing or alteration of the device's key)
- A GSM Module for communicating with the remote RPC server
- A GPS module
- A Temperature sensor

This simple architecture has been chosen to show the limited requirements of this approach. The implemented code is for sure hardware-dependent, but the general architecture is independent from the hardware chosen, and applicable to any kind of similar device.

Moreover, the simulation has been carried out using the Quadrans Testnet blockchain, accessed through its public RPC node, reachable at address `rpc.testnet.quadrans.io`. The communication with the remote RPC is based on HTTPS protocol, which ensures a higher layer of security within the interaction between the IoT device and the remote RPC server.

6.5. Monitoring the Cold-Chain of Fish Products

The boats of Fresh-Fish's fleet are equipped with our device, able to periodically record GPS coordinates. This makes it possible to have the complete history of the route navigated by every boat. When the boats arrive at the harbor, fishermen store the catch into the company's warehouses. Here, each batch of fish (e.g., one kilogram of mussels) is registered into Food-Chain's platform, through its web interface, and is identified with a unique ID that is written into an RFID tag. Then, according to fish shops daily requests, each batch is charged into refrigerator trucks to be delivered to its final destination. When a batch is charged on the truck, the device installed onto the truck scans the RFID tag on the batch package so that the device knows for which product it is going to start the monitoring process. Once the truck starts its delivery path, the IoT device starts to periodically monitor the temperature and the GPS position of the truck and sends data to the proper smart-contract for being registered, providing also the IDs of the batches under the current monitoring process. In the end, when the truck reaches its finals destination, batches' tags are scanned again for registering that the delivery has been completed and that the recipient retailer is now the owner of such products. At this point, the retailer shares the information with the customers by tagging its product with a specific QR code that represents an entry point to access the information stored in the BC. The customers can verify the complete story of the mussels and check if any violation of the cold-chain parameters has occurred.

A UML diagram of the use case is depicted in Figure 10.

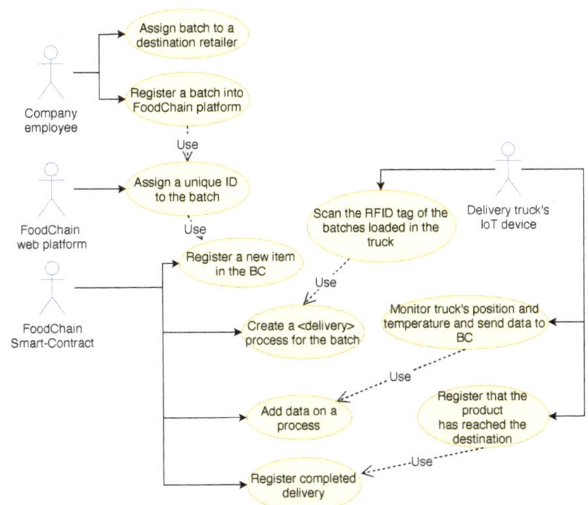

Figure 10. The UML use case diagram of the system (image created using https://draw.io).

A description of a scenario for the use-case involving the IoT device is given in Table 1.

Table 1. Use case scenario description

Step	Description
Pre-Condition	*The item X exists and a <delivery> process p has been activated for it*
1	Read Temperature from temp. sensor
2	Get the position from the GPS module
3	Encode data in the proper format required by the target Smart-Contract
4	Build the transaction object
5	Sign the transaction using the device's private key stored in the EEPROM
6	Send the signed transaction to the BC through an RPC call
8	Wait for the server to reply with the hash of the transaction submitted on the chain
Post-Condition	*Gathered data are now immutably stored in the BC for the item with id X.*

7. Results and Comments

The proposed system tackles the two fundamental issues in the food-chain domain: food integrity and food safety. These are the crucial parameters that efficiently protect customers from frauds or health-related problems, with the positive side effect of helping them to become loyal to a certain vendor. This permits a shift in paradigm, from a trust-based situation to a trust-less scenario. In the former, the buyer is led to put trust in a certain vendor because of an ad campaign, while in the latter the vendor shares with her all the information about the product that cannot be tampered thanks to the Blockchain technology. The use-case demonstrated how the fish tracking process, carried out combining both Blockchain and IoT devices, can enrich the amount of information that is shared with final consumers guaranteeing their authenticity and integrity without relying on a centralized trust authority.

Moreover, from the logistic company point-of-view, publishing data about their food shippings publicly and in transparency can become a good marketing tool for acquiring new customers interested in both shipping their products and, at the same time, sharing data about shipping-quality with their final consumers.

Last, but not least, the data-authenticity and data-immutability properties implemented by the blockchain can make the information registered by IoT devices valuable in legal trials, in case of debate about food-safety issues or even issues related to the shipment insurance.

Table 2 presents a structured comparison between the proposed solution and the other papers referred in the literature and described before. As the table shows, despite the fact that also other implementations in literature could be theoretically applied to the cold-chain monitoring task, the one presented here is the only one able to fulfill all the requirements together: actually working on resource-constrained IoT devices; the lack of need of a local full node of the BC; the lack of an intermediate and possibly insecure cloud platform; operating with a public domain blockchain, without need for trust-certifying parties; employing devices able to directly sign their own transactions.

Table 2. Comparison between the proposed paper and the cited literature, with a focus on cold-chain monitoring.

Reference	IoT hw Required	Full Node Required	Cloud Required	Kind of BC	On-Device Signing	Directly Applicable to Cold-Chain Monitoring
[21]	n/a	yes	yes	Custom, based on bitcoin	n/a	no
[22]	high-end/SBC	n/a	n/a	Ethereum	n/a	yes, but requires high-end device
[23]	high-end/SBC	yes	no	Ethereum	yes	yes
[24]	n/a	yes	no	Ethereum and Hyperledger	n/a	no
[25]	low-end	n/a	yes	n/a	n/a	no
[26]	high-end/SBC	yes	no	Ethereum	n/a	no
[27]	high-end	no	no	Hyperledger	yes	yes, but on a permissioned BC
This paper	yes	no	no	Quadrans	yes	yes

In order to estimate the performance and the effectiveness of the presented system, the following metrics have been considered:

- Hash Time (T_{hash}): the time needed by the microcontroller to hash the transaction, previously encoded as byte-stream using Recursive-Length-Prefix (RLP) algorithm. This metric is related to the amount of data that is included in the transaction.
- Signature time (T_{sign}): the time needed by the microcontroller for executing the ECDSA algorithm and generating a valid signature for the transaction's hash. Since the signature is always computed over the hash of the transaction (having a fixed length of 32 bytes), this metric is not related with the amount of data making up the transaction, but it gives a good index for evaluating the performance of the hardware on which the system is based.
- Confirmation time ($T_{confirmation}$): the time needed by the BC network to validate the transaction sent by the IoT device. This metric depends only on the blockchain layer and is strictly related to the average block-time of the network.

These three metrics have been chosen since their sum represents the least possible time interval that must elapse between the moment the data are gathered by the device (i.e., an on board sensor) and the moment these data become immutable in the blockchain.

From Table 3 it can be noticed that, as expected, the time required by the microcontroller for running the ECDSA algorithm is constant, and represents the major component of the total time needed by the gateway for preparing and issuing a transaction ($T_{transaction} = T_{sign} + T_{hash}$, in first approximation). At the moment, both hashing and ECDSA operations are implemented in software, but it would be worth exploring possible implementations of the system that can rely on hardware accelerators (maybe implemented on an FPGA connected to the micro-controller) which can speed up the execution of such tasks. Another important aspect is related to the time needed by the BC network for validating the transaction. It can be observed that such time goes in the order of 10–20 s, which reflects the average block-time of Quadrans Testnet that is about 15 seconds. Even if Quadrans Mainnet, the one used in production contexts, has a block-time of 5 seconds that would result in a 3x speed-up, it can still represent a barrier for some time-critical applications in automation and automotive fields, for example the control of a motor/robot or the automatic emergency braking of a car. However, in these particular mission critical applications, the use of a BC does not bring any benefit to the system architecture. This aspect goes beyond the goal of this paper, but trying to improve and reduce this time can become a valuable research case for the Quadrans Foundation because a further improvement in this direction will open the possibility to adapt the presented gateway also in scenarios where time constraints are more strict.

Table 3. Use case performance metrics.

	Min.	*Avg.*	*Max.*
T_{sig} **[ms]**		15	
T_{hash} **[ms]**	2	3	5
$T_{confirmed}$ **[s]**	12	16	18

8. Conclusions and Future Works

Summarizing, the integration between two emerging and disruptive technologies such as Internet-of-Things devices and BC-based infrastructure for decentralized applications can bring benefits in a very wide variety of fields, spanning far beyond the simple use case presented in this document. The proposed and developed system gives a proof-of-concept of a possible path to follow for a successful and efficient interaction between Internet-of-Things devices and decentralized applications based on a Blockchain ledger. The *blockchain-enabled gateway* presented in this document satisfies the essential requirements for performing secure and reliable data operation onto the BC infrastructure and, at the same time, provides both hardware and software interfaces for easy and seamless integration even with already developed applications. From the side of the decentralized applications, the possibility to add functionalities that can use data coming directly from the environment, without doubting over their integrity or authenticity, represents a powerful tool that can be exploited to enhance a lot of processes. Finally, for business actors, the services offered by the resulting kind of systems allow them to run a new form of marketing, focused on product and processes transparency. However, the possible application of this promising technologies combination can go even further from traceability and business processes, and be applied for example to healthcare, infrastructures, services, and so on.

Another important feature provided by the designed architecture, and which should be worth exploring better by developing proper use cases and DApps, is the one allowing Smart-Contracts running on the BC to asynchronously trigger some action on specific target IoT devices, identifiable on the network through their public address. Shortly, the idea would be to implement some logic on the IoT devices in order to periodically download the latest N blocks in the chain and search for some emitted event which is targeted to them. However, further studies should evaluate how these concepts can be applied in soft or hard real-time embedded systems, for example starting from the time-critical applications cited above. An exhaustive analysis should take into account not only both the delay of the network and the architecture itself, but also the delay related to the block-time parameter of the BC. Probably, explorations in this direction may lead to the development of IoT-oriented Blockchain infrastructures that optimize their parameters to be more adaptable to those contexts where timing constraints are critical and sometimes restrictive.

In conclusion, with respect to the currently applied solutions, the one presented in this document allows for developing decentralized applications of any kind that could interact directly with the Cyber-Physical-System, keeping costs low because they do not need to rely on any other service for accessing to the distributed network and, at the same time, satisfying the constraints about data-integrity and data-authenticity.

Author Contributions: J.G. provided the main idea and conducted the research and investigation process. E.G. and J.G. wrote and edited the paper. F.F. provided support during the development of the experimental use case. M.R. critically revised the paper. All authors have read and agreed to the published version of the manuscript.

Funding: This research received no external funding.

Conflicts of Interest: The authors declare no conflict of interest.

References

1. Lee, J.; Bagheri, B.; Kao, H.A. A Cyber-Physical Systems architecture for Industry 4.0-based manufacturing systems. *Manuf. Lett.* **2015**, *3*, 18–23. [CrossRef]
2. Lee, J.; Azamfar, M.; Singh, J. A blockchain enabled Cyber-Physical System architecture for Industry 4.0 manufacturing systems. *Manuf. Lett.* **2019**, *20*, 34–39. [CrossRef]
3. Ferrero, R.; Gandino, F.; Montrucchio, B.; Rebaudengo, M. A cost-effective proposal for an RFID-based system for agri-food traceability. *Int. J. Ad Hoc Ubiquitous Comput.* **2018**. [CrossRef]
4. Ashton, K. That Internet of Things Thing. *RFID J.* **2009**, *22*, 97–114.
5. Atzori, L.; Iera, A.; Morabito, G. The Internet of Things: A survey. *Comput. Netw.* **2010**, *54*, 2787–2805. [CrossRef]
6. Unlocking Opportunities in the Internet of Things-Bain & Company. Available online: https://www.bain.com/insights/unlocking-opportunities-in-the-internet-of-things (accessed on 29 November 2019).
7. Nakamoto, S. Bitcoin: A Peer-To-Peer Electronic Cash System. Technical Report. 2008. Available online: https://bitcoin.org/bitcoin.pdf (accessed on 29 November 2019)
8. Frizzo-barker, J.; Chow-white, P.A.; Adams, P.R.; Mentanko, J.; Ha, D.; Green, S. Blockchain as a disruptive technology for business : A systematic review. *Int. J. Inf. Manag.* **2019**, *51*, 102029. [CrossRef]
9. Tian, F. An agri-food supply chain traceability system for China based on RFID & blockchain technology. In Proceedings of the 2016 13th International Conference on Service Systems and Service Management, ICSSSM 2016, Kunming, China, 24–26 June 2016; Institute of Electrical and Electronics Engineers Inc.: Piscataway, NJ, USA, 2016. [CrossRef]
10. Salimitari, M.; Chatterjee, M. A Survey on Consensus Protocols in Blockchain for IoT Networks. *arXiv* **2018**, arXiv:1809.05613.
11. Rivest, R.L.; Shamir, A.; Adleman, L. A Method for Obtaining Digital Signatures and Public-Key Cryptosystems. *Commun. ACM* **1978**. [CrossRef]
12. Fedorov, A.K.; Kiktenko, E.O.; Lvovsky, A.I. Quantum computers put blockchain security at risk. *Nature* **2018**, *563*, 465–467. [CrossRef] [PubMed]
13. Koblitz, N. Elliptic Curve Cryptosystems. *Math. Comput.* **1987**, *48*, 201–209. [CrossRef]
14. Miller, V.S. Use of Elliptic Curves in Cryptography. In *Lecture Notes in Computer Science (Including Subseries Lecture Notes in Artificial Intelligence and Lecture Notes in Bioinformatics)*; Springer: Berlin/Heidelberg, Germany, 1986. [CrossRef]
15. Gallagher, P.D.; Romine, C. *Fips Pub 186-4 Digital Signature Standard (DSS) Category: Computer Security Subcategory: Cryptography*; Information Technology Laboratory, National Institute of Standards and Technology: Gaithersburg, MD, USA, 2013. [CrossRef]
16. Wood, G. Ethereum: a secure decentralised generalised transaction ledger. *Ethereum Proj. Yellow Pap.* **2014**. [CrossRef]

17. Buterin, V. Ethereum White Paper. Etherum. 2014 . Available online: https://ethereum.org/en/whitepaper/ (accessed on 29 November 2019).
18. Szabo, N. *Formalizing and Securing Relationships on Public Networks*; First Monday: Canton, TX, USA, 1997. [CrossRef]
19. Costa, D.; Fiori, F.; Milan, P.; Sala, M.; Vitale, A.; Vitale, M. *Quadrans White Paper*; Quadrans Foundation: Mendrisio, Switzerland, 2019.
20. Wu, M.; Wang, K.; Cai, X.; Guo, S.; Guo, M.; Rong, C. A Comprehensive Survey of Blockchain: From Theory to IoT Applications and beyond. *IEEE Internet Things J.* **2019**. [CrossRef]
21. Shafagh, H.; Burkhalter, L.; Hithnawi, A.; Duquennoy, S. Towards Blockchain-based Auditable Storage and Sharing of IoT Data. In Proceedings of the 2017 on Cloud Computing Security Workshop-CCSW '17, Dallas, TX, USA, 30 October–3 November 2017; ACM Press: New York, NY, USA, 2017; pp. 45–50. [CrossRef]
22. Huh, S.; Cho, S.; Kim, S. Managing IoT devices using blockchain platform. In Proceedings of the 2017 19th International Conference on Advanced Communication Technology (ICACT), Bongpyeong, Korea, 19–22 February 2017; pp. 464–467. [CrossRef]
23. Bahga, A.; Madisetti, V.K. Blockchain Platform for Industrial Internet of Things. *J. Softw. Eng. Appl.* **2016**, *9*, 533–546. [CrossRef]
24. Caro, M.P.; Ali, M.S.; Vecchio, M.; Giaffreda, R. Blockchain-based traceability in Agri-Food supply chain management: A practical implementation. In Proceedings of the 2018 IoT Vertical and Topical Summit on Agriculture-Tuscany, IOT Tuscany 2018, Siena, Italy, 8–9 May 2018; Institute of Electrical and Electronics Engineers Inc.: Piscataway, NJ, USA, 2018; pp. 1–4. [CrossRef]
25. Tsang, Y.P.; Choy, K.L.; Wu, C.H.; Ho, G.T.S.; Lam, H.Y. Blockchain-Driven IoT for Food Traceability with an Integrated Consensus Mechanism. *IEEE Access* **2019**, *7*, 129000–129017. [CrossRef]
26. Xu, Y.; Ren, J.; Wang, G.; Zhang, C.; Yang, J.; Zhang, Y. A blockchain-based nonrepudiation network computing service scheme for industrial iot. *IEEE Trans. Ind. Inform.* **2019**, *15*, 3632–3641. [CrossRef]
27. Dittmann, G.; Jelitto, J. A blockchain proxy for lightweight iot devices. In Proceedings of the 2019 Crypto Valley Conference on Blockchain Technology, CVCBT 2019, Rotkreuz, Switzerland, 24–26 June 2019; Institute of Electrical and Electronics Engineers Inc.: Piscataway, New Jersey, USA, 2019; pp. 82–85. [CrossRef]
28. Birrell, A.D.; Nelson, B.J. *Implementing Remote Procedure Calls*; Association for Computing Machinery: New York, NY, USA, 1984.
29. Technical Introduction to Events and Logs in Ethereum. Available online: https://media.consensys.net/technical-introduction-to-events-and-logs-in-ethereum-a074d65dd61e (accessed on 29 November 2019).
30. JSON-RPC 2.0 Specification. Available online: https://www.jsonrpc.org/specificationt (accessed on 29 November 2019).
31. Foodintegrity.org. *A Dangerous Food Disconnect When Consumers Hold You Responsible But Don't Trust You*; Technical Report; The Center for Food Integrity: Gladstone, MO, USA, 2018
32. Etienne, J.; Chirico, S.; McEntaggart, K.; Papoutsis, S.; Millstone, E. EU Insights—Consumer perceptions of emerging risks in the food chain. *EFSA Supporting Publ.* **2018**. [CrossRef]
33. Mipaaf-ICQRF-Report Attività 2018. Available online: https://www.politicheagricole.it/flex/cm/pages/ServeBLOB.php/L/IT/IDPagina/13602 (accessed on 29 November 2019).
34. The Economic Costs of IPR Infringement in Spirits and Wine. Available online: https://euipo.europa.eu/ohimportal/en/web/observatory/ipr_infringement_wines_and_spirits (accessed on 29 November 2019).
35. Hancoak, A. Younger Consumers Drive Shift to Ethical Products. *The Financial Times*, 22 December 2017.
36. Venkatesh, V.G.; Kang, K.; Wang, B.; Zhong, R.Y.; Zhang, A. System architecture for blockchain based transparency of supply chain social sustainability. *Robot. Comput. Integr. Manuf.* **2020**, *63*, 101896. [CrossRef]

energies

Article

A Multi-Source Harvesting System Applied to Sensor-Based Smart Garments for Monitoring Workers' Bio-Physical Parameters in Harsh Environments

Roberto de Fazio, Donato Cafagna, Giorgio Marcuccio, Alessandro Minerba and Paolo Visconti *

Department of Innovation Engineering, University of Salento, 73100 Lecce, Italy; roberto.defazio@unisalento.it (R.d.F.); donato.cafagna@unisalento.it (D.C.); giorgio.marcuccio@studenti.unisalento.it (G.M.); alessandro.minerba@studenti.unisalento.it (A.M.)

* Correspondence: paolo.visconti@unisalento.it; Tel.: +0832-297334

Received: 24 March 2020; Accepted: 23 April 2020; Published: 1 May 2020

Abstract: This paper describes the development and characterization of a smart garment for monitoring the environmental and biophysical parameters of the user wearing it; the wearable application is focused on the control to workers' conditions in dangerous workplaces in order to prevent or reduce the consequences of accidents. The smart jacket includes flexible solar panels, thermoelectric generators and flexible piezoelectric harvesters to scavenge energy from the human body, thus ensuring the energy autonomy of the employed sensors and electronic boards. The hardware and firmware optimization allowed the correct interfacing of the heart rate and SpO_2 sensor, accelerometers, temperature and electrochemical gas sensors with a modified Arduino Pro mini board. The latter stores and processes the sensor data and, in the event of abnormal parameters, sends an alarm to a cloud database, allowing company managers to check them via a web app. The characterization of the harvesting subsection has shown that $\approx$ 265 mW maximum power can be obtained in a real scenario, whereas the power consumption due to the acquisition, processing and BLE data transmission functions determined that a 10 mAh/day charge is required to ensure the device's proper operation. By charging a 380 mAh Lipo battery in a few hours by means of the harvesting system, an energy autonomy of 23 days was obtained, in the absence of any further energy contribution.

Keywords: wearable device; microcontroller; energy harvesting; piezoelectric harvester; thermo-electric generator; flexible solar panel

1. Introduction

Wearable devices are electronic systems, operating on the human body, kept continuously switched on to interact with the user or other devices. Nowadays, such devices are pervasive in several aspects of our lifestyle, enabling extensive and real-time monitoring of the wearer's psycho-physical state. Thus, several devices, including wireless communication modules, sensors, and computational units, have been developed and included inside a t-shirt [1] or connected to a belt [2] or integrated inside shoes [3]. Thanks to the recent scientific and technological advances in the biotechnology and bioengineering fields, considering the high computational capability of new processors with high level of integration and taking into account the availability of accurate, reliable and less invasive integrated sensors, the most recently manufactured wearable devices are smart, small and maintenance-free [4–8]. The main issue that limits the further development and use of wearable devices is the need for recharging and/or replacing the power supply source, usually represented by an energy storage device. If such technical obstacles were overcome, this would allow the manufacturing of energetically autonomous smart and miniaturized devices, thereby enabling easier integration of complex electronic systems into everyday objects. In the near future, this expected development will open the way to the so-called Internet of Everything (IoE), with billions of connected smart devices [9–11].

The design of wearable devices imposes several constraints to allow a perfect fit for the human body, to not hinder movements or to weigh down the user. Therefore, the dimensions, weight, flexibility, and comfort have to be taken into account for devising a smart wearable device design [12,13]. These constraints impose limits on the dimensions and capacity of the storage device used to supply power to the device. Moreover, current technologies don't permit the manufacture of batteries with high energy density, also by using new high-performance materials [14]. Recently, a smart approach for solving the aforementioned issues consists in the integration of energy harvesting systems into the wearable devices [15–17]. Harvesting the energy needed to feed the electronic device from the surrounding environment represents the most ecological and sustainable method. Specifically, wearable applications need harvesting systems able to scavenge energy from the sources typical of the human body, e.g., thermal energy, mechanical energy, luminous energy, etc... For these reasons, nowadays, the harvesting technologies are crucial to supply power for low power devices, thus making them energetically autonomous, ensuring a limitless operating time and eliminating service losses [18–20]. Different scientific works show how energy scavenging from environmental sources can integrate or totally eliminate the need for an external power supply, thanks to the continuous recharging of the employed storage device [21–24]. Generally, the approach to energy scavenging from environmental or artificial sources is a function of the typology and availability of energy sources. In fact, based on the power (AC or DC) supplied by source, on its amount and availability of multiple sources, different architectures have to be implemented. However, in its simplest form, the energy harvesting system basically requires an energy source (heat, luminous energy, mechanical energy) and the following key components:

- a harvester, that transduces the energy supplied by the source into electric energy; typical harvesters are the photovoltaic cells, thermoelectric generators (TEGs), antennas for the radio-frequency (RF) energy, the piezoelectric, electromagnetic, magneto-electric, electrostatic, or triboelectric transducers for mechanical energy (Figure 1);
- a conditioning electronic unit, that converts the variable signal produced by the harvester into a DC voltage for the electric load; the conditioning section includes a voltage regulator and complex control circuit able to manage the generated power as a function of the power load requirements and available power from the harvester (Figure 1). Also, this section has an impedance matching function for ensuring the maximum transfer of the harvested power;
- the energy storage device, a battery or super-capacitor (SC); it stores energy gathered by the harvesting unit in order to feed the electronic load in any operating condition (Figure 1).

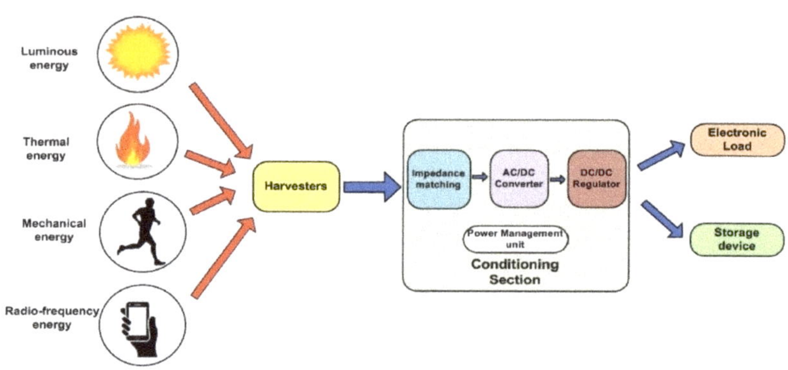

Figure 1. Block diagram of a generic energy harvesting system.

As regards wearable applications, the most common sources used for energy harvesting are the thermal gradient between the skin and environment, the luminous energy, mechanical energy related to walking, joint (i.e., knees and elbows) movements or arms and legs movements, that are repeated with a frequency of nearly 0.5–3Hz during the walking, thus providing continuous energetic input. Among the aforementioned energy sources, the thermal and luminous energies are the most used, followed by the mechanical one, due to the continuous availability, whereas RF sources are less frequently used, given the reduced amount of energy recoverable by such systems. Table 1 shows a comparison between the power densities related to the different energy sources both in indoor and outdoor environments [25]. A clever strategy for wearable applications consists of the combination of energy contributions provided by different transducers. This multi-source harvesting system ensures energy generations continuously, properly accumulated in a storage device able to power supply the electronic sections. This approach requires solutions for combining the contributions gathered from the different energy sources or simply selecting the source that instantly provides more energy [26–28].

Table 1. Comparison between the estimated power densities for different typologies of energy sources, both in the indoor and outdoor environments [25].

Source	Operating Conditions	Harvested Power
Light	Indoor Outdoor	4 μW/cm^2 4100 μW/cm^2
Thermal	Human (small temperature gradient) Industrial (high temperature gradient)	25 μW/cm^2 1–10 μW/cm^2
RF signals	Communication signals Industrial RF signals	0.1 μW/cm^2 1 μW/cm^2
Mechanical Vibrations	Human (Hz range) Industrial (kHz range)	40 μW/cm^2 800 μW/cm^2

The manuscript is structured as follows: the state of art of wearable technologies for energy harvesting and health monitoring is reported in sub-Section 1.1 whereas 1.2 describes the structure and functionalities of the developed energetically autonomous garment, highlighting its possible application scenarios. The second section describes the electronic devices included in both sensing and energy harvesting sub-systems, whereas the third one summarizes the results related to the characterization of each employed harvesting technology (i.e., solar, thermal, piezoelectric), and of the whole harvesting system in different operative scenarios; firmware development to interface sensors equipping the garment with Arduino Pro mini board is also described. In Section 3 the obtained experimental results are reported, whereas they are discussed in the following Section 4.

1.1. *Analysis of Wearable Technologies for Energy Harvesting and Health Monitoring*

The advances in sensing devices, ultra-low power microcontrollers, storage devices, and low power communication systems have led to a rapid evolution of wearable systems during the last decade [29,30]. Recent wearable applications have involved systems for managing business activities [31] as well as for preventing accidents and protecting workers' safety and security [32], and mainly in health monitoring, medical or fitness applications [33,34]. In this last context, the main users that could benefit from quasi-real-time monitoring of the health status are the elderly and children, but also users that wish to track their fitness performance. For instance, Carrè Technologies (Montreal, Quebec, Canada) developed a smart shirt for the Canadian space program, able to monitor and record the bio-physical parameters (heart rate, blood pressure and oxygenation, breath rate and body temperature); the acquired data are elaborated and transmitted to a smart device for detecting the onset of cardiovascular or respiratory diseases, as well as continuously monitoring the health status [35]. In [36], the authors reported a device including existing modules for monitoring user bio-physical parameters and warning the rescue unit in case of emergency by means of a communication module, able to also provide direct feedback to a medical center; in addition, the device can assist the patient in the prescribed medical treatment throughout the whole day. Similarly, Al-khafajiy et al. proposed a smart healthcare system able to monitor the health status of elderly people remotely [37]. The proposed system can track the user's bio-physical parameters for detecting the onset of a disease allowing prevention or rapid intervention in case of illness. The architecture is arranged in three related layers: the Patient layer, represented by the wearable device and smartphone sensors; the Data layer constituted by a cloud data center where the acquired data are stored; the Hospital layer, namely a doctors' platform synchronized in real-time, which monitors the data to efficiently coordinate a rescue team in the case of an emergency. Nakamura et al. have developed a belt-type wearable device (named *WaistonBelt X*) equipped with sensors and actuators; a magnetometer monitors the user's movements and body condition along with accelerometers and a gyroscope in order to recognize the wearer's lifestyle habits [38]. Statistical data from 17 types of signal (in the time and frequency domains) are calculated using a time window of 1.28 s; the information on the posture is obtained from the belt's angle by means of accelerometer data. If a bad posture is detected, a vibrator along with a message on a developed application warns the user to correct his position. In [39], the authors have developed two classification methods (named CM-I and CM-II) for detecting fall incidents in elderly users by using wearable altimeter sensors and a probability density function for both falls ($\mathcal{N}(\mu_F, \sigma_F)$) and lean over ($\mathcal{N}(\mu_B, \sigma_B)$) movements. The methods use the fall time and height data for classifying an incident as fall or lean over; the CM-II method was the best performing option, with a 98% detection capability.

As aforementioned, wearable devices are employed to assess user performances in the sport activity. For instance, Stetter et al. developed a method to measure the knee joint forces (KJF) during common movements, by using two inertial measurement units (IMUs), placed on the leg, and an artificial neural network (ANN) [40]. The IMU signals' matrix represents the input of a two-layer ANN as well as the KJF matrix is the ANN output; after initial training, the ANN-predicted results were compared with dynamic-calculated forces for several movements performed by 12 participants. The experimental results have shown that ANN-predicted KJFs have a good correlation for all tested movements. Besides, in [41], the authors presented a wearable inertial and magnetic platform enabling an end-to-end analysis of biomechanical parameters, usable for studying significant movement performance during sports activities. This platform consists of five IMU nodes placed in different points of the human body (the wrist, upper arm, forearm, chest, and the waist). These sensors were employed to collect data on the different body sections and thus to extract the motion patterns, which can be linked to injury risk. The timing sequence of the body intersegments can be extracted from the signals provided by the IMUs in a more accurate way than the detection through optical systems; furthermore, positional information can be extracted by employing magnetometer-augmented IMUs to provide information related to the position of the body during the analyzed athletic gesture.

A further application field of wearable devices is environmental monitoring aimed at the measurement of the spatiotemporal distribution of pollutants in the space strictly around the user, enabling one to detect parameters related to the environment healthiness and to correlate eventual diseases to the presence of a particular pollutant [42,43]. For instance, in [44], the authors presented the prototype of a smart t-shirt able to sense the air quality of the surrounding environment; the garment was an Arduino-based system including a TG2620 volatile organic compound (VOC) gas sensor (manufactured by Figaro, Arlington Heights, IL, USA), which provides a synthetic measure of the air quality. Furthermore, Spirjakin et al. developed a sensor-based wearable device for monitoring the working condition able to detect gas leakage [45]. The device is based on an ATxmega165E microcontroller and includes sensors for detecting the air temperature and combustible gases' concentration by using a catalytic gas sensor (manufactured by NTC IGD, Lyubertsy, Moscow, Russia). The system uses a measuring circuit based on the multi-stage pulse method for reducing the sensor's power consumption. The acquired data are transmitted to a base station by Zigbee transceiver module interfaced with the ATmega MCU by universal asynchronous receiver transmission (UART).

Furthermore, the device can be switched from the sleep condition to the measurement condition through an RF control signal over a distance of about 3 m.

An important challenge for wearable electronics is that the current batteries cannot provide enough energy for operations requiring a long time, inevitably leading to need to frequently replace of the energy storage device or its frequent recharging. Rechargeable Li-ion and Lipo batteries have been widely employed, thus promoting the development of portable electronics for quite a long time. A recent approach is the integration of energy harvesting systems and storage devices resulting in self-charging power systems (SCPSs) for low-power electronics and sensors. Sustainable energy sources to feed wearable devices depend heavily on their availability; different technologies for scavenging energy sources linked to the human body have been developed, such as piezoelectric (PENGs) and triboelectric nanogenerators (TENGs) for the kinetic energy related to movements/vibrations, photovoltaic panels for light energy, thermo-electric generators (TEGs) for thermal energy [46–48], etc.

The solar cells can be easily integrated with energy storage devices for providing energy to wearable systems [46]. Recent innovations have made available flexible solar cells with small dimensions, thus suitable for wearable applications; since inorganic materials (e.g., silicon cells) are not suitable, the main alternative is to use solar fiber (FSS) cells employing organic or hybrid materials. [49].

Organic solar cells (OSCs) featurey low weight, intrinsic flexibility of active layers, and low cost, making them most suitable to be fabricated in the form of flexible devices [50]. The OSCs' power conversion efficiency (PCE) was enhanced rapidly by the creation of new photovoltaic materials with conversion efficiencies higher than 13% [51]. Flexible semi-transparent OSCs have attracted a huge focus in the last few years, enabling new applications such as wearable energy harvesters and solar panels integrated into buildings. The crucial component for high-performance OSCs is cheap and mass-producible flexible bottom transparent electrodes (BTEs); these should have low sheet resistance, high transparency and durability, avoiding cracks and resistance increase with transducer bending [52]. Nowadays, perovskite solar cells (PSCs) are attracting huge interest due to their high conversion efficiency (i.e., $\approx$ 22%) and simple realization [49,53,54]. In [49], fiber-shaped PSCs are proposed with a flexible cathode realized by metallic twisted wires (i.e., stainless steel or titanium). Also, a thick layer and a porous one in titanium dioxide (TiO_2) act as hole blocking layer (HBL) and electron transport layer (ETL), respectively; then a perovskite layer is deposited on the wires. Afterward, a hole transport layer (HTL) made of 2,2′,7,7′-tetrakis(N,N-di-*para*-methoxy-phenylamine)-9,9-spirobifluorene (spiro-OMeTAD) is deposited; finally, an anode realized by a transparent conductive material, such as carbon nanotubes (CNTs) or a thin layer of gold, is deposited. The perovskite layer absorbs the incident light and the photogenerated electron-hole pairs are separated into the HTL and ETL resulting in a photovoltaic current. Such flexible PSCs can reach a PCE value of 3.3% also after 50 bending cycles. A further approach to increase PCE is the cathodic deposition technique [55], to obtain a uniform perovskite layer on curved supports. These PSCs include an ETL constituted by an array of Ti wires covered by a TiO_2 layer, radially grown by an anodization process, which acts as a mesoporous layer for electron extraction and transport. Afterward, a porous lead oxide (PbO) layer is deposited on TiO_2 nanorods by getting a lead iodide layer upon reaction with hydroiodic acid; then, the perovskite film is grown treating PbO layer with methylammonium iodide (CH_3NH_3I). The anode is constituted by a CNT thin film with high transmittance in the visible spectrum (more than 80%) realized by spin-coating an array of aligned CNTs mixed with isopropanol. The PSCs feature a 0.85 V open-circuit voltage (V_{OC}) and a 7.1% conversion efficiency, making them suitable for wearable applications, since they need to be close-fitting and conformable on the body.

A wearable sensor node that integrates a solar energy harvesting circuit and a Bluetooth Low Energy (BLE) transmission module is presented in [56], thus allowing the implementation of a self-powered wireless body area network (WBAN). Several nodes are distributed over different body areas to measure the vital parameters of the user wearing it, such as the temperature, heart rate, and also any falls suffered by user during the day. A web-based application allows the visualization of the acquired data by a smartphone. The energy harvesting system includes a flexible solar cell and the conditioning section with a buck converter and MPPT controller, so that sensor node can operate continuously and non-invasively, without requiring periodic battery recharging.

Mechanical energy related to body movements is an important source that can be exploited for feeding wearable devices. Several harvesters such as piezoelectric, triboelectric and electromagnetic generators have been developed to convert mechanical energy generated by different sources into electrical energy. The piezoelectric transducers represent an efficient solution to convert mechanical solicitations (walk, arms and legs movements, bending of elbows and knees' joints) into electrical energy. A further advance in this field is the development of piezoelectric nanogenerators (PENGs), realized by using aligned ZnO nanowires (NWs) encapsulated by a dielectric polymer film [46,57], able to harvest energy from tiny, random mechanical forces over a wide frequency range. Several energetically autonomous wearable applications were reported in literature employing piezoelectric harvesters to power supply sensing platforms for monitoring user health conditions [58–60]. Yang et al. studied the integration of flexible polyvinylidene fluoride (PVDF) transducers in clothes applied on different human body areas [61]. They evaluated the length variation of piezoelectric transducers and voltage values produced for specific movements (extension or compression of body joints, walking, push-up and torso flexion); higher voltages were obtained with the transducer placed at the elbow level during push-ups. During the walking with the harvester placed on the knee, a 3.1 V voltage was obtained whereas a maximum voltage of 4.4 V with 0.2 mW delivered power was reached during a squat with the harvester placed on the knee. Finally, the harvesting system was able to produce up to 1.42 mW, operating at a frequency of 1Hz and with a 3 MΩ load resistance.

In the last decade, triboelectric generators (TENGs) are gaining huge attention given their high conversion efficiency and constructive simplicity. The TENGs, based on the combination of contact electrification and electrostatic induction, represent a recent technology for energy harvesting and movement sensing. Different operating modes for TENGs have been developed depending on system implementation, excitation type and required outputs [62]. Thanks to their good performances, the TENGs have been used to scavenge energy from biomechanical activities, wind, water motions [63] and vibrations [64]. Different sensors based on TENGs, able to monitor force and pressure, motion, trajectory and biomechanical parameters, have also been proposed [65,66]. One of the most innovative applications involving TENGs is the self-powered electronic skin (or e-skin); try to mimic the sensory capabilities of the human skin by flexible electronic devices is turning out one of the most attractive research fields due to its wide promising applications such as wearable electronics, artificial intelligence and health monitoring [67]. To achieve this goal, it would be necessary to place on the body an enormous sensor network (BSN), but it would be hard to ensure concurrently the power supply. Despite this, the fast growth of innovative materials and micro/nano-manufacturing techniques makes the e-skin feasible by harvesting the mechanical energy related to the human movements. The e-skin can be applied everywhere on the body and employed in several applications, like wearable electronics [68], personal health monitoring, artificial prosthetics [69] and smart robots [70]. The electrodes and dielectric are essential components for building TENGs. The development of new flexible and extendible electrodes and dielectric structures are current issues [71,72]. The *e-skin* design requires bio-compatible and eco-friendly material, for these reasons organic materials were investigated. For instance, a silk fibroin was tested as TENG's dielectric, given its strong tendency to yield electrons and eco-compatibility [73]. Since TENG-based sensors are vulnerable to moisture, a chitosan-glycerol film was suggested to prevent the variation of the performance caused by moisture [73].

An energy harvesting method, that is receiving great attention for wearable applications, is the conversion of human biofluids, like sweat, into electrical energy [74,75]. Since many wearable devices operate on the human skin, epidermal biofuel cells (BFCs) have attracted great interest [76]. Sweat-based enzymatic BFCs represent a non-invasive harvesting solution [74], but they can provide power only if the user sweats, and because of the not-constant levels of both sweat and lactate biofuel, such harvesters can't provide steady output power. Given the low output voltage, the BFCs need an integrated conditioning unit (e.g., DC/DC converter) which could compromise the device flexibility [74].

The sweat-based BFCs harvest energy, through redox chemical reactions, from the lactate fuel contained in the sweat, and are constituted by a bio-anode and a cathode. The bio-anode is functionalized with the enzyme lactate oxidase (LOx) and a naphthoquinone (NQ) redox mediator, for catalyzing the oxidation of lactate and improving the power density, respectively. The active cathode is enriched with silver oxide thus, in the presence of sweat, the biofuel is enzymatically oxidized on the anode, producing an electrons flow toward the cathode (i.e., electrical power) [74]. Besides, in [76] a scalable low-cost screen-printing technology was employed to realize a textile-based BFC harvester by using engineered inks and electrodes with serpentine-shape. The experimental results suggest that at least 40 μL/cm^2 of sweat is needed to provide a stable output. An on-body test demonstrated the device operation in real conditions: after 37 min of sweating, the voltage on the storage SC was 0.4 V, suggesting that the harvesting solution doesn't provide enough power to feed an electronic section, but it represents a future promise. The experiments show that BFCs have good power density (252 μW/cm^2), durability, long-term stability, a wide operative range in terms of lactate concentrations, allowing to provide stable outputs for long charging times.

Thermal energy is widely exploited for energy harvesting in wearable applications since the human body heat causes a temperature difference between the skin and surrounding environment [59]. TEGs, based on the Seebeck effect, produce a difference of potential (ΔV) proportional to the temperature gradient on the two junctions (named hot and cold); namely $\Delta V = -S\Delta T$, where S is the Seebeck coefficient and ΔT the temperature gradient. The Seebeck coefficient varies with the temperature in conductors and depends on the used materials; for common materials (as Bi_2Te_3 and Sb_2Te_3), typical S values are between −100 μV/K and +1000 μV/K, while superconductors have a Seebeck coefficient equal to zero [77]. The TEG structure includes several p-n couples connected electrically in series and thermally in parallel, enclosed by two ceramic faces. Good thermoelectric materials have high electrical conductivity (σ) and very low thermal conductivity (κ); this property avoids the heat diffusion from a face to the other, which decreases the temperature gradient and so the generated voltage. The main disadvantage of TEGs is the low conversion efficiency: a module can convert about 7% of the available thermal power into electric power; this is due heat diffusion between the junctions, thus reducing the temperature gradient, and to the thermal resistance of human skin, much higher than TEG ones, so that heat hardly spreads on TEG hot side leading to a ΔT decrease.

In recent years, many advances have been done in the development of micro-TEGs (μ-TEGs), that can produce high output voltages with a very small volume. The working principle of μ-TEGs is the same as TEGs, but they present different structures or have differences in heat flow or electric behavior [78]. The μ-TEGs can be classified according to their structure (vertical, lateral or hybrid), materials and fabrication processes. The vertical structure, known as sandwich, is the conventional one of μ-TEGs (and TEGs). Generally, a μ-TEG has a simple design, high power density and PCE; nevertheless, the main issues of vertical-structured μ-TEGs are the manufacturing difficulties and high technological requirements. The lateral-structured μ-TEGs have easier manufacturing, thanks to their compatibility with IC planar technology [78], but they have lower performances compared to the vertical ones due to the high heat flux leakages towards the substrate and to package design given that the device operates with an in-plane temperature gradient. For these reasons, the lateral structure is commonly employed for sensor applications (e.g., flow, IR and power sensors) and rarely for energy harvesting. Also, μ-TEGs with hybrid structures (i.e., lateral/lateral, vertical/lateral and vertical/vertical) are possible with intermediate performances compared to the previous ones [79]; in a vertical-lateral hybrid structure, the heat flux flows vertically into one edge and passing planarly in the structure, is then emitted in the vertical direction from the other edge. Thereby, this structure exploits the vertical injection/ejection of heat flux for easier packaging as a vertical μ-TEG, keeping the compatibility of the manufacturing process with IC technologies as a laterally structured μ-TEG.

Different energetically autonomous wearable systems, power supplied by TEG harvesters, have been reported in the literature [80]. A BSN for health monitoring applications was proposed in [81], including a sensing section able to acquire, elaborate and wirelessly transmit the data concerning the electrocardiogram (ECG), electromyogram (EMG) and electroencephalogram (EEG). The device is fully fed by the body thermal energy harvested with a TEG and stored into a SC. A digital power management (DPM) system controls the nodes, manages the data flux and the elaborations; also, a low power RF transmitting section periodically transmits the elaborated data toward a gateway.

1.2. Description of the Proposed Energetically-Autonomous Wearable Device for Health and Environmental Monitoring Applications in the Workplace.

The problem of safety in the workplace is a has become a deeply felt issue in recent years. In civil society a deep awareness has spread as a consequence of several and recent events such as trauma associated to falls or bruises or intoxications due to leaks of dangerous gases, occurring in the workplace. Therefore, several solutions have been developed for preventing this kind of problem in order to preserve workers' health and safety. IoT technologies enable low-cost, efficient and not-invasive solutions for monitoring biophysical and environmental parameters directly on the user's body. The proposed solution is a garment for the upper body, like a jacket, that integrates several sensors to detect both environmental and biophysical quantities which can indicate a dangerous situation for user safety. Its main application is to monitor the conditions of workers operating into particularly dangerous workplaces for human safety, so obtaining detailed and real-time information. Particularly, the device is equipped with a low-power electrochemical carbon monoxide (CO) sensor (model ME2-CO, manufactured by Winsen Inc., Zhengzhou, China) for detecting the CO concentration in the environment around the human body, for instance due to gas leakages or abnormal emissions, an insidious danger since CO is odorless, colorless and particularly poisonous (Figure 2). The CO is produced in combustion phenomenon in conditions of oxygen deficiency, as occurs in stoves, boilers and furnaces. It can be found in numerous industrial areas since its use or emission is linked to several productive activities (e.g., food, chemical, metallurgical industries, etc). The smart garment includes a sulfur dioxide (SO_2) sensor (model 3SP_SO2_20, manufactured by Spec Inc, Newark, CA, USA) for obtaining an accurate and punctual detection of SO_2 concentration in dangerous workplaces (i.e., tanks for the food transport or in cramped environments), so avoiding accidents sometimes deadly. Given its anti-microbial properties, SO_2 is widely used in food industries (e.g., for sugar bleaching and conservation of wine and meat), but it can compromise (also at ppb levels) the respiratory function, leading to pharyngitis, fatigue and sensory disturbances (nose, eyes).

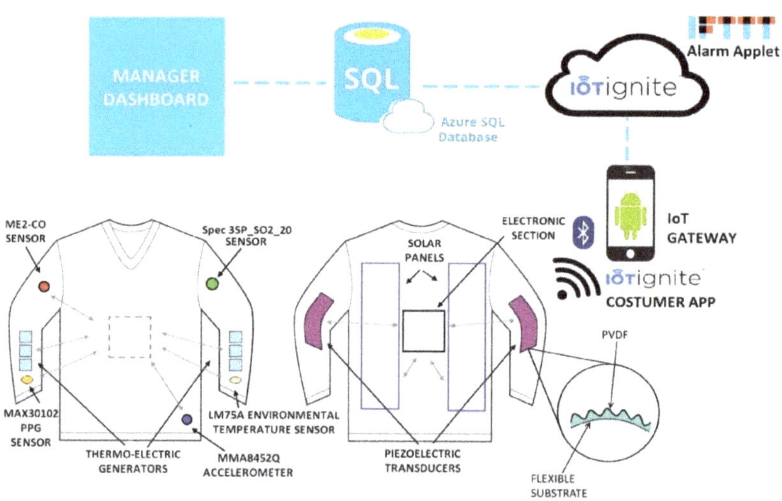

Figure 2. Schematic representation of the system architecture for monitoring environmental and biophysical parameters in workplaces based on the self-powered smart garment.

Furthermore, the sensing unit of the designed smart garment includes several sensors for monitoring the worker's biophysical conditions: a heart-rate, body temperature and SpO_2 (concentration of oxygenated hemoglobin) sensor (model MAX30102, manufactured by Maxim Integrated, San Jose, CA, USA) in optical technology (Photo-PlethysmoGram — PPG), placed at the jacket wrist, an integrated accelerometer (model MMA8452Q, manufactured by NXP Semiconductor, Eindhoven, The Netherlands), in correspondence of the user's waist for monitoring the physical activity level (PAL), the carried out steps number and eventual falls, and a temperature sensor (model LM75A, also manufactured by NXP Semiconductor) for detecting the environmental temperature. Given the application, the sensors were selected as a trade-off of power consumption, measurement accuracy, robustness and cost. The main board used to acquire and elaborate the data from the sensors, to coordinate the transmissions with the gateway and to optimize the device power consumption is a modified Arduino Pro mini (Arduino S.r.l, Torino, Italy).

It is worth noting that the proposed wearable device is self-powered since the garment is equipped with several harvesters for scavenging energy from sources related to the human body, as described below. Specifically, the garment includes two thin-film flexible photovoltaic cells placed on the user's back, which convert into electrical energy the incident luminous energy (solar or artificial), along with six TEGs for scavenging the thermal energy produced by the body, placed at the forearms to obtain a higher thermal gradient [82]. Furthermore, two piezoelectric harvesters are placed at the elbows to harvest the mechanical energy related to the joint movements (i.e., arms, forearms, shoulders). These transducers are constituted by a TPU (thermoplastic polyurethane) flexible support with a corrugated profile (Figure 2), similar to those reported in [61], on which is applied the series of two commercial polyvinylidene fluoride (PVDF) layers, each with 110 μm thickness, and finally the Ag and NiCu metallization on both sides. Specifically, each support featured by overall dimension 160 mm × 20 mm × 7 mm (thickness at the peak of the corrugation) has been realized by means of 3D printing using Flexismart TPU filament (manufactured by FFFworld, Cantabria, Spain), selected for its extreme flexibility (Shore A88) and resilience to bending and stretching. The support's corrugated profile allows both higher flexibility (in order not to be invasive for the user), to increase the surface on which to apply the piezoelectric layer, and the mechanical stress induced for a given movement, so getting a greater generated electric charge.

As aforementioned, suitable conditioning sections were used to efficiently extract energy from the harvesters. In particular, two LTC3108-based (manufactured by Linear Technology, Norwood, MA, USA) conditioning sections were used to scavenge energy both from solar cells and TEGs. The LTC3108, an integrated step-up converter with selectable DC output voltage, can be configured as a flyback converter through an external step-up transformer. By using a 1:90 step-up SMD transformer, a minimum input of 22 mV can be harvested, as detailed in the next section (Figure 3). The series of two flexible solar cells is connected to the input terminals of the solar conditioning section, as well as, for each arm the series of three TEGs is connected to the input terminals of the thermal conditioning section. The 5V DC output voltage was selected for all conditioning sections based on LTC3108 IC. Furthermore, two conditioning sections based on the LTC3588-2 buck converter (manufactured by Linear Technology) are employed to harvest energy from the two corrugated piezoelectric harvesters. The LTC3588-2, properly designed to harvest energy from alternate sources, includes an internal low-loss rectifying bridge and a programmable DC output voltage, set to 5 V for the two piezoelectric harvesting stages. The energy contributions provided by each conditioning section are combined through diodes (with 0.8 V forward voltage), employed to avoid the reverse current flow and reduce the output voltage to 4.2 V, maximum voltage value of the rechargeable storage device (380 mAh Lipo battery) (Figure 3).

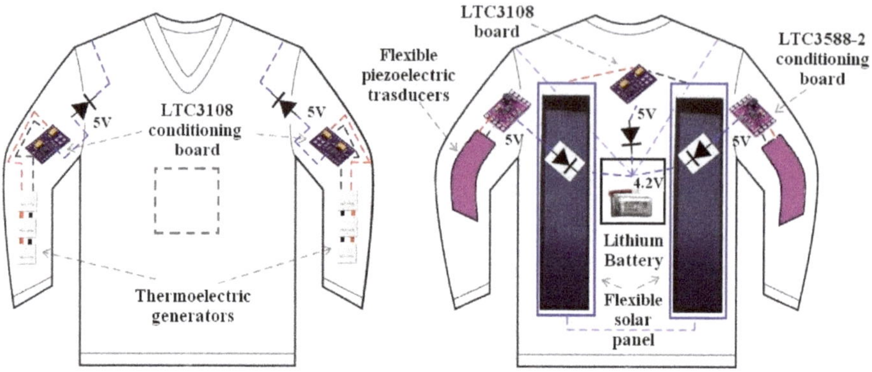

Figure 3. Schematic representation of the multi-source energy harvesting section of the developed smart garment; the electronic conditioning sections related to each energy harvester are shown.

As above described, an ultra-low-power electronic acquisition and elaboration section, based on ATmega328P microcontroller, periodically transmits, through a CC2541 BLE module (manufactured by Texas Instruments, Dallas, TX, USA), the acquired environmental and biophysical data towards a smartphone which acts as an IoT gateway (powered by the IoT-ignite® platform) for the Internet network (Figure 2). The acquired information, but also alarm messages, are periodically shared with a cloud-based Azure SQL database (DB), in which the information of all company employees is stored, along with the worker's actual GPS coordinates provided by their smartphones. Furthermore, if the wearable garment detects abnormal values of some environmental or biophysical parameters, by a suitable *IFTTT* applet, the system sends alarm messages, updating the related record in the storage DB, sending an e-mail/SMS to the security officers, with the last available GPS coordinates of the worker. The smart garment also includes signaling devices (LEDs) and a vibrations generator to immediately warn the user when abnormal parameters (e.g., high gas concentrations) are detected.

2. Materials and Methods

In this section, the energy harvesters (solar cells, TEGs, and piezoelectric transducers) and the sensing devices for detecting environmental and biophysical parameters applied to the designed smart garment are described.

Description of Employed Energy Harvesters and Sensors

The developed smart garment includes two commercial flexible solar panels (models 1w-rx and 0.125W 5V-45*25, manufactured by Dongguan City Xinliangguang New Energy Technology Co., Guangdong, China) placed on the back of the jacket (Figure 4). The used thin-film solar panels have a sandwich structure with three different elements, each tuned on distinct wavelength (red, green blue) to efficiently absorb the sunlight, thus improving the conversion efficiency also with low irradiance levels. The flexible cells are protected by ethylene vinyl acetate (EVA) and polyethylene terephthalate (PET) encapsulation layers laminated in order to make them impermeable and protected from the dust and gases. Two different typologies of solar cells, featured by different dimensions, and thus electrical parameters were employed:

- *Size 1*: 197 mm × 97 mm × 0.8 mm dimensions, 1 W maximum electrical power, operating current up to 666 mA, 1.5 V operating voltage, 800 mA short-circuit current and 2 V open-circuit voltage (Figure 4a,b).

- *Size 2*: 190 mm × 130 mm × 0.8 mm dimensions, 1.5 W maximum electrical power, operating current up to 1000 mA, 1.5 V operating voltage, 1200 mA short-circuit current and 2 V open-circuit voltage (Figure 4c).

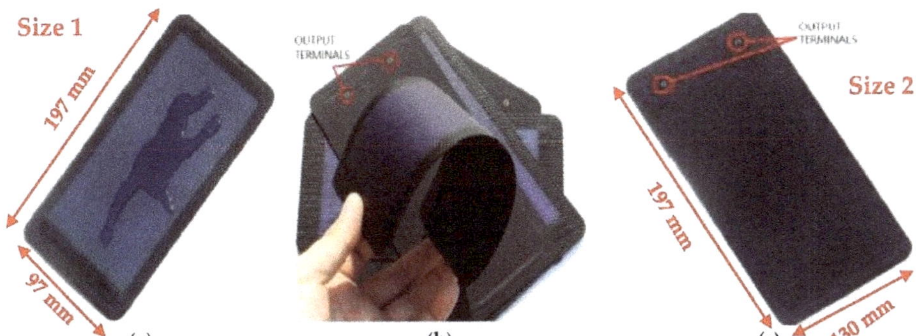

Figure 4. Top (**a**) and bottom (**b**) views of the flexible solar panel for the *Size 1* typology, and bottom view of the flexible solar panel for the *Size 2* typology (**c**).

Furthermore, three different TEG typologies have been tested, namely TEC1-12706 (manufactured by Thermonamic inc., Jiangxi, China), TMG-127-1.4-1.2 (manufactured by Ferrotec Nord Corp., Russia, Moscow), and TES1-03102 MINI (manufactured by Wellen Tech Inc., Shenzhen, China), shown in Figure 5a–c, respectively. The first two models have 40 mm × 40 mm dimensions and 127 p-n junctions, whereas the last one has 15 mm × 15 mm dimensions and 31 p-n junctions. They are commonly used as thermo-electric cooler (TEC) devices, based on the Peltier effect, but they can be also used as TEGs for converting thermal energy into electrical energy; in fact, the manufacturers provide their specifications as TECs and, for this reason, their characterization as TEGs was carried out. The output voltage provided by TEGs is variable over time due to variations of the temperature gradient between the two faces or to self-heating phenomenon (i.e., the diffusion of heat from the skin—hot face—the cold one—air). Also, the thermal resistance of skin is higher than that of TEG, leading to lower temperature gradient; hence, the TEGs require proper conditioning section to adapt the provided voltage values to those required by the electronic device to be powered and by the employed charge storage device.

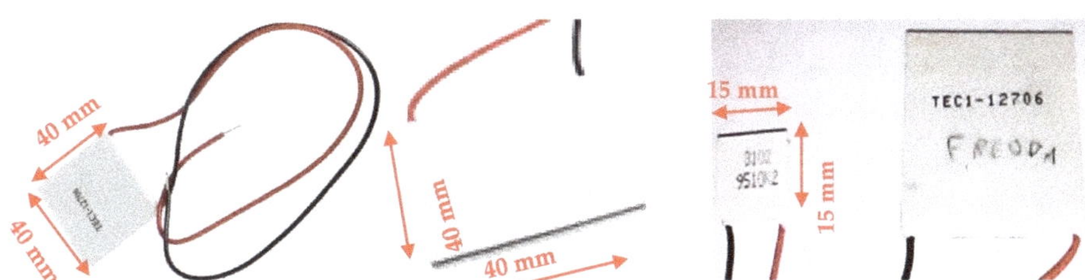

Figure 5. Top view of TEC1-12706 (**a**) (Thermonamic Inc.), TMG-127-1.4-1.2 (Ferrotec Nord Corp.) (**b**), and TES1-03102 MINI (Welletech Inc.) (**c**).

The electronic sections used to harvest energy from both solar panels and TEGs are based on LTC3108 IC (manufactured by Linear Technology), an integrated conditioning system designed to scavenge energy from low-level DC sources (e.g., TEGs, solar cells) (Figure 6). It is based on a DC/DC step-up converter, able to scavenge energy from sources with a 20 mV minimum input voltage (with a 1:100 step-up transformer placed in input to LTC3108 IC), configuring the board as a flyback converter.

Actually, a 1:90 step-up transformer (model LPR6235-752RMLB, manufactured by Coilcraft Inc., Cary, IL, USA) was used so increasing the minimum input voltage just to 22.2 mV. Also, the LTC3108 provides power management functionalities, selectable output voltage (2.35, 3.3, 4.1 and 5 V), whereas the impedance matching depends on the transformer's turn-ratio and input voltage level.

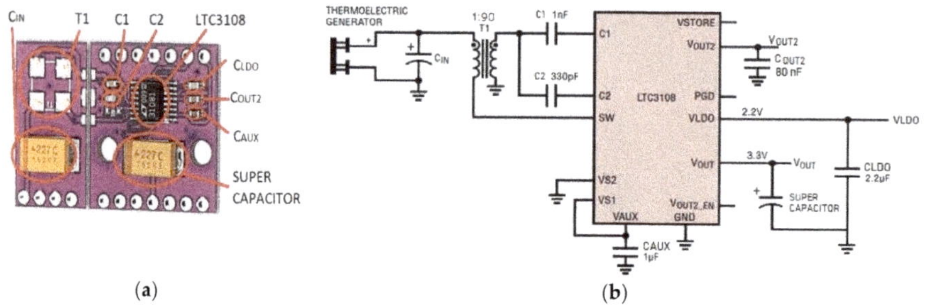

Figure 6. Top view of the conditioning board equipped with LTC3108 IC (**a**), schematic of the conditioning section based on LTC3108 for energy harvesting from TEGs and flexible solar cells (**b**).

As previously described, the smart garment is equipped with piezoelectric transducers to harvest energy from the joint movements. They were made with a TPU flexible support with a corrugated profile, realized by means of 3D printing, on which two PVDF layers connected in series have been applied for increasing the open-circuit voltage. A commercial PVDF layer with 110 μm thickness (manufactured by TE Connectivity Ltd., Schaffhausen, Switzerland, Figure 7) was employed, NiCu thin sputtered metallization, so ensuring good conductivity and oxidation resistance. The film was cut into strips with 160 mm × 20 mm dimensions and applied on the support's corrugated surface with silicone glue, to ensure strength and flexibility. The main specifications of the employed piezoelectric film are the following:

- Electro-Mechanical Conversion: (direction-1) 23×10^{-12} m/V, 700×10^{-6} N/V, (direction-3) -33×10^{-12} m/V;
- Mechano-Electrical Conversion: (direction-1) 12 mV per microstrain, 400 mV/μm, 14.4 V/N;
- Pyro-electrical Conversion: (direction 3) 13 mV/N, 8 V/K (@ 25 °C);
- Capacitance: 1.36 nF; Dissipation Factor of 0.018 @ 10 kHz; Impedance of 12 KΩ @ 10 kHz;
- Maximum Operating Voltage: DC) 280 V that yields a 7 μm displacement in the direction-1; AC) 840 V that yields a 21 μm displacement in the direction-1;
- Maximum applied Force (at the break, direction-1): 6–9Kg (yield voltage output: 830 V–1275 V).

Figure 7. PVDF thin film used to realize the piezoelectric harvester integrated into the smart garment.

The conditioning section used to scavenge energy from the piezoelectric harvester is based on LTC3588-2 IC (manufactured by Linear Technology, Figure 8a), an integrated solution designed and optimized for alternate sources, e.g the piezoelectric one. The LTC3588-2 relies on a high efficiency step-down (buck) DC/DC converter and includes a low losses full-wave rectifier, allowing the direct interfacing of the piezoelectric harvester with the IC (Figure 8b). Furthermore, an under-voltage lockout (UVLO) mechanism, featured by a wide hysteric window, optimizes the charge transfer from the input capacitor ($C_{STORAGE}$) to the storage devices (either a supercapacitor or lithium battery); the LTC3588-2 enters its UVLO modality, with an ultra-low quiescent current (450 nA), when an insufficient amount of charge is available from the energy source.

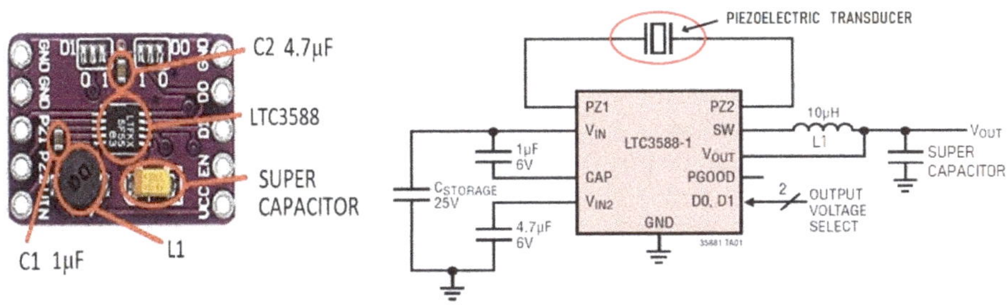

(a) (b)

Figure 8. Top view (**a**) and schematic (**b**) of the conditioning board based on LTC3588-2 IC, used to harvest energy from piezoelectric transducers.

The smart garment includes a MAX30102 PPG sensor (manufactured by Maxim Integrated), for monitoring the heart rate and SpO_2 level, which integrates two LEDs (i.e., a red LED at 660 nm and IR one at 880 nm), two photo-detectors for detecting the reflected light from the skin, optical elements and low-noise electronics to reduce the background signal due to ambient light (Figure 9a). It has a dual power supply, at 1.8 V (V_{DD}) for the electronic section and the other with a maximum value of 5.25 V (V_{LED+}), for the two LEDs. The communication with Arduino Pro mini board uses the I2C serial interface and supports standby mode with low power consumption. The MAX30102 sensor integrates a temperature sensor for measuring body temperature (Figure 9b); it is totally controllable through specific registers and the digital output can be saved in a FIFO register with 32 locations (for recording up to 32 samples), whose length depends on the active LED (data from each LED needs 3 bytes, so, if both LEDs are used, each location will be 6 bytes long). This register allows the sensor to be continuously connected with a processor or an external unit, sharing information through the communication bus. Some circuit changes were performed on the breakout board for reducing its power consumption, as discussed in Section 4. In addition, the surface of the board has been covered with a silicone coating to avoid short-cuts due to the contact of uncovered pins with skin. Since the MAX30102 sensor requires a direct contact with the human skin, it has been integrated into the jacket cuff, applied by an elastic band to guarantee a stable contact with the human body.

(a) (b) (c) (d)

Figure 9. MAX30102 sensor board (**a**), its pinout (**b**); the sensor board after the modifications (**c**) coating with an insulating layer for the protection of pins from possible short-circuits (**d**).

The software development has been performed for efficiently acquiring both the heart rate and the SpO_2, as described in the following section. Afterward, several test campaigns have been carried out by comparing the data acquired by the sensor with the measurements provided by a commercial pulse oximeter (model PR-10, manufactured by CocoBear, Shenzhen, Guangdong, China).

An LM75A digital temperature sensor (manufactured by NXP Semiconductor) is used to monitor the environmental temperature. It relies on a bandgap sensor and includes an 11-bit sigma-delta AD converter for ensuring 0.125 °C resolution on an operative range from −55 °C to +125 °C; the communication with the microcontroller is supported by I2C serial interface. The LM75A can be configured in two operating modes: (a) normal mode to periodically acquire the environmental temperature and (b) the shutdown mode to minimize the power consumption (3.5 μA absorbed current). The sensor has been sewn on the external surface of the jacket sleeve, in order to be in direct contact with the external environment and to measure the temperature regardless of the body temperature.

A monitoring system based on the use of accelerometers is a valid solution for estimating an individual's physical activity (e.g., pedometers are used in everyday life and in the workplace for discouraging a sedentary lifestyle). Another application of accelerometers concerns the fall detection using wearable devices, being falls very dangerous especially for elders and even fatal if no immediate help is provided. For these reasons, a pedometer and a fall detector have been implemented in the developed smart garment by using a MMA8452Q IC (manufactured by NXP Semiconductor) accelerometer. It is a low-energy three-axis MEMS (x, y, z) accelerometer with 12-bit resolution, embedded functions, flexible programmable options and two selectable interrupt sources. The default auto-sleep and wake up interrupt functions enable energy saving, avoiding that the main unit continuously processes data. The MMA8452Q sensor has a selectable full-scale value of ±2g/±4g/±8g with the possibility of data filtering by a high pass filter; it can be configured to generate an inertial interrupt signal for waking up from one of the possible sleep configurations, thus allowing the MMA8452Q to monitor events and remain in low-power mode during the inactivity periods.

A proper software development has been performed for implementing the aforementioned functionalities, exploiting the embedded functions provided by the MMA8452Q accelerometer, as described below. Furthermore, several tests have been carried out for validating the proper operation of both the implemented pedometer and fall detector under different operative conditions. The accelerometer has been securely fixed to a large elastic band sewn at the waist of the jacket, in order to keep the sensor in a fixed position respect the human body, thus reducing the detection of false falls induced by the motion of the garment respect to the body (Figure 2).

Electrochemical gas sensors have been employed to monitor the environmental concentration of dangerous gaseous species, given their reduced power consumption and small size. In fact, they are featured by lower operative temperature and power consumption than semiconductor sensors, and present lower cost and size compared to the optical ones. For detecting CO concentration, the ZE07-CO module (manufactured by Winsen Inc., Zhengzhou, China) was employed (Figure 10a); it is constituted by a ME2-CO sensor with a conditioning / acquisition board based on a potentiostatic circuit (Figure 10b).

The two-terminal ME2-CO sensor features high precision, low consumption, wide linearity range, and low cross-sensitivity respect to other gases; it has a working range from 0 to 1000 ppm (2000 ppm limit value), and sensitivity equal to 0.015 ± 0.005 μA/ppm. The conditioning board of ZE07-CO module also includes an acquisition section based on ST32F040F4 (manufactured by STMicroelectronics, Geneva, Switzerland) microcontroller, which acquires and elaborates the analog voltage provided by the conditioning section (Figure 10b), and then transmits the data to a master microcontroller by UART interface. This section was removed from the board and the analog voltage is acquired directly by the 10-bit ADC integrated into the ATmega 328P of Arduino Pro Mini board, in order to reduce the power consumption of the conditioning section.

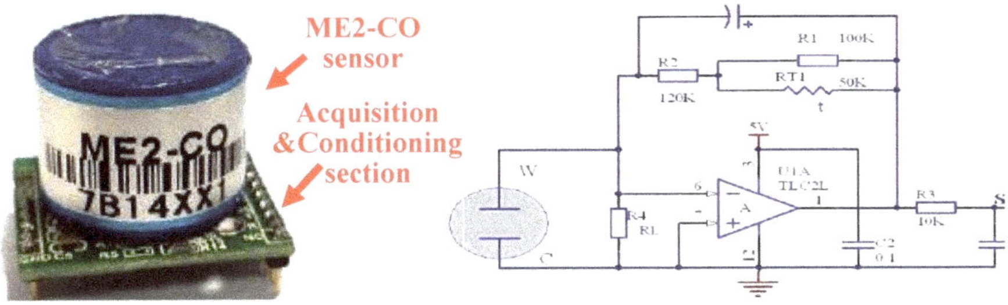

(a) (b)

Figure 10. ZE07-CO module constituted by the ME2-CO sensor and the relative acquisition and conditioning section (**a**), potentiostatic circuit included on the ZE07-CO acquisition and conditioning board (**b**).

For detecting the sulfur dioxide (SO_2) environmental concentration, an Ultra-Low Power Analog Module for Sulfur Dioxide (ULPSM-SO2) 968-006 sensor module (manufactured by Spec Inc.) has been employed (Figure 11a). The module is constituted by a 3SP_SO2_20 sensor and a conditioning board for converting the current signal provided by the sensor into an analog voltage between 0–3 V (Figure 11b). The 3SP_SO2_20 is a three-terminal electrochemical sensor with a measurement range from 0 to 20 ppm, sensitivity of 25 ± 10 nA/ppm, and power consumption between 10 and 50 μW, as a function of gas concentration. Both modules for gas detection were integrated into the garment by placing the conditioning sections in the jacket lining and letting out only the gas sensors from the jacket's outer fabric.

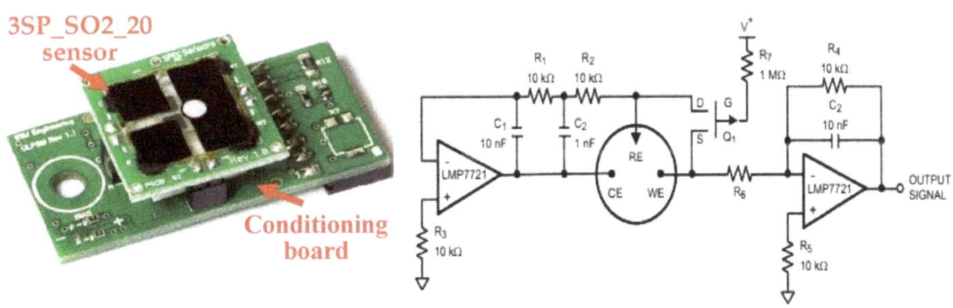

(a) (b)

Figure 11. ULPSM-SO2 968-006 module constituted by 3SP_SO2_20 sensor (Spec Inc.) and the relative conditioning section (**a**), and potentiostatic circuit integrated in the conditioning board (**b**).

3. Results

In this section, the results of characterization and functional tests carried out on both energy harvesting system and sensors are summarized, to evaluate the proper operation and performances.

3.1. Estimation of Solar Cells' Power Efficiency for Different Light Sources and Luminous Intensities

For the employed flexible solar cells, different behavior was observed when they were exposed to different light source typologies (e.g., sunlight and Ne source). In the following Table 2, the current (I_{IN}) and voltage (V_{IN}) values provided by the series of two *Size 2* solar cells to the LTC3108-based conditioning section (Figure 8a), equipped with 1:90 step-up transformer, are reported for the two light sources with different illuminance values (similar results are obtained for the *Size 1* cells model). The electrical quantities have been measured by two bench multimeters (model GDM-8351, manufactured by Gwinstek, Taipei, Taiwan), whereas the illuminance values by means of a digital luxmeter (model MT-4617, manufactured by Proskit, Taipei, Taiwan). As it can be noted (red box), for a given illuminance value, the tested solar cells have a better response when exposed to sunlight compared to Ne lamp; it can be explained with the different spectral distribution of the two sources. As evident from Figure 12, the solar spectrum covers uniformly the absorption spectrum of solar cells, whereas the Ne lamp one presents two main peaks, at 545 and 610 nm, resulting in a lower efficiency, as also reported in [83,84]. In fact, the smooth and continuous shape of the solar spectrum maximizes the conversion of light into electricity whereas the Ne lamp light concentrated in a few wavelengths determines a low efficiency due to saturation of photons converted into electric charge.

Table 2. Current (I_{IN}) and voltage (V_{IN}) values provided by the series of two *Size 2* solar cells to LTC3108-based conditioning section for different light sources (sunlight and Ne lamp) and illuminance values.

Illuminance [lux]										
		560	**640**	**1512**	**3000**	**5150**	**10,230**	**17,170**	**33,800**	**87,100**
Sunlight	V_{IN} [mV]	45.44	51.26	112.92	150.50	340.23	500.12	695.49	1570.89	2850.45
	I_{IN} [mA]	7.18	7.98	16.50	27.98	50.71	93.45	182.90	390.37	672.67
Ne lamp	V_{IN} [mV]	30.23	33.52	46.56	72.67					
	I_{IN} [mA]	3.81	4.05	7.57	10.42					

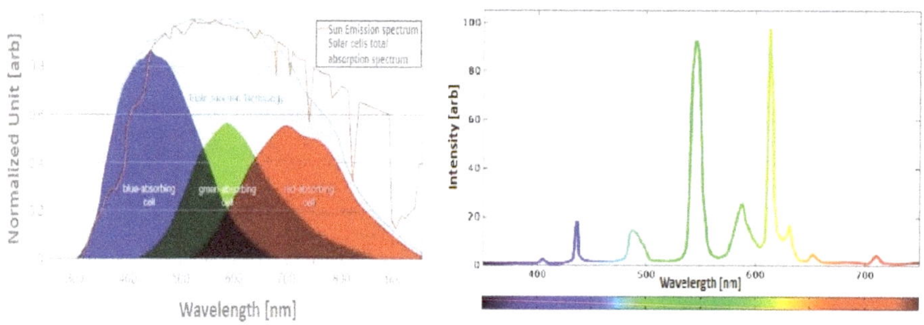

(a) (b)

Figure 12. Solar emission spectrum and solar cells absorption spectrum (**a**); Ne lamp emission spectrum (**b**).

In the following Figure 13, a histogram is shown with reported the power conversion efficiency (PCE) of the *Size 1* solar cell, exposed to sunlight (blue bars) or 2 × 14W T5 Ne lamp (red bars), for irradiance values between 0.22 W/m^2 and 33.99 W/m^2. The obtained data confirm, as previously discussed, that the PCE is higher when solar panels are exposed to the sunlight respect to Ne lamp, for a given irradiance value.

Furthermore, PCE decreases when the irradiance value is reduced regardless of the light source, due to internal parasitic shunt resistance of the solar cells, which dominates at low luminous intensities with consequent reduction of the open-circuit voltage and fill factor (FF) and thus of the power conversion efficiency [85].

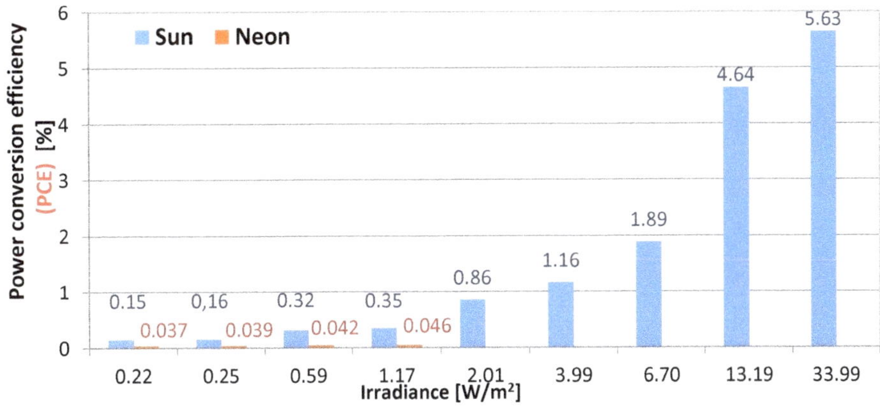

Figure 13. Histogram with reported the power conversion efficiency of employed flexible solar cells, as function of the irradiance level[1], when exposed to the sunlight (blue bars) and Ne lamp (red bars). [1]Irradiance values calculated from the illuminance ones [lux] in Table 2, using distinct conversion coefficients [Wm^{-2} lux^{-1}] for sunlight and Ne source (respectively 0.0079 and 0.0111), known the area of two *Size 2* cells.

3.2. Characterization of the Developed Piezoelectric Harvesters Applied to the Human Body

The flexible piezoelectric harvesters based on 110 μm PVDF layer, with the structure described in sub-Section 1.2 and overall dimensions of 160 mm × 20 mm × 7 mm (at the peak of the corrugated support), were applied by means of an elastic band inside and outside of the elbow and on the shoulder (as depicted in Figure 2) and characterized by performing common joint movements (e.g., arm bending and lifting). Each movement was repeated at 1 Hz frequency keeping them always the same; the maximum value of open-circuit voltage and relative root mean square (RMS) are reported in Table 3. The measurements were carried out employing a digital oscilloscope (model DSO5072P, manufactured by Hantek, Shandong, China) connected to the harvester terminals. In addition, the maximum power (P_{Max}) provided in output by the LTC3588-2 conditioning board was measured by a 100 kΩ load resistor.

Table 3. Table with reported the maximum open-circuit voltage and RMS values obtained from characterized piezoelectric harvesters for different joint movements repeated at 1 Hz frequency.

Movements	Description	$V_{OC,Max}$ [V]	$V_{OC,RMS}$ [V]	P_{Max} [μW]
1	Arm bending, transducer placed outside the elbow	31.25	6.87	256.12
2	Arm bending, transducer placed inside the elbow	27.57	5.98	234.83
3	Arm lifting, transducer placed outside the shoulder	28.61	6.26	245.39
4	Arm lifting, transducer placed inside the shoulder	26.78	5.97	238.65

3.3. Characterization of the Selected Thermo-Electric Generators and Relative Conditioning Section.

The characterization of TEG devices has been carried out using the experimental setup shown in Figure 14; the TEG series has been connected to the V_{IN} input of the LTC3108-based conditioning board (device 7 in Figure 14) with the output voltage set to 4.1 V and a 1 F SC (device 8) connected to its V_{OUT} pin; a bench multimeter (Gwinstek GDM 8351) (device 1) interfaced with PC (device 2) was used to measure the output current of the conditioning board, whereas two portable multimeters (Model GBC KDM-360CTF, manufactured by GBC Electronics, Milan, Italy, device 3) have been employed to measure the current (I_{IN}) and voltage (V_{IN}) values provided by TEGs.

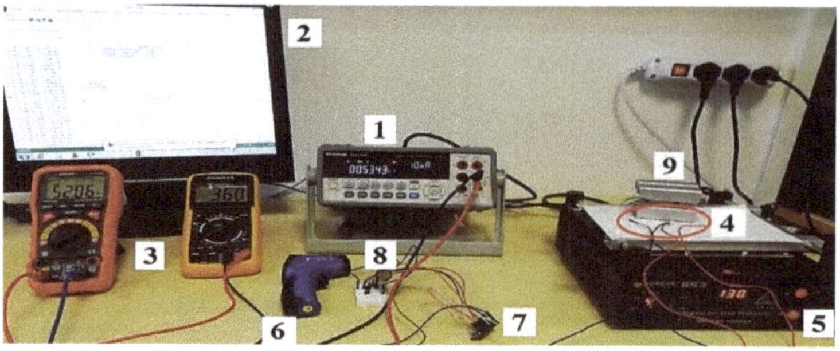

Figure 14. Experimental setup employed for characterization of TEGs.

An infrared heating plate (model 853, manufactured by Foshan Gordak Electric Co., Guangdong, China, device 5 in Figure 14) has been used to heat the hot junction of TEGs, whereas a heat sink has been placed on the cold junction (device 9) to keep the temperature gradient constant as long as possible; an infrared thermometer (model DEM100, manufactured by Velleman, Gavere, Belgium, device 6 in Figure 14), was used to measure the plate and cold joint temperatures at the beginning and end of each test. Each time, TEGs were placed on the hot plate for 10 s, the electrical quantities (V_{OC}, V_{IN}, I_{IN} and I_{OUT}) acquired and then the mean values determined; also, initial and final temperatures of the hot plate and cold joint were measured, calculating their mean values and thus the temperature gradient ΔT.

The following graphs depict the main electrical quantities' trends, measured for the series of two TEC1-12706 TEGs as function of the temperature gradient; the interpolation lines with relative angular coefficients are also reported (Figure 15). As can be noticed from the previous graphs, they show an almost linear trend, apart from a small deviation in the area around 15 °C–20 °C. Similarly, the electrical quantities have been acquired for two TMG-127-1.4-1.2 TEGs connected in series, by using the same operative modalities above described; the obtained trends are shown in Figure 16.

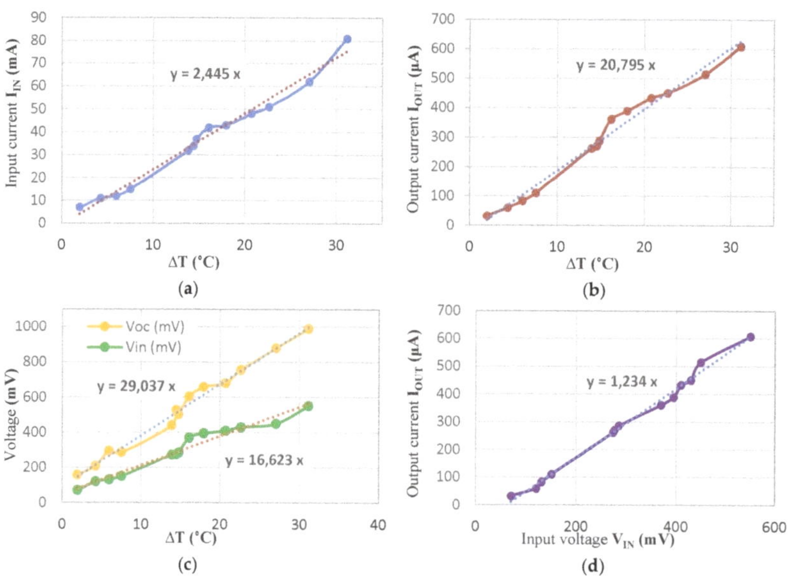

Figure 15. TEC1-12706 TEGs characterization: input current I_{IN} (**a**), output current I_{OUT} (**b**), open-circuit V_{OC} and input V_{IN} voltages (**c**) as function of ΔT, and output current I_{OUT} as function of the input voltage V_{IN} (**d**).

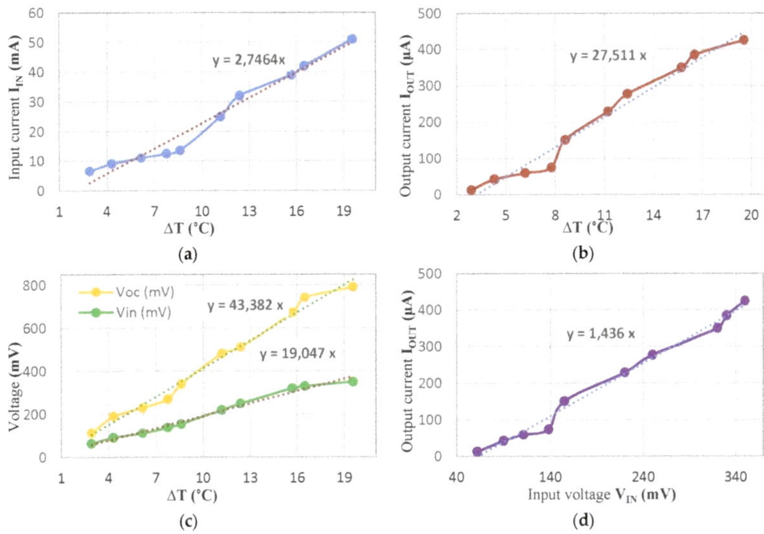

Figure 16. TMG-127-1.4-1.2 TEGs characterization: input current I_{IN} (**a**), output current I_{OUT} (**b**), open-circuit V_{OC} and input V_{IN} voltages (**c**) as function of ΔT, and output current I_{OUT} as function of the input voltage V_{IN} (**d**).

Afterward, the characterization of the TES1-03102 Mini TEG model has been carried out; this TEG device has a smaller area (15 mm × 15 mm × 3.8 mm) compared to the previous ones and only 31 p-n junctions (Figure 5c). The used experimental procedure for the characterization of two TEGs connected in series is the same used before but the monitoring time interval was increased to 20 s; the following graphs, shown in Figure 17, report the main electrical quantities' behavior.

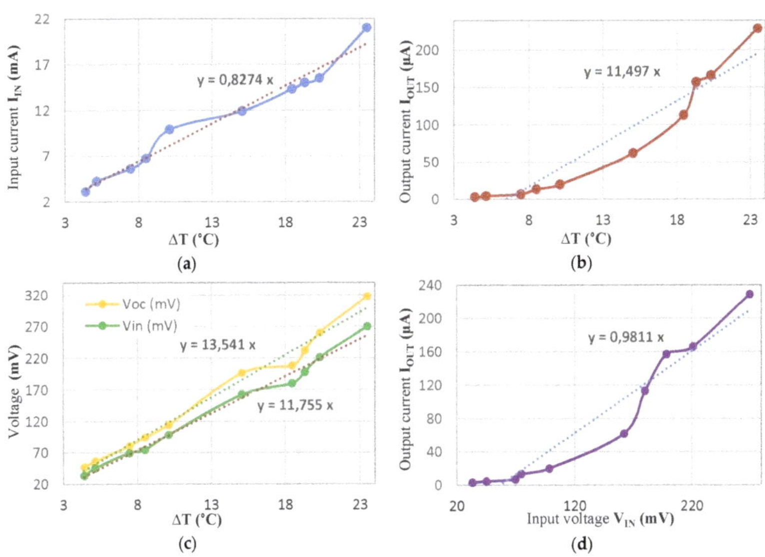

Figure 17. TES1-03102 Mini TEGs characterization: input current I_{IN} (**a**), output current I_{OUT} (**b**), open-circuit V_{OC} and input V_{IN} voltages (**c**) as function of ΔT, and output current I_{OUT} as function of the input voltage V_{IN} (**d**).

From the obtained experimental results, it is possible to draw very different conclusions from those concerning the first TEGs models; relatively to the open-circuit voltages V_{OC} and V_{IN} (Figure 15), the two trends are almost equivalent, since the input voltage values are obtained from related open-circuit voltage ones scaled by a factor of approximately 0.868 (calculated by the ratio of the angular coefficients of the interpolation lines in Figure 17); instead, for the larger models such ratio was equal to about 0.5 (Figure 15). This lower voltage reduction is ascribable to the smaller output impedance of the TES1-03102 Mini TEGs compared to the other tested models with larger area. However, the obtained voltage and current levels are much lower than previously tested models, due to the smaller size and lower number of p-n junctions. Relatively to the TES1-03102 Mini TEGs, considerable difficulties were encountered in carrying out the tests, due to their reduced area, to the uneven distribution of heat on the heating plate and a heterogeneous heat transfer to the TEGs' hot joint. It was determined that below a ΔT value of 4 °C (Figure 17a–c), the series of two Mini TEGs is unable to provide the minimum input voltage V_{IN} (22 mV) for LTC3108 conditioning module; since typical temperature difference between the skin and environment is in the range 2–8 °C, in the following tests three Mini TEGs connected in series have been employed.

By using the same operative modalities, the series of three TES1-03102 Mini TEGs have been characterized; the following graphs depict the trends of measured electrical quantities as function of ΔT, with also reported he interpolation lines and related angular coefficients (Figure 18).

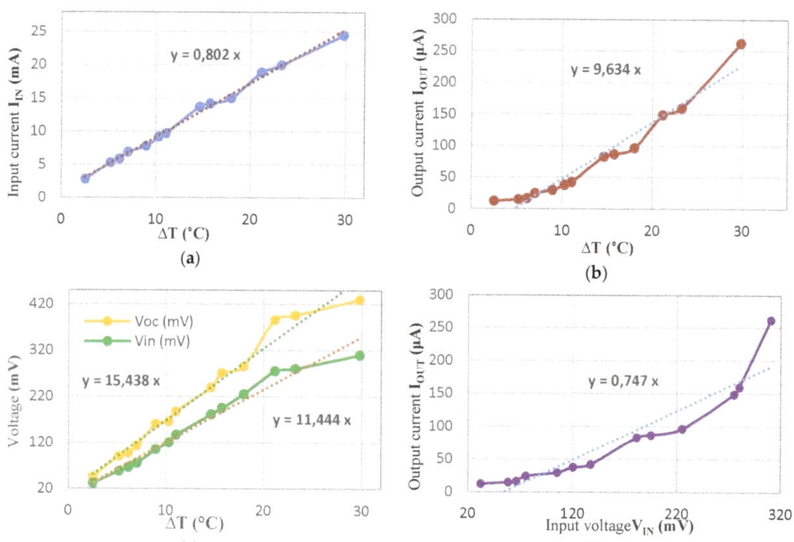

Figure 18. TES1-03102 Mini TEGs characterization: input current I_{IN} (**a**), output current I_{OUT} (**b**), open-circuit V_{OC} and input V_{IN} voltages (**c**) as function of ΔT, and output current I_{OUT} as function of the input voltage V_{IN} (**d**).

4. Discussion

From the results reported in Table 2, it can be noted that for a given irradiance level, the conversion efficiency of the tested thin-film solar cells is greater when exposed to solar light compared to an artificial light source (i.e., neon lamp). As afore described, this can be explained by the different spectral power distribution of the two light sources; nevertheless, the efficiency of the solar cells with the artificial neon source wasn't measured for irradiance values greater than 1.17 W/m^2 because of its limited luminous power. As shown in Figure 13, for both light sources, the power conversion efficiency of solar cells decreases greatly for low irradiance values, going from 5.63% to 0.15% with the irradiance changing from 33.99 W/m^2 up to only 0.22 W/m^2.

Relative to the designed piezoelectric harvesters, from the characterization results reported in Table 3, a higher open-circuit voltage and output power were obtained when they are applied outside the elbow due to the greater induced stress on a larger portion of the harvester. Further optimization of the shape of the flexible support for the piezoelectric layers and a more efficient positioning into the garment will improve their response. In addition, from results reported in paragraph 3.3, by placing three TES1-03102 Mini TEGs in series, the conditioning unit was activated also with low temperature gradient but, with the low current level provided by TEGs (of the order of mA), a long time is required for activating the LTC3108 internal circuitry due to the charge of C_{AUX} capacitor up to the activation voltage (2.5 V).

By comparing the input V_{IN} and open-circuit V_{OC} voltage values for the three Mini TEG series, it can be noted that their ratio (on average equal to 0.71) is lower compared with that obtained for the two TEGs in series (on average equal to 0.87); this is due to the higher impedance of the series of three TEGs with consequent higher internal voltage drop to thermo-electric source. In addition, by comparing the previously characterized TEGs (TMG-127-1.4-1.2 e TEC1-12706) with the TES1-03102 Mini TEGs, it is evident that despite the smaller size of the latter (about 1/7 respect to the larger model), they provide input V_{IN} and open-circuit V_{OC} voltages and input current I_{IN} values obviously lower but the reduction factors are only 1/3 for V_{OC} and 1/2 for V_{IN}; it can be concluded that the Mini TEGs are more performing, with the same ΔT, compared to the other larger TEG model.

Afterward, the energy harvesters and sensors were integrated inside a common jacket using the modality considered most appropriate for ensuring the functionality and efficiency of each device. Specifically, two *Size 2* photovoltaic cells connected in series were integrated on the back of the jacket by means of conductive snap buttons, so allowing to easily remove the photovoltaic cells for their replacement or cleaning. The connections with the conditioning section (LTC3108-based board, Figure 6) have been realized by conductive threads (model Electro-Fashion, manufactured by Kitronik Co., Nottingham, UK) soldered to the snap buttons and sewn on the back of the jacket.

The random variations of the incidence angle of the sun's rays during the body movements cause fluctuations of voltage and current values provided by the solar cells; however, the results obtained from the characterization of smart garment prototype, demonstrate a reduction of the mean power provided by the solar panels of only 20–25% compared to the results reported in sub-Section 3.1, due to the not always optimal orientation during the day of the cells respect to the sun.

Also, two developed piezoelectric transducers, described in the sub-Section 1.2, have been positioned at the elbows of the jacket by pasting them on elastic bands sewn onto the jacket; in this way, the mechanical coupling between the arm and transducer is strongly improved. Similarly, the connections between each harvester and the conditioning section (i.e., LTC3588-2 based board) have been carried out by means of the conductive threads sewn on the sleeve of the jacket. Besides, six TEC1-12706 TEGs were integrated into the smart garment, arranged in two series of three TEGs, each series positioned on the forearm of the jacket and equipped with the LTC3108-based conditioning board; the two TEG-based conditioning sections were connected in parallel through blocking diodes, as already detailed in sub-Section 1.2. Each TEG has been integrated within an elastic band sewn onto the internal surface of the jacket sleeve, in order to keep the hot junction constantly in contact with skin, whereas the cold junction is covered with a perforated fabric; in this way, the airflow on the jacket surface, also due to arm movements, can reduce and keep stable the temperature on the cold junction of TEGs. Because of the absence of a real heatsink and the not perfect contact of the TEGs' hot side with the skin during the measurement interval, also due to the body movements, a reduction of the provided power ranging between 20% and 30% was observed in the experimental tests of the garment compared to obtained results from the TEGs' laboratory characterization, reported in the sub-Section 3.3. The conditioning sections have been hidden inside some small pockets realized into the jacket lining, whereas the connections between the transducers and conditioning sections were realized using the conductive threads sewn directly onto the jacket fabric.

As previously described, the MAX30102 heart rate and SpO_2 sensor has been placed inside the jacket cuff employing an elastic band in order to keep firmly the sensor on the user's wrist, thus reducing the measurement errors. Furthermore, the MMA8452Q accelerometer has been embedded into a large elastic band sewn at the waist of the jacket and thus solidly fixed to the monitored body, for reducing errors in the steps counting or fall detection. The LM75A sensor has been sewn by using not conductive threads onto the jacket sleeve, in direct contact with the external environment in order to detect the air temperature.The modules for gas detection (i.e., the ZE8-CO and ULPSM-SO2 968-006) have been positioned inside the jacket lining, placing outside only the gas sensors. The acquisition and elaboration board (Arduino Pro mini) was positioned inside a small pocket on the back of the developed garment, along with the communication section (based on CC2541 IC) and the energy storage device (i.e., a 380 mAh Lipo battery).

Different measurement campaigns were performed to evaluate the performances of the multi-source harvesting system integrated into the smart garment when it is worn by a user, and in different operative scenarios, as following described:

- Scenario 1: a stationary user with a body temperature of 35.6 °C exposed to direct sunlight;
- Scenario 2: a user walking quickly (5 km/h) and exposed to direct sunlight with a body temperature of 36.1 °C;
- Scenario 3: a user walking quickly (5 km/h) and exposed to diffused sunlight with a body temperature of 36.3 °C;
- Scenario 4: a user walking quickly (5 km/h) and exposed to artificial light (neon lamp) with a body temperature of 36.3 °C;
- Scenario 5: a user performing pushups (0.5 Hz) and exposed to artificial light (neon lamp) with a body temperature of 35.9 °C.

An electronic load (model Tenma 72-13210, manufactured by Tenma®, Springboro, OH, USA), configured in constant resistance modality (100 kΩ), was connected to the output of the multi-source harvesting system, to monitor the maximum power provided to the load; the results in the different scenarios are summarized in the following Table 4. For each scenario, five different tests were carried out considering an observation interval of 30 min every time, during which the current values absorbed by the electronic load were acquired and recorded by means of a digital multimeter (Model PM8236, manufactured by Peakmeter®, Shenzhen, China), with a sampling time interval of 30 s. From the obtained time-domain trends, the maximum values of power provided by the multi-source harvesting system were obtained. In Table 4, the mean values of maximum power (P_{Max}) over the five tests for each scenario were reported together with the mean temperatures (T_{AIR} and T_{BODY}) and illuminance values. Furthermore, the mean value of the power ($\overline{P}$), provided by the harvesting section over the five tests, was reported in Table 4 for each scenario.

Table 4. Characterization of the multi-source energy harvesting system integrated into the designed smart garment in the five different scenarios.

Scenario	Activity	Illuminance [lux]	T_{BODY} [°C]	T_{AIR} [°C]	P_{Max} [mW]	$\overline{P}$[mW]
1	steady	27,918 (sunlight)	35.60	25.20	252.21	201.78
2	walking	29,322 (sunlight)	36.10	24.30	264.57	216.48
3	walking	530 (sunlight)	36.70	24.70	4.47	3.56
4	walking	530 (neon lamp)	36.30	23.60	4.87	3.87
5	pushups	530 (neon lamp)	35.90	24.50	4.25	3.54

The solar harvesting system provides the main contribution to the electrical power stored on the battery when the garment is exposed to sunlight, but it ensures a continuous charging of the

380 mAh Lipo battery also under diffused sunlight (Scenario 3) or artificial light (Scenarios 4–5), thanks to the continuous over-time contributions of the thermal and piezoelectric harvesting sub-sections. In the following section, it is demonstrated that the developed harvesting section, also in the worst case (Scenario 5), is able to guarantee the energy autonomy of the sensing and communication units. Following, additional tests were performed in order to determine the power contribution of each harvesting section, by placing a digital multimeter (Model PM8236) in series to each harvesting section. Specifically, for Scenarios 1 and 2, the contribution of the solar cells represents about 98% of the total power provided by the whole harvesting system, whereas the piezoelectric (0.1–0.2% only for Scenario 2) and thermal (1.5–2%) harvesters give some negligible contributions. Conversely, in the third scenario (Scenario 3), the power contributions of the solar and piezoelectric harvesting sections are much lower (respectively 6–7% and 5–6% of the total provided power) respect to the TEGs power contribution (high up to 87%). In Scenario 4, the contribution of the thermal harvesting section, including six TEC1-12706 TEGs, was equal to 4.46 mW (referred to the maximum provided power P_{Max}), whereas the contributions of the solar and piezoelectric ones were only 0.13 mW and 0.28 mW, respectively. Finally, in the Scenario 5, 3.77 mW was the contribution of the thermal harvesting section (due to a lower temperature gradient—precisely −1.3 °C—with respect to Scenario 4), whereas the photovoltaic and piezoelectric power contributions were only 0.13 mW and 0.35 mW, respectively. In this last case, the piezoelectric harvesters provide a higher power compared to Scenario 4, since push-ups excite more strongly the transducers, as already verified in sub-Section 3.2. These results confirm the afore reported concept, namely that also in the condition of low level of illumination (Scenarios 3, 4 and 5), the thermal harvesting section, based on six TEC1-12706 TEGs, is able to provide a continuous charge flow on the Lipo battery and therefore to power supply the entire designed system.

Testing and Characterization of the Sensors Included in the Smart Garment

In this section, the functional tests and the characterization carried out on the sensors, included in the developed smart garment, are described. A modified Arduino Pro mini, based on ATmega 328P MCU, represents the motherboard of the smart garment; this MCU offers six different sleep modalities to reduce its power consumption up to μA values. In particular, the built-in 3.3 V voltage regulator (78L05 made by STMicroelectronics) featured by 6 mA of maximum quiescent current was replaced with a more efficient one (model XC6206P332MRI, manufactured by Torex Semiconductor, Tokyo, Japan) with lower quiescent current (typical value 1 μA). Also, the MCU clock frequency was reduced from 16 MHz to 8 MHz provided by the internal oscillator, obtaining a drastic reduction of the absorbed current from the Arduino board in power-down mode, up to the value of only 23 μA.

Relative to the MAX30102 sensor, the Arduino code has been developed for performing the fundamental functions of signal processing provided by the two integrated photo-detectors (red and IR) and for interfacing the sensor with the MCU; subsequently, the code has been optimized for obtaining a low level of data memory occupation (an important aspect in this application). Also, the 3.3 V voltage regulator (model 10A45, manufactured by Saiertong, Shenzhen, China) was removed and bypassed, since the sensor is fed by using the 3.3 V Arduino output pin (yellow box in Figure 9c). Figure 19 shows the two anatomical areas (i.e., the wrist and fingertip) with the sensor of Figure 9d properly applied to detect the SpO_2 and HR. Relatively to the integration of MAX30102 sensor in the smart garment, it was positioned in correspondence of the wrist into the elastic jacket cuff.

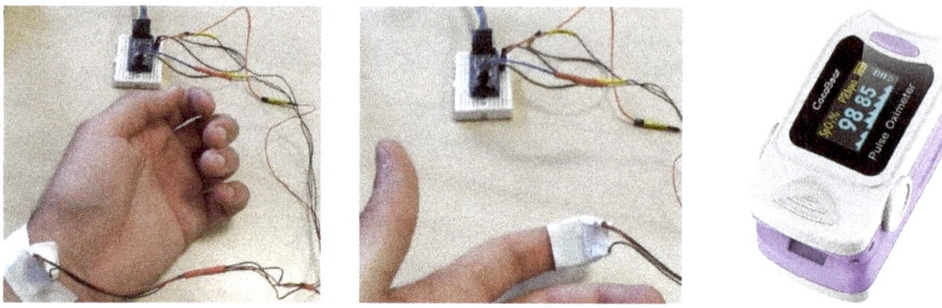

(a) (b) (c)

Figure 19. The positioning of the MAX30102 sensor on the wrist (**a**) and the fingertip for the correct detection of heart rate and blood oxygenation (SpO_2) (**b**); *CocoBear* pulse-oximeter used for the sensor characterization (**c**).

A commercial pulse-oximeter CocoBear (Figure 19c) has been used to measure the HR and SpO_2 values and compare them with those provided by the MAX30102 sensor (Table 5); the sensor was positioned as shown in Figure 19a and the values were detected every 30 s, simultaneously from both devices. From data reported in Table 5, after performed optimization of developed code for HR and SpO_2 calculation, a good agreement between the values obtained by the MAX30102 sensor and the CocoBear oximeter was obtained, with a maximum difference of 3 BPM for HR and 1% for SpO_2 values; moreover, the convergence time for getting correct SpO_2 values was less than 5 s. The power consumption of the MAX30102 module has been measured, both in active mode and power save mode, by acting on the shutdown control bit of the internal configuration register; the current absorbed by MAX30102 module was 1.25 mA in active mode and just 1.5 μA in power saving mode.

Table 5. Detected heart rate (HR) and blood oxygenation (SpO_2) values using the CocoBear oximeter and the MAX30102 sensor; differences reported in the fifth and sixth column are between HR and SpO2 values supplied by CocoBear and MAX30102 devices; also the skin temperature acquired by the MAX20102 sensor is reported.

Heart-Rate CocoBear (BPM)	Heart-Rate MAX30102 (BPM)	SpO_2 CocoBear (%)	SpO_2 MAX30102 (%)	Δ HR Coco Bear (BPM)	Δ SpO2 Coco Bear (%)	Temperature (°C)
78	75	97	98	3	1	34.7
79	76	97	97	3	0	34.6
83	81	96	95	2	1	33.7
84	83	95	96	1	-1	34.7
85	84	96	97	1	-1	34.8
86	85	98	98	1	0	34.8
87	86	97	98	1	-1	34.7
88	87	98	98	1	0	34.7
89	89	98	98	0	0	34.6
100	100	97	97	0	0	34.7
110	110	99	99	0	0	35.0
120	121	99	100	-1	-1	35.1
122	122	99	99	0	0	35.4

The developed garment is also equipped with a pedometer and fall detector, both implemented by means of the MMA8452Q 3-axis accelerometer. The firmware for the pedometer allows detecting a step by comparing the total acceleration value (i.e., combined acceleration on the three axes) with an experimentally derived threshold, whereas a hysterical detection mechanism prevents multiple increments for each step. The threshold value affects the algorithm sensitivity and its careful optimization has been carried out, in different test conditions, for avoiding an overestimation (for low threshold values) or underestimation (for high threshold values) of performed steps. The following Table 6 refers to the use of MMA8452Q accelerometer for a walk at normal pace, positioning it on the waist, as shown in Figure 20, to ensure high detection capability [86] and low influence by interpersonal differences in the body movements, since the sensor is not stressed by movements and accelerations, different from those generated by performed steps.

Table 6. Comparison of the number of steps calculated by the MMA8452Q accelerometer used as pedometer with the steps number counted manually by a second subject, during a walk at a normal pace.

Test	Threshold Value ([1] LSB)	Manually Counted Steps	Counted Steps by MMA8452Q Pedometer
1	150	50	70
2	150	100	127
3	200	100	109
4	200	110	118
5	200	125	137
6	220	70	72
7	220	100	101
8	220	110	112
9	225	100	102
10	225	110	112
11	230	100	98
12	230	105	102
13	230	108	105
14	240	110	103
15	240	125	115
16	250	100	86
17	250	108	96
18	300	115	100
19	300	136	116
20	350	125	100
21	400	125	95
22	500	125	90

[1] LSB is equal to 1 mg since the MMA8452Q accelerometer full scale was set to ± 2 g and 12 bit of resolution.

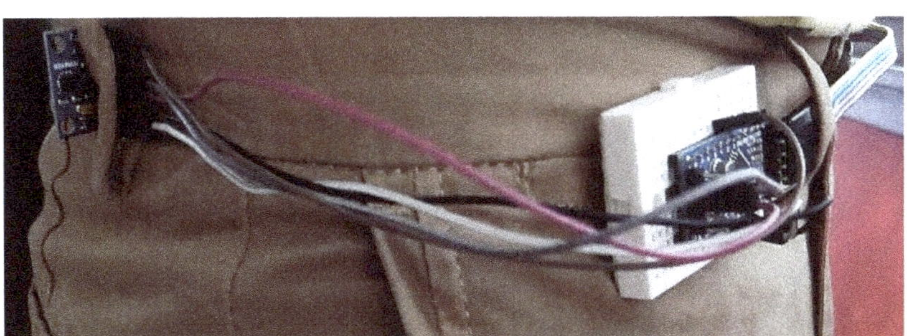

Figure 20. Positioning of the MMA8452Q acceleration sensor at the waist, in the belt loops, to keep it as firmly anchored as possible to the body and prevent the sensor from picking up fictitious accelerations.

The results reported in Table 6 demonstrate that, by lowering the threshold value, the sensor overestimates the number of performed steps, since even low acceleration values, not related to an actually executed step, are greater than the threshold and counted as a step. Conversely, with high threshold values, the accelerometer underestimates performed steps number, since low accelerations generated by a possible lighter step are not detected; more proper threshold values range from 220 to 230. Besides, to improve the accuracy of the developed pedometer during the run, a software mechanism was implemented for suppressing the signal spikes that could corrupt the steps count.

The MMA8452Q sensor implements a freefall detection algorithm based on the analysis of the acquired acceleration values along the three-axis, (by setting the *XEFE, YEFE, ZEFE* bits in the *FF_MT_CFG* register), verifying if they are all lower than a predetermined threshold (as detailed in the datasheet). If this event occurs for a time exceeding a set time duration (debounce time, set to 100 ms), then an interrupt is triggered; otherwise, events with duration lower than the debounce time are ignored. An accelerometer interrupt pin was set as interrupt source for the MCU normal routine; if a fall event occurs, an ISR (interrupt service routine) in the MCU is executed, so generating an alarm signal. The acceleration threshold value, set in the *FF_MT_THS* register of the MMA8452Q sensor and coded by seven bits (8 g full scale and consequently 0.063 g resolution), is crucial for the afore described mechanism; thus, its optimization has been performed to guarantee a high detection capability of the sensor. The MMA8452Q sensor was positioned on the waist (as shown in Figure 20) since this body area was identified as the most suitable to guarantee excellent fall detection [86]. In Table 7, a comparison between the number of falls performed by the user and falls detected by the sensor is reported, highlighting the false not-occurred falls as a function of the threshold value.

Table 7. Comparison between the number of falls performed by the user and number detected by MMA8452Q sensor for different threshold values; the last column reports false falls (not occurred) detected by the sensor.

Threshold Value	Number of Performed Falls	Number of Detected Falls	Number of False Falls
1	5	0	0
2	6	1	0
3	6	4	0
4	8	7	0
5	4	4	1
6	4	6	2
7	3	6	3

From data reported in Table 7, it is evident that the optimal threshold values for fall detection, regardless of the physical characteristics of the subject wearing the MMA8452Q sensor, are 4 and 5 (0.252 g or 0.315 g). Finally, the threshold value was set to 4, since even the employed commercial fall detector, placed on the user's waist, tends to underestimate the total number of performed falls.

Some modifications were carried out on the accelerometer breakout board; the 5 V built-in regulator (LG33, Micrel, San Jose, CA, USA) was removed since the sensor was fed directly to 3.3 V provided by the voltage regulator (XC6206P332MRI, Torex Semiconductor) placed on the Arduino Pro mini board. For reducing the power consumption, the built-in auto-wake/auto-sleep functionality was exploited, bringing the sensor in power-down mode if no motion is detected. In power-down mode, the data acquisition rate of the acceleration sensor is significantly lowered (Output Data Rate—ODR—reduced from 800 Hz to 56 Hz), and it is wake up, via an interrupt signal, when an acceleration higher than the set threshold is detected. The measurement of the current absorbed by the modified MMQ8452Q board demonstrates that the accelerometer draws 210 μA during the data acquisition phase and only 6 μA in the power-down modality.

As aforedescribed, the smart garment also includes some electrochemical gas sensors for detecting gaseous species dangerous for human health, for instance, a modified ZE07-CO module (Winsen Inc.) for detecting the CO concentration in the environment around the user. The module includes an acquisition and communication unit based on ST32F040F4 (made by STMicroelectronics) MCU and a potentiostatic conditioning unit based on SGM8042Y28 (made by SGMicro) operational amplifier (Figure 21a). For reducing the power consumption of the ZE07-CO module, the acquisition and communication unit has been removed from the board, along with the 3V voltage regulator LP2980 (made by National Semiconductor) and the onboard DAC (digital to analog) converter (Figure 21b). In this way, the analog signal produced by the conditioning unit was acquired directly by the 10-bit ATmega328P ADC of the Arduino Pro mini board and converted into the gas concentration (expressed in ppm-part per million) through the stage transfer function (1):

$$gas\ concentration = \frac{V - 0.2686\ Volt}{0.003\ V/ppm} \tag{1}$$

where the gas concentration is expressed in ppm and V, the acquired voltage, is in Volts.

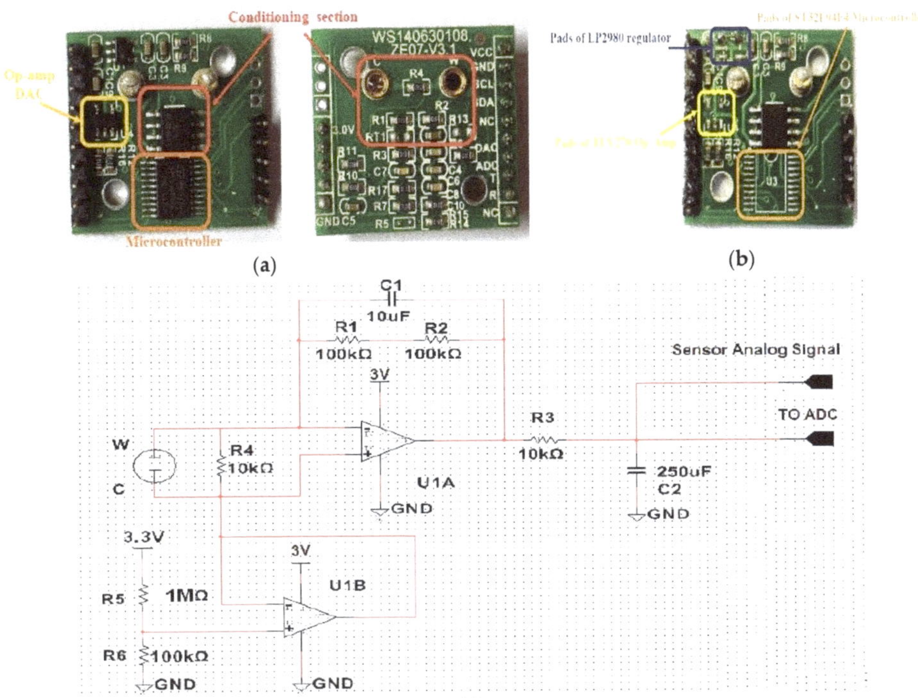

Figure 21. Top and bottom view (**a**) of the ZE07-CO (Winsen Inc.) conditioning board before the modification, and after the modification aimed to reduce the power consumption (**b**); schematic of the conditioning unit (**c**).

The measurement of the current absorbed by the modified ZE07-CO module demonstrates that a current of just 4.6 μA is required in the low gas concentration condition. By the tests carried out on the harvesting system, about 4 h of initial charging time in Scenario 2 (Table 4) are required to fully charge the 380 mAh Lipo battery. The operative strategy involves the acquisition, elaboration, and transmission of the environmental and biophysical parameters every 15 min, whereas the fall detector acts as an interrupt source for the MCU triggering an ISR, with consequent transmission of a warning message to the cloud app. Furthermore, the performed measurements on the electronic section indicate that the mean absorbed current in power-down mode is 88.74 μA, as well as in active mode just 7.46 mA; since the overall time duration of the measurement, elaboration and transmission phases was equal to 40 s, the charge required for the daily system operation was 9.99 mAh. Therefore, by supposing a 40% discharge margin of the Lipo battery, the autonomy of the smart garment in the total absence of any contribution from the harvesting section is given by (2):

$$Autonomy = \frac{Battery\ Capacity * (1 - Discharge\ Margin)}{Energy\ Consumption\ per\ day} = \frac{380\ mAh * (1 - 0.4)}{9.99\ mAh/day} = 22.82\ day \quad (2)$$

Thanks to the implementation of a proper operative strategy and power consumption minimization of the different sections into the smart garment, exploiting low power operating modes for the MCU and sensors, an autonomy of $\approx$ 23 days can be ensured in a total absence of the energetic contributions from the different harvesting sections included in the developed system. Finally, downstream of all performed hardware and software optimizations to obtain minimization of power consumption, considering that the sensing and communication units need about 0.42 mAh for every hour of operation (9.99 mAh/24 h), the energy harvesting section is able to provide, also in the worst scenario (Scenario 5 of Table 4), more than twice (0.850 mAh) compared to the needed charge (0.42 mAh), so ensuring the energy autonomy of the designed smart electronic garment.

After the integration of sensors inside the smart garment according to the modalities described in the fourth section, several tests were carried out exploiting the results obtained from the afore reported characterizations. For all sensors integrated into the smart garment, the correct operation was verified by comparing the measurements provided by each sensor with those obtained by external instruments, namely a thermometer, a pulse-oximeter, a step-counter and a gas detector. Only for the MMA8452Q-based steps counter has a slight (+20%) overestimation of the step counting been noticed, probably due to small movements of the elastic band where the accelerometer has been placed, issue solvable with a successive optimization of the acceleration threshold, involved in the step counter firmware, directly on the smart garment.

5. Conclusions

This manuscript aims to describe the development and characterization of a smart garment able to detect the environmental and biophysical parameters of a user wearing it, specially designed to monitor conditions in particularly dangerous workplaces and thus to prevent or reduce the consequences of worker accidents. The developed device is a joint application of low-power electronic sections and energy harvesting solutions, to ensure the energetic autonomy of the system; in fact, the smart garment includes flexible solar panels, TEGs and flexible piezoelectric harvesters to scavenge energy from sources closely related to the human body. The smart jacket shares acquired information and warning signaling with an on-cloud database, enabling company managers to check them through a web application. The results of the characterization of the selected energy harvesters with related conditioning sections have been reported in different real application scenarios. Moreover, the firmware to interface employed sensors (HR-SpO_2 sensor, accelerometers, environmental temperature and electrochemical gas sensors) with a modified Arduino Pro mini board has been developed; particular attention was paid to the implementation of hardware and software strategies aimed to reduce the overall energy consumption of the system. Furthermore, the characterization results derived from the experimental tests on different sensors included in the smart garment have been reported, to optimize the characteristic parameters of developed firmware.

The multi-source energy harvesting system integrated inside the smart garment was characterized in order to determine the maximum power provided in different real operative scenarios. The obtained results have demonstrated the predominance of solar contribution in the condition of direct solar illumination (Scenarios 1 and 2 with the user wearing the jacket, outdoors) compared to the thermal and piezoelectric effects; instead, the TEGs-based thermal harvesting section provides a continuous contribution, dominant respect to those others, in conditions of solar diffused or artificial light (Scenarios 3, 4 and 5), however ensuring the energy autonomy of the complete developed system in all the operative conditions. Besides, tests performed on the sensing section integrated inside the proposed wearable application indicated the correct operation of all included sensors, thanks to the parameters optimization previously performed. In addition, the power consumption measurements on acquisition, elaboration and communication subsections demonstrated that about a 10 mAh charge is needed to ensure the daily system functionality; in this way, by using a 380 mAh Lipo battery, as storage device charged by the designed harvesting section, an autonomy of more than 20 days was obtained by exploiting the usable battery charge (60%), if no energy contribution is available over time.

Author Contributions: Relatively to the present manuscript, the authors R.d.F., P.V., G.M. and A.M. carried out the experimental activities, as well as they wrote the manuscript. D.C. supervised the activities and data analysis, as well as he has contributed to the manuscript editing. All authors have read and agreed to the published version of the manuscript.

Funding: This research received no external funding.

Conflicts of Interest: The authors declare no conflict of interest.

References

1. Mathias, D.N.; Kim, S.-I.; Park, J.; Joung, Y.-H. Real Time ECG Monitoring Through a Wearable Smart T-shirt. *Trans. Electr. Electron. Mater.* **2015**, *16*, 16–19. [CrossRef]
2. Kamišalić, A.; Fister, I.; Turkanović, M.; Karakatič, S. Sensors and Functionalities of Non-Invasive Wrist-Wearable Devices: A Review. *Sensors* **2018**, *18*, 1714. [CrossRef] [PubMed]
3. Eskofier, B.M.; Lee, S.I.; Baron, M.; Simon, A.L.; Martindale, C.F.; Gassner, H.; Klucken, J. An Overview of Smart Shoes in the Internet of Health Things: Gait and Mobility Assessment in Health Promotion and Disease Monitoring. *Appl. Sci.* **2017**, *7*, 986. [CrossRef]
4. Prauzek, M.; Konecny, J.; Borova, M.; Janosova, K.; Hlavica, J.; Musilek, P. Energy Harvesting Sources, Storage Devices and System Topologies for Environmental Wireless Sensor Networks: A Review. *Sensors (Basel)* **2018**, *18*, 2446. [CrossRef]
5. Munir, B.; Dyo, V. On the Impact of Mobility on Battery-Less RF Energy Harvesting System Performance. *Sensors (Basel)* **2018**, *18*, 3597. [CrossRef]
6. Lee, H.G.; Chang, N. Powering the IoT: Storage-less and converter-less energy harvesting. In Proceedings of the 20th IEEE Asia and South Pacific Design Automation Conference, Chiba, Japan, 19–22 January 2015; pp. 124–129.
7. Tang, X.; Wang, X.; Cattley, R.; Gu, F.; Ball, A.D. Energy Harvesting Technologies for Achieving Self-Powered Wireless Sensor Networks in Machine Condition Monitoring: A Review. *Sensors (Basel)* **2018**, *18*, 4113. [CrossRef]
8. Eren, H.; Webster, J.G. *Telemedicine and Electronic Medicine*; CRC Press: Boca Raton, FL, USA, 2018; ISBN 978-1-351-23147-3.
9. Dimitrov, D.V. Medical Internet of Things and Big Data in Healthcare. *Healthc. Inform. Res.* **2016**, *22*, 156–163. [CrossRef]
10. Zeinab, K.A.M.; Elmustafa, S.A.A. Internet of Things Applications, Challenges and Related Future Technologies. *World Sci. News* **2017**, *67*, 126–148.
11. Visconti, P.; de Fazio, R.; Primiceri, P.; Cafagna, D.; Strazzella, S.; Giannoccarro, I.N. A Solar-Powered Fertigation System based on Low-Cost Wireless Sensor Network Remotely Controlled by Farmer for Irrigation Cycles and Crops Growth Optimization. *Int. J. Electron. Telecommun.* **2020**, *66*, 59–68.
12. Jokic, P.; Magno, M. Powering smart wearable systems with flexible solar energy harvesting. In Proceedings of the 2017 IEEE International Symposium on Circuits and Systems (ISCAS), Baltimore, MD, USA, 28–31 May 2017; pp. 1–4.
13. Haghi, M.; Thurow, K.; Stoll, R. Wearable Devices in Medical Internet of Things: Scientific Research and Commercially Available Devices. *Healthc. Inform. Res.* **2017**, *23*, 4–15. [CrossRef]

14. Abraham, K.M. Prospects and Limits of Energy Storage in Batteries. *J. Phys. Chem. Lett.* **2015**, *6*, 830–844. [CrossRef] [PubMed]
15. Visconti, P.; Primiceri, P.; Orlando, C. Solar Powered Wireless Monitoring System of Environmental Conditions for Early Flood Prediction or Optimized Irrigation in Agriculture. *ARPN J. Eng. Appl. Sci.* **2016**, *11*, 4623–4632.
16. Visconti, P.; Ferri, R.; Pucciarelli, M.; Venere, E. Development and Characterization of a solar-based energy harvesting and power management system for a WSN node applied to optimized goods transport and storage. *Int. J. Smart Sens. Intell. Syst.* **2016**, *9*, 1637–1667.
17. Visconti, P.; Primiceri, P.; Ferri, R.; Pucciarelli, M.; Venere, E. An Overview On State-of-Art Energy Harvesting Techniques and Choice Criteria: A WSN Node for Goods Transport and Storage Powered by a Smart Solar-Based EH System. *Int. J. Renew. Energy Res.* **2017**, *7*, 1281–1295.
18. Loncar-Turukalo, T.; Zdravevski, E.; da Silva, J.M.; Chouvarda, I.; Trajkovik, V. Literature on Wearable Technology for Connected Health: Scoping Review of Research Trends, Advances, and Barriers. *J. Med. Internet Res.* **2019**, *21*, 1–23. [CrossRef]
19. Iqbal, M.H.; Aydin, A.; Brunckhorst, O.; Dasgupta, P.; Ahmed, K. A review of wearable technology in medicine. *J. R. Soc. Med.* **2016**, *109*, 372–380. [CrossRef]
20. Al-Eidan, R.M.; Al-Khalifa, H.; Al-Salman, A.M. A Review of Wrist-Worn Wearable: Sensors, Models, and Challenges. *J. Sens.* **2018**, *2018*, 1–20. [CrossRef]
21. Ramadass, Y.K.; Chandrakasan, A.P. A Battery-Less Thermoelectric Energy Harvesting Interface Circuit With 35 mV Startup Voltage. *IEEE J. Solid-State Circuits* **2011**, *46*, 333–341. [CrossRef]
22. Kim, S.; Vyas, R.; Bito, J.; Niotaki, K.; Collado, A.; Georgiadis, A.; Tentzeris, M.M. Ambient RF Energy-Harvesting Technologies for Self-Sustainable Standalone Wireless Sensor Platforms. *Proc. IEEE* **2014**, *102*, 1649–1666. [CrossRef]
23. Thielen, M.; Sigrist, L.; Magno, M.; Hierold, C.; Benini, L. Human body heat for powering wearable devices: From thermal energy to application. *Energy Convers. Manag.* **2017**, *131*, 44–54. [CrossRef]
24. de Fazio, R.; Cafagna, D.; Marcuccio, G.; Visconti, P. Limitations and Characterization of Energy Storage Devices for Harvesting Applications. *Energies* **2020**, *13*, 783. [CrossRef]
25. Silva, F.A. Handbook of Energy Harvesting Power Supplies and Applications [Book News]. *IEEE Ind. Electron. Mag.* **2016**, *10*, 67–68. [CrossRef]
26. Cai, Y.; Deng, F.; Zhao, J.; Qiu, H.; Fan, X.; Liang, Z. The Distributed System of Smart Wearable Energy Harvesting Based on Human Body. In Proceedings of the 2018 IEEE 37th Chinese Control Conference (CCC), Wuhan, China, 25–27 July 2018; pp. 7450–7454.
27. Alhawari, M.; Tekeste, T.; Mohammad, B.; Saleh, H.; Ismail, M. Power management unit for multi-source energy harvesting in wearable electronics. In Proceedings of the 2016 IEEE 59th International Midwest Symposium on Circuits and Systems (MWSCAS), Abu Dhabi, UAE, 16–19 October 2016; pp. 1–4.
28. Kim, S.-E.; Kang, T.; Oh, K.-I.; Park, M.J.; Park, H.-I.; Lim, I.G.; Lee, J.-J. Energy Management Integrated Circuit for Multi-Source Energy Harvesters in WBAN Applications. *Appl. Sci.* **2018**, *8*, 1262. [CrossRef]
29. Visconti, P.; de Fazio, R.; Costantini, P.; Miccoli, S.; Cafagna, D. Arduino-Based Solution for In-Car- Abandoned Infants' Controlling Remotely Managed by Smartphone Application. *JCOMSS* **2019**, *15*, 89–100. [CrossRef]
30. Gaetani, F.; Primiceri, P.; Antonio Zappatore, G.; Visconti, P. Hardware design and software development of a motion control and driving system for transradial prosthesis based on a wireless myoelectric armband. *IET Sci. Meas. Technol.* **2019**, *13*, 354–362. [CrossRef]
31. Osswald, S.; Weiss, A.; Tscheligi, M. Designing wearable devices for the factory: Rapid contextual experience prototyping. In Proceedings of the 2013 IEEE International Conference on Collaboration Technologies and Systems (CTS), San Diego, CA, USA, 20–24 May 2013; pp. 517–521.
32. Alam, M.M.; Hamida, E.B. Surveying Wearable Human Assistive Technology for Life and Safety Critical Applications: Standards, Challenges and Opportunities. *Sensors* **2014**, *14*, 9153–9209. [CrossRef] [PubMed]
33. Takei, K.; Honda, W.; Harada, S.; Arie, T.; Akita, S. Toward Flexible and Wearable Human-Interactive Health-Monitoring Devices. *Adv. Healthc. Mater.* **2015**, *4*, 487–500. [CrossRef]
34. Mantua, J.; Gravel, N.; Spencer, R.M.C. Reliability of Sleep Measures from Four Personal Health Monitoring Devices Compared to Research-Based Actigraphy and Polysomnography. *Sensors* **2016**, *16*, 646. [CrossRef]
35. Bio-Monitor: Keeping an Eye on Astronauts' Vital Signs. Available online: https://www.asc-csa.gc.ca/eng/sciences/bio-monitor.asp (accessed on 12 April 2020).

36. Qaosar, M.; Ahmed, S.; Li, C.; Morimoto, Y. Hybrid Sensing and Wearable Smart Device for Health Monitoring and Medication: Opportunities and Challenges. In Proceedings of the AAAI Spring Symposium Series; Association for the Advancement of Artificial Intelligence, Palo Alto, CA, USA, 26–28 March 2018; Volume 1, pp. 269–274.
37. Al-khafajiy, M.; Baker, T.; Chalmers, C.; Asim, M.; Kolivand, H.; Fahim, M.; Waraich, A. Remote health monitoring of elderly through wearable sensors. *Multimed. Tools Appl.* **2019**, *78*, 24681–24706. [CrossRef]
38. Nakamura, Y.; Matsuda, Y.; Arakawa, Y.; Yasumoto, K. WaistonBelt X: A Belt-Type Wearable Device with Sensing and Intervention Toward Health Behavior Change. *Sensors* **2019**, *19*, 4600. [CrossRef]
39. Kyriakopoulos, G.; Ntanos, S.; Anagnostopoulos, T.; Tsotsolas, N.; Salmon, I.; Ntalianis, K. Internet of Things (IoT)-Enabled Elderly Fall Verification, Exploiting Temporal Inference Models in Smart Homes. *Int. J. Environ. Res. Public Health* **2020**, *17*, 408. [CrossRef] [PubMed]
40. Stetter, B.J.; Ringhof, S.; Krafft, F.C.; Sell, S.; Stein, T. Estimation of Knee Joint Forces in Sport Movements Using Wearable Sensors and Machine Learning. *Sensors* **2019**, *19*, 690. [CrossRef] [PubMed]
41. Lapinski, M.; Brum Medeiros, C.; Moxley Scarborough, D.; Berkson, E.; Gill, T.J.; Kepple, T.; Paradiso, J.A. A Wide-Range, Wireless Wearable Inertial Motion Sensing System for Capturing Fast Athletic Biomechanics in Overhead Pitching. *Sensors* **2019**, *19*, 3637. [CrossRef] [PubMed]
42. Mamun, M.A.A.; Yuce, M.R. Sensors and Systems for Wearable Environmental Monitoring Toward IoT-Enabled Applications: A Review. *IEEE Sens. J.* **2019**, *19*, 7771–7788. [CrossRef]
43. Haghi, M.; Stoll, R.; Thurow, K. A Low-Cost, Standalone, and Multi-Tasking Watch for Personalized Environmental Monitoring. *IEEE Trans. Biomed. Circuits Syst.* **2018**, *12*, 1144–1154. [CrossRef]
44. Kim, S.; Paulos, E.; Gross, M.D. WearAir: Expressive t-shirts for air quality sensing. In Proceedings of the TEI '10, Cambridge, MA, USA, 25–27 January 2010; pp. 295–296.
45. Spirjakin, D.; Baranov, A.; Akbari, S. Wearable Wireless Sensor System With RF Remote Activation for Gas Monitoring Applications. *IEEE Sens. J.* **2018**, *18*, 2976–2982. [CrossRef]
46. Pu, X.; Hu, W.; Wang, Z.L. Toward Wearable Self-Charging Power Systems: The Integration of Energy-Harvesting and Storage Devices. *Small* **2018**, *14*, 1–20. [CrossRef]
47. Lee, J.-H.; Kim, J.; Kim, T.Y.; Hossain, M.S.A.; Kim, S.-W.; Kim, J.H. All-in-one energy harvesting and storage devices. *J. Mater. Chem. A* **2016**, *4*, 7983–7999. [CrossRef]
48. Kim, J.H.; Kim, S.-W.; Wang, Z.L. Preface for Special Topic: Nanogenerators. *APL Mater.* **2017**, *5*, 073701. [CrossRef]
49. Varma, S.J.; Kumar, K.S.; Seal, S.; Rajaraman, S.; Thomas, J. Fiber-Type Solar Cells, Nanogenerators, Batteries, and Supercapacitors for Wearable Applications. *Adv. Sci.* **2018**, *5*, 1–32. [CrossRef]
50. Yao, H.; Ye, L.; Zhang, H.; Li, S.; Zhang, S.; Hou, J. Molecular Design of Benzodithiophene-Based Organic Photovoltaic Materials. *Chem. Rev.* **2016**, *116*, 7397–7457. [CrossRef] [PubMed]
51. Zhao, W.; Li, S.; Yao, H.; Zhang, S.; Zhang, Y.; Yang, B.; Hou, J. Molecular Optimization Enables over 13% Efficiency in Organic Solar Cells. *J. Am. Chem. Soc.* **2017**, *139*, 7148–7151. [CrossRef] [PubMed]
52. Li, Y.; Xu, G.; Cui, C.; Li, Y. Flexible and Semitransparent Organic Solar Cells. *Adv. Energy Mater.* **2018**, *8*, 1701791. [CrossRef]
53. Shin, S.S.; Yeom, E.J.; Yang, W.S.; Hur, S.; Kim, M.G.; Im, J.; Seo, J.; Noh, J.H.; Seok, S.I. Colloidally prepared La-doped BaSnO3 electrodes for efficient, photostable perovskite solar cells. *Science* **2017**, *356*, 167–171. [CrossRef] [PubMed]
54. Service, R.F. Perovskite solar cells gear up to go commercial. *Science* **2016**, *354*, 1214–1215. [CrossRef] [PubMed]
55. Qiu, L.; He, S.; Yang, J.; Deng, J.; Peng, H. Fiber-Shaped Perovskite Solar Cells with High Power Conversion Efficiency. *Small* **2016**, *12*, 2419–2424. [CrossRef] [PubMed]
56. Wu, T.; Wu, F.; Redouté, J.; Yuce, M.R. An Autonomous Wireless Body Area Network Implementation Towards IoT Connected Healthcare Applications. *IEEE Access* **2017**, *5*, 11413–11422. [CrossRef]
57. Zhang, Y.; Liu, C.; Liu, J.; Xiong, J.; Liu, J.; Zhang, K.; Liu, Y.; Peng, M.; Yu, A.; Zhang, A.; et al. Lattice Strain Induced Remarkable Enhancement in Piezoelectric Performance of ZnO-Based Flexible Nanogenerators. *ACS Appl. Mater. Interfaces* **2016**, *8*, 1381–1387. [CrossRef]

58. Leoni, A.; Stornelli, V.; Ferri, G.; Errico, V.; Ricci, M.; Pallotti, A.; Saggio, G. A human body powered sensory glove system based on multisource energy harvester. In Proceedings of the 2018 IEEE 14th Conference on Ph.D. Research in Microelectronics and Electronics (PRIME), Prague, Czech Republic, 2–5 July 2018; pp. 113–116.
59. Weddell, A.S.; Magno, M.; Merrett, G.V.; Brunelli, D.; Al-Hashimi, B.M.; Benini, L. A survey of multi-source energy harvesting systems. In Proceedings of the IEEE 2013 Design, Automation Test in Europe Conference Exhibition (DATE), Grenoble, France, 18–22 March 2013; pp. 905–908.
60. Dąbrowska, A.; Greszta, A. Analysis of the Possibility of Using Energy Harvesters to Power Wearable Electronics in Clothing. *Adv. Mater. Sci. Eng.* **2019**, *2019*, 1–14. [CrossRef]
61. Yang, J.-H.; Cho, H.-S.; Park, S.-H.; Song, S.-H.; Yun, K.-S.; Lee, J.H. Effect of garment design on piezoelectricity harvesting from joint movement. *Smart Mater. Struct.* **2016**, *25*, 1–15. [CrossRef]
62. Cao, X.; Jie, Y.; Wang, N.; Wang, Z. Triboelectric Nanogenerators Driven Self-Powered Electrochemical Processes for Energy and Environmental Science. *Adv. Energy Mater.* **2016**, *6*, 1–20. [CrossRef]
63. Ahmed, A.; Saadatnia, Z.; Hassan, I.; Zi, Y.; Xi, Y.; He, X.; Zu, J.; Wang, Z. Self-Powered Wireless Sensor Node Enabled by a Duck-Shaped Triboelectric Nanogenerator for Harvesting Water Wave Energy. *Adv. Energy Mater.* **2016**, 1–10. [CrossRef]
64. Chen, J.; Wang, Z.L. Reviving Vibration Energy Harvesting and Self-Powered Sensing by a Triboelectric Nanogenerator. *Joule* **2017**, *1*, 480–521. [CrossRef]
65. Li, Z.; Saadatnia, Z.; Yang, Z.; Naguib, H. A hybrid piezoelectric-triboelectric generator for low-frequency and broad-bandwidth energy harvesting. *Energy Convers. Manag.* **2018**, *174*, 188–197. [CrossRef]
66. Ahmed, A.; Zhang, S.L.; Hassan, I.; Saadatnia, Z.; Zi, Y.; Zu, J.; Wang, Z.L. A washable, stretchable, and self-powered human-machine interfacing Triboelectric nanogenerator for wireless communications and soft robotics pressure sensor arrays. *Extrem. Mech. Lett.* **2017**, *13*, 25–35. [CrossRef]
67. Chen, H.; Song, Y.; Cheng, X.; Zhang, H. Self-powered electronic skin based on the triboelectric generator. *Nano Energy* **2019**, *56*, 252–268. [CrossRef]
68. Hua, Q.; Sun, J.; Liu, H.; Bao, R.; Yu, R.; Zhai, J.; Pan, C.; Wang, Z.L. Skin-inspired highly stretchable and conformable matrix networks for multifunctional sensing. *Nat. Commun.* **2018**, *9*, 1–11. [CrossRef]
69. Chen, H.; Su, Z.; Song, Y.; Cheng, X.; Chen, X.; Meng, B.; Song, Z.; Chen, D.; Zhang, H. Omnidirectional Bending and Pressure Sensor Based on Stretchable CNT-PU Sponge. *Adv. Funct. Mater.* **2017**, *27*, 1–9. [CrossRef]
70. Moravčík, M.; Schmid, M.; Burch, N.; Lisý, V.; Morrill, D.; Bard, N.; Davis, T.; Waugh, K.; Johanson, M.; Bowling, M. DeepStack: Expert-level artificial intelligence in heads-up no-limit poker. *Science* **2017**, *356*, 508–513. [CrossRef]
71. Yu, X.; Mahajan, B.K.; Shou, W.; Pan, H. Materials, Mechanics, and Patterning Techniques for Elastomer-Based Stretchable Conductors. *Micromachines* **2017**, *8*, 7. [CrossRef]
72. Chen, J.; Guo, H.; He, X.; Liu, G.; Xi, Y.; Shi, H.; Hu, C. Enhancing Performance of Triboelectric Nanogenerator by Filling High Dielectric Nanoparticles into Sponge PDMS Film. *ACS Appl. Mater. Interfaces* **2016**, *8*, 736–744. [CrossRef] [PubMed]
73. Zhang, X.-S.; Brugger, J.; Kim, B. A silk-fibroin-based transparent triboelectric generator suitable for autonomous sensor network. *Nano Energy* **2016**, *20*, 37–47. [CrossRef]
74. Wang, J. Special Issue for Wearable Electrochemical Sensors. *Electroanalysis* **2016**, *28*, 1148. [CrossRef]
75. Gross, A.J.; Holzinger, M.; Cosnier, S. Buckypaper bioelectrodes: Emerging materials for implantable and wearable biofuel cells. *Energy Environ. Sci.* **2018**, *11*, 1670–1687. [CrossRef]
76. Lv, J.; Jeerapan, I.; Tehrani, F.; Yin, L.; Silva-Lopez, C.A.; Jang, J.-H.; Joshuia, D.; Shah, R.; Liang, Y.; Xie, L.; et al. Sweat-based wearable energy harvesting-storage hybrid textile devices. *Energy Environ. Sci.* **2018**, *11*, 3431–3442. [CrossRef]
77. Kanimba, E.; Tian, Z. *Modeling of a Thermoelectric Generator Device*, 1st ed.; IntechOpen: London, UK, 2016; Volume 1. [CrossRef]
78. Yan, J.; Liao, X.; Yan, D.; Chen, Y. Review of Micro Thermoelectric Generator. *J. Microelectromech. Syst.* **2018**, *27*, 1–18. [CrossRef]
79. Rojas, J.P.; Singh, D.; Inayat, S.B.; Sevilla, G.A.T.; Fahad, H.M.; Hussain, M.M. Review—Micro and Nano-Engineering Enabled New Generation of Thermoelectric Generator Devices and Applications. *ECS J. Solid State Sci. Technol.* **2017**, *6*, 3036–3044. [CrossRef]

80. Leonov, V. Thermoelectric Energy Harvesting of Human Body Heat for Wearable Sensors. *IEEE Sens. J.* **2013**, *13*, 2284–2291. [CrossRef]
81. Zhang, Y.; Zhang, F.; Shakhsheer, Y.; Silver, J.D.; Klinefelter, A.; Nagaraju, M.; Boley, J.; Pandey, J.; Shrivastava, A.; Carlson, E.J.; et al. A Batteryless 19µW MICS/ISM-Band Energy Harvesting Body Sensor Node SoC for ExG Applications. *IEEE J. Solid-State Circuits* **2013**, *48*, 199–213. [CrossRef]
82. Proto, A.; Bibbo, D.; Cerny, M.; Vala, D.; Kasik, V.; Peter, L.; Conforto, S.; Schmid, M.; Penhaker, M. Thermal Energy Harvesting on the Bodily Surfaces of Arms and Legs through a Wearable Thermo-Electric Generator. *Sensors* **2018**, *18*, 1927. [CrossRef]
83. Minnaert, B.; Veelaert, P. A Proposal for Typical Artificial Light Sources for the Characterization of Indoor Photovoltaic Applications. *Energies* **2014**, *7*, 1500–1516. [CrossRef]
84. Virtuani, A.; Lotter, E.; Powalla, M. Influence of the light source on the low-irradiance performance of Cu(In,Ga)Se2 solar cells. *Sol. Energy Mater. Sol. Cells* **2006**, *90*, 2141–2149. [CrossRef]
85. Benghanem, M.S.; Alamri, S.N. Modeling of photovoltaic module and experimental determination of serial resistance. *J. Taibah Univ. Sci.* **2009**, *2*, 94–105. [CrossRef]
86. Özdemir, A.T. An Analysis on Sensor Locations of the Human Body for Wearable Fall Detection Devices: Principles and Practice. *Sensors (Basel)* **2016**, *16*, 1161. [CrossRef] [PubMed]

energies

MDPI

Review

A Review of the Recent Developments in Integrating Machine Learning Models with Sensor Devices in the Smart Buildings Sector with a View to Attaining Enhanced Sensing, Energy Efficiency, and Optimal Building Management

Dana-Mihaela Petroşanu [1,*], George Căruţaşu [2,3], Nicoleta Luminiţa Căruţaşu [4] and Alexandru Pîrjan [2]

1 Department of Mathematics-Informatics, Faculty of Applied Sciences, University Politehnica of Bucharest, Splaiul Independenţei 313, 060042 Bucharest, Romania
2 Department of Informatics, Statistics and Mathematics, Romanian-American University, Expoziţiei 1B, 012101 Bucharest, Romania; carutasu.george@profesor.rau.ro (G.C.); alex@pirjan.com (A.P.)
3 Doctoral School, University Politehnica of Timişoara, Piaţa Victoriei 2, 300006 Timişoara, Romania
4 Department of Robotics and Production Systems, Faculty of Industrial Engineering and Robotics, University Politehnica of Bucharest, Splaiul Independenţei 313, 060042 Bucharest, Romania; nicoleta.carutasu@upb.ro
* Correspondence: danap@mathem.pub.ro; Tel.: +40-761-086-656

Received: 11 October 2019; Accepted: 9 December 2019; Published: 12 December 2019

Abstract: Lately, many scientists have focused their research on subjects like smart buildings, sensor devices, virtual sensing, buildings management, Internet of Things (IoT), artificial intelligence in the smart buildings sector, improving life quality within smart homes, assessing the occupancy status information, detecting human behavior with a view to assisted living, maintaining environmental health, and preserving natural resources. The main purpose of our review consists of surveying the current state of the art regarding the recent developments in integrating supervised and unsupervised machine learning models with sensor devices in the smart building sector with a view to attaining enhanced sensing, energy efficiency and optimal building management. We have devised the research methodology with a view to identifying, filtering, categorizing, and analyzing the most important and relevant scientific articles regarding the targeted topic. To this end, we have used reliable sources of scientific information, namely the Elsevier Scopus and the Clarivate Analytics Web of Science international databases, in order to assess the interest regarding the above-mentioned topic within the scientific literature. After processing the obtained papers, we finally obtained, on the basis of our devised methodology, a reliable, eloquent and representative pool of 146 papers scientific works that would be useful for developing our survey. Our approach provides a useful up-to-date overview for researchers from different fields, which can be helpful when submitting project proposals or when studying complex topics such those reviewed in this paper. Meanwhile, the current study offers scientists the possibility of identifying future research directions that have not yet been addressed in the scientific literature or improving the existing approaches based on the body of knowledge. Moreover, the conducted review creates the premises for identifying in the scientific literature the main purposes for integrating Machine Learning techniques with sensing devices in smart environments, as well as purposes that have not been investigated yet.

Keywords: internet of things; sensor networks; machine learning models; sensor devices; smart buildings; energy efficiency; optimal building management

1. Introduction

Globally nowadays, all types of buildings affect the environment to an overwhelming extent, by means ranging from the associated electricity consumption, through generated waste and pollution, up to natural habitat degradations, causing irreparable damages to the environment. Therefore, all over the world, concerted action is being carried out in order to limit these negative impacts. In addition to this, modern society faces issues regarding building safety along with comfort, and consequently, major efforts are being carried out all over the world in the direction of monitoring, identifying occupants' presence and activities in order to achieve enhanced sensing, energy efficiency and optimal building management, while at the same time minimizing or even eliminating the negative consequences imposed on the environment.

There is an increasing interest in the scientific literature in studies related to these topics; for example, there have been papers focusing on smart buildings [1–5], smart homes [6–10], smart hospitals [11], smart commercial buildings [12,13], sensor devices [9,14–17], supervised machine learning models for classification purposes [1,11,16–18] or for regression purposes [19–23], unsupervised machine learning models for clustering purposes [24–26], deep learning techniques [18,27,28], human activity recognition and classification with a view to assisted living [15,29–35], Internet of Things (IoT) [21,36–39], energy efficiency and an optimal building management [1,21,23,24,40–46], and the comfort and safety of the inhabitants [39,40,47–56].

In this context, a subject of utmost importance, which could lead to a wide range of advantages for the inhabitants of buildings, for constructors, for providers of different services, and even for society as a whole, is the analysis of recent developments in integrating machine learning models with sensor devices in the smart buildings sector with a view to attaining enhanced sensing, energy efficiency and optimal building management.

Therefore, this study aims to review the latest scientific articles that fuse emerging topics such as machine learning techniques, enhanced sensing, and smart buildings; hence attaining a proper categorization of a high number of scientific works in accordance with a well-defined encompassing taxonomy. In addition to providing a useful up-to-date overview to the researchers from different scientific fields who might be interested in devising project proposals or studying emerging complex topics like the analyzed ones, this review article sets its sights on providing scientists with valuable insights on enhancing existing methods from the current state of the art and on future research directions that have not yet been addressed by reviewing the recent advances that have been made with regard to integrating machine learning models with sensor devices in the smart buildings sector. Consequently, this review article aims to indicate the main purposes within the scientific literature for the integration of machine learning techniques with sensing devices in the smart buildings sector, thereby helping researchers identify possible novel purposes that have not been pursued up until now.

The review paper is structured as follows: the next section, namely "Research Methodology", presents the devised approach, developed with a view to identifying, filtering, classifying and analyzing the most important and relevant scientific articles related to the topic. The section also includes a flowchart of the developed survey, containing details regarding the steps of the devised research methodology. The Third Section, "Enhanced Sensing by Integrating Machine Learning Models with Sensor Devices in the Smart Buildings Sector" presents a review of the papers that were selected by applying the devised methodology, identifying through summarization tables and their analysis the machine learning models that are most suitable for integration with sensor devices in the smart buildings sector. The section also contains a review of the most highly cited scientific papers approaching the reviewed topics, as reported by the Elsevier Scopus and the Clarivate Analytics Web of Science International Databases. Afterwards, the Fourth Section, namely the "Discussion and Conclusions" Section, highlights the most important findings of the paper, presents an analysis of the conducted review research in perspective of previous surveys, highlighting a series of advantages offered by the devised approach, along with a few limitations of this study and future research directions targeted by the authors.

2. Research Methodology

The main purpose of our review is to survey the current state of the art with respect to recent developments in the integration of supervised and unsupervised machine learning models with sensor devices in the smart building sector with a view to attaining enhanced sensing, energy efficiency and optimal building management. We devised the research methodology with a view to identifying, filtering, classifying and analyzing the most important and relevant scientific articles related to the targeted topic.

We devised our review methodology in accordance to the SALSA (Search, AppraisaL, Synthesis and Analysis) framework, which was developed by Grant, M. J. and Booth, A. in their renowned paper [57], which had itself registered—at the time at which we devised our review methodology—a total of 1257 citations in the Clarivate Analytics Web of Science database and 1364 in Elsevier Scopus. Of the 14 review types and their associated methodologies, as depicted by Grant et al., we conducted our review in compliance with the "Literature Review" type. When developing the review methodology, we took into account the specifications corresponding to the "Literature Review" type provided by Grant et al. namely: the descriptive component characterizes "published materials that provide examination of recent or current literature; can cover wide range of subjects at various levels of completeness and comprehensiveness; may include research findings"; the search component of the SALSA framework for this type of review "may or may not include comprehensive searching"; the appraisal component "may or may not include quality assessment"; the synthesis component is "typically narrative"; the analysis component "may be chronological, conceptual, thematic, etc.".

To this end, we used reliable sources of scientific information, namely the Elsevier Scopus and Clarivate Analytics Web of Science international databases, in order to assess the interest in this topic within the scientific literature and to obtain a starting point for building a reliable, eloquent and representative database of scientific works that would be useful for developing our survey. We chose these two databases as we wanted to make sure that we were using globally accepted sources of information that distinctively select and index their contents in a uniformly consistent manner, backed up by decades of reliable, precise and comprehensive indexing. Furthermore, we took into account the fact that prestigious publishing groups categorize and promote their journals by highlighting the quality metrics of their journals as provided by the Web of Science Core Collection or the Elsevier Scopus databases. Therefore, we devised, based on the taxonomy of supervised and unsupervised machine learning techniques [58], custom search queries in order to assess the broad implementation and to identify which of the machine learning methods from the taxonomy represented in Figure 1 are most suitable for implementation with sensor devices in smart buildings with a view to achieving enhanced sensing, energy efficiency and optimal building management.

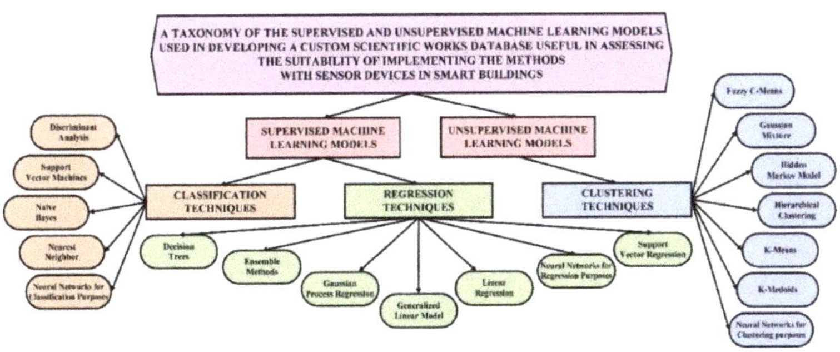

Figure 1. A taxonomy of the supervised and unsupervised machine learning models used in developing a custom scientific works database useful in conducting the survey.

After having tried several search patterns and criteria, we obtained custom search queries, with the terms smart, sensor, and at least one of the terms machine learning, artificial intelligence, supervised learning, and unsupervised learning along with their associated subcategories from the taxonomy depicted in Figure 1 being contained within the title, abstract or keywords. Consequently, according to the specific syntax of each scientific database, the search queries used for interrogating the databases are as follows:

- In the case of the Elsevier Scopus database: TITLE-ABS-KEY(Smart AND Sensor) AND TITLE-ABS-KEY("Machine Learning" OR "Artificial Intelligence" OR "Supervised Learning" OR "Classification" OR "Support Vector Machines" OR "SVM" OR "Discriminant Analysis" OR "DA" OR "Bayes" OR "NB" OR "Nearest Neighbor" OR "NNS" OR "Neural Networks" OR "ANN" OR "Regression" OR "Linear Regression" OR "LR" OR "Generalized Linear Model" OR "GLM" OR "Support Vector Regression" OR "SVR" OR "Gaussian Process Regression" OR "GPR" OR "Ensemble Methods" OR "EM" OR "Decision Tree" OR "DT" OR "Unsupervised Learning" OR "Clustering" OR "Fuzzy" OR "C-Means" OR "Gaussian Mixture" OR "Hidden Markov" OR "Hierarchical Clustering" OR "K-Means" OR "K-Medoids").
- In the case of the Clarivate Analytics Web of Science database: TS = (Smart AND Sensor) AND TS = (Machine Learning OR Artificial Intelligence OR Supervised Learning OR Classification OR Support Vector Machines OR SVM OR Discriminant Analysis OR DA OR Bayes OR NB OR Nearest Neighbor OR NNS OR Neural Networks OR ANN OR Regression OR Linear Regression OR LR OR Generalized Linear Model OR GLM OR Support Vector Regression OR SVR OR Gaussian Process Regression OR GPR OR Ensemble Methods OR EM OR Decision Tree OR DT OR Unsupervised Learning OR Clustering OR Fuzzy OR C-Means OR Gaussian Mixture OR Hidden Markov OR Hierarchical Clustering OR K-Means or K-Medoids).

The search queries were run, and two initial pools of scientific works were retrieved on the 14th of June 2019. Afterwards, the retrieved papers were filtered according to our devised methodology and synthesized into the following flowchart (Figure 2).

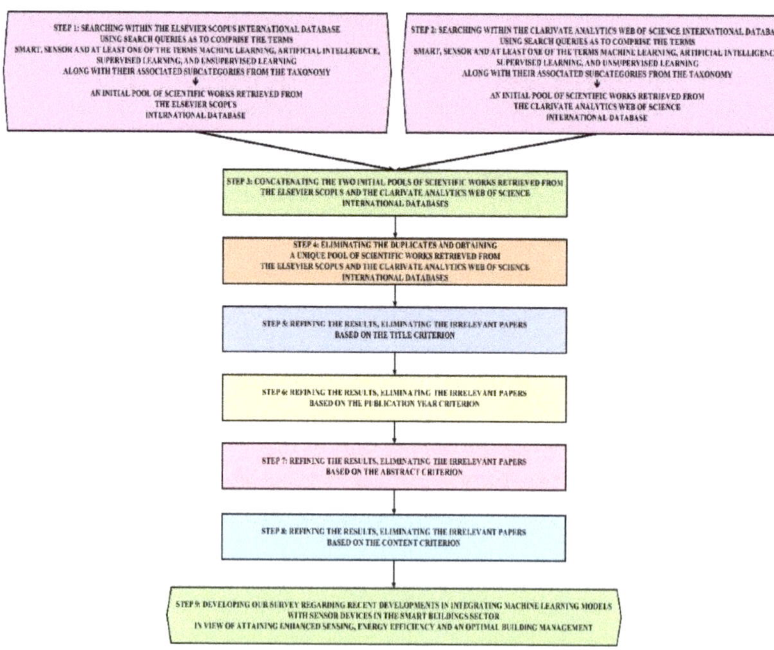

Figure 2. The flowchart of the developed survey.

Therefore, the first two steps of our methodology consist of searching the two international databases using the above-mentioned search queries, consequently obtaining two initial pools of scientific works useful for conducting the survey, consisting of 1255 papers retrieved from the Elsevier Scopus database and 381 papers from the Clarivate Analytics Web of Science database, that is, a total number of 1636 papers (with some papers being included in both databases).

The official data retrieved from the Web of Science and Scopus databases are unique to each database, meaning that the Web of Science database contains no duplicate items, and also that the Scopus database contains only unique entries. When concatenating the scientific articles retrieved from the two international databases, we took into account the fact that some scientific articles might be indexed in both the Web of Science and Scopus databases, thus resulting in duplicate entries, while other scientific works may only be indexed in one of the databases. Consequently, in Step 4 of the review methodology, after having concatenated the works retrieved from the two scientific databases, we eliminated any duplicate entries, retaining only a single instance of each scientific paper.

The particular reason for distinguishing between the two databases is the sheer fact that the two internationally renowned databases have different contents with regard not only to the indexed scientific works, but also with regard to the categories of classification by domain of interest of the papers, and this is why we had to represent the charts depicting the data corresponding to each particular indexing database in different graphics. One can therefore observe that there has been an increasing interest in the literature over the years in the topic targeted by this review, as is clearly depicted by the official data retrieved from the individual databases and distinctly graphically represented for each of database, in accordance with the official records for each database in the absence of duplicate entries.

In order to obtain an initial image regarding the number and content of the scientific papers retrieved from the two databases, we computed, for both the Elsevier Scopus and Clarivate Analytics Web of Science international databases, a series of plots highlighting the number of publications per year (Figure 3), the number of publications by type (Figure 4) and the number of publications per subject area (Figure 5).

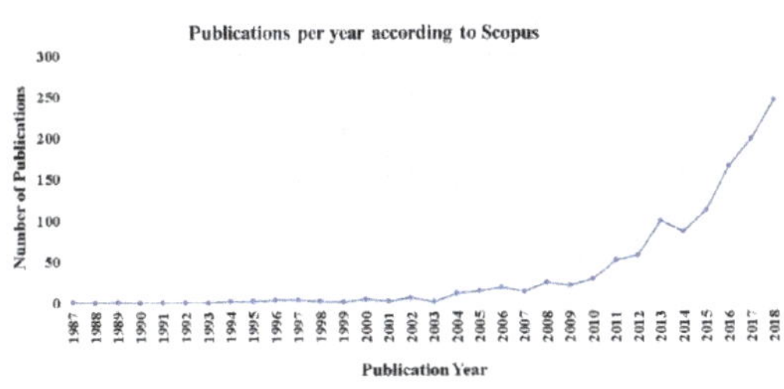

(**a**) Publications per year according to Elsevier Scopus international database

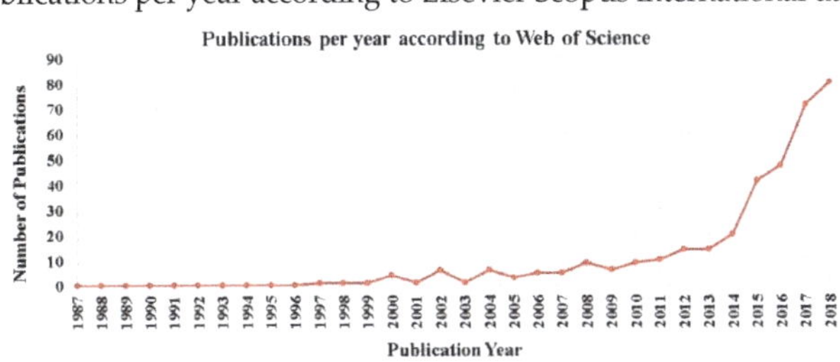

(**b**) Publications per year according to Clarivate Analytics Web of Science international database

Figure 3. The number of publications per year according to the two used databases.

By analyzing Figure 3, we noticed that during the last 5 years, the targeted subjects have been the focus of the research activity of an exponential growing number of papers indexed in both of the used databases, reflecting not only the interest of the authors of these papers but also the development of machine learning models and their integration with sensor devices in smart buildings during the analyzed period of time.

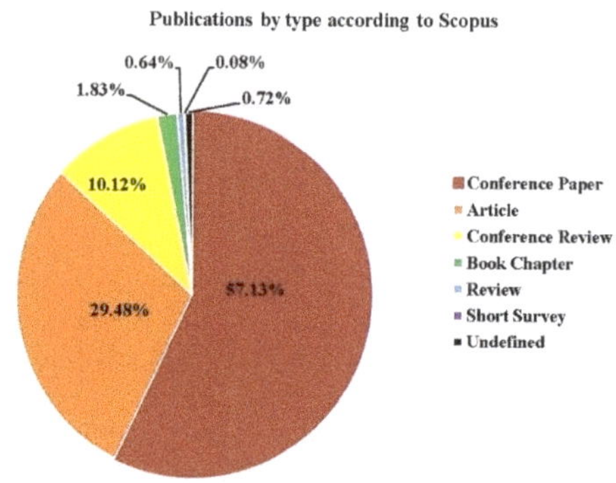

(**a**) Publications by type according to Elsevier Scopus international database

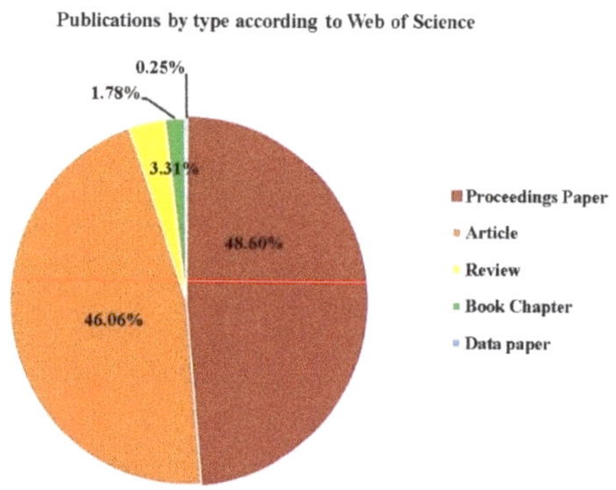

(**b**) Publications by type according to Clarivate Analytics Web of Science international database

Figure 4. The number of publications by type according to the two used databases.

Analyzing Figure 4, it can be remarked that the searches performed across the two databases returned a wide range of publication types. Therefore, even if the two consulted databases had returned different search results, the statistics regarding the number of publications by type according to the two databases would be similar, to a large extent, with respect to the hierarchy of the types, if not the percentages. Even though the two databases structure their searches into slightly different categories, the order of the categories of publications returned (in descending order by number of papers) by the searches performed within the two databases are highly similar. With respect to the percentages of different types of publications within the returned results, by analyzing Figure 4, it can be observed that in the case of the Elsevier Scopus international database, the "Article" type of paper represents a percentage of 29.48, while in the case of the Clarivate Analytics Web of Science international database, this type of paper represents a percentage of 46.06 of the total number of published scientific works. With respect to the papers of the "Review" type, they represent a percentage of 0.64 in the case of the Elsevier Scopus international database and a percentage of 3.31 in the case of the Clarivate Analytics Web of Science international database. With respect to "Book Chapters", the search within the Elsevier

Scopus database returned a percentage of 1.83 of the total number of retrieved scientific works, while the Clarivate Analytics Web of Science database returned a percentage of 1.78.

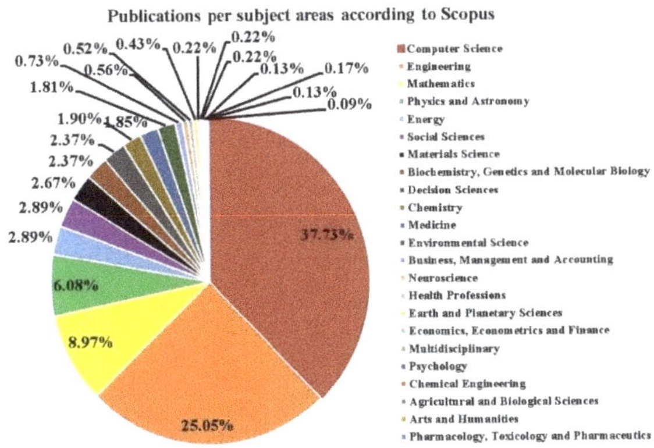

(**a**) Publications per subject area according to Elsevier Scopus international database

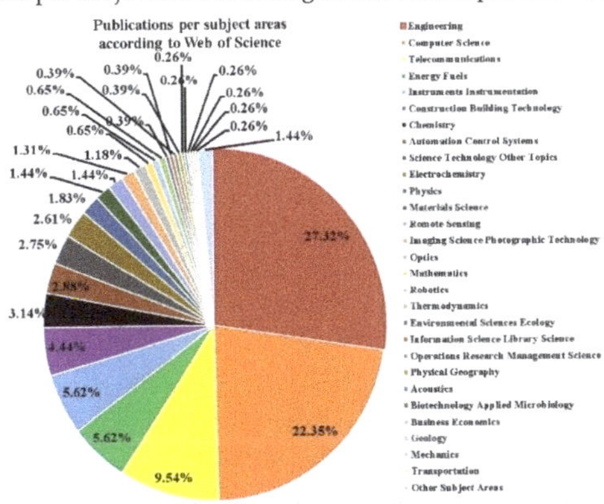

(**b**) Publications per subject area according to Clarivate Analytics Web of Science international database

Figure 5. The number of publications per subject area according to the two used databases.

Examining Figure 5, it can be observed that the searches returned an extensive assortment of subject areas based on the search terms of the queries. One interesting aspect of the results depicted in Figure 5 is the fact that, in the cases of both Elsevier Scopus and Clarivate Analytics Web of Science international databases, some papers are considered to belong to more than one subject area.

Even if the results returned are structured by the two databases into slightly different types and subject areas, it is still possible to observe a series of similarities regarding the statistics of the returned results. Therefore, in the case of the Elsevier Scopus database, the most frequently approached subject areas are: Computer Science, Engineering and Mathematics (representing percentages of 37.73, 25.05 and 8.97, respectively, of the returned results), while in the case of the Clarivate Analytics Web of Science database, the hierarchy of the three most frequently approached subject areas is: Engineering, Computer Science, and Telecommunications (with percentages of 27.32, 22.35 and 9.54, respectively).

In the third step of the devised approach, by concatenating the two initial pools of scientific works retrieved from the Elsevier Scopus and Clarivate Analytics Web of Science international databases, we obtained a raw custom scientific works database. However, the raw set of scientific papers obtained still required further refinement, due to the fact that at the end of the third step, the constructed set contained duplicate copies of some papers. Therefore, during the fourth step, we eliminated the duplicates from the set of scientific papers.

Afterwards, in order to make further improvements to the obtained set of scientific papers, in the fifth, sixth, seventh, and eighth steps, we successively refined the obtained set of scientific works by taking into account the following criteria: title, year of publication, abstract, and content of the paper. Regarding the year of publication, we decided not to plot papers published in 2019 in Figure 3 (as only half of the year had passed at the point at which we retrieved the papers used for our survey), or those papers scheduled to be published in the following year, 2020, because in these two cases, there would be further papers still to be published, and therefore the actual numbers of published papers from these two years would not be able to be taken into account when computing statistics regarding the number of publications per year according to the two used databases. However, in the subsequent analyses, in Figures 4 and 5 and throughout the whole developed survey, for reasons of consistency, we took into account papers whose publication year is (or is scheduled to be) up to 2020. Regarding the earliest year of publication taken into consideration when devising our survey, as we were targeting recent developments in integrating machine learning models with sensor devices in the smart buildings sector with a view to attaining enhanced sensing, energy efficiency, and optimal building management, in our review article we focused mainly on scientific papers published after the year 2012. Moreover, the topic that we are addressing in our survey actually began to soar after this year, as can be seen from Figure 3a,b.

Regarding the filtering performed in the eighth step, when refining the results based on the content criterion, we also eliminated documents published in conference proceedings from the custom database, on account of the fact that the most prominent proceedings papers have also been published in extenso in prestigious journals as scientific articles or reviews, while the remainder, being proceedings, do not contain comprehensive details regarding the developed methodologies and their implementations. Therefore, at this point our database contained a total number of 146 papers.

In the last step of the devised methodology, based on the final form of the custom tailored database of scientific papers, we developed our survey regarding recent developments in the integration of machine learning models with sensor devices in the smart buildings sector with a view to attaining enhanced sensing, energy efficiency, and optimal building management.

In the following, we present a review of the papers that were identified by applying the devised methodology, identifying on the basis of summarization tables and their analysis the machine learning models that are most suitable for integration with sensor devices in the smart buildings sector.

3. Enhanced Sensing by Integrating Machine Learning Models with Sensor Devices in the Smart Buildings Sector

In the following, we conducted a review of the most recent scientific articles, on the basis of the devised research methodology. For each of the identified supervised or unsupervised machine learning models, we summarize, according to the search criteria and methodology, the papers addressing those respective models. A selection of the most recent papers (sorted in descending order of publication year) is presented in the following sections, while comprehensive summarization tables are presented in the Supplementary Materials (Tables S1–S16).

3.1. Supervised Learning

3.1.1. Classification

Based on the devised methodology, we selected and summarized scientific papers that implement the Support Vector Machines (SVM) method integrated with sensor devices in smart buildings. A summary of 25 articles from the scientific papers pool that address Support Vector Machine approaches integrated with sensor devices in smart buildings can be found in Table S1 in the Supplementary Materials file, while a selection of five of the most recent papers is presented in Table 1.

Table 1. Five of the most recent scientific articles addressing the Support Vector Machines method integrated with sensor devices in smart buildings.

Reference	Publication Year	Type of Smart Building	Type of Sensors	Reason for Using the SVM Method with Sensor Devices	SVM Only or Hybrid	Performance Metrics
[1]	2019	smart building	indoor environment sensors: thermocouple TX-FF-0.32-1P (FUKUDEN) for the temperature; photosensor HD2021T AA-SP (Deltaohm) for the illuminance; OPUS20 TCO (Lufft) sensor for the relative humidity and CO_2 concentration; occupancy information sensor: PN1500 (Botem); electricity meters: PR300 (Yokogawa) for the lighting power; Enertalk Plug (Encored Technologies) for the PC electricity consumption and EHP electricity meter	assessing occupancy status information in order to improve the energy prediction performance of a building energy model	Support Vector Machine compared with Decision Tree and Artificial Neural Networks	Overall Accuracy and Standard Deviation
[6]	2018	smart home	motion sensors, item sensors (kitchen items), door sensor, temperature sensor, electricity usage, burner, cold water, hot water sensors	human activity recognition in order to help disabled persons	Support Vector Machines with a polynomial kernel of degree 3 (P-SVM); a comparison with other four classifiers: Radial Basis Function kernel—Support Vector Machine (RBF-SVM), Naïve-Bayes, Logistic Recognition, Recurrent Neural Network (RNN)	True Positives, False Positives, Precision, Recall, the F-Measure, the Receiver-Operating-Characteristic (ROC) Curve
[7]	2018	smart home	smart phones' built-in three-axis acceleration sensors and Kinect motion sensors	human fall detection	Support Vector Machine (SVM)	True Positive (TP), True Negative (TN), False Positive (FP), False Negative (FN), Sensitivity or True Positive Rate (TPR), Specificity (SPC) or True Negative Rate (TNR), Accuracy (ACC)
[8]	2018	smart home	passive radar-based sensor to achieve multiple level activities detection by adjusting Doppler resolution	human activity recognition and classification	Support Vector Machine (SVM) in order to classify the feature vectors into corresponding activity groups	Confusion Matrices, Classification Accuracy
[2]	2018	smart building	thermal sensor	human behavior recognition	Support Vector Regression (SVR) and Recurrent Neural Network (RNN)	Average Error, Error Rate

Examining the 25 papers selected and summarized in Table S1, presented in the Supplementary Materials file, it can be observed that 32% of them take into consideration smart buildings in general (including smart care houses, smart hospitals, smart offices), while the remaining percentage of scientific papers refer solely to smart homes. With respect to the publication year, 60% of the identified articles were published during the last 5 years.

In their research, the authors of these papers implement various types of sensors, according to their purposes, namely: indoor sensors [1], occupancy information sensors [1], electricity meters [1,6,44], motion sensors [6,7,30,59,60], item kitchen sensors [6], door sensors [6,59,61,62], temperature sensors [1,2,6,59,63], photosensors [1,3,63], status of water and burner sensors [6,59], acceleration sensors [4,7], Kinect motion sensors [7], modern smartphone sensors [4,7,60], passive radar-based sensors [8], unobtrusive sensors [9,14], infrared sensors [15,30], wireless sensor networks [61,62], accelerometers [5,63], altimeters [63], gyroscopes [63], barometers [63], heart rate monitor [63], embedded sensors [4,10,32,60,63], binary sensors [29,31,59,61], sensors installed in everyday objects [62], ubiquitous sensors [29], building management systems [44], weather stations [44], video systems [52], multi-appliance recognition systems [64], sensors for the Heating, Ventilation, and Air Conditioning (HVAC) technology [65].

With respect to the reasons for using the SVM method with sensor equipment in smart buildings, it can be observed that the recognition of human activity is at the forefront, as this is addressed in most of the papers [3,4,6,8–10,14,15,29–32,59,60,62,63]. Assisted living was a strong motivation for using the SVM method with sensor devices in the smart buildings sector; seven of the identified papers focusing on the recognition of human activity did so in order to provide appropriate assisted living [6,14,15,30–32,63], while other papers aimed to achieve assisted living by focusing on human fall detection [7], human behavior recognition [2], assessment of occupancy status information, and identification of human behavior [61]. Other reasons for applying SVM with sensors in smart buildings include measuring the occupancy status of a building's inhabitants in order to improve the energy prediction performance of the building's energy model [1], classifying the gender of occupants [5], forecasting electricity consumption [44], detecting and classifying human behavior with a view to maximizing comfort with optimized energy consumption [52], recognizing household appliances in order to assess their usage and develop habits of power preservation [64], and selecting optimal sensors for use in complex system monitoring problems such as HVAC chillers [65].

With respect to the devised methods, in [1], the authors made use of the Support Vector Machine technique and compared the obtained results with those obtained using Decision Tree and Artificial Neural Networks. In [6], the Support Vector Machine approach was implemented with a polynomial kernel of degree 3 (P-SVM), and afterwards, a comparison was conducted with other four classifiers: Radial Basis Function kernel–Support Vector Machine (RBF-SVM), Naïve Bayes, logistic recognition, and Recurrent Neural Network (RNN). The authors of [7,8,32,52,59,60] developed their research based solely on the Support Vector Machine technique. In [2], the Support Vector Regression (SVR) and Recurrent Neural Network (RNN) approaches were used. In [9] the Support Vector Machine technique was implemented for classification purposes, along with two different feature extraction methods: a manually defined method, and a Convolutional Neural Network (CNN). The authors of [3] implemented the Support Vector Machine (SVM), Convolutional Neural Network-Hidden Markov Model (CNN-HMM) and Long Short-Term Memory networks (LSTM) learning algorithms. In [10], the authors developed a hybrid approach combining the Beta Process Hidden Markov Model (BP-HMM) and the Support Vector Machine (SVM). In [4], the authors developed a Coordinate Transformation and Principal Component Analysis (CT-PCA) scheme and compared the results obtained using the K-Nearest Neighbor (KNN), Decision Tree (DT), Artificial Neural Network (ANN), Support Vector Machine (SVM) techniques. The authors of [14] used a hybrid approach, combining the Neural Network, C4.5 Decision Tree, Bayesian Network and Support Vector Machine techniques. Also based on a hybrid approach, the authors of [15] made use of SVM, Linear Kernel, Multinomial Kernel, and Radial Basis Function (RBF) kernel, and compared their results with those obtained using the K-Nearest Neighbor,

Gaussian Mixture Hidden Markov Model (GM-HMM), and Naïve Bayes approaches. The hybrid approach developed in [61] combines resampling methods such as oversampling and undersampling with Support Vector Machines and Linear Discriminant Analysis (LDA). In [5], the authors combined Bagged Decision Tree, Boosted Decision Tree, Support Vector Machines (SVMs) and Neural Networks in order to carry out gender classification. In [30], the authors used a series of learning classification algorithms, namely Naïve Bayesian (NB), Support Vector Machine (SVM), and Random Forest (RF). The authors of [63] developed their research using the Multilayer Perceptron Neural Network (MLP), Radial Basis Function Neural Network (RBF), and Support Vector Machine (SVM) techniques. In [31], the authors made use of the Support Vector Machine (SVM), Evidence-Theoretic K-Nearest Neighbor (ET-KNN), Probabilistic Neural Network (PNN), K-Nearest Neighbor (KNN), and Naïve Bayes (NB) techniques. The authors of [62] conducted their research using various methods of feature extraction, including Principal Component Analysis (PCA), Independent Component Analysis (ICA), and Linear Discriminant Analysis (LDA); afterwards, the new features selected by each method were used as inputs for a Weighted Support Vector Machines (WSVM) classifier. In [29], a hybrid method was developed by combining the Synthetic Minority Oversampling Technique (SMOTE) with Cost-Sensitive Support Vector Machines (CS-SVM). The authors of [44] developed a model based on Support Vector Regression (SVR). In [64], the authors developed a hybrid method by combining the Support Vector Machine with the Gaussian Mixture Model (SVM/GMM) classification model with a view to classifying electric appliances. In [65], the authors compared the Support Vector Machines (SVMs), Principal Component Analysis (PCA), and Partial Least Squares (PLS) methods.

The performance metrics considered in the scientific papers that used Support Vector Machines integrated with sensor devices in smart buildings include: Accuracy [1,3,5,7–10,29,31,59,61,62,64]; Standard Deviation [1,63]; True Positive Rate [6,7,59]; False Positive Rate [6,7,59]; Precision [6,29,30,59,61,62]; Recall [6,29,59,61,62]; F-measure [6,29,30,59,61,62]; True Negative rate [7,59]; False Negative Rate [7,59]; Sensitivity [7,14,30]; Specificity [7,14,30,59]; Recognition Rate [10,60,64]; Receiver-Operating-Characteristic (ROC) Curve [6,14]; Confusion Matrix [8,15]; Average Error and Error Rate [2]; Root Mean Square Error (RMSE) [9] and Mean Squared Error (MSE) [3]; Classification Rate [15,52]; Absolute Mean, Variance, Median Absolute Deviation, Maximum, Minimum, Signal Magnitude range, Power, Interquartile range for computing the time and the Maximum, Mean, Skewness, Kurtosis, and Power of the frequency [4]; Matthews Correlation Coefficient [59]; Similarity Degree [32]; Mean, Standard Deviation (STD), Maximum, Minimum, Median, Mode, Kurtosis, Skewness, Intensity, Difference, Root-Mean-Square (RMS), Energy, Entropy, and Key Coefficient [63]; Coefficient of Variation (CV) and Standard Error [44]; and Success Rate [64].

With regard to the five most recent scientific articles making use of Support Vector Machines with sensor devices in smart buildings (Table 1), it can be seen that in [1], Kim et al. aimed to enhance the accuracy of energy forecasting for buildings that were not under construction, by means of assessing occupancy status information using a machine learning approach consisting of applying Support Vector Machines, Decision Tree and Artificial Neural Networks to process the data recorded by different types of sensors. The authors gathered the necessary data using indoor environmental sensors like the thermocouple TX-FF-0.32-1P manufactured by Fukuden with a view to measuring the temperature, a Deltaohm HD2021T AA-SP photosensor for measuring the illuminance level, a Lufft OPUS20 TCO sensor for measuring the relative humidity and CO_2 concentration, a PN1500 occupancy status sensor built by Botem, a Yokogawa PR300 electricity meter along with an Enertalk Plug produced by Encored Technologies for measuring the electricity consumption of the Personal Computer (PC), and an Electric Heat Pump (EHP). After carrying out the training and validation processes, the authors noticed that all of the tested machine learning algorithms provided their best results during the summer and their worst results during the spring, whereas the Support Vector Machine approach provided an increased level of accuracy compared with the other two approaches. In light of the promise of the obtained results, the authors aimed to extend their research by addressing open office spaces, which are frequently encountered in office buildings, overcoming the limitation of using only a single private office.

In [6], Machot et al. proposed a method making use of Support Vector Machines with a Polynomial Kernel of Degree 3 (P-SVM) for the recognition of human activity in order to help persons with disabilities in smart homes. The authors put forward a windowing technique relying on data recorded by different types of sensors used for motion, kitchen items, doors, temperature measurements, electricity metering, burner state determination, and cold and hot water usage. In addition to the data recorded from smart homes, available from the Center for Advanced Studies in Adaptive Systems (CASAS) dataset, Machot et al. performed experimental tests on data simulated by the Human Behavior Monitoring and Support (HBMS) software tool, identifying a set of temporal and spatial characteristics that were then used in order to compute, assess and build a conclusive feature vector. The authors compared their proposed method with the Radial Basis Function kernel–Support Vector Machine (RBF-SVM), Naïve Bayes, Logistic Recognition, and Recurrent Neural Network (RNN) approaches, obtaining improved results, as highlighted by the applied performance metrics, which included True Positives, False Positives, Precision, Recall, F-Measure, and the Receiver-Operating-Characteristic Curve.

Acknowledging the importance of accurate human fall detection and the numerous challenges arising due to the plethora of possible activities carried out by a person within a residential environment, in [7], Li et al. propounded a collaborative platform for detecting human falls. The platform comprises two sub-systems: one that uses a smart phone's built-in three-axis acceleration sensors and another that processes, using an SVM approach, the recorded data from a Kinect's motion sensors. The developed platform identifies a fall by combining the data provided by the two sub-systems based on two approaches: a logical rules process and a Dempster–Shafer theory-based method. In terms of performance, Li et al. computed and analyzed the True Positive (TP), True Negative (TN), False Positive (FP), False Negative (FN), Sensitivity/True Positive Rate (TPR), Specificity (SPC)/True Negative Rate (TNR) and Accuracy (ACC) metrics, concluding that the proposed approach was promising when taking into account the rapid development, diversification and integration of sensors.

In [8], Li et al. proposed a passive radar-based human activity recognition and classification method that was able to distinguish the particular body movements, physical activity patterns, and respiration of a person. A wireless energy transmitter device, such as a WiFi access point, was used to provide the signals necessary to identify the residents' activity in the smart home. The method devised by the authors comprises two stages: the Doppler data is obtained and subsequently processed by means of SVM classification in order to recognize human physical activity, while in order to detect the respiration process, a micro Doppler extraction is performed upon a Doppler spectrogram followed by the application of a Savitzky–Golay noise removal filter. The analysis of the performance metrics, which included Confusion Matrices and Classification Accuracy, confirmed that the proposed method offered satisfactory performance levels for the two analyzed situations, namely, physical activity recognition and breathing detection. The authors concluded by stating that the obtained results were promising in the healthcare field, with one advantage being the fact that no wearables or intrusive sensors were needed, meaning that the proposed system could therefore prove useful when the monitoring is being carried out over longer periods of time. The authors remarked that the developed system targets single user scenarios, and that implementing it in real-world working environments would necessitate the development of enhanced methods for separating multiple signals and behavior patterns.

Simulated sensor data related to temperature and heat were used by Zhao et al. in [2] with the aim of recognizing human behavior in smart buildings. Using the EnergyPlus software, the authors simulated different time-series of building-related data samples on which they subsequently applied two methods, one based on Support Vector Regression (SVR) and the other based on Recurrent Neural Networks (RNNs). The results obtained after conducting the experimental tests indicated that the two approaches provided similar levels of performance, as shown by the registered performance metrics, namely the Average Error and the Error Rate. This study confirmed that the Support Vector Regression approach was more flexible, and made it possible to add or remove features from the model without significantly affecting the model's accuracy; meanwhile, the Recurrent Neural Network approach provides a higher level of accuracy when the model's features do not change much over the course of time.

Then, from the obtained pool of scientific articles resulting from applying the devised review methodology, we identified, analyzed and summarized those that make use of the Discriminant Analysis technique integrated with sensor devices in smart buildings for classification purposes. A complete summarization table (Table S2) is provided in the Supplementary Materials file, while Table 2 presents five of the most recent papers that address this subject.

Analyzing the papers in Table S2 in the Supplementary Materials file, it can be observed that 83% of them refer to smart homes, while the remainder deal with any type of smart buildings (like smart offices, smart hospitals, smart foster care houses, smart retirement homes).

In these papers, the authors make use of a variety of different types of sensors. In [17], Brennan et al. considered a scalable wireless sensor network with CO_2-based estimation. In [61], Abidine et al. used a wireless sensor network comprising binary sensors like reed switches to determine the open-closed state of the doors and cabinets, pressure mats to determine whether the subject was lying down in the bed or on the couch, and float sensors to determine whether the toilet had been flushed. In [62], Abidine et al. analyzed sensor networks in a pervasive environment, with sensors installed in everyday objects such as doors, cupboards, the refrigerator, and the toilet flush to record activation/deactivation events (opening/closing events). Liao et al. based their study in [66] on sensors for motion detection. In [16], Tian et al. used a wearable accelerometer, which provided inertial information of human activity. In [33], Alam et al. considered four kinds of biosensors: Electro-Dermal Activity sensors (EDA), Electrocardiogram sensors (ECG), Blood Volume Pulse sensors (BVP) and surface Electromyography sensors (EMG).

In the identified papers, the reasons for using the Discriminant Analysis method with sensor devices in smart buildings were equally distributed between human activity recognition/classification [16,17,62] and the detection of human behavior in the context of assisted living [33,61,66].

With respect to the devised methods, in [16], the authors used the Kernel Fisher Discriminant Analysis (KFDA) technique and the Extreme Learning Machine (ELM) and performed a comparison between Best Base ELM, SVM, Bagging, AdaBoost and the proposed method. In [17], the authors compared Gradient Boosting, K-Nearest Neighbor (KNN), Linear Discriminant Analysis, and Random Forest. In [61], the authors used a hybrid method, combining resampling methods like Oversampling and Undersampling with Support Vector Machines and Linear Discriminant Analysis (LDA). The authors of [66] implemented the Discriminant Analysis technique. In [33], the authors implemented a Hidden Markov Model (HMM), Viterbi path counting, and a scalable Stochastic Variational Inference (SVI)-based training algorithm, along with Generalized Discriminant Analysis. In [62], the authors made use of various methods of feature extraction (Principal Component Analysis (PCA), Independent Component Analysis (ICA), and Linear Discriminant Analysis (LDA)) and the new features selected by each method were subsequently used as the inputs for a Weighted Support Vector Machines (WSVM) classifier.

The performance metrics considered in the scientific papers that use the Discriminant Analysis technique integrated with sensor devices in Smart Buildings include: Accuracy [16,17,33,61,62,66]; Precision [61,62]; Recall [16,61,62] and F-measure [33,61,62]; Root-Mean-Square Error (RMSE) [17]; Coefficient of Variance (CV) [17]; Normalized Root-Mean-Square Error (NRMSE) [17]; Coefficient of Variation of the RMSD (CV) [17]; Sensitivity (Sen) [33], Specificity (Spe) [33]; and Area Under the Receiver Operating Characteristic (ROC) Curve (AUC) [33].

Regarding five of the most recent scientific articles that make use of the Discriminant Analysis technique with sensor devices in smart buildings (Table 2), it can be observed that in [16], Tian et al. put forward a method for human activity recognition in a smart home. The proposed approach makes use of a wearable tri-axial accelerometer that provides inertial data related to the resident's activity. The collected data from the sensors are further processed using the Kernel Fisher Discriminant Analysis (KFDA) technique in order to refine and improve the feature vectors that were to be used in the subsequent processing step, which consisted of applying the Extreme Learning Machine classifier trained using the bootstrap method. After comparing the proposed method with the Best Base ELM, SVM, Bagging and AdaBoost approaches, the authors stated that their obtained results were superior, as confirmed by the Accuracy and Recall performance metrics.

Human activity recognition in smart buildings was also addressed in another recent paper [17], in which Brennan et al. studied the performance of several machine learning models, namely, Linear Discriminant Analysis, Gradient Boosting, K-Nearest Neighbor and Random Forest, with data gathered from a scalable wireless sensor network with CO_2-based estimation, with a view to accurately recognizing human activity without having to make use of expensive and privacy intrusive equipment such as computer vision and smart video cameras. In order to compare the results obtained using each of the models, the authors computed performance metrics which included Accuracy, Root-Mean-Square Error (RMSE), Normalized Root-Mean-Square Error (NRMSE) and Coefficient of Variance (CV), thereby concluding that all of the models were able to provide increased levels of performance when the training dataset comprised information regarding the sensor data in terms of structure and magnitude.

In [61], Abidine et al. aimed to assess the occupancy status information and detect human behavior within a smart home with a view to providing assisted living health care. The authors recorded the data using a wireless sensor network comprising binary sensors like reed switches to determine the open-closed state of the doors and cabinets, pressure mats to determine whether someone was lying down in the bed or on the couch, and float sensors to identify whether the toilet had been flushed. The collected data were processed using a hybrid approach, obtained by combining resampling methods like Oversampling and Undersampling with Linear Discriminant Analysis (LDA) and Support Vector Machines (SVM). The authors compared the obtained results in terms of accuracy, precision, recall and F-measure with other methods from the scientific literature that rely on the Hidden Markov Model (HMM) and the Conditional Random Field (CRF) statistical modeling technique, concluding that Oversampling with Linear Discriminant Analysis offers the best performance level.

Another scientific work that uses the Discriminant Analysis technique with sensing equipment in a smart home is that of Liao et al. [66], in which the authors aimed to overcome the limitations of existing human fall detection methods in terms of both accuracy detection and privacy intrusion issues. To this end, the authors collected data using motion detection sensors and made use of the Discriminant Analysis method to extract certain features corresponding to a resident's behavior, and to build an associated feature vector, which was then compared with features representing the state of having fallen down. After performing the experimental tests with respect to the robustness of the proposed approach, the authors stated that the results obtained confirmed the performance of the devised method.

Acknowledging the numerous benefits that assisted living brings to a patient's health and wellbeing, in [33], Alam et al. proposed a framework for Ambient Assisted Living (AAL) with a view to predicting emergencies concerning the psychiatric states of patients in a smart home environment. In order to record the different symptoms of psychiatric patients, the authors made use of four types of biosensors, namely Electro-Dermal Activity (EDA) sensors, Electrocardiogram (ECG) sensors, Blood Volume Pulse (BVP) sensors, and surface Electromyography (EMG) sensors. The recorded data were processed using a method that made use of several machine learning techniques, specifically the Hidden Markov Model (HMM) for modeling the psychiatric states, the Viterbi algorithm and the Stochastic Variational Inference (SVI) scalable algorithm for approximating the model's parameters, and Generalized Discriminant Analysis (GDA) in order to focus better on the characteristics belonging to the same psychiatric state class. After conducting an experimental study and analyzing the results in terms of prediction Accuracy (Acc), Sensitivity (Sen), Specificity (Spe), F-Measure (FM) and Area Under the ROC Curve (AUC), the authors concluded that their proposed approach was able to supplement existing psychiatric care in residential spaces.

Subsequently, taking into consideration the devised methodology, we identified and summarized scientific papers that implemented the Naïve Bayes method integrated with sensor devices in smart buildings. The research articles that address Naïve Bayes approaches integrated with sensor devices in smart buildings are summarized in Table S3 in the Supplementary Materials file, while a selection of five of the most recent papers is presented in Table 3.

Table 2. Five of the most recent scientific articles addressing Discriminant Analysis integrated with sensor devices in smart buildings.

Reference	Publication Year	Type of Smart Building	Types of Sensor	Reason for Using the Discriminant Analysis Method with Sensor Devices	Discriminant Analysis Only or Hybrid	Performance Metrics
[16]	2019	smart home	wearable sensor, accelerometer providing inertial information of human activity	human activity recognition	Kernel Fisher Discriminant Analysis (KFDA) technique, Extreme Learning Machine (ELM); comparison among Best Base ELM, SVM, Bagging, AdaBoost and the proposed method	Accuracy, Recall
[17]	2018	smart buildings	a scalable wireless sensor network with CO_2-based estimation	human activity recognition	comparison of Gradient Boosting, K-Nearest Neighbor (KNN), Linear Discriminant Analysis, and Random Forest	Accuracy, Root-Mean-Square Error (RMSE), Normalized Root-Mean-Square Error (NRMSE), Coefficient of Variance (CV)
[61]	2016	smart home	wireless sensor network comprising binary sensors like reed switches to determine the open-closed state of the doors and cabinets; pressure mats to determine if one is staying laid down in the bed or on the couch; float sensors to determine if the toilet has been flushed	assessing the occupancy status information and detecting the human behavior with a view to assisted living	hybrid, combining resampling methods like Oversampling and Undersampling with Support Vector Machines and Linear Discriminant Analysis (LDA)	Accuracy, Precision, Recall and F-measure
[66]	2016	smart home	sensors for motion detection	human fall detection	Discriminant Analysis	Accuracy
[33]	2016	smart home	four kinds of biosensors: Electro-Dermal Activity sensor (EDA), Electrocardiogram sensor (ECG), Blood Volume Pulse sensor (BVP) and surface Electromyography sensor (EMG)	ambient assisted living framework for emergency psychiatric state prediction	Hidden Markov Model (HMM), Viterbi path counting, scalable Stochastic Variational Inference (SVI)-based training algorithm Generalized Discriminant Analysis	Prediction Accuracy (Acc), Sensitivity (Sen), Specificity (Spe), F-Measure (FM) and Area Under the ROC Curve (AUC)

Table 3. Five of the most recent scientific articles addressing the Naïve Bayes integrated with sensor devices in smart buildings.

Reference	Publication Year	Type of Smart Building	Type of Sensors	Reason for Using the Naïve Bayes Method with Sensor Devices	Bayes Only or Hybrid	Performance Metrics
[11]	2019	smart hospital	biomedical sensors, providing medical data (based on physiological signals), behavioral patterns (e.g., smoking, drinking alcoholics, taking medications, etc.), ambient data (e.g., humidity, temperature, noise, etc.), contextual information (e.g., location, activity, etc.)	achieving remote monitoring of patients outside the hospital in real time	a hybrid algorithm of Naïve Bayes (NB) and Whale Optimization Algorithm (WOA); a comparison between six classifiers: Decision tree (J48), Random Forest (RF), Ripper (JRip), Naïve Bayes (NB), Nearest Neighbor (IBK), Support Vector Machine (SVM)	Accuracy, Recall, Precision, F-Measure
[67]	2018	smart home	acoustic sensor network	accurate knowledge of the positions of surrounding objects useful for autonomous systems and smart devices	Bayesian filter	Mean Value and Standard Deviation
[68]	2018	smart home	carbon dioxide, total volatile organic compounds, air temperature, and air relative humidity sensors	occupancy detection in smart homes	comparison of the supervised learning models: Naïve Bayes (NB), C4.5 Decision Tree, Logistic Regression, K-Nearest Neighbor, Random Forest	For occupancy: Accuracy, True Positive Rate, True Negative Rate; For the number of occupants: Mean Absolute Error, Root Mean Square Error
[36]	2018	smart home	WiFi-enabled sensors for food nutrition quantification, and a smart phone application that collects nutritional facts of the food ingredients	Internet of Things (IoT)-based fully automated nutrition monitoring system	Bayesian algorithms and 5-layer Perceptron Neural Network method for diet monitoring	Accuracy of classification of food items and meal prediction
[34]	2016	smart home	Passive Infrared Sensor (PIR) and environmental sensors to measure pressure, temperature, humidity, and the light intensity in a particular area of the home	human presence identification and location with sub room accuracy in the context of home-based assisted living	Bayes filter algorithm	Error Rate

Analyzing the papers in Table S3, it can be observed that, according to the authors of these papers, all of the studies focused on smart homes. The authors of these scientific articles made use in their analyses of different types of sensors, including: biomedical sensors [11]; ambient data sensors [11,34,68]; acoustic sensor networks [67]; WiFi-enabled sensors [36]; Passive Infrared (PIR) sensors [30,34]; binary sensors [31,69]; and motion sensors [30,70].

With respect to the reasons for using the Naïve Bayes method with sensor equipment in smart buildings, one can observe that the recognition of human activity was the main subject of the identified papers summarized in Table S3, being addressed in papers [11,30,31,34,68–70]. Meanwhile, several of the above-mentioned scientific papers that use the Naïve Bayes integrated with sensor devices in Smart Buildings also addressed issues regarding assisted living [11,30,31,34,36]. Other reasons for applying the Naïve Bayes method with sensors in smart buildings include obtaining accurate information regarding the positions of surrounding objects, an aspect especially useful for autonomous systems and smart devices [67] or in developing an Internet of Things (IoT)-based fully automated nutrition monitoring system [36].

With respect to the devised methods, in [11], the authors made use of a hybrid approach based on the Naïve Bayes (NB) Algorithm and the Whale Optimization Algorithm (WOA), subsequently presenting a comparison among six classifiers: Decision tree (J48), Random Forest (RF), Ripper (JRip), Naïve Bayes (NB), Nearest Neighbor (IBK), Support Vector Machine (SVM). In [67], the authors implemented the Bayesian filter in order to estimate the trajectories of source positions using an acoustic sensor network. In [68], a comparison of the supervised learning models was presented: Naïve Bayes (NB), C4.5 Decision Tree, Logistic Regression, K-Nearest Neighbor, and Random Forest were used in order to detect and estimate occupancy in smart homes. In [36], the authors developed a hybrid approach by combining Bayesian algorithms and a 5-layer Perceptron Neural Network method for diet monitoring purposes; the authors of [34] used the Bayes filter algorithm to locate people. In [30], the authors made use of learning classification algorithms, including Naïve Bayes (NB), Support Vector Machine (SVM) and Random Forest (RF). The authors of [31] made use of the Naïve Bayes (NB), Support Vector Machine (SVM), Evidence-Theoretic K-Nearest Neighbor (ET-KNN), Probabilistic Neural Network (PNN), and K-Nearest Neighbor (KNN) methods. In [69], the Dempster–Shafer theory was implemented, and was subsequently compared with the Naïve Bayes classifier and J48 Decision Tree. In [70], the authors applied a hybrid approach based on the Naïve Bayes classifier, Hidden Markov Model and Viterbi algorithm.

The performance metrics that were chosen by the authors of the scientific papers that use the Naïve Bayes method integrated with sensor devices in smart buildings include: Accuracy [11,31,34,36,68,70]; Precision [11,30,69]; Recall [11,69]; F–measure [11,30,69]; Mean Value and Standard Deviation [67]; Accuracy, True Positive Rate, True Negative Rate with a view to assessing the performance in detecting the occupancy, along with the Mean Absolute Error and the Root Mean Square Error, for establishing the number of occupants [68]; and Error Rate [34].

Regarding five of the most recent scientific articles that make use of the Naïve Bayes machine learning classifiers with sensor devices in smart buildings (Table 3), it can be observed that in [11], Hassan et al. proposed a hybrid approach, consisting of a hybrid algorithm combining Naïve Bayes (NB) and Whale Optimization Algorithm (WOA) in order to achieve real-time remote monitoring in a smart hospital of patients affected by chronic illnesses who reside outside of a hospital, thereby increasing the number and quality of monitored patients while reducing the associated hospitalization costs. The datasets were recorded by means of biomedical sensors for acquiring medical data based on physiological signals, behavioral patterns (e.g., smoking, drinking alcoholic beverages, taking medications), ambient data (e.g., humidity, temperature, noise), and contextual information (e.g., location, activity). After comparing the obtained results of their proposed hybrid approach with those recorded by using six machine learning classifiers, namely, Decision tree (J48), Random Forest (RF), Ripper (JRip), Naïve Bayes (NB), Nearest Neighbor (IBK) and Support Vector Machine (SVM), the authors concluded that the performance metrics Accuracy, Recall, Precision and F-Measure confirmed the superiority of their proposed approach.

With a view to acquiring accurate knowledge of the positions of surrounding objects in a smart home, an aspect that is useful for both autonomous systems and smart devices, in [67], Evers et al. used a Bayesian filter in order to approximate the position trajectories of sources by acquiring data using a network of acoustic sensors. The authors aimed to overcome the challenges implied by approximating the direction of arrival for the source positions, directions that become more difficult to approximate due to the sound field becoming more diffuse as the distance from the sensor increases, causing an increase in reverberations and noises. The authors proposed using a coherent to diffuse ratio to measure the reliability of a direction of arrival in the case of localizing a single source, and showed that it is possible to triangulate the positions of a source by probabilistic means, taking advantage of the spatial diversity of network nodes.

In [68], Zimmerman et al. made use of environmental sensors that record data related to carbon dioxide, total volatile organic compounds, air temperature, and relative air humidity in order to determine the occupancy level within smart homes. The datasets retrieved from sensors were categorized using a correlation method, and the authors subsequently compared several supervised learning models: Naïve Bayes (NB), C4.5 Decision Tree, Logistic Regression, K-Nearest Neighbor, and Random Forest. These were used to detect and estimate the occupancy level. On the basis of the Accuracy, True Positive Rate and True Negative Rate for assessing the occupancy, along with the Mean Absolute Error and Root Mean Square Error for evaluating the number of occupants, the authors evaluated the performance of various classifiers (ZeroR, JRip, Naïve Bayes, J48, Logistic, K-Nearest Neighbor, Random Forest), concluding that the best performance metrics were registered when using the NB machine learning technique.

Taking into account how important the correct nutritional intake is for people, especially for infants, in [36], Sundaravadivel et al. put forward an automated nutrition monitoring system based on the Internet of Things (IoT) concept, aiming to achieve smart nutritional healthcare in smart homes. The authors' proposed system comprises WiFi-enabled sensors for food nutrition quantification, a smart phone application that collects nutritional facts regarding food ingredients, a five-layer perceptron ANN, and an algorithm based on a Bayesian Artificial Neural Network for predicting and monitoring meals. After performing the experimental tests, the authors concluded, on the basis of the Accuracy for the classification of food items and meal prediction, that their proposed system was a reliable tool for monitoring one's diet, having the potential to become an indispensable tool for childcare and for household residents.

In order to accurately identify human presence and to locate residents with sub-room accuracy in a smart home for assisted living purposes, in [34], Ballardini et al. proposed a probabilistic method that relied on the Bayes filter algorithm. In order to collect the necessary data, the authors made use of a Passive Infrared Sensor (PIR) and environmental sensors to measure pressure, temperature, humidity, and light intensity in a particular area of the home. After having analyzed the obtained results and the obtained Error Rate, the authors concluded that their developed system provided a high level of performance, with its only limitation being the fact that the system was only suitable for situations in which the smart home is inhabited by only a single resident.

Afterwards, using the devised methodology, we selected and summarized scientific papers that implement the Nearest Neighbor method integrated with sensor devices in smart buildings. A summary of the papers that address the Nearest Neighbor approaches integrated with sensor devices in smart buildings is presented in Table S4 in the Supplementary Materials file, while a selection containing five of the most recent papers is presented in Table 4.

Table 4. Five of the most recent scientific articles addressing the Nearest Neighbor method integrated with sensor devices in smart buildings.

Reference	Publication Year	Type of Smart Building	Type of Sensors	Reason for Using the Nearest Neighbor Method with Sensor Devices	Nearest Neighbor Only or Hybrid	Performance Metrics
[17]	2018	smart buildings	a scalable wireless sensor network with CO_2-based estimation	human activity recognition	comparison of Gradient Boosting, K-Nearest Neighbor (KNN), Linear Discriminant Analysis, and Random Forest	Accuracy, Root-Mean-Square Error (RMSE), Normalized Root-Mean-Square Error (NRMSE), Coefficient of Variance (CV)
[68]	2018	smart home	carbon dioxide, total volatile organic compounds, air temperature, and air relative humidity sensors	occupancy detection in smart homes	comparison of the supervised learning models: Naïve Bayes (NB), C4.5 Decision Tree, Logistic Regression, K-Nearest Neighbor, Random Forest	For occupancy: Accuracy, True Positive Rate, True Negative Rate; For the number of occupants: Mean Absolute Error, Root Mean Square Error
[71]	2018	smart home	a single point Electromagnetic Interference (EMI) smart sensor	detect and track the operation of the information technology (IT) appliances (such as desktops and printers), operating in non-working hours in office buildings	Nearest Neighbor only	Precision and Recall
[72]	2015	smart home	an accelerometer in order to indicate a potential fall and the Kinect sensor in order to authenticate the eventual fall alert	human activity recognition and fall detection	the k-Nearest Neighbor (k-NN) classifier and comparison with the results obtained using linear SVM	Sensitivity, Specificity, Precision, Classification Accuracy
[31]	2015	smart home	binary sensors	human activity recognition and classification in home-based assisted living	Support Vector Machine (SVM), Evidence-Theoretic K-Nearest Neighbor (ET-KNN), Probabilistic Neural Network (PNN), K-Nearest Neighbor (KNN), Naïve Bayes (NB)	The Classification Accuracy (the Classification Error Results)

80% of the scientific papers selected and summarized in Table S4, presented in the Supplementary Materials file, present research exclusively focused on smart homes, while the remaining 20% take into consideration smart buildings in general. In these papers, the authors make use of different types of sensors. In [17], a scalable wireless sensor network with CO_2-based estimation was used. In [68], carbon dioxide, total volatile organic compounds, air temperature, and air relative humidity sensors were employed. In [71], a single-point Electromagnetic Interference (EMI) smart sensor was used. In [72], an accelerometer was used. In [31], binary sensors were used.

In these papers, the reasons for using the Nearest Neighbor integrated with sensor devices in smart buildings were mainly related to human activity recognition/classification [17,31,68,72], the detection of human behavior in the context of assisted living [31,72], and the detection and tracking of the operation of information technology (IT) appliances (such as desktops and printers) operating during non-working hours in office buildings [71].

With regard to the devised research methods, in [17], Brennan et al. compared the Gradient Boosting, K-Nearest Neighbor (KNN), Linear Discriminant Analysis, and Random Forest methods. In [68], Zimmermann et al. compared a series of supervised learning models, including Naïve Bayes (NB), C4.5 Decision Tree, Logistic Regression, K-Nearest Neighbor, Random Forest. In [71], Gulati et al. developed a Nearest Neighbor-based classification algorithm for the statistical features extracted from histograms of the measured common mode electromagnetic emissions. In [72], Kwolek et al. made use of the K-Nearest Neighbor (K-NN) classifier and compared the results with those obtained using linear SVM. In [31], Fahad et al. used the Support Vector Machine (SVM), Evidence-Theoretic K-Nearest Neighbor (ET-KNN), Probabilistic Neural Network (PNN), K-Nearest Neighbor (KNN) and Naïve Bayes (NB) techniques.

The performance metrics considered in the scientific papers that use the Nearest Neighbor method integrated with sensor devices in smart buildings include: Accuracy [17,68,72]; Root-Mean-Square Error (RMSE), Normalized Root-Mean-Square Error (NRMSE) and Coefficient of Variance (CV) [17]; Precision [71,72]; True Positive Rate, True Negative Rate, Mean Absolute Error, and Root Mean Square Error [68]; Recall [71]; Classification Accuracy [31,72]; and Sensitivity and Specificity [72].

With respect to the five most recent scientific articles addressing the Nearest Neighbor method integrated with sensor devices in smart buildings (Table 4), it can be observed that in [17], Brennan et al. developed a Wireless Sensor Network (WSNs) prototype based on CO_2 measurements in order to estimate the occupancy estimation in a smart building. With a view to improving the developed method, the authors compared the performance provided by four learning models, namely Gradient Boosting, K-Nearest Neighbor (KNN), Linear Discriminant Analysis and Random Forest, using as performance metrics the Accuracy, Root-Mean-Square Error (RMSE), Normalized Root-Mean-Square Error (NRMSE), and Coefficient of Variance (CV), finally concluding that the KNN model had produced the best results.

In [68], Zimmerman et al. made use of environmental sensors (carbon dioxide, total volatile organic compounds, air temperature, and air relative humidity sensors) in order to assess the occupancy detection in smart homes. Data retrieved from sensors were classified using a correlation method, and the authors subsequently compared a few supervised learning models: Naïve Bayes (NB), C4.5 Decision Tree, Logistic Regression, K-Nearest Neighbor, and Random Forest. These were used in order to detect and estimate occupancy. Based on the Accuracy, True Positive Rate and True Negative Rate for assessing the occupancy, along with the Mean Absolute Error and Root Mean Square Error for evaluating the number of occupants, the authors evaluated the performance of different classifiers (ZeroR, JRip, Naïve Bayes, J48, Logistic, k-Nearest Neighbor, Random Forest) and concluded that the best performance metrics were registered when using the NB technique.

In paper [71], the authors analyzed the case in which a single-point Electromagnetic Interference (EMI) smart sensor is used in order to detect and track the operation of the information technology (IT) devices, operating during non-working hours in office buildings. To this end, Gulati et al. developed a Nearest Neighbor-based classification algorithm for the statistical features extracted from histograms of the measured common mode electromagnetic emissions. Based on the developed experiments, and computing in each case the Precision and Recall performance metrics, the authors concluded that their proposed approach was extremely useful in practice.

In paper [72], Kwolek et al. aimed to improve fall detection using an accelerometer (in order to indicate a potential fall) and a Kinect sensor (in order to authenticate the eventual fall alert) as sensors. The authors used the K-Nearest Neighbor (K-NN) classifier, and subsequently compared the results obtained with those obtained using the linear SVM approach by computing and comparing the Sensitivity, Specificity, Precision, and Classification Accuracy performance metrics. The authors concluded that in the case of their dataset, the K-NN approach outperformed the linear SVM one from a classification performance point of view.

In [31], Fahad et al. made use of binary sensors in order to analyze human activity recognition and classification in home-based assisted living. The authors carried out a comparative analysis by taking into consideration five different learning models, namely the Support Vector Machine (SVM), Evidence-Theoretic K-Nearest Neighbor (ET-KNN), Probabilistic Neural Network (PNN), K-Nearest Neighbor (KNN) and Naïve Bayes (NB) models. Based on Classification Accuracy, the authors noted that the SVM and ET-KNN registered an improved performance when compared to the other three analyzed learning models (PNN, KNN and NB).

Afterwards, of the obtained pool of scientific articles obtained based on the devised review methodology, we identified, analyzed and summarized those that made use of Neural Networks for classification purposes integrated with sensor devices in smart buildings. A complete summarization table (Table S5) is provided in the Supplementary Materials file, while Table 5 presents five of the most recent papers addressing this subject.

Table 5. Five of the most recent scientific articles addressing Neural Networks for classification purposes integrated with sensor devices in smart buildings.

Reference	Publication Year	Type of Smart Building	Type of Sensors	Reason for Using the Neural Networks for Classification Method with Sensor Devices	Neural Networks for Classification Only or Hybrid	Performance Metrics
[18]	2019	smart home	wearable hybrid sensor system comprising motion sensors and cameras	human activity recognition in medical care, smart homes, and security monitoring	hybrid approach, combining Long Short-Term Memory (LSTM) and Convolutional Neural Network (CNN) methods	Confusion Matrices, F1 Accuracy
[27]	2019	smart home	a two-dimensional acoustic array	human activity recognition	Convolutional Neural Networks compared with traditional recognition approaches such as K-Nearest Neighbor and Support Vector Machines	Overall Accuracy
[23]	2019	smart building	Wireless Sensor Network (WSN)	energy consumption forecasting	Multilayer Perceptron (MLP) compared with: Linear Regression (LR), Support Vector Machine (SVM), Gradient Boosting Machine (GBM) and Random Forest (RF)	Coefficient of Determination (R^2), Root Mean Square Error (RMSE), Mean Absolute Error (MAE), Mean Absolute Percentage Error (MAPE)

Table 5. *Cont.*

Reference	Publication Year	Type of Smart Building	Type of Sensors	Reason for Using the Neural Networks for Classification Method with Sensor Devices	Neural Networks for Classification Only or Hybrid	Performance Metrics
[1]	2019	smart building	indoor environment sensors: thermocouple TX-FF-0.32-1P (FUKUDEN) for the temperature; photosensor HD2021T AA-SP (Deltaohm) for the illuminance; OPUS20 TCO (Lufft) sensor for the relative humidity and CO_2 concentration; occupancy information sensor: PN1500 (Botem); electricity meters: PR300 (Yokogawa) for the lighting power; Enertalk Plug (Encored Technologies) for the PC electricity consumption and EHP electricity meter	assessing the occupancy status information in order to improve the energy prediction performance of a building energy model	Support Vector Machine compared with Decision Tree and Artificial Neural Networks	Overall Accuracy and Standard Deviation
[73]	2018	smart home	Environmental sensors: Passive Infrared (PIR) and temperature sensors	human activity recognition	Deep Convolutional Neural Network (DCNN) compared with Naïve Bayes (NB), Back-Propagation (BP) algorithms	Precision, Specificity, Recall, F1 Score, Accuracy, Total Accuracy, Confusion Matrix

Analyzing the papers from the Table S5, it can be observed that 79% of them refer to smart homes, while the remainder take into consideration the more general case of smart buildings. The authors of these scientific articles make use of different types of sensors in their analyses. These include wearable sensors [18,74]; environmental sensors [73,74]; motion sensors [18,75]; a two-dimensional acoustic array [27]; a Wireless Sensor Network (WSN) [23] and sensor networks [76]; temperature sensors [1,63,73,77]; photosensors [1,63]; Passive Infra-Red Sensors (PIR) [73,75]; sensors for humidity and for evaluating the carbon dioxide concentration [1,77]; microphones [77]; cameras [18]; occupancy information sensors [1]; electricity meters [1,75]; accelerometers [5,63]; sensors of IoT devices [38]; an altimeter, a gyroscope and a barometer [63]; sensors mounted on different objects [75]; an unobtrusive sensing module [14]; and binary and ubiquitous sensors [29].

With respect to the reasons for implementing Neural Networks for classification integrated with sensor devices in smart buildings, these are mainly related to the recognition/classification of human activity in the papers [1,5,14,18,23,27,29,63,73–77]. In some of these papers, human activity recognition has as a final purpose the detection and prediction of abnormal behavior [75], monitoring the activities of elderly who are living alone [14,63], classification of the gender of occupants in a building [5], and monitoring the activities of elderly who are living in smart homes care [18,77]. In addition to these purposes, in other papers, the authors target the study of energy consumption forecasting [1,23] or achieving advanced connectivity between devices, systems, and services that continuously record enormous amounts of data from the sensors of IoT devices [38].

With respect to the devised methods, in the paper [18], the authors made use of a hybrid approach, combining Long Short-Term Memory (LSTM) and Convolutional Neural Network (CNN) methods. In [27], the authors implemented Convolutional Neural Networks, comparing them with traditional recognition approaches such as K-Nearest Neighbor and Support Vector Machines. In [23], the authors used the Multilayer Perceptron (MLP) method and compared it with Linear Regression (LR),

Support Vector Machine (SVM), Gradient Boosting Machine (GBM) and Random Forest (RF). The authors of [1] made use of the Support Vector Machine technique and compared it with the Decision Tree and the Artificial Neural Networks techniques. In [73], a Deep Convolutional Neural Network (DCNN) approach was implemented, and this was compared with the Naïve Bayes (NB) and the Back-propagation (BP) algorithms. In [38], the authors made use of a Bayesian Network approach that was subsequently compared with the Decision Tree and Monolithic Bayesian Network methods. In [77], the authors developed an Artificial Neural Network based on the Levenberg–Marquardt algorithm (LMA). In [14], an approach was used combining Neural Network, C4.5 Decision Tree, Bayesian Network and Support Vector Machine techniques. The authors of [5] implemented the Bagged Decision Tree, Boosted Decision Tree, Support Vector Machines (SVMs), and Neural Networks methods in order to classify gender. In [74], Recurrent Neural Networks (RNNs) were used for the activity recognition process. In [63], the authors used the Multilayer Perceptron Neural Network (MLP), Radial Basis Function (RBF) Neural Network and Support Vector Machine (SVM) methods. The authors of [29] used a hybrid method, combining Synthetic Minority Oversampling Technique (SMOTE) with Cost-Sensitive Support Vector Machines (CS-SVM). In [76], the authors developed a Bayesian Belief Network (BBN), which was improved using an Edge-Encode Genetic Algorithm (EEGA) approach and afterwards; they compared the developed approach with the Naïve Bayesian Network (NBN) and Multiclass Naïve Bayes Classifier (MNBC). In [75], the authors made use of the Echo State Network (ESN), Back Propagation Through Time (BPTT) and Real Time Recurrent Learning (RTRL) methods.

The performance metrics considered in the scientific papers that use Neural Networks for classification purposes integrated with sensor devices in smart buildings include: Confusion Matrix [18,38,73]; F1 Score [18,73,74,76]; Accuracy [1,5,18,27,29,38,73,74,76]; Root Mean Square Error (RMSE) [23,75,77]; Precision [29,38,73,74,76]; Recall [29,38,73,74,76]; Standard Deviation (STD) [1,63]; Mean Absolute Percentage Error (MAPE) [23,77]; Mean Squared Error (MSE) [77]; Coefficient of Determination (R^2) and Mean Absolute Error (MAE) [23]; Specificity [14,73]; Sensitivity (SN), Area Under the Receiver Operating Characteristic Curve (AUC) [14]; and Maximum, Minimum, Median, Mode, Kurtosis, Skewness, Intensity, Difference, Root-Mean-Square (RMS), Energy, Entropy and Key Coefficient [63].

With regard to the five most recent scientific articles that make use of neural networks for classification purposes with sensor devices in smart buildings (Table 5), it can be observed that in [18], Yu et al. aimed to enhance human activity recognition in medical care and smart homes and to ensure secure monitoring by means of a hybrid approach, combining the Long Short-Term Memory (LSTM) and Convolutional Neural Network (CNN) methods. The authors recorded the necessary data using a wearable hybrid sensor system comprising motion sensors for identifying and categorizing the different states of the performed activities, along with cameras that recorded photo streams to finalize the human activity recognition within the different groups of identified states. After carrying out the experimental tests and computing the performance metrics, which included Confusion Matrices and F1-Accuracy, the authors concluded that their devised approach had managed to optimally fuse the data from the motion sensors with those from the cameras' photo streams, thereby increasing the performance when compared with a direct fusing approach.

In [27], Guo et al. proposed a method that made use of Convolutional Neural Networks for human activity recognition in smart homes in reliance on the data recorded by a two-dimensional sensor array. The authors aimed to overcome the limitations of traditional methods that make use of ultrasonic sensors with respect to the numerous operations needed for extracting features from a recorded data stream by using a single feature for recognizing human activity. The authors compared their proposed method with traditional recognition approaches such as K-Nearest Neighbor and Support Vector Machines, obtaining improved results, as highlighted by the Overall Accuracy performance metric.

Considering the numerous benefits and the importance attached to accurate electricity consumption forecasting in smart buildings and the numerous prediction methods arising from the literature due to the evolution of wireless sensing devices and IoT equipment, in [23], Chammas et al. proposed a Multilayer Perceptron (MLP) approach for forecasting the electricity consumption in a building. The authors recorded the necessary data using a Wireless Sensor Network (WSN) comprising sensors for measuring temperature, humidity, and ambient light, along with the information regarding the weather and timestamp data. Chammas et al. compared their proposed approach with the Linear Regression (LR), Support Vector Machine (SVM), Gradient Boosting Machine (GBM) and Random Forest (RF) machine learning methods with respect to the Coefficient of Determination (R^2), Root Mean Square Error (RMSE), Mean Absolute Error (MAE) and the Mean Absolute Percentage Error (MAPE) performance metrics, concluding that the developed approach was efficient.

Paper [1] was reviewed previously, when analyzing the most recent scientific articles that integrate Support Vector Machine approaches with sensor devices in smart buildings (Table 1).

Passive Infrared (PIR) and temperature environmental sensors were used by Tan et al. [73] with a view to recognizing and classifying, in an unobtrusive manner, the activity of multiple inhabitants within the same smart home. The authors proposed a method based on analyzing the sensor-acquired Red-Green-Blue (RGB) images by means of a Deep Convolutional Neural Network (DCNN), which was trained and tested using the Cairo open dataset. The results obtained after conducting the experimental tests indicated a higher level of performance than those achieved using the Naïve Bayes (NB) and the Back-Propagation (BP) algorithms, as confirmed by the Precision, Specificity, Recall, Confusion Matrix, F1 Score, Accuracy, and Total Accuracy performance metrics. The authors concluded that the devised method could be used for practical purposes in cases of smart homes inhabited by two or three residents, and that the enhancement of the Deep Convolutional Neural Network for the classification of more intricated human activities would be worth investigating in a future study.

3.1.2. Regression

Subsequently, from the obtained pool of scientific articles obtained based on the devised review methodology, we identified, analyzed and summarized those making use of Decision Tree integrated with sensor devices in smart buildings. A complete summarization table (Table S6) is presented in the Supplementary Materials file, while Table 6 presents five of the most recent papers addressing this subject.

Table 6. Five of the most recent scientific articles addressing the Decision Tree integrated with sensor devices in smart buildings.

Reference	Publication Year	Type of Smart Building	Type of Sensors	Reason for Using the DT Method with Sensor Devices	DT Only or Hybrid	Performance Metrics
[19]	2019	smart building	smartphone sensors and Bluetooth beacons data	group activity detection and recognition	a framework for indoor Group Activity Detection and Recognition (GADAR) and Hierarchical Clustering, along with Decision Tree classifier, K-Neighbors classifier, Deep Neural Network, Gaussian Process classifier, Logistic regression, Support Vector Machine, Linear Discriminant Analysis, Gaussian Naïve Bayes (comparison)	Confusion Matrix, Accuracy (Mean), Accuracy (Variation), Precision, Recall, F1 Score

Table 6. *Cont.*

Reference	Publication Year	Type of Smart Building	Type of Sensors	Reason for Using the DT Method with Sensor Devices	DT Only or Hybrid	Performance Metrics
[1]	2019	smart building	indoor environment sensors: thermocouple TX-FF-0.32-1P (FUKUDEN) for the temperature; photosensor HD2021T AA-SP (Deltaohm) for the illuminance; OPUS20 TCO (Lufft) sensor for the relative humidity and CO_2 concentration; occupancy information sensor: PN1500 (Botem); electricity meters: PR300 (Yokogawa) for the lighting power; Enertalk Plug (Encored Technologies) for the PC electricity consumption and EHP electricity meter	assessing the occupancy status information in order to improve the energy prediction performance of a building energy model	Support Vector Machine compared with Decision Tree and Artificial Neural Networks	Overall Accuracy and Standard Deviation
[50]	2019	smart office buildings	air temperature, relative humidity, air speed, CO_2	personal thermal comfort	comparison between Decision Tree, Random Forest, Boosted Trees	the Overall Prediction Accuracy, the On-State Accuracy, the Present State Accuracy, Confusion Matrix, the Mean Squared Error (MSE), the Root-Mean-Squared Error (RMSE) and the Average Test Accuracy
[21]	2019	smart building	wireless sensor networks	forecasting Packet Delivery Ratio (PDR) and Energy Consumption (EC) in Internet of Things (IoT)	comparison between Linear Regression, Gradient Boosting, Random Forest, Baseline and Deep Learning Neural Networks	Root Mean Square Error (RMSE), Mean Percentage Error (MPE), and Mean Absolute Percentage Error (MAPE)
[78]	2019	smart office building	common sensors: motion detection, power consumption, CO_2 concentration	estimating the number of occupants	Decision Tree C4.5, parameterized rule-based classifier	Average Error of Occupancy Estimation

It can be seen that 32% of the scientific papers selected and summarized in Table S6, presented in the Supplementary Materials file, analyze smart buildings in general, while 53% target exclusively smart homes, 11% take into consideration smart office buildings, and the remaining 4% analyze smart spaces. The authors of these papers make use of different types of sensors, including wireless sensor networks [17,21,53,79]; sensors for detecting carbon dioxide concentration [1,17,50,53,68,78]; sensors for detecting total volatile organic compounds [68]; air temperature and humidity sensors [1,50,53,68,80]; pressure sensors [5,80]; wind speed sensors [50,80]; motion sensors [30,78,81]; Passive Infrared (PIR) sensors [30,82]; electricity meters [1,78,81]; smartphone sensors and Bluetooth beacon data [19]; indoor environment sensors [1]; occupancy information sensors [1]; sensors measuring the visibility outside the building [80]; sensors embedded in the environment [81]; wearable and environmental sensors [53,74]; binary infrared sensors [83]; unobtrusive sensing modules, including a gateway and a set of passive sensors [14]; simple non-intrusive sensors, door sensors and occupancy sensors [82]; high-sensitivity underfloor mounted accelerometers [5]; binary sensors installed in doors, cupboards, and toilet flushes [69]; and cameras, microphones, accelerometers, multisensor board and PC monitoring, and external sensors integrated in the user's home automation system [84].

In these papers, the reasons for using the Decision Tree integrated with sensor devices in smart buildings were mainly related to human activity recognition [1,5,14,17,19,21,30,50,53,68,69,74,78–84]. In some of these papers, human activity recognition was just a first step, subsequently focusing on: analyzing and improving the energy prediction performance [1,80]; analyzing and ensuring the thermal comfort of the occupants [50,53]; forecasting energy consumption [21]; estimating the number of occupants [78]; identifying behavioral patterns [79]; detecting deviating human behavior [82]; monitoring the activities of elderly people living alone [14]; classifying the gender of occupants [5]; and improving home-based assisted living [30].

With respect to the devised research methods, in [19], Chen et al. made use of a hybrid approach, combining a framework for indoor group activity detection/recognition and hierarchical clustering, along with the Decision Tree classifier, the K-Neighbors classifier, Deep Neural Network, the Gaussian Process classifier, Logistic Regression, Support Vector Machine, Linear Discriminant Analysis, and Gaussian Naïve Bayes, drawing a comparison among these techniques. In [79], Zamil et al. used the ordered Decision Tree and compared their results with those obtained using the ClaSP and CMCla methods. In [81], Malazi et al. made use of the Emerging Patterns and Random Forest (CARER) method, comparing it with the Hidden Markov Model, Bayesian Network, Naïve Bayes, SVM, Decision Tree, and Random Forest. In [84], Bjelica et al. implemented the Decision-Tree technique only. In [69], Sebbak et al. made use of the Dempster–Shafer theory, comparing it with the Naïve Bayes classifier and J48 Decision Tree. In [17], Brennan et al. compared Gradient Boosting, K-Nearest Neighbor (KNN), Linear Discriminant Analysis, and Random Forest. In [82], Lundström et al. made use of a hybrid approach, combining Random Forest and the third-order Markov chain. In [1], Kim et al. used the Support Vector Machine and compared their results with those obtained using Decision Tree and Artificial Neural Networks. In [78], Amayri et al. made use of Decision Tree C4.5, a parameterized rule-based classifier. In [30], Nef et al. used a series of learning classification algorithms, namely Naïve Bayesian (NB), Support Vector Machine (SVM), and Random Forest (RF). In [68], Zimmermann et al. presented a comparison of the following supervised learning models: Naïve Bayes (NB), C4.5 Decision Tree, Logistic Regression, K-Nearest Neighbor, and Random Forest. These were used to detect and estimate occupancy. In [5], Bales et al. combined Bagged Decision Tree, Boosted Decision Tree, Support Vector Machines (SVMs) and Neural Networks in order to classify gender. In [74], Palumbo et al. made use of Recurrent Neural Networks (RNNs) for the activity recognition process. In [50], Shetty et al. compared the Decision Tree, Random Forest and Boosted Trees methods. In [21], Ateeq et al. compared the Linear Regression, Gradient Boosting, Random Forest, Baseline and Deep Learning Neural Networks. In [53], Li et al. compared Logistic Regression, K-Nearest Neighbor, Support Vector Machine, and Random Forest. In [80], Fong et al. made use of an improved version of the Very Fast Decision Tree (VFDT) classification algorithms and compared their results with those obtained with CART Decision Tree version 4.8, the Active Learning classifier for evolving data streams, Fast Incremental Model Trees with Drift Detection (FIMT-DD), Hoeffding Tree or VFDT, the K-Nearest Neighbor algorithm, Naïve Bayes, Online Regression Tree with Options, and Stochastic Gradient Descent. In [83], Zhao et al. implemented the Fuzzy Decision Tree method. In [14], Kim et al. used a hybrid approach, combining the Neural Network, C4.5 Decision Tree, Bayesian Network and Support Vector Machine techniques.

The performance metrics chosen by the authors of the scientific papers that used the Decision Tree method integrated with sensor devices in smart buildings included Accuracy [1,5,17,19,50,53,68,74,80]; Confusion Matrix [19]; Precision [19,30,69,74,80]; Recall [19,69,74,80]; F1 Score [19,74]; Standard Deviation [1]; Root Mean Square Error (RMSE) [17,21,68,80]; Mean Percentage Error (MPE) and Mean Absolute Percentage Error (MAPE) [21]; Average Error of Occupancy Estimation [78]; Normalized

Root-Mean-Square Error and Coefficient of Variation of the RMSD (CV) [17]; True Positive Rate and True Negative Rate [68,80]; Mean Absolute Error (MAE) [68,80,84]; Runtime [79]; the Receiver Operating Characteristic (ROC) curve [80]; the F-Measure [30,69,80,81]; Recognition Success Rate [83,84]; Sensitivity (SN), Specificity (SP) and Area Under the Receiver Operating Characteristic Curve (AUC) [14]; Local Outlier Factor (LOF), Z-Score values, and Cluster Transition Probability [82]; Average Specificity, Sensitivity [30].

With respect to five of the most recent scientific articles making use of Decision Tree along with sensor devices in smart buildings (Table 6), it can be observed that in [19], Chen et al. put forward a framework for indoor group activity detection and recognition (GADAR), achieving hierarchical clustering in smart buildings by using a Decision Tree classifier and data collected from smartphone sensors and Bluetooth beacons. The developed framework was designed to contain four layers: one for the user, one for the data package, one for processing, and one for output. The selection of the Decision Tree classifier was based on the experimental results obtained after comparing several machine learning approaches, namely Decision Tree, the K-Neighbors classifier, Deep Neural Network, the Gaussian Process classifier, Logistic Regression, Support Vector Machine, Linear Discriminant Analysis, and Gaussian Naïve Bayes. A group activity recognition system was developed based on the devised framework and tasked with distinguishing different types of educational group activities. The best results were obtained when using the DT classifier, as confirmed by the Confusion Matrix, Accuracy (Mean), Accuracy (Variation), Precision, Recall and F1 Score performance metrics. The most important result was the Accuracy of 89% in the cases of both group activity detection and group activity recognition.

The Decision Tree classifier was employed and compared with Support Vector Machines and artificial neural networks in paper [1], which was previously analyzed when reviewing the most recent scientific articles that integrate SVM approaches with sensor devices in smart buildings (Table 1).

Ensuring the wellbeing of inhabitants in smart office buildings in terms of personal thermal comfort is a topic that has been approached in a recent paper [50], in which Shetty et al. analyzed and compared the performance of several machine learning approaches, namely Decision Tree, Random Forest, and Boosted Trees with data recorded from sensors measuring the air temperature, relative humidity, air speed and CO_2, with to the aim of classifying a desk fan's state and forecasting its speed in accordance with individual preferences regarding desk fan usage. In order to compare the results obtained for each of the machine learning approaches, the authors computed the Overall Prediction Accuracy, the On State Accuracy, the Present State Accuracy, the Confusion Matrix, the Mean Squared Error (MSE), the Root-Mean-Squared Error (RMSE), and the Average Test Accuracy performance metrics, concluding that the Random Forest approach registered the highest performance level.

In article [21], Ateeq et al. aimed to forecast the Packet Delivery Ratio (PDR) and Energy Consumption (EC) of wireless sensor networks, given their paramount importance for Internet of Things (IoT) devices, which are increasingly being employed in small- to medium-sized smart buildings. The authors compared the results obtained after applying the Linear Regression, Gradient Boosting, Random Forest, Single Hidden Layer, and Deep Learning Neural Networks approaches to predict the PDR and EC, using an open dataset regarding the IEEE 802.15.4 technical standard. After conducting the experimental study and analyzing the results in terms of Root Mean Square Error (RMSE), Mean Percentage Error (MPE), and Mean Absolute Percentage Error (MAPE), the authors concluded that the Deep Learning Neural Networks registered the best level of performance, followed closely by the Random Forest approach.

Estimating the number of people within a smart office environment with a minimum number of interactions through video stream acquisition, so as not to disturb the occupants and avoid invading their privacy, was the topic of interest in [78], where Amayri et al. studied Decision Tree C4.5 and a Parameterized Rule-Based Classifier using data recorded from commonly available sensors for motion detection, power consumption, and CO_2 concentration. Analyzing the obtained results, the authors concluded that the C4.5 DT algorithm provided the highest level of performance after approximately

14 interaction spaces, while the Parameterized Rule-Based approach performed better at the beginning but, due to having only two parameters, in the end the C4.5 DT assessed the number of people within the smart office environment with a higher degree of accuracy, as determined on the basis of the Average Error of Occupancy Estimation performance metric.

Subsequently, from the obtained pool of scientific articles resulting from the application of the devised review methodology, we identified, analyzed and summarized those making use of Ensemble Methods integrated with sensor devices in smart buildings for classification purposes. A complete summarization table (Table S7) is presented in the Supplementary Materials file, while Table 7 presents five of the most recent papers addressing this subject.

Table 7. Five of the most recent scientific articles addressing Ensemble Methods integrated with sensor devices in smart buildings.

Reference	Publication Year	Type of Smart Building	Type of Sensors	Reason for Using the Ensemble Methods with Sensor Devices	Ensemble Methods Only or Hybrid	Performance Metrics
[20]	2019	smart building	smartphone sensors (acceleration, gyroscope)	human activity recognition	Extreme Learning Machine (ELM) for ensemble learning, compared with Artificial Neural Networks (ANN), Extreme Learning Machine (ELM), Support Vector Machine (SVM), Random Forest (RF), and deep Long Short-Term Memory (LSTM) approaches	Accuracy
[16]	2019	smart home	wearable sensor, accelerometer providing inertial information of human activity	human activity recognition	Kernel Fisher Discriminant Analysis (KFDA) technique, Extreme Learning Machine (ELM); comparison among Best Base ELM, SVM, Bagging, AdaBoost and the proposed method	Accuracy, Recall
[3]	2018	smart building	Light-Emitting Diode (LED) luminaires used as light sensors	human activity recognition	Support Vector Machine (SVM), Convolutional Neural Network-Hidden Markov Model (CNN-HMM), Long Short-Term Memory networks (LSTM) learning algorithms	Accuracy and Mean Square Error (MSE)
[85]	2014	smart home	wireless sensors associated with different objects, monitoring the activities	human activity recognition	Cluster-Based Classifier Ensemble (ensemble method)	Confusion Matrix presenting number of True Positives, True Negatives, False Positives and False Negatives, Precision, Recall and F-Measure
[86]	2013	smart home	embedded sensors: stove-sensor, refrigerator-sensor, door-sensor	Activity recognition	ensemble method, combining one of the methods: Artificial Neural Networks (ANN), Hidden Markov Model (HMM), Conditional Random Fields (CRF) with the Genetic Algorithm (GA) approach	Precision, Recall, F-measure and Accuracy

Analyzing the scientific articles summarized in Table S7, presented in the Supplementary Materials file, it can be observed that 40% of them analyze smart buildings in general, while the remaining 60% take smart homes into consideration. The authors of these scientific articles make use of different types of sensors in their analyses, including smartphone sensors [16,20]; accelerometers providing inertial information of human activity [16]; Light-Emitting Diode (LED) luminaires used as light sensors [3]; and sensors associated with different objects [85,86]. In all of the papers selected and summarized in Table S7, the reason for using the Ensemble Methods integrated with the sensor devices in smart buildings was the recognition of human activity.

Regarding the devised research methods, in [20], Chen et al. made use of the Extreme Learning Machine (ELM) for ensemble learning, and compared it with the Artificial Neural Network (ANN), Extreme Learning Machine (ELM), Support Vector Machine (SVM), Random Forest (RF), and deep Long Short-Term Memory (LSTM) approaches. In [16], Tian et al. implemented the Kernel Fisher Discriminant Analysis (KFDA) technique, along with the Extreme Learning Machine (ELM), and compared their proposed method with Best Base Extreme Learning Machine (ELM), Support Vector Machine (SVM), Bagging, and AdaBoost. In [3], Hao et al. made use of the Support Vector Machine (SVM), Convolutional Neural Network–Hidden Markov Model (CNN-HMM), and Long Short-Term Memory (LSTM) networks learning algorithms. In [85], Jurek et al. implemented the Cluster-Based Classifier Ensemble as an ensemble method. In [86], Fatima et al. developed an ensemble approach, combining each of Artificial Neural Networks (ANN), Hidden Markov Model (HMM), and Conditional Random Fields (CRF) with the Genetic Algorithm (GA) approach.

The performance metrics chosen by the authors of the scientific papers that use Ensemble Methods integrated with sensor devices in smart buildings include Accuracy [3,16,20,86]; Recall [16,85,86]; Precision and F-measure [85,86]; Mean Squared Error (MSE) [3]; and Confusion Matrix presenting a number of true Positives, True Negatives, False Positives and False Negatives [85].

With respect to the scientific articles making use of Ensemble Methods along with sensor devices in smart buildings (Table 7), after applying the devised review methodology, five recent scientific works were identified. In [20], Chen et al. proposed an ensemble Extreme Learning Machine (ELM) approach using Gaussian Random Projection to initialize the input weights with a view to achieving accurate recognition of a diversity of human activities in smart buildings using non-intrusively recorded data by means of smartphone sensors, namely accelerometers and gyroscopes. The authors compared the results provided by their approach with those obtained by using the Artificial Neural Networks (ANNs), Extreme Learning Machine (ELM) that didn't use Gaussian Random Projection to initialize the input weights, Support Vector Machine (SVM), Random Forest (RF), and deep Long Short-Term Memory (LSTM) approaches. They concluded that their proposed approach was superior in terms of recognition accuracy when compared to other existing methods.

An ensemble Extreme Learning Machine method was devised by Tian et al. in [16] and compared with Best Base ELM, SVM, Bagging and AdaBoost. This paper was previously analyzed when reviewing the most recent scientific articles that use Discriminant Analysis approaches with sensor devices in smart buildings (Table 2).

Human activity recognition while the persons are moving in smart buildings is a topic addressed in a recent paper [3], in which Hao et al. proposed an ensemble learning approach consisting of the Support Vector Machine (SVM), Convolutional Neural Network-Hidden Markov Model (CNN-HMM) and Long Short-Term Memory (LSTM) networks learning algorithms. The authors used light-emitting diode luminaires as light sensors and applied a forward sequential pruning technique to improve the performance of their proposed ensemble method. The results obtained from the experimental tests were analyzed in terms of the Accuracy and Mean Squared Error (MSE) performance metrics, with results of 88% and 0.13 MSE, respectively, for the dynamical occupancy dataset.

In article [85], Jurek et al. aimed to recognize human activity in smart homes by proposing a cluster-based classifier ensemble method, using numeric and binary data collected by means of wireless sensors attached to different objects. After conducting the experimental tests and analyzing the results in terms of the Confusion Matrix presenting the number of True Positives, True Negatives, False Positives and False Negatives, Precision, Recall and F-Measure, the authors concluded that their proposed approach offered a higher level of performance than a range of state-of-the-art single clustering algorithms.

Achieving reliable human activity recognition in the context of the many distinctive features that different smart homes may exhibit is a topic addressed in [86], where Fatima et al. studied an ensemble method developed by combining one of the Artificial Neural Networks (ANN), Hidden Markov Model (HMM) or Conditional Random Fields (CRF) approaches with the Genetic Algorithm (GA) approach, using data recorded from embedded sensors mounted on refrigerators, stoves and doors. Analyzing the obtained results, the authors concluded that their proposed approach offered a higher level of performance than single classifiers and classical multi-class models, as reflected in the Precision, Recall, F-Measure and Accuracy performance metrics.

Subsequently, from the pool of scientific articles obtained based on the devised review methodology, we identified, analyzed and summarized those making use of the Gaussian Process Regression (GPR) integrated with sensor devices in smart buildings. A complete summarization table (Table S8) is presented in the Supplementary Materials file, while Table 8 presents five of the most recent papers addressing this subject.

Table 8. Five of the most recent scientific articles addressing the Gaussian Process Regression (GPR) integrated with sensor devices in smart buildings.

Reference	Publication Year	Type of Smart Building	Type of Sensors	Reason for Using the Gaussian Process Regression with Sensor Devices	Gaussian Process Regression Only or Hybrid	Performance Metrics
[20]	2019	smart building	smartphone sensors (acceleration, gyroscope)	human activity recognition	Extreme Learning Machine (ELM) for ensemble learning, compared with Artificial Neural Networks (ANN), Extreme Learning Machine (ELM), Support Vector Machine (SVM), Random Forest (RF), and deep Long Short-Term Memory (LSTM) approaches	Accuracy
[87]	2017	smart home	smart meter	human activity monitoring	Non-Intrusive Load Monitoring (NILM) algorithm, Dempster—Shafer theory compared with the Gaussian Mixture model	Score for test events
[88]	2017	smart home	smart phones as sensors to capturing voice signals, Electroglottography (EGG) electrodes as sensors to capture EGG signals	voice pathology assessment	Gaussian Mixture model-based classifier, using different numbers of Gaussian mixtures	Accuracy
[89]	2017	smart home	wearable sensors providing inertial data, environment sensors and data processed video streams that anonymize the individual	machine monitoring of human health	linear-Gaussian transition model with hard boundaries, nonlinear-Gaussian observation model, post-regularized particle filter (C-ERPF), compared to other methods: Extended Kalman Filter (EKF), constrained-EKF, and Extended Regularized Particle Filtering (ERPF) without transition constraints	Average Error
[35]	2017	smart home	The smart meter or another third-party device	ambient assisted living	the developed PQD-PCA Classifier along with the Gaussian Mixture Mode (GMM) and the Dempster—Shafer Theory (DST) compared with other classifiers (K-Nearest-Neighbors KNN, Gaussian Naïve Bayes GNB, Logistic Regression Classifier LGC, Decision Tree DTree and Random Forest Rforest)	True Positive Percentage (TPP), False Positive Percentage (FPP), Precision, Recall, F1 Score, F2 Score

A total of 83% of the scientific papers selected and summarized in Table S8, presented in the Supplementary Materials file, focus their research exclusively on smart homes, while the remaining 17% analyze both smart homes and smart buildings in general. In these papers, the authors make use of different types of sensors, including smartphone sensors [88]; electroglottography (EGG) electrodes [88]; smart meters [35,87]; wearable sensors providing inertial data, environment sensors and data processed video streams [89]; electricity, water and natural gas consumption sensors [90]; and multi-appliance recognition systems, designing a single smart meter using a current sensor and a voltage sensor in combination with a microprocessor to meter multi-appliances [64].

With respect to the reasons for implementing the GPR integrated with sensor devices in smart buildings, these are mainly related to human activity recognition/monitoring [35,87–89]; voice pathology assessment [88]; monitoring of human health [89]; ambient assisted living [35]; recognizing household appliances in order to assess their usage and develop habits of power preservation [64]; and developing a framework for automatic leakage detection in smart water and gas grids [90].

With respect to the devised research methods, in [87], Alcalá et al. implemented the Non-Intrusive Load Monitoring (NILM) algorithm and the Dempster–Shafer theory and compared them with the Gaussian Mixture model. In [88], Muhammad et al. used the Gaussian Mixture model-based classifier, using different numbers of Gaussian Mixtures. In [89], Villeneuve et al. made use of the linear-Gaussian transition model with hard boundaries, the nonlinear-Gaussian observation model, and post-regularized particle filter (C-ERPF), and compared these to other methods, including Extended Kalman Filter (EKF), constrained-EKF, and Extended Regularized Particle Filtering (ERPF) without transition constraints. In [35], Alcalá et al. implemented a PQD-PCA Classifier along with the Gaussian Mixture Mode (GMM) and the Dempster–Shafer Theory (DST) and compared their approach with other classifiers (K-Nearest-Neighbor (KNN), Gaussian Naïve Bayes (GNB), Logistic Regression Classifier (LGC), Decision Tree (DTree) and Random Forest (Rforest)). In [90], Fagiani et al. compared Gaussian Mixture Model (GMM), Hidden Markov Model (HMM) and One-Class Support Vector Machine (OC-SVM). In [64], Lai et al. developed a hybrid approach, combining Support Vector Machine with Gaussian Mixture Model (SVM/GMM) with a view to classifying electric appliances.

The performance metrics chosen by the authors of papers using Gaussian Process Regression (GPR) integrated with sensor devices in smart buildings included Score for test events [87]; Accuracy [88]; Average Error [89]; True Positive Percentage (TPP), False Positive Percentage (FPP), Precision, Recall, F1 Score, and F2 Score [35]; the probability of correctly detecting an anomaly, the probability of erroneously detecting an anomaly, the Receiver Operating Characteristic (ROC) curve, and Area Under the ROC Curve (AUC) [90]; and Accuracy, the Success Rate and the Recognition Rate [64].

Regarding the five most recent scientific articles retrieved according to the review methodology (Table 8), in [20], Chen et al. put forward an ensemble Extreme Learning Machine (ELM) approach using Gaussian Random Projection to initialize the input weights. This paper was reviewed previously when analyzing the most recent scientific works using Ensemble Methods approaches with sensor devices in smart buildings (Table 7).

Acknowledging the importance of human activity monitoring in ensuring a certain level of independence for the elderly without sacrificing their wellbeing, in [87], Alcalá et al. aimed to overcome the challenges arising from the rejection of intrusive monitoring techniques due to privacy issues by the residents of smart homes. To this end, the authors proposed a Non-Intrusive Load Monitoring (NILM) algorithm developed based on the Dempster–Shafer theory using only the data retrieved from a smart metering device, and compared this with the Gaussian Mixture model using the Score for Test Events as a performance metric. Based on the obtained results, the authors stated that their proposed method offered a higher level of performance than the model based on the Gaussian Mixture approach.

Considering the numerous disabilities that affect people's overall quality of life by limiting their movements, senses, or activities, in [88], Muhammad et al. put forward a system for assessing voice pathological features within smart homes by means of processing the data, which consisted of voice signals recorded using smartphone sensors and electroglottography (EGG) electrodes for capturing EGG signals, through different numbers of Gaussian mixtures. The authors performed the experimental tests on the open Saarbrucken public database, which consists of a variety of voice samples, concluding the viability of the proposed system on the basis of the Accuracy performance metric, as well as the processing speed. Muhammad et al. remarked that in the case of acute pathological voice features, the information obtained after processing only the electroglottography data was insufficient; for moderate cases, the use of either the EGG or voice recorded signals offered similar levels of performance, while the highest accuracy level was obtained through a fusion of both sources.

Machine monitoring of human health in smart homes is the topic of another recent scientific article [89], in which Villeneuve et al. devised a system based on the Linear-Gaussian transition model with hard boundaries, the Nonlinear-Gaussian observation model, and the Post-Regularized Particle Filter (C-ERPF). This system was designed to process data, recorded by wearable inertial sensors, environmental sensing devices and video streams, that had been anonymized with respect to the residents' identity. The authors compared the results obtained with their proposed approach with those obtained when using the extended Kalman Filter (EKF), the constrained-EKF, and the Extended Regularized Particle Filtering (ERPF) without transition constraints in terms of Average Error as a performance metric, concluding that two wearable wrist accelerometer sensors were sufficient to predict the kinematics of the arm.

In the scientific article [35], Alcalá et al. aimed to achieve ambient assisted living for the elderly in smart homes by proposing a Power Quality Disturbances (PQD)–Principal Component Analysis (PCA) classifier along with the Gaussian Mixture Mode (GMM) and the Dempster–Shafer Theory (DST) using data recorded by means of a smart meter or another single third-party sensing device. After conducting the experimental tests and analyzing the results with respect to True Positive Percentage (TPP), False Positive Percentage (FPP), Precision, Recall, F1 Score, F2 Score, the authors concluded that their devised method was a viable option for the elderly population who live alone.

Subsequently, from the obtained pool of scientific articles resulting from the application of the devised review methodology, we identified, analyzed and summarized those making use of the Linear Regression integrated with sensor devices in smart buildings. A complete summarization table (Table S9) is presented in the Supplementary Materials file, while Table 9 presents five of the most recent papers addressing this subject.

Table 9. Five of the most recent scientific articles addressing Linear Regression integrated with sensor devices in smart buildings.

Reference	Publication Year	Type of Smart Building	Type of Sensors	Reason for Using the Linear Regression Method with Sensor Devices	Linear Regression Only or Hybrid	Performance Metrics
[21]	2019	smart building	wireless sensor networks	forecasting Packet Delivery Ratio (PDR) and Energy Consumption (EC) in Internet of Things (IoT)	comparison between Linear Regression, Gradient Boosting, Random Forest, Baseline and Deep Learning Neural Networks	Root Mean Square Error (RMSE), Mean Percentage Error (MPE), and Mean Absolute Percentage Error (MAPE)
[91]	2018	smart building	three virtual sensors: temperature, airflow, and fan speed	improving electricity consumption by correctly identifying faults within a smart building's ventilation system	Linear Regression compared with Autoregressive Moving Average with Exogenous Variables (ARMAX) models, Support Vector Machine (SVM), Artificial Neural Network (ANN).	the Coefficient of Determination (for linear models) and Acceptable Ranges (for non-linear ones)
[92]	2018	smart home	wireless sensor networks	adaptive interference suppression	Linear Regression only	range of power savings, ratio of received packet

Table 9. *Cont.*

Reference	Publication Year	Type of Smart Building	Type of Sensors	Reason for Using the Linear Regression Method with Sensor Devices	Linear Regression Only or Hybrid	Performance Metrics
[41]	2017	smart home	temperature and humidity sensors from a Wireless Sensor Network	forecasting the energy use of appliances	comparing: Multiple Linear Regression, Support Vector Machine with Radial Kernel, Random Forest, Gradient Boosting Machines (GBM)	Root Mean Square Error (RMSE), Coefficient of Determination, Mean Absolute Error (MAE), Mean Absolute Percentage Error (MAPE)
[93]	2014	smart home	passive radio-frequency identification antennas various sensors: ultrasonic, infrared, load cells	gesture recognition	Linear regression only	Accuracy

Analyzing the scientific articles summarized in Table S9, presented in the Supplementary Materials file, it can be observed that 50% of these scientific papers analyze smart buildings in general, while the remaining 50% take smart homes into consideration. The authors of these scientific articles make use of different types of sensors in their analyses, including wireless sensor networks [21,41,92,94]; temperature, airflow, and fan virtual sensors [91]; temperature and humidity sensors [41]; and Passive Radio-frequency identification antennas along with various sensors such as ultrasonic, infrared, load cells [93].

With respect to the reasons for implementing the Linear Regression integrated with sensor devices in smart buildings, these were related to the analysis of forecasting Packet Delivery Ratio (PDR) and Energy Consumption (EC) in the Internet of Things (IoT) [21]; improving electricity consumption by correctly identifying faults within a smart building's ventilation system [91]; analyzing Adaptive Interference Suppression [92]; forecasting the energy use of appliances [41]; gesture recognition [93]; and controlling smart lighting [94].

Regarding the devised research methods, in [21], Ateeq et al. compared Linear Regression, Gradient Boosting, Random Forest, Baseline and Deep Learning Neural Networks. In [91], Mattera et al. made use of Linear Regression compared with Autoregressive Moving Average With Exogenous Variables (ARMAX), Support Vector Machine (SVM) and Artificial Neural Network (ANN) methods. In [92], Lynggaard implemented Linear Regression only. In [41], Candanedo et al. compared the Multiple Linear Regression, Support Vector Machine with Radial Kernel, Random Forest, and Gradient Boosting Machines (GBM) methods. In [93], Bouchard et al. made use of Linear Regression only. In [94], Basu et al. made use of the Linear Regression and Support Vector Regression (SVR) models.

The authors of the scientific papers using Linear Regression integrated with sensor devices in smart buildings chose various performance metrics, including the Root Mean Squared Error (RMSE) Mean Absolute Percentage Error (MAPE) [21,41,94]; Mean Percentage Error (MPE) [21]; Coefficient of Determination [41]; Mean Absolute Error (MAE) [41]; range of power savings, ratio of received packet [92]; Accuracy [93]; Normalized Mean Square Error (NMSE) [94]; and Coefficient of Determination (for linear models) and Acceptable Ranges (for non-linear ones) [91].

Concerning the five most recent scientific articles retrieved according to the review methodology (Table 9), in [21], Ateeq et al. proposed a method for predicting Packet Delivery Ratio and energy consumption, and compared the results obtained using the Linear Regression, Gradient Boosting, Random Forest, Baseline and Deep Learning neural networks approaches. This paper was reviewed previously when analyzing the most recent scientific works that use Decision Tree approaches with sensor devices in smart buildings (Table 6).

Considering the major negative impacts that faulty ventilation units can have on the electricity consumption of a building, in [91], Mattera et al. proposed a method for correctly identifying faults that might occur within a smart building's ventilation system by means of developing temperature, airflow and fan speed virtual sensors based on the data provided by existing physical sensors, thereby overcoming the expense and space conditions needed to install supplementary hardware sensing devices. To identify the moments in which virtual sensors were operating outside the correct parameters of a hardware sensor, the authors used and compared Linear Regression, Autoregressive Moving Average with Exogenous Variables (ARMAX), Support Vector Machine (SVM), and Artificial Neural Network (ANN) approaches in terms of the Coefficient of Determination (for linear models) and Acceptable Ranges (for nonlinear ones). Analyzing the obtained results, the authors concluded that their proposed approach yielded satisfactory results, thereby offering the possibility of reducing costs and equipment expenditure while ensuring an appropriate reliability level.

Acknowledging the problems that will arise due to limited radio spectrum availability in the context of IoT devices, which are increasingly present in smart homes, in [92], Lynggaard put forward an adaptive interference suppression system based on the Linear Regression method in order to correctly forecast in wireless sensor networks, using the information related to the radio channels' states, the power needed to successfully transmit a data package. The author performed comprehensive experimental tests using data retrieved from wireless sensor networks in smart homes, and concluded that the savings in terms of power ranged from 42% to 82%, while the receive ratio of a data packet was greater than or equal to 92%.

In the scientific article [41], Candanedo et al. aimed to forecast the electricity usage of appliances in smart homes by comparing the results obtained after applying Multiple Linear Regression, Support Vector Machine with Radial Kernel, Random Forest and Gradient Boosting Machines (GBM) approaches on data recorded by means of temperature and humidity sensors in a wireless sensor network. After conducting the experimental tests and analyzing the results in terms of the Root Mean Square Error (RMSE), Coefficient of Determination, Mean Absolute Error (MAE), and Mean Absolute Percentage Error (MAPE), the authors concluded that for all of the machine learning approaches, the timestamps were the most significant information for accurately forecasting the electricity consumption of appliances.

Gesture recognition of the elderly in smart home environments was studied in [93], in which Bouchard et al. devised an algorithm based on the Linear Regression in order to distinguish movement direction and segment the datasets in order to identify a gesture's starting and ending points with a view to recognizing gestures in situations that exhibit a high degree of uncertainty by processing data recorded through means of a Passive Radio-frequency identification antennas system, along with load cells and ultrasonic and infrared sensors. The authors analyzed the results obtained using their proposed approach in terms of the Accuracy performance metric and concluded that even though the accuracy level was low, the passive radio-frequency identification system was a promising tool for the recognition of human activity. The authors intended to enhance the system in the future by means of fuzzy inference methods.

Subsequently, from the pool of scientific articles obtained based on the devised review methodology, we identified, analyzed and summarized those that make use of the Neural Networks for Regression Purposes integrated with sensor devices in smart buildings. A complete summarization table (Table S10) is presented in the Supplementary Materials file, while Table 10 presents five of the most recent papers addressing this subject.

Table 10. Five of the most recent scientific articles addressing the Neural Networks for regression purposes integrated with sensor devices in smart buildings.

Reference	Publication Year	Type of Smart Building	Type of Sensors	Reason for Using the ANN Regression Method with Sensor Devices	ANN Regression Only or Hybrid	Performance Metrics
[22]	2019	smart buildings	sensors for registering the electricity consumption	forecasting the electricity consumption	ANN compared with Linear Regression (LR), Auto-Regressive Integrated Moving Average (ARIMA), Evolutionary Algorithms (EAs) for Regression Trees (EVTree), Generalized Boosted Regression Models (GBM), Random Forest (RF), Ensemble, Recursive Partitioning and Regression Trees (Rpart), Extreme Gradient Boosting (XGBoost)	Mean Absolute Error (MAE) and the Root Mean Square Error (RMSE)
[23]	2019	smart building	Wireless Sensor Network (WSN)	energy consumption forecasting	Multilayer Perceptron (MLP) compared with: Linear Regression (LR), Support Vector Machine (SVM), Gradient Boosting Machine (GBM) and Random Forest (RF)	Coefficient of Determination (R^2), Root Mean Square Error (RMSE), Mean Absolute Error (MAE), Mean Absolute Percentage Error (MAPE)
[95]	2018	smart home	smart metering system and sensors installed at a residential consumer, corresponding to 15 individual appliances (water heater, refrigerator, microwave, furnace, master bedroom, front bedroom, kitchen stove wall, dishwasher disposal, kitchen sink wall, family room, kitchen half-bath foyer, washing machine, guest bedroom, dryer, basement)	forecasting the electricity consumption	mixed Artificial Neural Network (ANN) approach using both Non-Linear Autoregressive with Exogenous Input (NARX) ANNs and Function Fitting Neural Networks (FITNETs)	Mean Squared Error (MSE), Correlation Coefficient (R), the differences between the real consumption and the forecasted ones
[12]	2018	smart commercial and residential buildings	weather sensors	forecasting the electricity consumption	deep Recurrent Neural Network (RNN) models	Root Mean Square Error relative to Root Mean Squared (RMS) average of electricity consumption in test data, Root Mean Square Error relative to Root Mean Squared (RMS) average of electricity consumption in training data, Pearson Coefficient
[43]	2018	smart home	flowmeter sensor	identifying the occurrence of a specific pattern in a Water Management System (WMS)	three types of ANN for Multi-Step-Ahead (MSA) forecasting: "Multi-Input Multi-Output (MIMO), Multi-Input Single-Output (MISO), and Recurrent Neural Network (RNN)"	Accuracy, Precision, Recall, and F-Measure

A total of 45.5% of the scientific articles summarized in Table S10, presented in the Supplementary Materials file, analyzed smart buildings in general; the same percentage of papers considered smart homes, while the remaining 9% analyzed both smart homes and smart buildings. The authors of these scientific papers make use of different types of sensors in their analyses, including sensors for registering the electricity consumption [22]; Wireless Sensor Networks (WSNs) [23,45,96]; Passive Infrared (PIR) sensors or motion detectors [75,97]; smart metering systems and sensors installed by the residential consumer, corresponding to 15 individual appliances [95]; weather sensors [12]; flowmeter sensors [43]; temperature sensors, external humidity sensors, solar radiation sensors [98]; thermal sensors [2]; and door/window entry point sensors, electricity power usage sensors, bed/sofa pressure sensors, and flood sensors [75].

With respect to the reasons for implementing the Neural Networks for regression purposes integrated with sensor devices in smart buildings, these were mainly related to forecasting electricity consumption [12,22,23,45,95]; identifying the occurrence of a specific pattern in a Water Management System (WMS) [43]; indoor temperature monitoring and forecasting [96,98]; human behavior recognition [2,75]; and short-term prediction of occupancy [97].

With respect to the devised research methods, in [22], Divina et al. made use of an Artificial Neural Network (ANN) approach, and compared this with Linear Regression (LR), Auto-Regressive Integrated Moving Average (ARIMA), Evolutionary Algorithms (EAs) for Regression Trees (EVTree), Generalized Boosted Regression Models (GBM), Random Forest (RF), Ensemble, Recursive Partitioning and Regression Trees (Rpart), and Extreme Gradient Boosting (XGBoost). In [23], Chammas et al. developed a Multilayer Perceptron (MLP) Neural Network approach and compared it with Linear Regression (LR), Support Vector Machine (SVM), Gradient Boosting Machine (GBM), and Random Forest (RF). In [95], Oprea et al. made use of a mixed Artificial Neural Network (ANN) approach using both Nonlinear Autoregressive with Exogenous Input (NARX) ANNs and Function Fitting Neural Networks (FITNETs). In [12], Rahman et al. implemented deep Recurrent Neural Network (RNN) models. In [43], Khan et al. used three types of ANN for Multi-Step-Ahead (MSA) forecasting methods: Multi-Input Multi-Output (MIMO), Multi-Input Single-Output (MISO), and Recurrent Neural Network (RNN). In [98], Attoue et al. made use of an Artificial Neural Network (ANN) with Multilayer Perceptron (MLP) structure. In [2], Zhao et al. implemented the Support Vector Regression (SVR) and Recurrent Neural Network (RNN) methods. In [97], Li et al. used an ANN approach and compared the obtained results with the Traditional inhomogeneous Markov chain model, the New Markov chain model, the Probability Sampling model, and Support Vector Regression (SVR). In [45], Collotta et al. developed a hybrid method, combining the Bluetooth Low-Energy Home Energy Management System (BluHEMS) and an Artificial Neural Network (ANN) approach. In [96], Pardo et al. developed two ANNs: a linear model and a Multilayer Perceptron (MLP) model with one hidden layer, comparing the results with the Bayesian standard model. In [75], Lotfi et al. made use of different types of recurrent Neural Networks, such as Echo State Network (ESN), Back Propagation Through Time (BPTT), and Real-Time Recurrent Learning (RTRL).

The performance metrics considered in the scientific papers using the Neural Networks for Regression Purposes integrated with sensor devices in smart buildings included Mean Absolute Error (MAE) and Root Mean Squared Error (RMSE) [12,22,23,45]; Coefficient of Determination (R^2) [23]; Mean Absolute Percentage Error (MAPE) [23,45]; Mean Squared Error (MSE) [45,95,98]; Correlation Coefficient (R) [95,98]; the differences between the real consumption and the forecasted ones [95]; Pearson Coefficient [12]; Accuracy [43,97]; Precision, Recall, and F-Measure [43]; Average Error and Error Rate [2]; Mean Absolute Error (MAE) [96]; and Root Mean Squared Error (RMSE) [75].

With respect to the five most recent scientific articles retrieved according to the review methodology (Table 10), in [22], Divina et al. addressed issues regarding the prediction of smart buildings' electricity consumption, using data retrieved from sensors that registered electricity consumption. To this end, the authors analyzed a series of prediction methods, comparing the ANN approach with Linear Regression (LR), Auto-Regressive Integrated Moving Average (ARIMA), Evolutionary Algorithms

(EAs) for Regression Trees (EVTree), Generalized Boosted Regression Models (GBM), Random Forest (RF), Ensemble, Recursive Partitioning and Regression Trees (Rpart), Extreme Gradient Boosting (XGBoost). Based on this comparison, the authors observed that the methods based on machine learning models were the most suitable for task under consideration.

Article [23] was previously detailed when analyzing the most recent scientific articles that integrate Neural Networks for Classification Purposes with sensor devices in smart buildings (Table 5).

In [95], Oprea et al. presented a forecasting method for providing accurate predictions of electricity consumption at the residential level, refined to the electrical devices level. The authors considered smart home complexes that were capable of partially sustaining their electricity consumption based on renewable energy resources. The authors stated that, in contrast to other existing studies, their approach did not require supplementary meteorological datasets. The devised method was based on an ANN approach that combined the Nonlinear Autoregressive with Exogenous Input (NARX) model and Function Fitting Neural Networks (FITNETs). The input dataset was retrieved from a smart metering system and from sensors installed in the residence, corresponding to a selection of the electrical devices. In the case of the NARX model, they also used a timestamp dataset as exogenous variables. In order to validate the developed prediction method, the authors computed the Mean Squared Error (MSE), the Correlation Coefficient (R), and the differences between the real consumption and the forecasted ones and used these as performance metrics. Subsequently, they compared the obtained results with those found in the scientific literature. The authors concluded that the developed approach was a practical and efficient alternative to the existing approaches in the literature.

To obtain medium-to-long term predictions of aggregated hourly electricity consumption in both commercial and residential buildings, in [12], Rahman et al. presented a Recurrent Neural Network approach. Using the Root Mean Square Error relative to Root Mean Squared (RMS) average of electricity consumption in test data, Root Mean Square Error relative to Root Mean Squared (RMS) average of electricity consumption in training data, and the Pearson Coefficient as performance metrics, the authors evaluated the performance of their developed approach and compared it with that provided by the multilayered perceptron model. The authors compared their results to those obtained in the case of the Multilayered Perceptron Model, and the authors concluded that in the case of commercial buildings, their approach registered a lower relative error, while in the case of residential buildings, the results registered by the two methods were comparable.

In [43], Khan et al. addressed issues regarding real-time analysis of data retrieved from sensors in order to develop a process for making decisions by automated means, without any human involvement, in smart homes based on Internet of Things. To identify the patterns in a Water Management System (WMS), the authors made use of three types of ANNs: Multi-Input Multi-Output (MIMO), Multi-Input Single-Output (MISO), and Recurrent Neural Network (RNN). These were compared in order to achieve multi-step-ahead forecasting based on flowmeter sensors. Conducting a series of experiments, using Accuracy, Precision, Recall, and F-Measure as performance metrics, the authors remarked that the Recurrent Neural Network approach provided the best performance, and using its prediction, the implementation of an automated decision-making system provided an accuracy of 86%.

Subsequently, from the pool of scientific articles resulting from the application of the devised review methodology, we identified, analyzed and summarized those making use of the Support Vector Regression (SVR) integrated with sensor devices in smart buildings. A complete summarization table (Table S11) is presented in the Supplementary Materials file, while Table 11 presents five of the most recent papers approaching this subject.

Most of the scientific articles summarized in Table S11, presented in the Supplementary Materials file, analyze smart buildings in general (75%), while 12.5% consider smart homes, and the remaining 12.5% consists of studies regarding commercial buildings. The authors of these scientific articles make use of different types of sensors in their analyses, including wireless sensor networks [23,41,51,94]; thermal sensors [2]; passive infrared motion detecting sensors [97]; temperature and humidity sensors [41]; occupancy and light sensors [13]; and energy smart meters, building management systems, and weather stations [44].

The reasons for implementing the Support Vector Regression (SVR) integrated with sensor networks in smart buildings were mainly related to forecasting electricity consumption [13,23,41,44]; controlling smart lighting [94]; human behavior recognition [2]; thermal comfort optimization [51]; and short-term prediction of occupancy [97].

Regarding the devised research methods, in [23], Chammas et al. developed a Multilayer Perceptron (MLP) Neural Network approach and compared it with Linear Regression (LR), Support Vector Machine (SVM), Gradient Boosting Machine (GBM), and Random Forest (RF). In [2], Zhao et al. implemented the Support Vector Regression (SVR) and Recurrent Neural Network (RNN) methods. In [51], Viani et al. implemented the Support Vector Regression method. In [97], Li et al. used an ANN approach and compared the obtained results with the traditional inhomogeneous Markov chain model, the New Markov chain model, Probability Sampling model, and Support Vector Regression (SVR). In [41], Candanedo et al. compared the Multiple Linear Regression, Support Vector Machine with Radial Kernel, Random Forest, and Gradient Boosting Machines (GBM) methods. In [13], Caicedo et al. implemented the Support Vector Regression method. In [44], Jain et al. developed a model based on the Support Vector Regression (SVR) method. In [94], Basu et al. made use of the Linear Regression and Support Vector Regression (SVR) models.

The authors of the scientific papers using the Support Vector Regression (SVR) method integrated with sensor devices in smart buildings chose various performance metrics, including Coefficient of Determination (R^2), Root Mean Square Error (RMSE), Mean Absolute Error (MAE), and Mean Absolute Percentage Error (MAPE) [23,41]; Average Error and Error Rate [2]; Prediction Error [51]; Accuracy [97]; comparison between the actual energy consumption per day and predicted energy consumption per day [13]; Coefficient of Variation (CV) and Standard Error [44]; and Root Mean Square Error (RMSE) along with Normalized Mean Square Error (NMSE) [94].

With respect to the five most recent scientific articles addressing the Support Vector Regression (SVR) method integrated with sensor devices in smart buildings (Table 11) it can be observed that paper [23] was previously reviewed when analyzing the most recent scientific articles that integrate Neural Networks for classification purposes with sensor devices in smart buildings (Table 5); paper [2] was reviewed previously when analyzing the most recent scientific articles that integrate Support Vector Machines with sensor devices in smart buildings (Table 1); article [41] was reviewed previously when analyzing the most recent scientific articles that integrate Linear Regression with sensor devices in smart buildings (Table 9).

Table 11. Five of the most recent scientific articles addressing the Support Vector Regression (SVR) integrated with sensor devices in smart buildings.

Reference	Publication Year	Type of Smart Building	Type of Sensors	Reason for Using the Support Vector Regression Method with Sensor Devices	Support Vector Regression Only or Hybrid	Performance Metrics
[23]	2019	smart building	Wireless Sensor Network (WSN)	energy consumption forecasting	Multilayer Perceptron (MLP) compared with: Linear Regression (LR), Support Vector Machine (SVM), Gradient Boosting Machine (GBM) and Random Forest (RF)	Coefficient of Determination (R^2), Root Mean Square Error (RMSE), Mean Absolute Error (MAE), Mean Absolute Percentage Error (MAPE)
[2]	2018	smart building	thermal sensor	human behavior recognition	Support Vector Regression (SVR) and Recurrent Neural Network (RNN)	Average Error, Error Rate
[51]	2017	smart building	Wireless Sensor Networks	thermal comfort optimization	Support Vector Regression	Prediction Error
[97]	2017	smart building	passive infrared motion detecting sensors	short-term prediction of occupancy	ANN compared with traditional inhomogeneous Markov Chain model, New Markov Chain model, Probability Sampling model, Support Vector Regression (SVR)	Accuracy (Correctness)
[41]	2017	smart home	temperature and humidity sensors from a Wireless Sensor Network	forecasting the energy use of appliances	comparing: Multiple Linear Regression, Support Vector Machine with Radial Kernel, Random Forest, Gradient Boosting Machines (GBM)	Root Mean Square Error (RMSE), Coefficient of Determination, Mean Absolute Error (MAE), Mean Absolute Percentage Error (MAPE)

In [51], Viani et al. addressed issues regarding the thermal comfort forecasting in smart buildings in order to improve the management of the Heating, Ventilation, Air Conditioning (HVAC) systems, to fulfill the users' requirements and to obtain reduced energy costs. Using a Wireless Sensor Network in order to evaluate the indoor conditions, the authors developed a customized SVR technique in order to determine the indoor temperature necessary to ensure the comfort of the inhabitants. Subsequently, the authors conducted a series of experiments in order to evaluate the performance of their prediction and concluded that the forecasting error was lower than 1 degree Celsius, and that their approach was therefore proved to be useful for ensuring the thermal comfort of the smart building's inhabitants.

In paper [97], Li et al. made use of passive infrared motion detection sensors in order to provide a short-term prediction of occupancy based on an inhomogeneous Markov model. The proposed approach was subsequently compared to existing models such as Probability Sampling, Artificial Neural Network, and Support Vector Regression. With the aim of evaluating the prediction accuracy of their method, the authors took into account various forecasting time intervals, including a quarter of hour, half an hour, one hour, and 24 h. In order to assess the precision of the devised approach at the spatial level, the authors evaluated the forecasting accuracy at both room and house level. The authors observed that their approach outperformed the existing models analyzed, especially when considering the quarter of an hour prediction timeframe, while for the day—ahead prediction, the differences were insignificant.

3.2. *Unsupervised Learning*

Clustering

Subsequently, from the obtained pool of scientific articles obtained based on the devised review methodology, we identified, analyzed and summarized those that make use of the Fuzzy C-Means method integrated with sensor devices in smart buildings. A complete summarization table (Table S12) is provided in the Supplementary Materials file, while Table 12 presents five of the most recent papers addressing this subject.

Examining the papers selected and summarized in Table S12, presented in the Supplementary Materials file, it can be observed that 53% of them focus on smart homes and smart houses, 37% refer to smart buildings in general, and the remaining 10% are equally divided among smart structures, residential buildings and smart spaces. With respect to the publication year, 63% of the identified articles were published during the last 5 years. The authors of these scientific articles made use in their analyses of different types of sensors, including sensors and actuators related to the primary heating circuits and power generation systems [24]; telecare medicine information systems (TMIS) comprising specialized sensors that provide key health data parameters [99]; distributed sensors [100]; temperature, humidity and flame sensors [101]; string-type strain gauges [49]; temperature and occupancy sensors [54]; wireless sensors [47,102]; environment sensors for measuring indoor illuminance, temperature-humidity, carbon dioxide concentration and outdoor rain and wind direction [103]; sensors for measuring the indoor and outdoor temperature and the humidity [39]; vision sensors [55]; sensor networks [56,104]; binary infrared sensors [83]; motion detectors, light sensors, meteorological sensors for the wind and solar radiation data [105]; light and motion sensors [106]; environmental sensors [107]; in-house and city sensors [108]; meteorological stations [46]; smart home sensors, remote monitoring systems, and data and video review systems [102]; temperature and infrared sensors [109]; temperature sensors [110]; inside and outside home sensors [111]; different sensors and effectors [112]; smart systems for controlling the vibration of building structures by means of smart dampers [113]; virtual sensor based on a fisheye video camera [48]; and indoor and outdoor light sensors [114].

Table 12. Five of the most recent scientific articles addressing the Fuzzy C-Means integrated with sensor devices in smart buildings.

Reference	Publication Year	Type of Smart Building	Type of Sensors	Reason for Using the Fuzzy C-Means Method with Sensor Devices	Fuzzy C-Means Only or Hybrid	Performance Metrics
[24]	2019	smart building	more than 450 sensors and actuators related to the primary heating circuits and power generation system, managed by a Supervisory Control and Data Acquisition (SCADA) system	appropriate energy management	a state-of-the-art scalable distributed genetic fuzzy system (GFS) based on scalable fuzzy rule learning through evolution for regression (S-FRULER)	Root Mean Square Error (RMSE), Rules, Time
[99]	2019	smart home	Telecare Medicine Information System (TMIS) comprising specialized sensors that provide key health data parameters	identifying the patients based on their biometric data using a fuzzy extractor within a proposed security protocol	Fuzzy Extractor	the performance is assessed at the level of the whole developed protocol, taking into account the computational costs, user anonymity, mutual authentication, off-line password guessing attacks, impersonation attacks, replay attacks, and the assurance of formal security
[100]	2018	smart home	distributed sensors	object localization	Fuzzy logic techniques compared with similar approaches from other papers: Wireless Network, Radio-Frequency Identification (RFID), Visional Approach	Inaccuracy Rate, Experiment Environment Dimension and Root-Mean-Square Error (RMSE), the dependency of the localization approach to the number of wireless nodes (topology), which are employed to localize the objects
[101]	2018	smart buildings	temperature, humidity and flame sensors	fire monitoring and warning	Fuzzy Logic	Accuracy
[49]	2018	smart building	string-type strain gauge	integrity of the building, assuring public safety	Fuzzy Theory	Coefficient of Determination (R^2)

In these papers, the reasons for using the Fuzzy C-Means with the sensor devices in smart buildings were mainly related to monitoring and controlling energy management processes [24,39,46,47,54,55,106,109]; monitoring building integrity, thus ensuring public safety [49,101,111]; human activity recognition in the context of assisted living [83,99,102,114]; improving indoor environments [48,56,103,105,108,110]; object localization [100]; identifying user location within the smart home [107]; assessing the behavior of a smart home sensor network's nodes [104]; passive Radio-Frequency Identification (RFID) localization in smart homes [112]; and identifying and isolating sensors faults [113].

With respect to the devised research methods, in [24], Rodriguez-Mier et al. developed a state-of-the-art scalable distributed Genetic Fuzzy System (GFS) based on scalable Fuzzy rule learning through evolution for regression (S-FRULER). In [100], Amirjavid et al. made use of Fuzzy Logic techniques and compared them with similar approaches from other papers, including the wireless network, Radio-Frequency Identification (RFID), and Visional approaches. In [83], Zhao et al. implemented the Fuzzy Decision Tree method. In [48], Anthierens et al. made use of the Fuzzy

Logic algorithm. In [112], Fortin-Simard et al. implemented a hybrid approach, using the elliptical trilateration and the Fuzzy Logic method. In [102], Yuan et al. made use of the pervasive healthcare system, the Context-Aware Real-time Assistant (CARA), which combines a case-based reasoning engine and the Fuzzy Logic method. In [101], Sarwar et al. implemented a Fuzzy Logic approach. In [103], Wang et al. made use of a Fuzzy microcontroller implemented by Arduino UNO. In [114], Chen et al. used Fuzzy Logic and Neuro-Fuzzy systems. In [109], Panna et al. implemented a fuzzy temperature controller. In [110], Wang et al. used a Fuzzy Cognitive Map (FCM) in order to develop a genetic algorithm with a view to identifying the connection matrix of the FCM. In [106], Liu et al. implemented a Fuzzy Logic controller. In [49], Chang et al. made use of the Fuzzy theory. In [39], Meana-Llorián et al. used the Fuzzy Logic approach. In [54], Ain et al. implemented a Fuzzy inference system. In [104], Usman et al. used Fuzzy Logic, with the same method being used by Motamed et al. in [55] and by Ulpiani in [56]. In [99], Khatoon et al. made use of the Fuzzy Extractor. In [113], Sharifi et al. developed a semi-active nonlinear Fuzzy Control System. In [107], Ahvar et al. made use of the Fuzzy Set theory. In [111], Sang-Hyun et al. developed an Adaptive Network Fuzzy Inference System (ANFIS). In [105], Kıyak et al. developed a Fuzzy Expert System for testing the light. In [47], Keshtkar et al. developed a Fuzzy Logic Decision-Making algorithm. In [46], Jabłonski made use of a Fuzzy Controller that generates the output settings for the building actuators according to a general Fuzzy Set processing scheme. In [108], a set of concepts and their Fuzzy Semantic relations were defined, extracted and used by Vlachostergiou et al.

The performance metrics considered in the scientific papers that use the Fuzzy C-Means integrated with sensor devices in smart buildings were evaluated based on experiments and simulations [46,47,103,107–109,111,114]; Root Mean Square Error (RMSE) [24]; computational cost, user anonymity, mutual authentication, off-line password guessing attacks, impersonation attacks, replay attacks, and the assurance of formal security [99]; Inaccuracy Rate, experiment environment dimension and Root-Mean-Square Error (RMSE), and the dependency of the localization approach on the number of wireless nodes (topology) employed to locate the objects [100]; Accuracy [101,110]; Coefficient of Determination (R^2) [49]; energy consumption, Electricity Cost, Peak-to-Average Ratio (PAR) [54]; energy saving percentage in different working scenarios [39]; Standard Error of Mean (SEM), Horizontal Illuminance, Daylight Glare Probability, paper-based Landolt test, Freiburg Visual Acuity Test (FrACT), Electric Lighting Energy Consumption, total number of shading and lighting commands [55]; turbulence intensity, draught rates, operative temperature, Predicted Mean Vote (PMV) and Percentage of People Dissatisfied (PPD) [56]; Identification Rate [83]; Energy Consumption and illumination level [105]; energy savings [106]; Detection Accuracy, Energy Consumption, Memory Consumption, Processing Time Estimation [104]; True Positive, False Positive, True Negative, False Negative, and Accuracy [102]; Accuracy and a comparison with the results presented in related works (based on Ultrasonic, Ultrasonic/RFID, ZigBee, Active RFID, Passive RFID) [112]; Fault Detection Index values for certain fault magnitudes, residual values for individual sensors corresponding to different fault magnitudes [113]; and comfort level [48].

With respect to the five most recent scientific articles addressing the Fuzzy C-Means method integrated with sensor devices in smart buildings (Table 12), it can be observed that in [24], Rodriguez-Mier et al. developed a Genetic Fuzzy system designed to build a scalable information database, useful in forecasting smart buildings' energy consumption. To this end, the authors developed a state-of-the-art scalable distributed Genetic Fuzzy System (GFS) based on Scalable Fuzzy Rule Learning through Evolution for Regression (S-FRULER). The authors subsequently carried out experiments based on real data and concluded that the developed approach provided a high level of accuracy.

In [99], Khatoon et al. proposed a secure and efficient authentication method, along with a key agreement protocol for the Telecare Medicine Information System (TMIS), offering healthcare services to patients, particularly to those who were elderly and vulnerable, and were unable to go to hospitals. The developed protocol was based on a Fuzzy Method in order to identify the patients, making use of their biometric data. To ensure the security of the proposed approach and the privacy of the users, the authors made use of the elliptic curves' theory. Subsequently, the authors stated that "the performance is assessed at the level of the whole developed protocol, taking into account the computational costs, user anonymity, mutual authentication, off-line password guessing attacks, impersonation attacks, replay attacks, and the assurance of formal security".

In [100], Amirjavid et al. addressed issues regarding the tracking of objects within smart homes, proposing a method that did not require the attachment of sensors to the targeted objects, making use only of distributed sensors (among which were included visual sensors). The authors developed a series of simulations and, comparing the obtained results with those provided by other state-of-art methods, they concluded that their approach offered an improved performance, as highlighted by the following performance metrics: Inaccuracy Rate, the experiment environment dimension and Root-Mean-Square Error (RMSE), and the dependency of the localization approach on the number of wireless nodes (topology) employed to locate the objects.

In their paper [101], Sarwar et al. presented a Fire Monitoring and Warning System (FMWS), developed based on a Fuzzy Logic approach, that was designed to detect the actual existence of fire and to send alarms to a system providing a complete infrastructure for fire safety management, namely, the Fire Management System (FMS), using the Global System for Mobile (GSM) Communication technology. The authors made use of temperature, humidity and flame sensors in their study. The performance of the developed method was assessed by computing the Accuracy as a performance metric, then it was compared with similar existing methods, with the authors ultimately concluding that their approach had the potential to reduce the rate of false alarms, providing an increased potential to save lives and reduce material damage.

In [49], Chang et al. approached a subject related to both the civil engineering and automatic control fields, analyzing issues regarding the detection in real time of the falling of the tiles that cover building exteriors in Taiwan, endangering public safety. The authors combined the micro-resistance approach and the Fuzzy Theory, implementing string-type strain gauges as sensors, the Coefficient of Determination as a performance metric. They concluded that their developed method represented a feasible approach that could be further utilized with a view to assessing the status of the tiles in real time.

Subsequently, from the obtained pool of scientific articles resulting from the application of the devised review methodology, we identified, analyzed and summarized those making use of the Hidden Markov Model integrated with sensor devices in smart buildings for classification purposes. A complete summarization table (Table S13) is presented in the Supplementary Materials file, while Table 13 presents five of the most recent papers addressing this subject.

Table 13. Five of the most recent scientific articles addressing the Hidden Markov Model integrated with sensor devices in smart buildings.

Reference	Publication Year	Type of Smart Building	Type of Sensors	Reason for Using the Hidden Markov Model with Sensor Devices	Hidden Markov Model Only or Hybrid	Performance Metrics
[25]	2019	smart home	34 sensors (3 door and 31 motion sensors)	sensor-based activity recognition and abnormal behavior detection	Convolutional Neural Networks (CNNs) for detecting abnormal behavior related to dementia, the results being compared with methods such as Naïve Bayes (NB), Hidden Markov Models (HMMs), Hidden Semi-Markov Models (HSMM), Conditional Random Fields (CRFs)	Precision, Recall, F-Measure and Accuracy, Sensitivity, Specificity
[115]	2019	smart building	Wireless Sensor Network	presence detection in a building	Hidden Markov Model (DS-HMM)	Accuracy
[116]	2019	smart home	unobtrusive sensing infrastructures, environmental sensors monitoring the interaction of the inhabitant with home artifacts, context conditions (e.g., temperature) and presence in certain locations	human activity recognition	the developed newNECTAR framework, based on Markov Logic Network compared with state-of-the-art techniques such as Multilayer Perceptron, Random Forest, Support Vector Machine, Naïve Bayes	Average F1 Score, Confusion Matrix
[117]	2019	smart home	passive infrared motion sensors and door sensors	human activity recognition	Hidden Markov Models and Regression Models	Average Accuracy using real data, synthetic data and randomly generated data; Accuracy first using only the real data and then Accuracy using the real data enlarged with a month of synthetically generated data
[118]	2018	smart home	motion sensors, beacons, switches, thermometers	determining the risk of an anomaly related to the healthcare of a resident happening and provide adequate actions to be taken so that a real anomaly does not occur	Markov Logic Network	Precision, Recall, and Correctness

Analyzing the papers selected and summarized in Table S13, it can be observed that 78% of them exclusively analyze smart homes, 16% take into consideration smart buildings in general, 3% analyze both smart homes and buildings, while the remaining 3% of the selected papers refer to smart workplace environments. The authors of these scientific articles make use of different types of sensors in their analyses, including wireless sensor networks [70,115,119–124]; passive infrared motion sensors [82,97,117,118,122,125,126]; motion sensors [25,70,81,118,120,127,128]; environmental sensors [10,25,81,82,116–118,123,127–132]; temperature sensors [116,118,120,123,125,131–133]; humidity sensors [123,131–133];

pressure sensors [128,130,131,133]; light sensors [3,123,132]; unobtrusive sensing infrastructures [116]; real and virtual sensors [134]; radar sensors [135]; accelerometers [127,136]; light-emitting diodes (LED) [3]; electricity and electrical sensors [81,131,132]; smartphone sensors [127,131]; microphones [125, 129]; distributed sensor networks [137]; simple non-intrusive sensors [82]; infrared sensors [124,129–131]; actuators and home automation equipment [125]; shelf binary sensors [128]; biosensors [33]; smart meters [138]; acoustics and CO_2 sensors [133]; non-wearable ambient sensors [131].

With respect to the reasons for using the Hidden Markov Model with sensor equipment in smart buildings, it can be observed that the recognition of human activity is the main subject of the identified papers summarized in Table S13, and is addressed in papers [3,10,25,70,81,82,116,117,120–123,125–127,130–132,135,136]. Additional applications include abnormal behavior detection [25,82,118,126]; presence detection in a building [115]; fault-tolerant maintenance of a networked environment in the domain of the Internet of Things [134]; providing proximity services in smart home and building automation [119]; forecasting the presence of residents at the room and house level [97]; modeling the decision process in the context of a voice-controlled smart home [129]; event recognition in cyber-physical systems [137]; the detection of visits in the home of older adults living alone [128]; emergency psychiatric state prediction [33]; load disaggregation [138]; occupancy detection with a view to energy saving [133]; state estimation for a special class of flag Hidden Markov Models [124].

With respect to the devised methods, the authors of papers [115,122,124,126,133,136,138] implemented solely the Hidden Markov Model, while in other papers, a hybrid approach was used, based on: hidden Markov models and regression models [117]; continuous-time Markov chains, together with a cooperative control algorithm [134]; two layers of classifiers: a first-level Bayesian classifier whose inferential results are used as inputs for the second level Hidden Markov Model (HMM) [135]; Support Vector Machine (SVM), Convolutional Neural Network-Hidden Markov Model (CNN-HMM), and Long Short-Term Memory (LSTM) networks learning algorithms [3]; Beta Process Hidden Markov Model (BP-HMM) and Support Vector Machine (SVM) [10]; Hidden Markov Model and Conditional Random Field model [120]; Random Forest and third-order Markov chain [82]; Hidden Markov Model (HMM), Conditional Random Fields (CRF) and a sequential Markov Logic Network (MLN), the obtained results of which were compared to those of three non-sequential models: a Support Vector Machine (SVM), a Random Forest (RF) and a non-sequential MLN [125]; Hidden Markov Model (HMM), Viterbi path counting, scalable Stochastic Variational Inference (SVI)-based training algorithm, and Generalized Discriminant Analysis [33]; Naïve Bayes classifier, Hidden Markov Model and Viterbi algorithm [70]; Coupled Hidden Markov Model (CHMM) and Factorial Conditional Random Field (FCRF) [123]. Other methods implemented by the authors of the papers selected and summarized in Table S13 include Convolutional Neural Networks (CNNs) for detecting abnormal behavior related to dementia, with the results being compared to methods such as Naïve Bayes (NB), Hidden Markov Models (HMMs), Hidden Semi-Markov Models (HSMM), and Conditional Random Fields (CRFs) [25]; the developed newNECTAR framework, based on Markov Logic Network compared with state-of-the-art techniques such as Multilayer Perceptron, Random Forest, Support Vector Machine, and Naïve Bayes [116]; the Markov Logic Network [118]; the Markov chain model [119]; the Inhomogeneous Markov model compared with the Probability Sampling (PS), Artificial Neural Network (ANN) and Support Vector Regression approaches [97]; the Complex Activity Recognition using Emerging patterns and Random Forest (CARER) compared with Hidden Markov Model, Bayesian Network, Naïve Bayes, SVM, Decision Tree, and Random Forest [81]; the Markov Logic Network [129]; an original proposed model, compared with the results obtained when using the Hidden Markov Model and the Conditional Random Field Model [131]; semi-supervised learning algorithms and Markov-based models [132]; the Markov modulated multidimensional non-homogeneous Poisson process (M3P2) compared with the classical Markov modulated Poisson process (MMPP) [128]; a coupled Hidden Markov Model [127]; semantical Markov Logic Network [137]; Markov Logic Network (MLN) compared with Artificial Neural Network (ANN), Support Vector Machine, Bayesian Network (BN) and Hidden Markov Model [121]; two different approaches: a factorial Hidden Markov model for modeling two separate chains corresponding to two residents, and nonlinear Bayesian tracking for decomposing the observation space into the number of residents.

The performance metrics that chosen by the authors of the scientific papers using the Hidden Markov Model integrated with sensor devices in smart buildings included: Accuracy [3,10,25,33,70,115,117,120,122,123,125,127,131,133,136,138]; Precision [25,118,128,133,135,137]; Recall [25,118,128,135]; F-Measure [25,81,121,130,133]; Sensitivity and Specificity [25,33,133]; F1 Score [116,133]; Confusion Matrix [116,127,129]; and Correctness [97,118]. In addition to the above-mentioned performance metrics, other methods that were used to assess the performance of the developed methods by the authors of the scientific papers selected and summarized in Table S13 included: a numerical case study highlighting the efficiency of the developed model [134]; thread latency [119]; evaluation of energy savings [135]; memory and response time requirements [136]; Mean Squared Error (MSE) [3]; Receiver Operating Characteristic (ROC) scores computed based on the True Positive Rates against the False Positive ones [97]; Mean Recognition Rate [10]; Leave-One-Subject-Out-Cross-Validation (LOSOCV) [129]; execution speed [127]; Local Outlier Factor (LOF), the Z-Score values, cluster transition probability [82]; the APL: Average Path Length, LTA: Location and Time Accuracy, PRDOS: Pressure of Receiving Data On Sink Node, and APRDOS: average PRDOS of sink node [122]; the probability of error [124]; a series of experiments along with the F-Value [128]; simulation tests in order to compare the Generalized Version Space (GVS) algorithm with a simple method using an epsilon greedy mechanism [132]; the Area Under the ROC Curve (AUC) [33]; Correlation Factors depicting the similarities between simulated and real displacement activities [126]; and the heuristic merit of a sensor feature subset S containing k features [123].

With respect to the five most recent scientific articles addressing the Nearest Neighbor method integrated with sensor devices in smart buildings (Table 13), it can be observed that in [25], Arifoglu et al. analyzed the possibility of detecting abnormal behavior in elderly people in order to identify early indicators and symptoms associated with a decline in memory, indicating dementia or brain disease, by making use of Convolutional Neural Networks. After identifying patterns within the daily activity and abnormal activities within them, the authors compared the performance of their approach with those obtained when using other methods, such as Naïve Bayes, Hidden Markov Models (HMMs), Hidden Semi-Markov Models, and Conditional Random Fields (computing the Precision, Recall, F-measure and Accuracy, Sensitivity, Specificity), and concluded that the developed approach was comparable with the state-of-art methods.

In [115], Papatsimpa et al. addressed issues regarding the human presence in a smart building equipped with a Wireless Sensor Network, making use of various Hidden Markov Models (HMMs). The authors proposed a method based on an efficient transmission strategy along with a blending algorithm that was designed to combine data from various Hidden Markov Models perceiving the same Markovian process. To evaluate their approach, the authors analyzed a series of experimental results and stated that these results confirmed the functionality and benefits of their developed method. Taking into account the accuracy of their scheme, along with the reduction in terms of communication requirements, the authors concluded that their method was suitable and applicable for many situations requiring information merging in wireless sensor devices.

In [116], Civitarese et al. focus on human activity recognition with a view to developing an affordable ambient assisted living approach, ensuring the individual's data privacy. To this end, the authors developed a hybrid approach, combining collaborative active learning with probabilistic and knowledge-based reasoning. The authors developed the newNECTAR framework, which was based on the Markov Logic Network, and compared it with state-of-the-art techniques (such as Multilayer Perceptron, Random Forest, Support Vector Machine, Naïve Bayes). The authors concluded that their developed learning solution improved recognition rates, generated a reduced number of feedback requests, and was comparable and sometimes even better than other existing activity recognition methods based on the performance metrics used (the Average F1 Score and Confusion Matrix).

In [117], Dahmen et al. analyzed methods for "testing machine learning techniques for healthcare applications", aiming to overcome the limitations related to the complexity and lack of applicability of many actual approaches. To this end, the authors developed a synthetic data generation method based on Machine Learning techniques, SynSys. The authors made use of Hidden Markov Models and regression models, and afterwards, they tested the generated set of synthetic data on a dataset recorded from a real smart home. To evaluate the developed approach, the authors made use of the following performance metrics: the Average Accuracy using real data, synthetic data and randomly generated data; the Accuracy first using only the real data, and then the Accuracy using the real data enlarged by a month of synthetically generated data. The authors concluded that their data generation method had the ability to provide a higher human activity recognition accuracy than that obtained when solely using real data.

In paper [118], Sfar et al. developed an approach for early detection of abnormal behavior in elderly people living in smart homes, in order to prevent risks related to their health, based on identifying and extracting anomalous causes from datasets, making use of causal association rules mining. These causes were subsequently used in order to detect the risks of anomalies occurring by using the Markov Logic Network Machine Learning method. The authors evaluated their approach by using real datasets, concluding that the devised method proved to be efficient in terms of the computed performance metrics (Precision, Recall, Recognition Rate and Correctness).

Subsequently, from obtained pool of scientific articles obtained based on the devised review methodology, we identified, analyzed and summarized those making use of Hierarchical Clustering integrated with sensor devices in smart buildings. A complete summarization table (Table S14) is presented in the Supplementary Materials file, while Table 14 presents the most recent papers targeting this subject.

Table 14. The most recent scientific articles addressing the Hierarchical Clustering integrated with sensor devices in smart buildings.

Reference	Publication Year	Type of Smart Building	Type of Sensors	Reason for Using the Hierarchical Clustering Approach with Sensor Devices	Hierarchical Clustering Only or Hybrid	Performance Metrics
[19]	2019	smart building	smartphone sensors and Bluetooth beacons data	group activity detection and recognition	a framework for indoor Group Activity Detection and Recognition (GADAR) and Hierarchical Clustering, along with Decision Tree classifier, K-Neighbors classifier, Deep Neural Network, Gaussian Process classifier, Logistic regression, Support Vector Machine, Linear Discriminant Analysis, Gaussian Naïve Bayes (comparison)	Confusion Matrix, Accuracy (Mean), Accuracy (Variation), Precision, Recall, F1 Score
[37]	2019	smart building	WiFi-enabled IoT device-user	Personalized location-based service	hybrid: Hierarchical Clustering and Location Similarity Matching	Accuracy
[139]	2014	smart building	smart meters organized into clusters	data collection in hierarchical smart building networks	hybrid hierarchical clustering containing a two-layer transmission process	simulations scenarios, comparison of the proposed scheme's performance with the performance of the Uniform Algorithm

Analyzing the papers selected and summarized in Table S14, presented in the Supplementary Materials file, it can be observed that all of them analyze smart buildings in general. In these papers, the authors make use of different types of sensors, for example: smartphone sensors and Bluetooth beacons data [19]; WiFi-Enabled IoT Device-User [37]; smart meters organized into clusters [139]. In these papers, the reasons for using the Hierarchical Clustering approach with the sensor devices in smart buildings are related to group activity detection and recognition [19]; Personalized Location-Based Services [37]; and data collection in hierarchical smart building networks [139].

Regarding the devised research methods, in [19], Chen et al. made use of a hybrid approach, combining a framework for indoor group activity detection/recognition and hierarchical clustering, along with the Decision Tree classifier, K-Neighbors classifier, Deep Neural Network, Gaussian Process Classifier, Logistic Regression, Support Vector Machine, Linear Discriminant Analysis, Gaussian Naïve Bayes, making a comparison between these techniques. In [37], Zou et al. developed a hybrid approach, combining Hierarchical Clustering and location similarity matching. In [139], Luan et al. made use of a hybrid Hierarchical Clustering containing a two-layer transmission process.

The performance metrics considered in the scientific papers that use the Hierarchical Clustering method integrated with sensor devices in smart buildings include: the Confusion Matrix, Accuracy (Mean), Accuracy (Variation), Precision, Recall, F1 Score [19]; Accuracy [37]; and the development of simulated scenarios and a comparison of the proposed scheme's performance with that of the uniform algorithm, in which the cluster heads are uniformly distributed and the resources are uniformly allocated [139].

With respect to the most recent scientific articles addressing the Hierarchical Clustering method integrated with sensor devices in smart buildings (Table 14), it can be observed that in [37], Zou et al. addressed personalized location-based services in smart buildings. To this end, the authors developed a method that used a non-intrusive device, based on WiFi technology, and an association scheme based on an unsupervised learning algorithm. The authors developed a hybrid approach, combining Hierarchical Clustering and location similarity matching. To test the performance of the developed approach, the authors conducted a series of experiments and, using Accuracy as a performance metric, concluded that their method had the potential to be implemented in real-world situations, "for practical personalized context-aware and location-based services in the era of IoT".

The scientific paper [19] was reviewed previously when analyzing the most recent scientific articles that integrate Decision Tree approaches with sensor devices in smart buildings (Table 6).

In [139], Luan et al. proposed a hybrid cooperation scheme useful in collecting data in hierarchical smart buildings networks, making use of machine-to-machine communication. In this study, the authors used smart meters organized into clusters as sensors, sending information to the cluster-heads. The authors developed hybrid Hierarchical Clustering, containing a two-layer transmission process. In the first-layer transmission, the distributed smart meters send the data to their respective cluster heads. In the second-layer transmission, the cluster-heads forward all of the data to the base station. With a view to highlighting the advantages and properties of their developed scheme, the authors developed a series of simulated scenarios and compared the proposed scheme's performance with that of the uniform algorithm, whereby the cluster heads were uniformly distributed and the resources were uniformly allocated.

Subsequently, from the obtained pool of scientific articles resulting from the application of the devised review methodology, we identified, analyzed and summarized those making use of the K-Means integrated with sensor devices in smart buildings for classification purposes. A complete summarization table (Table S15) is presented in the Supplementary Materials file, while Table 15 presents the most recent papers addressing this subject.

Examining the papers selected and summarized in Table S15, presented in the Supplementary Materials file, it can be observed that 67% of them take into consideration smart buildings in general, while the remaining 33% refer to smart homes. The authors of these scientific articles made use of different types of sensors in their analyses, including binary sensors [26]; sensor networks [140]; smart meters, Personal Weather Stations (PWS), and sensors providing data useful in computing the mean values of: hourly indoor temperature, hourly outdoor temperature, hourly value of precipitation, hourly value of wind direction, hourly value of solar radiation, hourly value of ultraviolet index, hourly value of humidity, hourly value of pressure [42].

Table 15. The most recent scientific articles addressing the K-Means integrated with sensor devices in smart buildings.

Reference	Publication Year	Type of Smart Building	Type of Sensors	Reason for Using the K-Means Method with Sensor Devices	K-Means Only or Hybrid	Performance Metrics
[26]	2018	smart home	binary sensors	extraction of behavioral patterns	hybrid: K-Means Algorithm combined with Nominal Matrix Factorization method	comparison with existing methods based on both synthetic and publicly available real smart home datasets
[140]	2018	smart buildings	sensor network	discovering electricity consumption patterns	Cluster Validation Indices (CVIs) for establishing the optimal number of clusters k for the dataset, combined with the Parallelized Version of K-Means Clustering Algorithm for discovering patterns from the dataset	Cluster Analysis, Centroids of the electricity consumption clusters, Centroids of the clusters with lower consumptions, Computing times
[42]	2018	smart building	Smart meters, Personal Weather Stations (PWS), sensors providing data useful in computing the mean values of: hourly indoor temperature, hourly outdoor temperature, hourly value of precipitation, hourly value of wind direction, hourly value of solar radiation, hourly value of ultraviolet index, hourly value of humidity, hourly value of pressure	managing energy consumption	Data Mining Engine, METATECH (METeorological data Analysis for Thermal Energy CHaracterization)	Support, Confidence and Lift

In these papers, the reasons for using the K-Means method with the sensor devices in smart buildings were related to extraction of behavioral patterns [26]; determining electricity consumption patterns [140]; and managing energy consumption [42]. With respect to the devised research methods, in [26], Li et al. made use of a hybrid approach, combining the K-Means algorithm with Nominal Matrix Factorization method. In [140], Pérez-Chacón et al. used the Cluster Validation Indices (CVIs) method to establish the optimal number of clusters for the dataset, combined with the parallelized version of K-Means clustering algorithm for discovering patterns from the dataset. In [42], Di Corso et al. implemented the data mining engine, METATECH (METeorological data Analysis for Thermal Energy CHaracterization), which computes the similarity between two objects by using the Euclidean distance, and integrates a partitional algorithm, the K-Means algorithm.

The performance metrics considered in the scientific papers using the K-Means integrated with sensor devices in smart buildings were evaluated based on a comparison with existing methods based on both synthetic and publicly available real smart home datasets [26]; cluster analysis, centroids of the electricity consumption clusters, centroids of the clusters with lower consumptions, and computing times [140]; and support, confidence and lift [42].

Regarding the most recent scientific articles that make use of the K-Means method along with sensor devices in smart buildings (Table 15), it can be observed that in [26], Li et al. aimed to devise a methodology for the automatic detection of the behavioral patterns of elderly people living in smart homes. The authors made use of binary sensors and devised a hybrid approach, combining the K-Means algorithm with Nominal Matrix Factorization method in order to obtain the daily routines. To assess the performance and suitability of their method, the authors compared their developed approach with existing methods based on both synthetic and publicly available real smart home datasets and considered their obtained results to be promising.

In [140], Pérez-Chacón et al. proposed a method for identifying patterns in big data time series with respect to energy consumption in smart buildings, making use of sensor networks. The authors based their approach on Cluster Validation Indices (CVIs) for establishing the optimal number of clusters for the dataset, combined with the parallelized version of K-Means clustering algorithm (from the Apache Spark's Machine Learning Library) in order to discover patterns from the dataset. The devised method was tested using a large dataset, representing the energy consumption of eight smart buildings over a seven-year period (2011–2017). As performance metrics, the authors used cluster analysis, centroids of the electricity consumption cluster, and centroids of the clusters with lower consumptions, along with computing times, and concluded that their devised approach represented a valuable tool for the optimization of energy usage.

In paper [42], Di Corso et al. proposed a data mining engine, METeorological Data Analysis for Thermal Energy CHaracterization (METATECH), which computes the similarity between two objects by using the Euclidean distance, and integrates a partitional algorithm, the K-Means algorithm. The authors made use of various types of sensors, including Smart meters, Personal Weather Stations (PWS), and sensors providing data useful in computing the mean values of: hourly indoor temperature, hourly outdoor temperature, hourly value of precipitation, hourly value of wind direction, hourly value of solar radiation, hourly value of ultraviolet index, hourly value of humidity, and hourly value of pressure [42]. The devised method aimed to develop models for correlating meteorological conditions and the energy consumption in smart buildings at various levels of granularity. To validate the devised approach, the authors performed a series of experimental tests using real datasets and concluded that these tests highlighted the effectiveness of their method in the process of data mining.

3.3. Deep Learning Techniques

Taking into account recent increases in the computational power of hardware processing architectures (especially parallel processing ones), which have led to the widespread application of Deep Learning techniques, in addition to the above-mentioned categories, we also identified, analyzed and summarized, with respect to the obtained pool of scientific papers, those that make use of Deep Learning techniques with sensor devices in the smart building sector. A selection of the most recent papers (sorted in descending order of publication year) is presented in Table 16, while a comprehensive summarization table can be found in the Supplementary Materials file (Table S16).

It can be observed that 78% of the scientific papers selected and summarized in Table S16, presented in the Supplementary Materials file, focused their research exclusively on smart homes, while 17% focused on smart buildings in general, and the remaining 5% focused on smart commercial and residential buildings.

In these papers, the authors made use of different types of sensors, including motion sensors [18,25,28,141–145]; temperature sensors [28,40,73,143,144]; wireless sensor networks [21,40,141,145]; door sensors [25,143]; smartphone inertial sensors [146] and a smartphone application [36]; cameras [18]; a two-dimensional acoustic array [27]; daily activity recognition sensors [28]; actuators [143]; tactile sensors, power meters, and microphones in the ceiling [144]; non-wearable sensors [147]; unobtrusive sensors [9]; environmental sensors [73,142]; weather sensors [12]; WiFi-enabled sensors for food nutrition quantification [36]; and binary sensors [148].

Table 16. Five of the most recent scientific articles addressing Deep Learning techniques integrated with sensor devices in smart buildings.

Reference	Publication Year	Type of Smart Building	Type of Sensors	Reason for Using the Deep Learning Method with Sensor Devices	Deep Learning Only or Hybrid	Performance Metrics
[18]	2019	smart home	wearable hybrid sensor system comprising motion sensors and cameras	human activity recognition in medical care, smart homes, and security monitoring	hybrid approach, combining Long Short-Term Memory (LSTM) and Convolutional Neural Network (CNN) methods	Confusion Matrices, F1 Accuracy
[27]	2019	smart home	a two-dimensional acoustic array	human activity recognition	Convolutional Neural Networks compared with traditional recognition approaches such as K-Nearest Neighbor and Support Vector Machines	Overall Accuracy
[28]	2019	smart home	daily activities recognition sensors, infrared motion and temperature sensors	human activity recognition	hybrid, using Term Frequency-Inverse Document Frequency (TF-IDF), along with each of the Support Vector Machine (SVM), Sequential Minimal Optimization (SMO), and Random Forest (RF), Long Short-Term Memory (LSTM) methods and comparison between them	Accuracy, Precision, and F-Measure
[25]	2019	smart home	34 sensors (3 door and 31 motion sensors)	sensor-based activity recognition and abnormal behavior detection	Convolutional Neural Networks (CNNs) for detecting abnormal behavior related to dementia, the results being compared with methods such as Naïve Bayes (NB), Hidden Markov Models (HMMs), Hidden Semi-Markov Models (HSMM), Conditional Random Fields (CRFs)	Precision, Recall, F-Measure and Accuracy, Sensitivity, Specificity
[21]	2019	smart building	wireless sensor networks	forecasting Packet Delivery Ratio (PDR) and Energy Consumption (EC) in Internet of Things (IoT)	comparison between Linear Regression, Gradient Boosting, Random Forest, Baseline and Deep Learning Neural Networks	Root Mean Square Error (RMSE), Mean Percentage Error (MPE), and Mean Absolute Percentage Error (MAPE)

In the scientific papers selected and summarized in Table S16, the reasons for using Deep Learning techniques integrated with sensor devices in smart buildings were mainly related to human activity recognition [9,18,25,27,28,73,142,143,145–148]; ensuring health care [18,25,142]; forecasting Packet Delivery Ratio (PDR) and Energy Consumption (EC) in Internet of Things (IoT) [21]; realizing small and big data management [141]; adaptive decision-making in smart homes [144]; thermal comfort modeling [40]; forecasting the electricity consumption [12]; and Internet of Things (IoT)-based fully automated nutrition monitoring systems [36].

With respect to the devised methods, in paper [18], the authors made use of a hybrid approach, combining the Long Short-Term Memory (LSTM) networks with the Convolutional Neural Network (CNN) approach. In [27], the authors implemented Convolutional Neural Networks, and compared them with traditional recognition approaches such as K-Nearest Neighbor and Support Vector Machines. The authors of [28] developed a hybrid approach, using Term Frequency-Inverse Document Frequency (TF-IDF), along with the Support Vector Machine (SVM), Sequential Minimal Optimization (SMO), and Random Forest (RF), Long Short-Term Memory (LSTM) methods and compared them. In [25], the authors made use of Convolutional Neural Networks (CNNs) for detecting abnormal behavior related to dementia, the results were compared with methods such as Naïve Bayes (NB), Hidden Markov Models (HMMs), Hidden Semi-Markov Models (HSMM), and Conditional Random Fields (CRFs). In [21], Ateeq et al. compared the Linear Regression, Gradient Boosting, Random Forest, Baseline and Deep Learning Neural Networks. The authors of [141] used Deep Neural Networks for system monitoring and optimization. In [146], the authors implemented a Deep Belief Network (DBN), comparing it with Support Vector Machine (SVM) and Artificial Neural Network (ANN) approaches. In [142], the authors developed a hybrid Deep Learning-based gesture/locomotion recognition model, integrating CNN and RNN. In [143], the authors made use of different Deep Learning (DL) models based on Long Short-Term Memory (LSTM), comparing their approach with the Hidden Markov Model (HMM), Conditional Random Field (CRF), and Naïve Bayes (NB) approaches. In [144], the authors developed a hybrid method, namely the Adaptive Reinforced Context-Aware Deep Decision System (ARCADES), combining Deep Neural Networks and Reinforcement Learning (RL). In [145], the authors compared Recurrent Neural Networks (Long Short-Term Memory, Gated Recurrent Units), Convolutional Neural Network, Behavior Explanatory Models, and Sensor Profiles. In [147], the authors developed a Deep Learning technique, namely the Recurrent Neural Network (RNN), using the Long Short-Term Memory (LSTM) architecture. In [9], the authors made use of the SVM classifier along with two different feature extraction methods: a manually defined method and a Convolutional Neural Network (CNN). The authors of [40] developed an intelligent Thermal Comfort Management (iTCM) system black-box neural network (ITCNN), whose performance was compared with the Fanger's Predicted Mean Vote (PMV) model and six classical machine learning approaches: three traditional white-box machine learning approaches and three classical black-box machine learning methods. In [73], the authors made use of a Deep Convolutional Neural Network (DCNN), comparing it with the Naïve Bayes (NB) and Back-Propagation (BP) algorithms. In [12], the authors used deep Recurrent Neural Network (RNN) models. In [36], the authors implemented Bayesian algorithms and the 5-layer Perceptron Neural Network method for diet monitoring. In [148], the authors developed an Activity Recognition (AR) model based on Deep Learning for two cases: one-layer Denoising Autoencoder (DAE) and two-layer Stacked Denoising Autoencoder (SDAE). The results obtained were compared with those obtained by five commonly used baselines: Naïve Bayes (NB), Hidden Markov Model (HMM), Hidden Semi-Markov Model (HSMM), K-Nearest-Neighbor (KNN), and Support Vector Machine (SVM) with linear kernel.

The performance metrics chosen by the authors of the papers focusing on Deep Learning techniques integrated with sensor devices in smart buildings included Confusion Matrices and F1 Accuracy [18]; Overall Accuracy [27]; Accuracy, Precision and F-Measure [28]; Precision, Recall, F-Measure, Accuracy, Sensitivity, and Specificity [25]; Root Mean Squared Error (RMSE), Mean Percentage Error (MPE), and Mean Absolute Percentage Error (MAPE) [21]; Overall Accuracy and Mean Recognition Rate [146]; Accuracy, Precision, Recall and F1 Score [142,143]; reward per episode, Precision, Recall, F1 Score [144]; methods discussed and evaluated on the basis of real-life data and the Confusion Matrix [145]; Accuracy and Root Mean Square Error (RMSE) [9]; energy cost savings [40]; Precision, Specificity, Recall, F1 Score, Accuracy, Total Accuracy, and Confusion Matrix [73]; Root Mean Squared Error relative to Root Mean Squared (RMS) average of electricity consumption in test data, Root Mean Squared Error relative to Root Mean Squared (RMS), average of electricity consumption in training data, and Pearson

Coefficient [12]; Accuracy of classification of food items and meal prediction [36]; and time-slice accuracy and class accuracy [148].

With respect to the most recent scientific articles that make use of Deep Learning techniques along with sensor devices in smart buildings (Table 16), it can be observed that papers [18] and [27] were reviewed previously when analyzing the most recent scientific articles integrating Neural Networks for classification purposes with sensor devices in smart buildings (Table 5). Paper [25] was reviewed when analyzing the most recent scientific articles integrating the Hidden Markov Model with sensor devices in smart buildings (Table 13). Article [21] was detailed when analyzing the most recent scientific articles integrating Decision Tree with sensor devices in smart buildings (Table 6).

In paper [28], Guo et al. aimed to achieve human activity recognition based on a non-invasive method in order to improve residents' lives. In their research, the authors made use of daily activity recognition sensors, and infrared motion and temperature sensors, and developed a hybrid approach using Term Frequency–Inverse Document Frequency (TF-IDF), along with the Support Vector Machine (SVM), Sequential Minimal Optimization (SMO), Random Forest (RF), and Long Short-Term Memory (LSTM) methods, carrying out a comparison between them. By computing the Accuracy, Precision and F-Measure performance metrics, the authors evaluate the Machine Learning methods and Deep Learning technique, thereby concluding that their strategy, based on the Term Frequency-Inverse Document Frequency (TF-IDF) approach, has the potential to improve the performance of human activity recognition systems.

In the following, we review the most frequently cited articles from the scientific papers pool addressing the reviewed topics, as reported by the two considered international databases.

3.4. Frequently Cited Scientific Papers Addressing the Reviewed Topics, as Reported by the Elsevier Scopus and the Clarivate Analytics Web of Science International Databases

We devised our research methodology and conducted our review with a view to identifying, filtering, categorizing, and analyzing the most important and relevant scientific articles with respect to recent developments in the integration of machine learning models with sensor devices in the smart buildings sector with a view to attaining enhanced sensing, energy efficiency, and optimal building management. Therefore, we focused our attention on the most recent scientific papers, meanwhile being aware of the fact that these topics represent an important subject, and that new research is disseminated day by day throughout the scientific literature. In addition to this, the choice to review the most recent scientific works addressing developments concerning the integration of machine learning models with sensor devices in the smart buildings sector offers the possibility of grasping the recent advancements in technology and sensing equipment.

Another criterion that can be addressed when devising a review paper is based on the visibility of the papers in the scientific literature, evaluated on the basis of their number of citations. Nevertheless, this approach has its disadvantages, due to the fact that in this way, the most recent papers may not be taken into account, as they have not had the chance to be cited as frequently as those published at an earlier date, as sufficient time has not yet elapsed since their publication. However, in order to highlight the most visible papers in the scientific literature that address the reviewed topics, in addition to the above-mentioned analysis, we also identified, analyzed and summarized from the obtained scientific papers pool the most frequently cited scientific papers, as reported by the Clarivate Analytics Web of Science (WoS) and the Elsevier Scopus (ES) international databases. These papers are summarized in Table 17, sorted into descending order of number of citations.

Table 17. The most frequently cited scientific articles addressing Machine Learning Models integrated with sensor devices in smart buildings as reported by the WoS and the ES international databases.

Reference	Publication Year	Number of Citations According to		Type of Smart Building	Type of Sensors	Reason for Using the Machine Learning Models with Sensor Devices	Machine Learning Models Only or Hybrid	Performance Metrics
		WoS	ES					
[44]	2012	170	197	smart building	energy smart meters, building management systems, and weather stations	energy consumption forecasting	a model based on Support Vector Regression (SVR) using the Scikit-learn module, which provides a Python front-end to LIBSVM, a widely cited Support Vector Machine library	Coefficient of Variation (CV) and Standard Error In %
[75]	2011	79	118	smart home	Passive Infra-Red (PIR) sensors or motion detectors; door/window entry point sensors; electricity power usage sensors; bed/sofa pressure sensors; flood sensors	human activity recognition for detecting and predicting abnormal behavior	Echo State Network (ESN), Back Propagation Through Time (BPTT) and Real Time Recurrent Learning (RTRL)	Root Mean Square Error (RMSE)
[123]	2011	60	76	smart home	wireless sensor network highlighting the user movement (i.e., both hands), user location, human-object interaction (i.e., objects touched and sound), human-to-human interaction (i.e., voice), environmental information (i.e., temperature, humidity and light)	human activity recognition	Coupled Hidden Markov Model (CHMM) and Factorial Conditional Random Field (FCRF)	Accuracy, the heuristic merit of a sensor feature subset S containing k features
[65]	2007	54	66	smart home	sensors for HVAC Chillers	optimal sensor selection in complex system monitoring problems	comparison of: Support Vector Machines (SVMs), Principal Component Analysis (PCA), and Partial Least Squares (PLS)	Recognition Rate
[138]	2007	53	64	smart home	smart meter	load disaggregation	Nonintrusive Load Monitoring (NILM), Hidden Markov Models	Accuracy

Analyzing the papers selected and summarized in Table 17, it can be observed that 80% of them focus exclusively on smart homes, while the remaining 20% take into consideration smart buildings in general. The authors of these scientific articles make use of different types of sensors in their analyses, including energy smart meters, building management systems, and weather stations [44]; Passive Infra-Red (PIR) sensors or motion detectors; door/window entry point sensors; electricity power usage sensors; bed/sofa pressure sensors; flood sensors [75]; wireless sensor network highlighting user movement, user location, human-object interaction, human-to-human interaction, environmental information [123]; sensors for HVAC chillers [65]; and smart meters [138].

In these papers, the reasons for using Machine Learning Models with sensor equipment in the smart buildings are mainly related to the recognition of human activity [75,123]; forecasting of energy consumption [44]; optimal sensor selection in complex system monitoring problems [65]; and load disaggregation [138].

With respect to the devised methods, in [44], Jain et al. developed a model based on Support Vector Regression (SVR). In [75], Lotfi et al. made use of the Echo State Network (ESN), Back Propagation Through Time (BPTT) and Real-Time Recurrent Learning (RTRL) methods. In [123], Wang et al. made use of the Coupled Hidden Markov Model (CHMM) and Factorial Conditional Random Field (FCRF) methods. In [65], Namburu et al. compared Support Vector Machines (SVMs), Principal Component Analysis (PCA), and Partial Least Squares (PLS) methods. In [138] Egarter et al. solely implemented the Hidden Markov Model.

The performance metrics chosen by the authors of the most frequently cited scientific articles addressing Machine Learning Models integrated with sensor devices in smart buildings reported by the WoS and the ES International Databases included the Coefficient of Variation (CV) and Standard Error [44]; Root Mean Square Error (RMSE) [75]; Accuracy, the heuristic merit of a sensor feature subset containing a certain number of features [123]; Recognition Rate [65]; and Accuracy [138].

By analyzing the most frequently cited scientific articles addressing Machine Learning Models integrated with sensor devices in smart buildings reported by the Clarivate Analytics Web of Science and the Elsevier Scopus international databases (Table 17), it can be observed that in [44], Jain et al. started their study by highlighting the importance of the accurate forecasting of a building's energy consumption in order to achieve appropriate, efficient urban energy management. To this end, the authors developed a forecasting model based on the Support Vector Regression method, and applied it to a residential building in New York City, endowed with various types of sensors such as weather stations, smart meters and building management systems. The authors analyzed the impact of spatial and temporal granularity on forecasting accuracy by taking into consideration several parts of the building and a variety of time intervals. By comparing the obtained results, using the Coefficient of Variation (CV) and the Standard Error as performance metrics, the authors concluded that the best results were those registered when forecasting the energy consumption at the floor level, with an hourly timeframe.

In [75], Lotfi et al. proposed a method for monitoring the activities of elderly people living alone in homes equipped with sensor networks (comprising motion and door sensors) by detecting and predicting any abnormal behavior. The authors presented methods for analyzing the large datasets retrieved from the sensors, representing them in formats that were suitable for grouping the abnormalities. Subsequently, they used recurrent neural networks in order to predict potential upcoming values of the activities monitored by each implemented sensor. Thereby, if an abnormal behavior were forecasted to take place, health professionals could be informed. The authors compare their Echo State Network (ESN) approach with those based on other recurrent neural network techniques such as the Back Propagation Through Time (BPTT) and Real-Time Recurrent Learning (RTRL), using the Root Mean Square Error (RMSE) and the training time as performance metrics, concluding that the forecasting results provided by the ESN approach were better than those of the other two approaches with respect to training time. The developed forecasting method was evaluated by implementing it in a smart home inhabited by elderly people suffering from brain diseases.

A wireless sensor network highlighting environmental information, user location, user movement, human-to-human interactions, and human-object interactions was used by Wang et al. in [123] with the aim of multi-user activity recognition in smart homes. The authors made use of a wearable sensor platform in order to retrieve data from multiple users, modeling the interaction processes by the means of two models, namely, the Coupled Hidden Markov Model (CHMM) and the Factorial Conditional Random Field (FCRF). The authors conducted a series of experiments in order to assess the performance of the two developed probabilistic models, concluding that the CHMM model provided an accuracy of 96.41%, while the FCRF model registered an accuracy of 87.93% with respect to multi-user activity recognition.

Acknowledging the importance of the Chillers as components in Heating, Ventilating and Air-Conditioning (HVAC) systems, and the fact that they involve significant energy consumption, in [65], Namburu et al. proposed a generic Fault Detection and Diagnosis (FDD) scheme for centrifugal chillers and "a nominal data-driven model of the chiller" that could be useful in forecasting the system response under changing loading conditions. The authors made use of sensors for HVAC Chillers in order to achieve "an optimal sensor selection in complex system monitoring problems", and compared the Support Vector Machines (SVMs), Principal Component Analysis (PCA), and Partial Least Squares (PLS) classification techniques using the Recognition Rate as a performance metric. Using an approach based on a genetic algorithm, the authors selected the sensor suite that was most suitable for forecasting system response in the context of new loading conditions and also assessed the performance provided by the above-mentioned classification techniques when using the identified sensor suite. Using the loading conditions obtained through the nominal model, the authors forecast the responses of the sensor suite. Afterwards, the authors used real HVAC equipment in order to obtain a benchmark dataset for use in validating the developed approach.

In [138], Egarter et al. addressed issues regarding Particle Filter-Based Load Disaggregation (PALDi) in smart homes. The authors commenced their study by highlighting the fact that smart meters provide information that can be used in order to disaggregate appliance consumption by means of Nonintrusive Load Monitoring (NILM), a method that analyzes the consumption provided by the smart meter device within the smart home and identifies the appliances that are being used in the house, along with their individual associated consumption. The authors made use of the NILM method and estimated the appliance states using the particle filtering approach. Using Hidden Markov Models for modeling the appliances and their combinations, the authors obtained a description of the household power demand. Afterwards, in order to evaluate the developed approach, the authors made use of generated and real datasets and concluded that their method registered an accuracy of 90% when detecting the appliance states in the real dataset case.

4. Discussion and Conclusions

The conducted review focused on recent developments in the scientific literature with respect to the integration of Machine Learning models with sensor devices in the smart buildings sector with a view to attaining enhanced sensing, energy efficiency, and optimal building management. To ensure the quality and reliability of the reviewed works, prominent scientific databases (the Elsevier Scopus and the Clarivate Analytics Web of Science) were used as a means to devise custom tailored queries.

In contrast to other, previously existing review papers, our approach was focused on recent scientific articles, highlighting and comparing, for these papers, the details regarding publication year, type of smart building, types of sensor device implemented, reason for using the respective method with sensor devices, developed approach, and the performance metrics implemented in the study. We first conducted an overall comparative analysis of the pool of scientific papers identified according to the devised review methodology with respect to a previously identified and constructed taxonomy. Subsequently, for each taxonomy branch, the most recent scientific articles were analyzed separately, emphasizing the details of the implementation, along with the specific aspects pertaining to the respective papers.

A review of the most recent scientific articles that deal with emergent topics like machine learning, sensor devices and smart buildings offers a series of undeniable advantages in terms of categorizing a high number of scientific articles according to a clear, comprehensive taxonomy. This review article offers a useful up-to-date overview for researchers from different fields who may wish to submit a project proposal or study complex topics like those reviewed.

At the same time, by reviewing recent advancements in the integration of Machine Learning models with sensor devices in the smart buildings sector, the current study offers scientists the possibility of identifying future research directions that have not yet been addressed in the scientific literature or of improving the approaches that already exist within the body of knowledge. The conducted review provides the possibility of identifying the main applications for which approaches have been developed in the literature integrating Machine Learning techniques with sensing devices in smart environments, as well as those applications that have not yet been pursued.

An important challenge that still remains after decades of evolving research in the semiconductors field is the need to develop novel low-power sensing equipment, considering that the vast majority of sensing devices rely for their operation on different power sources, thereby incurring power consumption costs for the acquisition, processing and transmission of the data streams in addition to the physical wiring installation and maintenance costs when using them at the level of an entire smart building. As can be seen from the results of the performed survey, several methods process the data locally, while others adopt a cloud-based approach. Both of these proposed approaches raise important challenges with regard to data processing, power consumption and data transmission power consumption costs. While local processing of the acquired data consumes computational and power resources on the long run, uploading the data into the cloud raises several security-related challenges, including confidentiality, authenticity, integrity, non-repudiation, and accountability. In addition, there is a need for future studies to focus on developing optimized compression algorithms and uploading schemes for the acquired data into the cloud systems, considering that this process is consumes resources from an energy requirements point of view. It is the authors' opinion that the integration of machine learning techniques in sensing equipment benefits not only enhanced sensing, but the development of optimized processing and uploading strategies, in the end leading to a reduction in the overall energy consumption.

When analyzing the pool of scientific works obtained after applying the devised review methodology, we noticed an important aspect that had not been taken into consideration by the scientific papers focusing on human-centric society and on the improvement of the life quality, namely, the perceived notion of "comfort". According to the International Organization for Standardization (ISO) and the International Electrotechnical Commission (IEC) 25010 specifications [149], comfort is defined as the "degree to which the user is satisfied with physical comfort", and this physical comfort can often be a matter of individual perception, being dependent to some extent to a human being's acoustic, visual, thermal and sensorial traits, while also being influenced by gender, age, and overall health status.

An important aspect that should be further studied by researchers and implemented in practice is improving the data security and privacy of IoT systems, due to the fact that most of the data that resulting from the processes highlighted by our review paper, in which machine learning models are integrated with sensor devices in the smart buildings sector, contain sensitive, personal information related to the inhabitants of the respective buildings. These data must therefore be protected. In addition to this, the entire ecosystem of hardware and software components is also vulnerable, and threat protection must therefore evolve accordingly. The above-mentioned vulnerabilities could be overcome by means of appropriate technologies designed to protect data, networks, systems, and devices from malicious attacks, implementing cryptography, securing both the hardware and software components, and ensuring communication protection in order to prevent unauthorized access to private information, avoid the interruption of communications, and guarantee the accuracy of information managed by the respective system.

Even if the developed review covers the most relevant and important actual scientific articles dealing with the above-mentioned research topics, we are aware of the fact that, as with any other review paper, this is affected by the rapid development of the body of knowledge with regard to the reviewed topics, which is strongly correlated with the extremely rapid evolution of the technology, of sensor devices, and of machine learning approaches.

With respect to future work, we will aim to conduct a review of the most relevant patents awarded, along with those that are pending, that propose methods and devices related to the fusion of machine learning techniques with sensor devices in the smart buildings sector. In our opinion, this is an aspect worth being studied and reviewed, considering the numerous existing patents that have not been disseminated yet as scientific articles in the literature.

Supplementary Materials: The following are available online at http://www.mdpi.com/1996-1073/12/24/4745/s1. Table S1: Scientific articles addressing the Support Vector Machines integrated with sensor devices in smart buildings; Table S2: Scientific articles addressing the Discriminant Analysis integrated with sensor devices in smart buildings; Table S3: Scientific articles addressing the Naïve Bayes integrated with sensor devices in smart buildings; Table S4: Scientific articles addressing the Nearest Neighbor integrated with sensor devices in smart buildings; Table S5: Scientific articles addressing the Neural Networks for Classification Purposes integrated with sensor devices in smart buildings; Table S6: Scientific articles addressing the Decision Tree integrated with sensor devices in smart buildings; Table S7: Scientific articles addressing the Ensemble Methods integrated with sensor devices in smart buildings; Table S8: Scientific articles addressing the Gaussian Process Regression (GPR) integrated with sensor devices in smart buildings; Table S9: Scientific articles addressing the Linear Regression integrated with sensor devices in smart buildings; Table S10: Scientific articles addressing the Neural Networks for Regression Purposes integrated with sensor devices in smart buildings; Table S11: Scientific articles addressing the Support Vector Regression (SVR) integrated with sensor devices in smart buildings; Table S12: Scientific articles addressing the Fuzzy C-Means integrated with sensor devices in smart buildings; Table S13: Scientific articles addressing the Hidden Markov Model integrated with sensor devices in smart buildings; Table S14: Scientific articles addressing the Hierarchical Clustering integrated with sensor devices in smart buildings; Table S15: Scientific articles addressing the K-Means integrated with sensor devices in smart buildings; Table S16: Scientific articles addressing the Deep Learning techniques integrated with sensor devices in smart buildings.

Author Contributions: Conceptualization, D.-M.P., G.C., A.P., and N.L.C.; Methodology, D.-M.P., G.C., A.P., and N.L.C.; Investigation, D.-M.P., G.C., A.P., and N.L.C.; Resources, D.-M.P., G.C., A.P., and N.L.C.; Data Curation, D.-M.P., G.C., A.P., and N.L.C.; Writing—Original Draft Preparation, D.-M.P., G.C., A.P., and N.L.C.; Writing—Review & Editing, D.-M.P., G.C., A.P., and N.L.C.; Visualization, D.-M.P., G.C., A.P., and N.L.C.; Supervision, A.P.; Funding Acquisition, A.P.

Funding: The article processing charge (APC) was discounted integrally by the Multidisciplinary Digital Publishing Institute (MDPI).

Conflicts of Interest: The authors declare no conflict of interest.

References

1. Kim, S.; Song, Y.; Sung, Y.; Seo, D. Development of a consecutive occupancy estimation framework for improving the energy demand prediction performance of building energy modeling tools. *Energies* **2019**, *12*, 433. [CrossRef]
2. Zhao, H.; Hua, Q.; Chen, H.-B.; Ye, Y.; Wang, H.; Tan, S.X.-D.; Tlelo-Cuautle, E. Thermal-Sensor-Based Occupancy Detection for Smart Buildings Using Machine-Learning Methods. *ACM Trans. Des. Autom. Electron. Syst.* **2018**, *23*, 54. [CrossRef]
3. Hao, J.; Yuan, X.; Yang, Y.; Wang, R.; Zhuang, Y.; Luo, J. Visible Light Based Occupancy Inference Using Ensemble Learning. *IEEE Access* **2018**, *6*, 16377–16385. [CrossRef]
4. Chen, Z.; Zhu, Q.; Soh, Y.C.; Zhang, L. Robust Human Activity Recognition Using Smartphone Sensors via CT-PCA and Online SVM. *IEEE Trans. Ind. Inform.* **2017**, *13*, 3070–3080. [CrossRef]
5. Bales, D.; Tarazaga, P.A.; Kasarda, M.; Batra, D.; Woolard, A.G.; Poston, J.D.; Malladi, V.V.N.S. Gender Classification of Walkers via Underfloor Accelerometer Measurements. *IEEE Internet Things J.* **2016**, *3*, 1259–1266. [CrossRef]
6. Al MacHot, F.; Mosa, A.H.; Ali, M.; Kyamakya, K. Activity Recognition in Sensor Data Streams for Active and Assisted Living Environments. *IEEE Trans. Circuits Syst. Video Technol.* **2018**, *28*, 2933–2945. [CrossRef]
7. Li, X.; Nie, L.; Xu, H.; Wang, X. Collaborative Fall Detection Using Smart Phone and Kinect. *Mob. Netw. Appl.* **2018**, *23*, 775–788. [CrossRef]

8. Li, W.; Tan, B.; Piechocki, R. Passive Radar for Opportunistic Monitoring in E-Health Applications. *IEEE J. Transl. Eng. Heal. Med.* **2018**, *6*, 2800210. [CrossRef]
9. Chen, Z.; Wang, Y.; Liu, H. Unobtrusive sensor-based occupancy facing direction detection and tracking using advanced machine learning algorithms. *IEEE Sens. J.* **2018**, *18*, 6360–6368. [CrossRef]
10. Lu, L.; Qing-ling, C.; Yi-Ju, Z. Activity Recognition in Smart Homes. *Multimed. Tools Appl.* **2017**, *76*, 24203–24220. [CrossRef]
11. Hassan, M.K.; El Desouky, A.I.; Elghamrawy, S.M.; Sarhan, A.M. A Hybrid Real-time remote monitoring framework with NB-WOA algorithm for patients with chronic diseases. *Futur. Gener. Comput. Syst.* **2019**, *93*, 77–95. [CrossRef]
12. Rahman, A.; Srikumar, V.; Smith, A.D. Predicting electricity consumption for commercial and residential buildings using deep recurrent neural networks. *Appl. Energy* **2018**, *212*, 372–385. [CrossRef]
13. Caicedo, D.; Pandharipande, A. Sensor Data-Driven Lighting Energy Performance Prediction. *IEEE Sens. J.* **2016**, *16*, 6397–6405. [CrossRef]
14. Kim, J.Y.; Liu, N.; Tan, H.X.; Chu, C.H. Unobtrusive Monitoring to Detect Depression for Elderly with Chronic Illnesses. *IEEE Sens. J.* **2017**, *17*, 5694–5704. [CrossRef]
15. Guan, Q.; Yin, X.; Guo, X.; Wang, G. A novel infrared motion sensing system for compressive classification of physical activity. *IEEE Sens. J.* **2016**, *16*, 2251–2259. [CrossRef]
16. Tian, Y.; Wang, X.; Chen, L.; Liu, Z. Wearable sensor-based human activity recognition via two-layer diversity-enhanced multiclassifier recognition method. *Sensors* **2019**, *19*, 2039. [CrossRef]
17. Brennan, C.; Taylor, G.W.; Spachos, P. Designing learned CO2-based occupancy estimation in smart buildings. *IET Wirel. Sens. Syst.* **2018**, *8*, 249–255. [CrossRef]
18. Yu, H.; Pan, G.; Pan, M.; Li, C.; Jia, W.; Zhang, L.; Sun, M. A hierarchical deep fusion framework for egocentric activity recognition using a wearable hybrid sensor system. *Sensors* **2019**, *19*, 546. [CrossRef]
19. Chen, H.; Cha, S.H.; Kim, T.W. A framework for group activity detection and recognition using smartphone sensors and beacons. *Build. Environ.* **2019**, *158*, 205–216. [CrossRef]
20. Chen, Z.; Jiang, C.; Xie, L. A Novel Ensemble ELM for Human Activity Recognition Using Smartphone Sensors. *IEEE Trans. Ind. Inform.* **2019**, *15*, 2691–2699. [CrossRef]
21. Ateeq, M.; Ishmanov, F.; Afzal, M.K.; Naeem, M. Multi-parametric analysis of reliability and energy consumption in IoT: A deep learning approach. *Sensors* **2019**, *19*, 309. [CrossRef] [PubMed]
22. Divina, F.; Torres, M.G.; Vela, F.A.G.; Noguera, J.L.V. A comparative study of time series forecasting methods for short term electric energy consumption prediction in smart buildings. *Energies* **2019**, *12*, 1934. [CrossRef]
23. Chammas, M.; Makhoul, A.; Demerjian, J. An efficient data model for energy prediction using wireless sensors. *Comput. Electr. Eng.* **2019**, *76*, 249–257. [CrossRef]
24. Rodriguez-Mier, P.; Mucientes, M.; Bugarín, A. Feature Selection and Evolutionary Rule Learning for Big Data in Smart Building Energy Management. *Cognit. Comput.* **2019**, *11*, 418–433. [CrossRef]
25. Arifoglu, D.; Bouchachia, A. Detection of abnormal behaviour for dementia sufferers using Convolutional Neural Networks. *Artif. Intell. Med.* **2019**, *94*, 88–95. [CrossRef]
26. Li, C.; Cheung, W.K.; Liu, J.; Ng, J.K. Automatic extraction of behavioral paterns for elderly mobility and daily routine analysis. *ACM Trans. Intell. Syst. Technol.* **2018**, *9*, 54. [CrossRef]
27. Guo, X.; Su, R.; Hu, C.; Ye, X.; Wu, H.; Nakamura, K. A single feature for human activity recognition using two-dimensional acoustic array. *Appl. Phys. Lett.* **2019**, *114*, 214101. [CrossRef]
28. Guo, J.; Mu, Y.; Xiong, M.; Liu, Y.; Gu, J.; Garcia-Rodriguez, J. Activity Feature Solving Based on TF-IDF for Activity Recognition in Smart Homes. *Complexity* **2019**, *2019*, 5245373. [CrossRef]
29. Abidine, B.M.; Fergani, B.; Oussalah, M.; Fergani, L. A new classification strategy for human activity recognition using cost sensitive support vector machines for imbalanced data. *Kybernetes* **2014**, *43*, 1150–1164. [CrossRef]
30. Nef, T.; Urwyler, P.; Büchler, M.; Tarnanas, I.; Stucki, R.; Cazzoli, D.; Müri, R.; Mosimann, U. Evaluation of three state-of-the-art classifiers for recognition of activities of daily living from smart home ambient data. *Sensors* **2015**, *15*, 11725–11740. [CrossRef]
31. Fahad, L.G.; Ali, A.; Rajarajan, M. Learning models for activity recognition in smart homes. *Inf. Sci. Appl.* **2015**, *339*, 819–826.
32. Amirjavid, F.; Bouzouane, A.; Bouchard, B. Data driven modeling of the simultaneous activities in ambient environments. *J. Ambient Intell. Humaniz. Comput.* **2014**, *5*, 717–740. [CrossRef]

33. Alam, M.G.R.; Abedin, S.F.; Al Ameen, M.; Hong, C.S. Web of objects based ambient assisted living framework for emergency psychiatric state prediction. *Sensors* **2016**, *16*, 1431. [CrossRef] [PubMed]
34. Ballardini, A.L.; Ferretti, L.; Fontana, S.; Furlan, A.; Sorrenti, D.G. An Indoor Localization System for Telehomecare Applications. *IEEE Trans. Syst. Man Cybern. Syst.* **2016**, *46*, 1445–1455. [CrossRef]
35. Alcala, J.M.; Urena, J.; Hernandez, A.; Gualda, D. Sustainable Homecare Monitoring System by Sensing Electricity Data. *IEEE Sens. J.* **2017**, *17*, 7741–7749. [CrossRef]
36. Sundaravadivel, P.; Kesavan, K.; Kesavan, L.; Mohanty, S.P.; Kougianos, E. Smart-Log: A Deep-Learning Based Automated Nutrition Monitoring System in the IoT. *IEEE Trans. Consum. Electron.* **2018**, *64*, 390–398. [CrossRef]
37. Zou, H.; Zhou, Y.; Yang, J.; Spanos, C.J. Unsupervised WiFi-Enabled IoT Device-User Association for Personalized Location-Based Service. *IEEE Internet Things J.* **2019**, *6*, 1238–1245. [CrossRef]
38. Yang, K.; Cho, S.B. A context-aware system in Internet of Things using modular Bayesian networks. *Int. J. Distrib. Sens. Netw.* **2017**, *13*, 155014771770898. [CrossRef]
39. Meana-Llorián, D.; González García, C.; Pelayo G-Bustelo, B.C.; Cueva Lovelle, J.M.; Garcia-Fernandez, N. IoFClime: The fuzzy logic and the Internet of Things to control indoor temperature regarding the outdoor ambient conditions. *Futur. Gener. Comput. Syst.* **2017**, *76*, 275–284. [CrossRef]
40. Hu, W.; Wen, Y.; Guan, K.; Jin, G.; Tseng, K.J. ITCM: Toward Learning-Based Thermal Comfort Modeling via Pervasive Sensing for Smart Buildings. *IEEE Internet Things J.* **2018**, *5*, 4164–4177. [CrossRef]
41. Candanedo, L.M.; Feldheim, V.; Deramaix, D. Data driven prediction models of energy use of appliances in a low-energy house. *Energy Build.* **2017**, *140*, 81–97. [CrossRef]
42. Di Corso, E.; Cerquitelli, T.; Apiletti, D. METATECH: METeorological data analysis for thermal energy characterization by means of self-learning transparent models. *Energies* **2018**, *11*, 1336. [CrossRef]
43. Khan, N.S.; Ghani, S.; Haider, S. Real-time analysis of a sensor's data for automated decision making in an IoT-based smart home. *Sensors* **2018**, *18*, 1711. [CrossRef]
44. Jain, R.K.; Smith, K.M.; Culligan, P.J.; Taylor, J.E. Forecasting energy consumption of multi-family residential buildings using support vector regression: Investigating the impact of temporal and spatial monitoring granularity on performance accuracy. *Appl. Energy* **2014**, *123*, 168–178. [CrossRef]
45. Collotta, M.; Pau, G. An Innovative Approach for Forecasting of Energy Requirements to Improve a Smart Home Management System Based on BLE. *IEEE Trans. Green Commun. Netw.* **2017**, *1*, 112–120. [CrossRef]
46. Jabłoński, I. Smart transducer interface—From networked on-site optimization of energy balance in research-demonstrative office building to smart city conception. *IEEE Sens. J.* **2015**, *15*, 2468–2478. [CrossRef]
47. Keshtkar, A.; Arzanpour, S.; Keshtkar, F.; Ahmadi, P. Smart residential load reduction via fuzzy logic, wireless sensors, and smart grid incentives. *Energy Build.* **2015**, *104*, 165–180. [CrossRef]
48. Anthierens, C.; Leclercq, M.; Bideaux, E.; Flambard, L. A smart sensor to evaluate visual comfort of daylight into buildings. *Int. J. Optomech.* **2008**, *2*, 413–434. [CrossRef]
49. Chang, C.Y.; Hung, S.S.; Liu, L.H.; Lin, C.P. Innovative strain sensing for detection of exterior wall tile lesion: Smart skin sensory system. *Materials* **2018**, *11*, 2432. [CrossRef]
50. Shetty, S.S.; Hoang, D.C.; Gupta, M.; Panda, S.K. Learning desk fan usage preferences for personalised thermal comfort in shared offices using tree-based methods. *Build. Environ.* **2019**, *149*, 546–560. [CrossRef]
51. Viani, F.; Polo, A. A forecasting strategy based on wireless sensing for thermal comfort optimization in smart buildings. *Microw. Opt. Technol. Lett.* **2017**, *59*, 2913–2917. [CrossRef]
52. Ahmed, H.S.; Faouzi, B.M.; Caelen, J. Detection and classification of the behavior of people in an intelligent building by camera. *Int. J. Smart Sens. Intell. Syst.* **2013**, *6*, 1317–1342. [CrossRef]
53. Li, D.; Menassa, C.C.; Kamat, V.R. Personalized human comfort in indoor building environments under diverse conditioning modes. *Build. Environ.* **2017**, *126*, 304–317. [CrossRef]
54. Ain, Q.-u.; Iqbal, S.; Khan, S.A.; Malik, A.W.; Ahmad, I.; Javaid, N. IoT operating system based fuzzy inference system for home energy management system in smart buildings. *Sensors* **2018**, *18*, 2802. [CrossRef] [PubMed]
55. Motamed, A.; Deschamps, L.; Scartezzini, J.L. On-site monitoring and subjective comfort assessment of a sun shadings and electric lighting controller based on novel High Dynamic Range vision sensors. *Energy Build.* **2017**, *149*, 58–72. [CrossRef]

56. Ulpiani, G. Overheating phenomena induced by fully-glazed facades: Investigation of a sick building in Italy and assessment of the benefits achieved via fuzzy control of the AC system. *Sol. Energy* **2017**, *158*, 572–594. [CrossRef]
57. Grant, M.J.; Booth, A. A typology of reviews: An analysis of 14 review types and associated methodologies. *Health Info. Libr. J.* **2009**, *26*, 91–108. [CrossRef]
58. Shobha, G.; Rangaswamy, S. Machine Learning. In *Handbook of Statistics*; Rao, C.R., Ed.; Elsevier: Amsterdam, The Netherlands, 2018; Volume 38, pp. 197–228, ISBN 9780444640420.
59. Gillani Fahad, L.; Khan, A.; Rajarajan, M. Activity recognition in smart homes with self verification of assignments. *Neurocomputing* **2015**, *149*, 1286–1298. [CrossRef]
60. Kim, T.S.; Cho, J.H.; Kim, J.T. Mobile motion sensor-based human activity recognition and energy expenditure estimation in building environments. *Smart Innov. Syst. Technol.* **2013**, *22*, 987–993. [CrossRef]
61. Abidine, M.B.; Fergani, B.; Ordóñez, F.J. Effect of over-sampling versus under-sampling for SVM and LDA classifiers for activity recognition. *Int. J. Des. Nat. Ecodyn.* **2016**, *11*, 306–316. [CrossRef]
62. Abidine, M.B.; Fergani, B. News schemes for activity recognition systems using PCA-WSVM, ICA-WSVM, and LDA-WSVM. *Information* **2015**, *6*, 505–521. [CrossRef]
63. Chernbumroong, S.; Cang, S.; Yu, H. Genetic algorithm-based classifiers fusion for multisensor activity recognition of elderly people. *IEEE J. Biomed. Heal. Inform.* **2015**, *19*, 282–289. [CrossRef] [PubMed]
64. Lai, Y.X.; Lai, C.F.; Huang, Y.M.; Chao, H.C. Multi-appliance recognition system with hybrid SVM/GMM classifier in ubiquitous smart home. *Inf. Sci.* **2013**, *230*, 39–55. [CrossRef]
65. Namburu, S.M.; Azam, M.S.; Luo, J.; Choi, K.; Pattipati, K.R. Data-driven modeling, fault diagnosis and optimal sensor selection for HVAC chillers. *IEEE Trans. Autom. Sci. Eng.* **2007**, *4*, 469–473. [CrossRef]
66. Liao, R.; Changqing, S. Smart Home Design Based on Cloud Computing and Internet of Things. *J. Comput. Theor. Nanosci.* **2016**, *13*, 8075–8080. [CrossRef]
67. Evers, C.; Habets, E.A.P.; Gannot, S.; Naylor, P.A. DoA reliability for distributed acoustic tracking. *IEEE Signal Process. Lett.* **2018**, *25*, 1320–1324. [CrossRef]
68. Zimmermann, L.; Weigel, R.; Fischer, G. Fusion of nonintrusive environmental sensors for occupancy detection in smart homes. *IEEE Internet Things J.* **2018**, *5*, 2343–2352. [CrossRef]
69. Sebbak, F.; Benhammadi, F.; Chibani, A.; Amirat, Y.; Mokhtari, A. Dempster–Shafer theory-based human activity recognition in smart home environments. *Ann. Des Telecommun. Telecommun.* **2014**, *69*, 171–184. [CrossRef]
70. Fang, H.; Srinivasan, R.; Cook, D.J. Feature selections for human activity recognition in smart home environments. *Int. J. Innov. Comput. Inf. Control* **2012**, *8*, 3525–3535.
71. Gulati, M.; Ram, S.S.; Majumdar, A.; Singh, A. Single Point Conducted EMI Sensor with Intelligent Inference for Detecting IT Appliances. *IEEE Trans. Smart Grid* **2018**, *9*, 3716–3726. [CrossRef]
72. Kwolek, B.; Kepski, M. Improving fall detection by the use of depth sensor and accelerometer. *Neurocomputing* **2015**, *168*, 637–645. [CrossRef]
73. Tan, T.H.; Gochoo, M.; Huang, S.C.; Liu, Y.H.; Liu, S.H.; Huang, Y.F. Multi-resident activity recognition in a smart home using RGB activity image and DCNN. *IEEE Sens. J.* **2018**, *18*, 9718–9727. [CrossRef]
74. Palumbo, F.; Gallicchio, C.; Pucci, R.; Micheli, A. Human activity recognition using multisensor data fusion based on Reservoir Computing. *J. Ambient Intell. Smart Environ.* **2016**, *8*, 87–107. [CrossRef]
75. Lotfi, A.; Langensiepen, C.; Mahmoud, S.M.; Akhlaghinia, M.J. Smart homes for the elderly dementia sufferers: Identification and prediction of abnormal behaviour. *J. Ambient Intell. Humaniz. Comput.* **2012**, *3*, 205–218. [CrossRef]
76. Zeng, X.H.; Chen, X.T.; Ye, C.Y. An EEGA-based bayesian belief network model for recognition of human activity in smart home. *J. Donghua Univ.* **2012**, *29*, 497–500.
77. Vanus, J.; Belesova, J.; Martinek, R.; Nedoma, J.; Fajkus, M.; Bilik, P.; Zidek, J. Monitoring of the daily living activities in smart home care. *Hum. Cent. Comput. Inf. Sci.* **2017**, *7*, 30. [CrossRef]
78. Amayri, M.; Ploix, S.; Bouguila, N.; Wurtz, F. Estimating occupancy using interactive learning with a sensor environment: Real-time experiments. *IEEE Access* **2019**, *7*, 53932–53944. [CrossRef]
79. Al Zamil, M.G.H.; Samarah, S.M.J.; Rawashdeh, M.; Hossain, M.A. An ODT-based abstraction for mining closed sequential temporal patterns in IoT-cloud smart homes. *Clust. Comput.* **2017**, *20*, 1815–1829. [CrossRef]

80. Fong, S.; Li, J.; Song, W.; Tian, Y.; Wong, R.K.; Dey, N. Predicting unusual energy consumption events from smart home sensor network by data stream mining with misclassified recall. *J. Ambient Intell. Humaniz. Comput.* **2018**, *9*, 1197–1221. [CrossRef]
81. Tabatabaee Malazi, H.; Davari, M. Combining emerging patterns with random forest for complex activity recognition in smart homes. *Appl. Intell.* **2018**, *48*, 315–330. [CrossRef]
82. Lundström, J.; Järpe, E.; Verikas, A. Detecting and exploring deviating behaviour of smart home residents. *Expert Syst. Appl.* **2016**, *55*, 429–440. [CrossRef]
83. Zhao, Q.; Tsai, C.M.; Chen, R.C.; Huang, C.Y. Resident activity recognition based on binary infrared sensors and soft computing. *Int. J. Mach. Learn. Cybern.* **2019**, *10*, 291–299. [CrossRef]
84. Bjelica, M.Z.; Mrazovac, B.; Papp, I.; Teslic, N. Context-aware platform with user availability estimation and light-based announcements. *IEEE Trans. Syst. ManCybern. Part A Syst. Hum.* **2013**, *43*, 1228–1239. [CrossRef]
85. Jurek, A.; Nugent, C.; Bi, Y.; Wu, S. Clustering-based ensemble learning for activity recognition in smart homes. *Sensors* **2014**, *14*, 12285–12304. [CrossRef] [PubMed]
86. Fatima, I.; Fahim, M.; Lee, Y.K.; Lee, S. A genetic algorithm-based classifier ensemble optimization for activity recognition in smart homes. *KSII Trans. Internet Inf. Syst.* **2013**, *7*, 2853–2873. [CrossRef]
87. Alcalá, J.M.; Ureña, J.; Hernández, Á.; Gualda, D. Assessing human activity in elderly people using non-intrusive load monitoring. *Sensors* **2017**, *17*, 351. [CrossRef] [PubMed]
88. Muhammad, G.; Alhamid, M.F.; Shamim Hossain, M.; Almogren, A.S.; Vasilakos, A.V. Enhanced living by assessing voice pathology using a co-occurrence matrix. *Sensors* **2017**, *17*, 267. [CrossRef]
89. Villeneuve, E.; Harwin, W.; Holderbaum, W.; Janko, B.; Sherratt, R.S. Reconstruction of angular kinematics from wrist-worn inertial sensor data for smart home healthcare. *IEEE Access* **2017**, *5*, 2169–3536. [CrossRef]
90. Fagiani, M.; Squartini, S.; Gabrielli, L.; Severini, M.; Piazza, F. A statistical framework for automatic leakage detection in smart water and gas grids. *Energies* **2016**, *9*, 665. [CrossRef]
91. Mattera, C.G.; Quevedo, J.; Escobet, T.; Shaker, H.R.; Jradi, M. A Method for Fault Detection and Diagnostics in Ventilation Units Using Virtual Sensors. *Sensors* **2018**, *18*, 3931. [CrossRef]
92. Lynggaard, P. Using Machine Learning for Adaptive Interference Suppression in Wireless Sensor Networks. *IEEE Sens. J.* **2018**, *18*, 8820–8826. [CrossRef]
93. Bouchard, K.; Giroux, S.; Bouchard, B.; Bouzouane, A. Regression analysis for gesture recognition using passive RFID technology in smart home environments. *Int. J. Smart Home* **2014**, *8*, 245–260. [CrossRef]
94. Basu, C.; Caubel, J.J.; Kim, K.; Cheng, E.; Dhinakaran, A.; Agogino, A.M.; Martin, R.A. Sensor-based predictive modeling for smart lighting in grid-integrated buildings. *IEEE Sens. J.* **2014**, *14*, 4216–4229. [CrossRef]
95. Oprea, S.-V.; Pîrjan, A.; Căruțașu, G.; Petroșanu, D.-M.; Bâra, A.; Stănică, J.-L.; Coculescu, C. Developing a Mixed Neural Network Approach to Forecast the Residential Electricity Consumption Based on Sensor Recorded Data. *Sensors* **2018**, *18*, 1443. [CrossRef] [PubMed]
96. Pardo, J.; Zamora-Martínez, F.; Botella-Rocamora, P. Online learning algorithm for time series forecasting suitable for low cost wireless sensor networks nodes. *Sensors* **2015**, *15*, 9277–9304. [CrossRef] [PubMed]
97. Li, Z.; Dong, B. A new modeling approach for short-term prediction of occupancy in residential buildings. *Build. Environ.* **2017**, *121*, 277–290. [CrossRef]
98. Attoue, N.; Shahrour, I.; Younes, R. Smart building: Use of the artificial neural network approach for indoor temperature forecasting. *Energies* **2018**, *11*, 395. [CrossRef]
99. Khatoon, S.; Rahman, S.M.M.; Alrubaian, M.; Alamri, A. Privacy-Preserved, Provable Secure, Mutually Authenticated Key Agreement Protocol for Healthcare in a Smart City Environment. *IEEE Access* **2019**, *7*, 47962–47971. [CrossRef]
100. Amirjavid, F.; Spachos, P.; Plataniotis, K.N. 3-D Object Localization in Smart Homes: A Distributed Sensor and Video Mining Approach. *IEEE Syst. J.* **2018**, *12*, 1307–1316. [CrossRef]
101. Sarwar, B.; Bajwa, I.S.; Ramzan, S.; Ramzan, B.; Kausar, M. Design and application of fuzzy logic based fire monitoring and warning systems for smart buildings. *Symmetry* **2018**, *10*, 615. [CrossRef]
102. Yuan, B.; Herbert, J. Context-aware hybrid reasoning framework for pervasive healthcare. *Pers. Ubiquitous Comput.* **2014**, *18*, 865–881. [CrossRef]
103. Wang, J.M.; Yang, M.T.; Chen, P.L. Design and implementation of an intelligent windowsill system using smart handheld device and fuzzy microcontroller. *Sensors* **2017**, *17*, 830. [CrossRef] [PubMed]

104. Usman, M.; Muthukkumarasamy, V.; Wu, X.W. Mobile agent-based cross-layer anomaly detection in smart home sensor networks using fuzzy logic. *IEEE Trans. Consum. Electron.* **2015**, *61*, 197–205. [CrossRef]
105. Kıyak, İ.; Oral, B.; Topuz, V. Smart indoor LED lighting design powered by hybrid renewable energy systems. *Energy Build.* **2017**, *148*, 342–347. [CrossRef]
106. Liu, J.; Zhang, W.; Chu, X.; Liu, Y. Fuzzy logic controller for energy savings in a smart LED lighting system considering lighting comfort and daylight. *Energy Build.* **2016**, *127*, 95–104. [CrossRef]
107. Ahvar, E.; Lee, G.M.; Han, S.N.; Crespi, N.; Khan, I. Sensor network-based and user-friendly user location discovery for future smart homes. *Sensors* **2016**, *16*, 969. [CrossRef]
108. Vlachostergiou, A.; Stratogiannis, G.; Caridakis, G.; Siolas, G.; Mylonas, P. User Adaptive and Context-Aware Smart Home Using Pervasive and Semantic Technologies. *J. Electr. Comput. Eng.* **2016**, *2016*, 4789803. [CrossRef]
109. Panna, R.; Thesrumluk, R.; Chantrapornchai, C. Development of energy saving smart home prototype. *Int. J. Smart Home* **2013**, *7*, 47–66.
110. Wang, K.J.; Wu, C.Y.; Ning, W.L. Fuzzy cognitive map control on room temperature in a smart house. *J. Chin. Soc. Mech. Eng.* **2013**, *34*, 431–440.
111. Sang-Hyun, L.; Lee, J.G.; Kyung-Il, M. Smart home security system using multiple ANFIS. *Int. J. Smart Home* **2013**, *7*, 121–132.
112. Fortin-Simard, D.; Bouchard, K.; Gaboury, S.; Bouchard, B.; Bouzouane, A. Accurate passive RFID localization system for smart homes. *Netw. Embed. Syst. Enterp. Appl.* **2012**, *1*, 391–399. [CrossRef]
113. Sharifi, R.; Kim, Y.; Langari, R. Sensor fault isolation and detection of smart structures. *Smart Mater. Struct.* **2010**, *9*, 105001. [CrossRef]
114. Chen, S.Y.; Chiu, M.L. Designing Smart Skins for Adaptive Environments: A fuzzy logic approach to smart house design. *Comput. Aided. Des. Appl.* **2007**, *4*, 751–760. [CrossRef]
115. Papatsimpa, C.; Linnartz, J.-P. Distributed Fusion of Sensor Data in a Constrained Wireless Network. *Sensors* **2019**, *19*, 1006. [CrossRef] [PubMed]
116. Civitarese, G.; Bettini, C.; Sztyler, T.; Riboni, D.; Stuckenschmidt, H. newNECTAR: Collaborative active learning for knowledge-based probabilistic activity recognition. *Pervasive Mob. Comput.* **2019**, *56*, 88–105. [CrossRef]
117. Dahmen, J.; Cook, D. SynSys: A synthetic data generation system for healthcare applications. *Sensors* **2019**, *19*, 1181. [CrossRef]
118. Hela, S.; Amel, B.; Badran, R. Early anomaly detection in smart home: A causal association rule-based approach. *Artif. Intell. Med.* **2018**, *91*, 57–71. [CrossRef]
119. Lan, D.; Pang, Z.; Fischione, C.; Liu, Y.; Taherkordi, A.; Eliassen, F. Latency Analysis of Wireless Networks for Proximity Services in Smart Home and Building Automation: The Case of Thread. *IEEE Access* **2019**, *7*, 2169–3536. [CrossRef]
120. Al Zamil, M.G.; Rawashdeh, M.; Samarah, S.; Hossain, M.S.; Alnusair, A.; Rahman, S.M.M. An Annotation Technique for In-Home Smart Monitoring Environments. *IEEE Access* **2017**, *6*, 1471–1479. [CrossRef]
121. Gayathri, K.S.; Easwarakumar, K.S.; Elias, S. Probabilistic ontology based activity recognition in smart homes using Markov Logic Network. *Knowl. Based Syst.* **2017**, *121*, 173–184. [CrossRef]
122. Wang, C.; Peng, Y.; De, D.; Song, W.Z. DPHK: Real-time distributed predicted data collecting based on activity pattern knowledge mined from trajectories in smart environments. *Front. Comput. Sci.* **2016**, *10*, 1000–1011. [CrossRef]
123. Wang, L.; Gu, T.; Tao, X.; Chen, H.; Lu, J. Recognizing multi-user activities using wearable sensors in a smart home. *Pervasive Mob. Comput.* **2011**, *7*, 287–298. [CrossRef]
124. Doty, K.; Roy, S.; Fischer, T.R. Explicit State-Estimation Error Calculations for Flag Hidden Markov Models. *IEEE Trans. Signal Process.* **2016**, *64*, 4444–4454. [CrossRef]
125. Chahuara, P.; Fleury, A.; Portet, F.; Vacher, M. On-line human activity recognition from audio and home automation sensors: Comparison of sequential and non-sequential models in realistic Smart Homes. *J. Ambient Intell. Smart Environ.* **2016**, *8*, 399–422. [CrossRef]
126. Noury, N.; Hadidi, T. Computer simulation of the activity of the elderly person living independently in a Health Smart Home. *Comput. Methods Programs Biomed.* **2012**, *108*, 1216–1228. [CrossRef]
127. Roy, N.; Misra, A.; Cook, D. Ambient and smartphone sensor assisted ADL recognition in multi-inhabitant smart environments. *J. Ambient Intell. Humaniz. Comput.* **2016**, *7*, 1–19. [CrossRef]

128. Nait Aicha, A.; Englebienne, G.; Kröse, B. Unsupervised visit detection in smart homes. *Pervasive Mob. Comput.* **2017**, *34*, 157–167. [CrossRef]
129. Chahuara, P.; Portet, F.; Vacher, M. Context-aware decision making under uncertainty for voice-based control of smart home. *Expert Syst. Appl.* **2017**, *75*, 63–79. [CrossRef]
130. Alemdar, H.; Ersoy, C. Multi-resident activity tracking and recognition in smart environments. *J. Ambient Intell. Humaniz. Comput.* **2017**, *8*, 513–529. [CrossRef]
131. Chikhaoui, B.; Wang, S.; Pigot, H. ADR-SPLDA: Activity discovery and recognition by combining sequential patterns and latent Dirichlet allocation. *Pervasive Mob. Comput.* **2012**, *8*, 845–862. [CrossRef]
132. Karami, A.B.; Fleury, A.; Boonaert, J.; Lecoeuche, S. User in the loop: Adaptive smart homes exploiting user feedback-State of the art and future directions. *Information* **2016**, *7*, 35. [CrossRef]
133. Von Bomhard, T.; Wörner, D.; Röschlin, M. Towards smart individual-room heating for residential buildings. *Comput. Sci. Res. Dev.* **2016**, *31*, 127–134. [CrossRef]
134. Casado-Vara, R.; Vale, Z.; Prieto, J.; Corchado, J.M. Fault-tolerant temperature control algorithm for IoT networks in smart buildings. *Energies* **2018**, *11*, 3430. [CrossRef]
135. Papatsimpa, C.; Linnartz, J.P.M.G. Propagating sensor uncertainty to better infer office occupancy in smart building control. *Energy Build.* **2018**, *179*, 73–82. [CrossRef]
136. Khan, W.M.; Zualkernan, I.A. SensePods: A ZigBee-Based Tangible Smart Home Interface. *IEEE Trans. Consum. Electron.* **2018**, *64*, 145–152. [CrossRef]
137. Mohammed, A.W.; Xu, Y.; Liu, M.; Hu, H. Semantical Markov Logic Network for Distributed Reasoning in Cyber-Physical Systems. *J. Sens.* **2017**, *2017*, 4259652. [CrossRef]
138. Egarter, D.; Bhuvana, V.P.; Elmenreich, W. PALDi: Online load disaggregation via particle filtering. *IEEE Trans. Instrum. Meas.* **2015**, *64*, 467–477. [CrossRef]
139. Luan, X.; Zheng, Z.; Wang, T.; Wu, J.; Xiang, H. Hybrid cooperation for machine-to-machine data collection in hierarchical smart building networks. *IET Commun.* **2015**, *9*, 421–428. [CrossRef]
140. Pérez-Chacón, R.; Luna-Romera, J.M.; Troncoso, A.; Martínez-Alvarez, F.; Riquelme, J.C. Big data analytics for discovering electricity consumption patterns in smart cities. *Energies* **2018**, *11*, 683. [CrossRef]
141. Zeiler, W.; Labeodan, T. Human-in-the-loop energy flexibility integration on a neighbourhood level: Small and Big Data management. *Build. Serv. Eng. Res. Technol.* **2019**, *40*, 305–318. [CrossRef]
142. Zhu, H.; Chen, H.; Brown, R. A sequence-to-sequence model-based deep learning approach for recognizing activity of daily living for senior care. *J. Biomed. Inform.* **2018**, *84*, 148–158. [CrossRef] [PubMed]
143. Liciotti, D.; Bernardini, M.; Romeo, L.; Frontoni, E. A sequential deep learning application for recognising human activities in smart homes. *Neurocomputing* **2019**, 1–13. [CrossRef]
144. Brenon, A.; Portet, F.; Vacher, M. ARCADES: A deep model for adaptive decision making in voice controlled smart-home. *Pervasive Mob. Comput.* **2018**, *49*, 92–110. [CrossRef]
145. Mora, N.; Matrella, G.; Ciampolini, P. Cloud-based behavioral monitoring in smart homes. *Sensors* **2018**, *18*, 1951. [CrossRef] [PubMed]
146. Hassan, M.M.; Uddin, M.Z.; Mohamed, A.; Almogren, A. A robust human activity recognition system using smartphone sensors and deep learning. *Futur. Gener. Comput. Syst.* **2018**, *81*, 307–313. [CrossRef]
147. Hsiu-Yu, L.; Yu-Ling, H.; Wen-Nung, L. Convolutional recurrent neural networks for posture analysis in fall detection. *J. Inf. Sci. Eng.* **2018**, *34*, 577–591. [CrossRef]
148. Chen, G.; Wang, A.; Zhao, S.; Liu, L.; Chang, C.Y. Latent feature learning for activity recognition using simple sensors in smart homes. *Multimed. Tools Appl.* **2018**, *77*, 15201–15219. [CrossRef]
149. ISO/IEC 25010:2011 Systems and Software Engineering—Systems and Software Quality Requirements and Evaluation (SQuaRE)—System and Software Quality. Available online: https://www.iso.org/standard/35733.html (accessed on 10 October 2019).

energies

MDPI

Article

IoT-Based Smart Plug for Residential Energy Conservation: An Empirical Study Based on 15 Months' Monitoring

Jooseok Oh

Department of Architecture, Korea University, Seoul 02841, Korea; ohjooseok@korea.ac.kr

Received: 6 July 2020; Accepted: 29 July 2020; Published: 4 August 2020

Abstract: The study examines the implications of educating prosumers regarding Internet of Things (IoT) use and monitoring to reduce power consumption in the home and encourage energy conservation, sustainable living, and behavior change. Over 15 months, 125 households and household owners received training regarding IoT plug equipment, usage monitoring, and energy reduction. A face to face survey was then conducted regarding power consumption reductions, frequency of monitoring, and user satisfaction compared to the previous year. The study found that participating households used around 5% less energy compared to average households. The reduction rate was found to have increased when more appliances were connected to smart plugs and their power usage was monitored more frequently. Power usage also fell in a greater level when participants were more satisfied with being given smart plugs and related education. Moreover, energy reduction rates increase when smart plugs were used for cooling and heating appliances as well as video, audio, and related devices. The results suggest that this program can be used to reduce energy use, which can be beneficial for smart homes and smart cities. The study demonstrates the importance of education from the perspective of energy conservation and related policies.

Keywords: smart home; IoT; energy conservation; smart city; energy education

1. Introduction

Over the past few decades, the electric power system has faced unprecedented demand growth [1,2]. Buildings are one of the main consumers of total electricity; many scholars have discussed the energy management problem and have presented a number of power management schemes in both the domestic and residential sectors [3]. In countries outside the United States, various programs for demand response (DR) in the commercial and industrial field have been implemented [2]. Fully automated DR is the most well-known automation type realized by the home energy management (HEM) system, and research in related fields has produced tangible results.

Since the 1990s, several studies have been conducted on the residential sector, from both the perspective of power production and management, and from residents who are the primary power consumers. Among consumers, the real-time electricity pricing model has a high economic-environmental value than the conventional common flat rates [4]. Real-time monitoring of power consumption is an environmentally sustainable strategy for power usage reduction, helping the user reduce unnecessary power consumption and expenditure.

The advent of the smart city era in the 2000s highlighted the importance of energy management through various methods in smart homes, which were considered the end point. "Smart home" refers to a building equipped with devices that benefit the end users [5–7] with the development of application models and services for home networks recognized as a business model for the private sector [8]. These technologies and services increase the user's satisfaction with the adaptive home service [9]. The emergence of the Internet of Things (IoT) appliances, devices, and applications, especially in the energy sector, is believed to increase the convenience of citizens' daily lives, bringing about sustainable development in homes and cities [10]. Accordingly, IoT equipment and applications are being considered from various perspectives [11].

The IoT-based smart plug is a fundamental alternative to energy reduction from the user's point of view. It provides selective access, has an appropriate level of technology and price competitiveness, which is why many companies offer solutions that connect to a single product or other equipment. The market has expanded from $0.71 billion in 2016 to an expected $2.59 billion in 2021 [12].

Points of improvement for the smart plug have been suggested because it is voluntarily chosen, purchased, and utilized in homes where smart home equipment is not installed [13], but it plays a unique role at the midpoint of the energy prosumer era that we currently predict. It has been pointed out, however, that the smart plug by itself is insignificant for energy saving as "you use energy to save energy" [14]. Ultimately, a change in user and household behavior is also required to reduce energy usage [15,16]. Thus, educating consumers on the efficient use and reduction of energy is needed for effective utilization of sustainable energy [17]. This is also a means to enhance the economy, society, and quality of life [18]. From this point of view, the usage of equipment or devices, which are easily accessible and usable by households for energy conservation, and related education on those items are expected to serve as a stepping stone to the production and the management of sustainable energy. In order to ensure the sustainability of related technologies and policy measures, it is necessary to monitor equipment and devices that have been developed and utilized, subsequent changes of users' behaviors and related education programs.

As such, this study examined apartment residents over 15 months, wherein they used both IoT smart plugs and received education about energy saving. Based on the results, the study analyzed methods the participants used to reduce electricity consumption, factors that affected the reduction, users' satisfaction level, and the possibility for their sustainable utilization in the future.

1.1. Literature Review

1.1.1. Transition from Consumer to Manager and Prosumer

So far, studies of the energy sector and residential space have been conducted primarily from the perspective of public energy supply and management stability based on power demand. For example, Ref. [19] conducted a study based on game-theory which predicted the balance of power loads and daily schedules for residents and the pricing tariffs of the power supply. Ref. [20] emphasized the importance of scheduling for cost-benefit while presenting and verifying algorithms for the balance and optimization of households' minimum price and maximum comport. In particular, various technologies, hypotheses, and research models were established and verified to realize the aforementioned smart home and energy management system. However, the reality is that most countries and cities rely on the top-down method of public or large companies for the stable supply and demand of energy. The creation of the energy market's equilibrium condition is important in the long run to overcome this [21].

From this perspective, the growth and spread of smart city and smart grid concepts since the 2000s have redefined both the role of users, who were previously regarded as consumers of power, and new methods of power supply. The word "prosumer" was coined to exemplify the combination of producer and consumer. Many scholars predict that in the future, communities or individuals will act as major agents of energy production, management, and sales; in some ways, this is already being accomplished [22]. Thus, related research areas will be needed to predict and redefine users' roles and behaviors, which will expand in the future [23]. The public sector is required to serve its role of changing smart energy management, while simultaneously supplying and monitoring various related goods and services [24].

Except for a few standardized and commercialized technologies, however, there is still a lack of fundamental, foundational, and realistic research from a user's behavioral point of view, i.e., the most accessible device or service that can change the user's fundamental energy-saving behavior, and the education needed regarding energy conservation.

At a time when energy consumers have increasingly become "prosumers," multiple researches have focused on users themselves as well as effects of their voluntary energy reduction from the perspectives of smart home or smart city.

The first point of such discussions is users' voluntary improvement regarding their energy demand through the revision of energy supply or charging schemes [25,26]. They also emphasize that advanced information and knowledge could lead to structural changes in users' behaviors [27,28]. Secondly, they highlight control and monitoring led by users with a focus on the development and monitoring of smart home-related technologies such as smart plugs and IoT-based home appliances. Given that IoT, which can actually be controlled by users, is one of the efficient means for saving energy [28,29], the introduction of smart plug/socket and other similar items could reduce the energy consumption in cities and households [30,31]. As related technologies and equipment are still in their development stage, researches on their impacts and utilities have just begun.

Case in point is Ref. [32], who analyzed the survey on the use of smart plugs, which found that the quality of information and the usefulness of control applications, among other things, could enhance users' satisfaction with the device. However, there still have not been enough researches that review real effects that IoT equipment, such as real smart plugs, has on energy saving as well as their impacts on users in socio-scientific terms.

1.1.2. Importance of Evaluating User Experiences

From the perspective discussed in the previous section, recent studies have focused on measuring citizens' perceptions in the energy saving and renewable energy field, and suggested the direction of relevant policies based on the results.

In particular, as the field has come to include renewable energies [33] and new energy technologies [26], as well as willingness to pay (WTP) [34] and quality of life (Qol) [35], related research is going beyond the development of existing energy mechanisms and technologies that respond to user demand; it is recognizing citizens as agents of energy consumption and conservation by listening to and analyzing their opinions. In particular, public opinion must be considered when formulating and implementing policies, with an even greater emphasis in democratic societies [36].

Thus, the socio-demographic characteristics, ideology, and implementation of surveys based on individual value judgments in civil society is the basis for energy-related policymaking, which is the basis for public and private investment in related fields [37]. In addition, given that each individual's energy saving behavior is an important factor for reducing energy consumption [38], and that education is an important tool to achieve this goal [39], monitoring these items is an important factor in diagnosing the present and predicting and preparing for the future.

From this point of view, a number of studies are being carried out to create a framework for the user's experience of relevant policies, technologies, and equipment, or to monitor the results. In particular, the importance of IoT in transportation, energy, and housing for user experience in smart cities is being emphasized [40], but there is a lack of discussion on practical research about the relationship between users' perceptions, acceptance, and consumption and the importance of IoT [41]. Moreover, given that only relevant discussions are considered in detail, such as education, monitoring of its effectiveness [42], or reviews of related products and services [43,44], studies that combine user experience and the evaluation of related products or services with various factors that influence power consumption may be necessary in future research.

2. Materials and Methods

2.1. Research Process and Method

The smart plug, which has become more popular recently, is one of the most active and accessible user-led energy saving means but requires education. Study participants were selected in the following manner. They were then provided with smart plugs and energy conservation education, and the results were observed for 15 months.

The selected participants included those 20 years old or older who lived in apartment complexes. Complexes were chosen due to Seoul's residential characteristics, as apartments are one of the most common housing types in Korea. Selected households (144) had lived in one of four apartment complexes for two years or more. Data on monthly electricity usage for the past year was obtained from either the participants or the complex management office so that usage could be compared with the previous year.

Second, three IoT-based smart multi-tabs, two plugs, and two smart switches were given to participants. The products provided were IoT-based smart plugs and switches, and their functions were power control, scheduling, and timer from far and near distances using 5G, LTE, Wi-Fi, and Bluetooth (offline-based), all of which are currently available on the market. The participants installed the smart plugs and switches as they saw fit and notified the researchers of the type, number, and location of the connecting device.

The apartment complex was between 20 years and 30 years old and used hot water heating supplied by CHP (combined heat and power) plant. Because of this, electricity use in winter was low, except when using electric heaters. However, since air conditioning is used separately in each household, a lot of electricity tends to be used in the summer.

Third, participants received an hour-long education session when they received the smart plug and at one month, two months, four months, six months, eight months, and 12 months after they received the plug. The content of the training included: (1) How to use the IoT-based smart plug and how to use the control application; (2) real-time power monitoring, other methods of use, and comparison workshops on current power consumption compared to the previous year; (3) understanding and reducing the power consumption of home appliances; (4) how to use air-conditioning and heating devices to increase cooling and heating efficiency; (5) two workshops comparing current power consumption compared to the previous year; and (6) purchasing high-efficiency home appliances and government subsidies and reviews.

Fourth, participants' monthly power usage was collected for the 15 months of the experiment, starting in, March 2019, when the devices were supplied, to May 2020, when the experiment ended. Prior to the experiment, a preliminary survey was conducted on users' gender, age, number of household members, and area of residence; during and at the end of the experiment, a survey was conducted on the number of times the device control application was monitored, satisfaction level, and willingness to participate in the program.

2.2. Research Models and Variables

A total of 125 survey results who attended more than 80% of the education programs were collected and then analyzed. First, a pre- and post-use comparative evaluation was conducted. Participant expectations about the product before use were measured and compared with the satisfaction level after use. In terms of actual power usage, the results and effects were analyzed by comparing and analyzing the average monthly power usage for 15 months before using the product with the power usage for 15 months during the experiment. For the two before and after comparisons, the paired-samples *t*-test was performed.

Second, the factors affecting the increase and decrease in power usage were analyzed. The reduction in power consumption based on pre- and post-use power usage was set up as the dependent variable, while the number of installations of the supplied device, the number of times power and status were controlled and monitored using the application, and the socio-economic characteristics were set up as independent variables. Given that users' energy conservation can be derived from the results of relevant devices supplied and training, the study comprehensively examined which factors affected the increase or decrease of electricity consumption by adding the satisfaction level measured in the survey as an independent variable. The reason socio-economic variables such as gender, age, and income were included in the regression model is to provide a basis for future public policymaking or support [45,46]. Additionally, since residential area variables tend to increase the demand for electricity as the size of the dwelling increases, education regarding the use of, and then using, smart plugs was expected to reduce power usage.

Third, with the dependent variable as the amount of power reduced, the study examined which appliances affected participants' reduced power usage through an additional regression analysis that used the supplied plug and the connected home appliances as independent variables. The reduced power usage depended on the user's lifestyle or the amount of power consumed by the home appliance, but it was expected that this analysis would provide insight into the user's tendency to install smart plugs and the impact of home appliances on power reduction, and that the results would provide a basis for the improvement of the related curriculum.

The dependent variables and the types and contents of each of the independent variables, which combine the contents of these two regression models, are shown in Tables 1 and 2.

Table 1. Dependent (*D.V.*) and Independent (*I.V.*) Variables of Research Model I.

Class	Variable	Scales
D.V.	Average Reduction Rate of Electricity Consumption [1]	Percentage
I.V. (Socio-economic Characteristics)	Gender (Male/Female)	Selective
	Age of Respondent	Number
	Area of Housing	m^2
	Monthly Income per Household	USD
I.V. (Observed Variables)	Frequency of Monitoring Using an Application	Number
	Number of home Appliances Connected with Smart Plugs	Number
	Overall Satisfaction of Smart Plugs and Education	Likert (11pts)

[1] Mean of reduction rate for 15 months compared to the same month last year.

Table 2. Dependent (*D.V.*) and Independent (*I.V.*) Variables of Research Model II.

Class	Variable	Scales
D.V.	Average reduction Rate of Electricity Consumption [1]	Percentage
I.V. (Home Appliances Connected with Smart Plugs)	Lighting Fixtures	Selective ($Y = 1$, $N = 0$)
	Washing and Drying Machine	
	Kitchen Appliances	
	Air Conditioner	
	Living Appliances: (de)Humidifier, Air purifier, etc.	
	Electric Heater	
	Desktop (or Laptop)	
	TV, AV devices, Set-top box	
	Etc.	

[1] Average value for 15 months compared to the same month last year.

This study used a multiple regression model with several independent variables, but since the measurement scale is different, all the variables included were converted to Z-score and used for the regression model. The equation for the regression model is as follows:

$$Y_i = \beta_0 + \beta_1 X_{1i} + \beta_2 X_{2i} + \cdots\cdots + \beta_k X_{ki} + \varepsilon_i \tag{1}$$

where Y_i is the $_i$th observation of the $D.V.$, β_0 is the intercept, $\beta_1 \cdots \beta_k$ is the slope coefficient. For each of the $I.V.$ ε_i is the error term for the $_i$th observation. F for measuring the overall significance of the regression model was calculated as follows:

$$F = \frac{R^2/k}{(1-R^2)/(n-l-1)} = \frac{MS_R}{MS_E} \tag{2}$$

where R^2 is for RSS is residual sum of squares and SSE is for the sum of squared estimate of errors. Additionally, MSR is for regression mean square and MSE is for mean square error. Since the value increases the more $I.V.$ there are in the multiple regression, R^2 was used as the criterion for selecting $I.V.$, whose value is the following. Variance inflation factor (VIF) was used to measure multicollinearity between $I.V.$ and the equation is shown below to the right.

$$R^2_{adj.} = R^2 - \frac{k(1-R^2)}{n-k-1}, \quad VIF_i = \frac{1}{1-R^2_{1,2\cdots k}} \tag{3}$$

where R^2 is for sample R-square, $'n'$ is for total sample size, and k is for number of predictors.

3. Results

3.1. Overview

In this study, 144 people participated in the training and smart plug and switch usage monitoring programs. The results were monitored over 15 months. However, for quantitative analysis, 19 households were excluded because they either replaced or installed new living and kitchen appliances or electric heaters during the period, moved to other areas, or whose attendance was less than 60%. Results from the final 125 households were analyzed. The socio-economic characteristics for the 125 people are shown in Table 3 below. The respondents had a mean age of 40.56, a mean residential area of 87.89 m^2, and an average of 2.60 people per household. This was slightly higher than the average, according to government statistics, of 2.4 people per household [47], and higher than the mean residential area of 84.2 m^2 for complex residential areas.

In this study, 144 people participated in the training and smart plug and switch usage monitoring programs. The results were monitored over 15 months. However, for quantitative analysis, 19 households were excluded because they either replaced or installed new living and kitchen appliances or electric heaters during the period, moved to other areas, or whose attendance was less than 60%. Results from the final 125 households were analyzed. The socio-economic characteristics for the 125 people are shown in Table 3 above. The respondents had a mean age of 40.56, a mean residential area of 87.89 m^2, and an average of 2.60 people per household. This was slightly higher than the average, according to government statistics, of 2.4 people per household [47], and higher than the mean residential area of 84.2 m^2 for complex residential areas.

Prior to the regression model analysis, the effects of the corresponding training program and the use of smart plugs and switches were reviewed. The results showed that for the 15 months, participants had used an average of 204.19 kWh per month during the previous year, but toward the end of the experiment they were using an average of 196.57 kWh per month, showing a reduction of power consumption by about 3.53%. Figure 1 shows the decrease rates for the 125 participating households.

Participants were found to have connected at least two and up to 11 appliances to the smart plugs and switches, averaging 5.18. As shown in Figure 2, the satisfaction level for the six training sessions and smart plug supply programs conducted over a year averaged 5.72 points out of 10. On average, participants' willingness to participate in further smart plug usage and related education programs was 6.18, which is higher than the satisfaction level.

Table 3. Result of General Characteristics of Respondents (n = 125).

Variable	Class	Number (n)	Percentage
Gender (Over 20)	Male	81	64.80%
	Female	44	35.20%
Age Group	20–30	30	24.00%
	31–40	39	31.00%
	41–50	29	23.00%
	51–60	13	10.00%
	Over 60	14	11.00%
Education	Under Secondary Education	0	0.00%
	Under Higher Education	51	40.80%
	Associates Degree or Higher	74	59.20%
Average Income [1]	1250–2083 USD	11	8.80%
	2083–2500 USD	23	18.40%
	2500–3333 USD	52	41.60%
	3333–4167 USD	28	22.40%
	Over 4166 USD	11	8.80%
Number of People in a Households	1–2	14	11.20%
	3	44	35.20%
	4	45	36.00%
	5	22	17.60%
	Over 5	0	0.00%
Area of Residential Area	52–59 m^2	34	27.20%
	59–80 m^2	25	20.00%
	81–110 m^2	31	24.80%
	111–130 m^2	9	7.20%
	Over 130 m^2	26	20.80%

[1] Average value for 15 months compared to the same month last year.

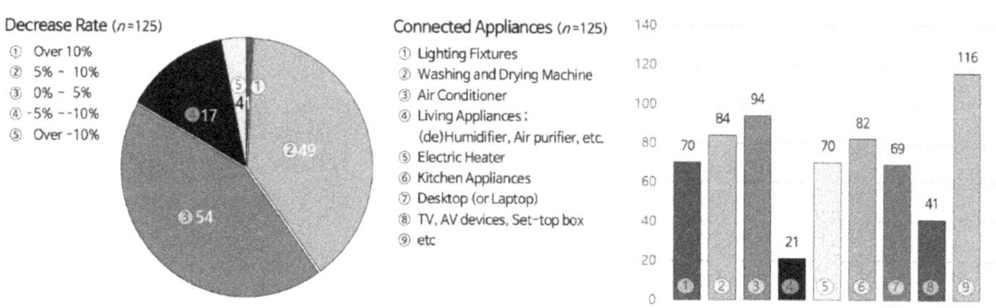

(a) (b)

Figure 1. Result of electric power decrease rate of the program participants for 15 months compared to: (a) the same month last year; (b) connected appliances with smart plugs and switches.

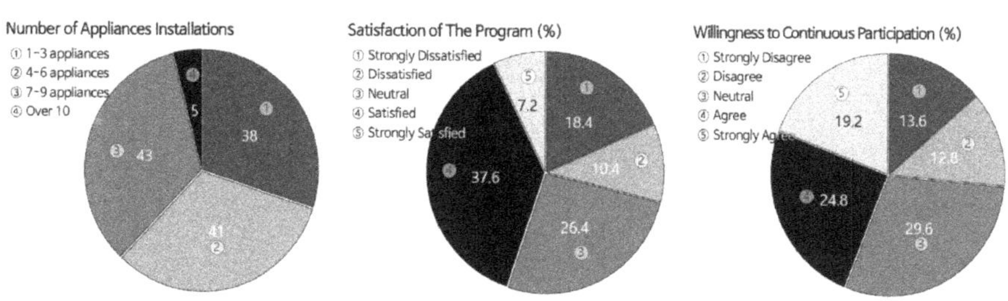

(a) (b) (c)

Figure 2. Results: (a) Number of installed smart plugs and switches; (b) satisfaction with the program; and (c) willingness to continue to participate.

3.2. Effects of Home Energy Savings by The Program

A paired-samples t-test was conducted to verify the quantitative difference between the amount of power reduction and the year-on-year reduction in power usage, measured over the 15 months experiment (see Table 4); the t-value came to 10.666, meaning that the reduced amount of power usage before and after this program shows a statistically significant difference. Moreover, given that the average power consumption of apartment complexes where program participants reside increased by about 1.69% year-on-year, participants in this program reduced power usage through relevant devices and training.

Table 4. Comparison of Home Energy Saving Rate Before and After Program Participation.

Variables	Mean	Std.	t-Value	p
Before program participation	204.19 kWh	28.411	10.666	0.000 [1]
After program participation	196.57 kWh	25.474		

[1] $p < 0.000$.

A paired-samples t-test was conducted to verify the quantitative difference between the amount of power reduction and the year-on-year reduction in power usage, measured over the 15 months experiment; the t-value came to 10.666, meaning that the reduced amount of power usage before and after this program shows a statistically significant difference. Moreover, given that the average power consumption of apartment complexes where program participants reside increased by about 1.69% year-on-year, participants in this program reduced power usage through relevant devices and training.

In order to measure the actual impact on the power reduction, this study selected a control group involving a total of 375 households and compared their average power consumption with that of program participants over the same period. The control group was chosen from the same four apartment complexes as those at which the participants reside. Samples from each complex were selected first in consideration of the distribution of the participants, and the final members were randomly chosen among the residents whose living space is the same as that of the participants. The number of households in the control group was three times larger than that of the program participants. In order to compare the average power consumption between the two groups, this study conducted independent two-sample t-test. The result showed that there found statistical differences between them, just as shown in Table 5.

Table 5. Comparison of Energy Saving Rate Before and After Program with control group.

Variables	Mean		S.D. [2]		t-Value	p
	Participants	C.G. [1]	Participants	C.G. [1]		
Average power usage before program	204.19 kWh	203.20 kWh	28.411	33.505	0.250	0.614
Average power usage after program	196.58 kWh	206.84 kWh	31.474	36.591	−2.528	0.012 *

[1] control group, [2] standard deviation, * $p < 0.05$.

In details, the average power consumption for the 15 months before the launch of the program came to 204.18 kWh and 203.20 kWh, respectively, which showed no statistical differences.

However, the Table 5 for 15 months after the program stood at 196.56 kWh and 206.84 kWh, respectively, which shows a 10-kWh gap, or about 4.97%. The t value is −2.528, which means that the reduction of power consumption between the participants and the control group shows differences in the statistically significant level. This can be translated into that the distribution of related equipment and providing education have a significant impact on the reduction of power consumption.

The analysis results of the two regression models that identify factors affecting participants' power usage are as follows. First, the socio-economic variables, smart plug and switch usage characteristics, and the program's satisfaction level, including training, were input as independent variables (see Tables 6 and 7) to measure factors affecting power usage reduction.

Table 6. Result of Regression Model I.

Variables (Std. Values)	Std. Error	β	t-Value	p-Value [1]	Tolerance
(constant)	0.165	–	21.456	0.000	–
Gender	0.207	0.079	1.476	0.143	0.637
Age of Respondent	0.222	−0.062	−1.082	0.282	0.556
Area of Housing	0.205	0.088	1.658	0.100	0.647
Monthly Income per Household	0.169	−0.028	−0.635	0.527	0.954
Frequency of Monitoring	0.275	0.168	2.355	0.020 *	0.361
No. of Appliances with Smart Plugs	0.264	0.211	3.072	0.003 **	0.391
Overall satisfaction of Program	0.258	0.451	6.715	0.000 ***	0.409

[1] * $p < 0.05$, ** $p < 0.01$, *** $p < 0.001$.

Table 7. Summary of Regression Model I.

R	R^2	Adjusted R^2	F-Statistics	p-Value [1]	Durbin–Watson
0.886	0.784	0.772	60.833	0.000 ***	1.709

[1] *** $p < 0.001$.

The adjusted R^2 showed a high explanatory power of 77.2%. The Durbin–Watson value was 1.709, which is close to 2 and not close to 0 or 4. In short, there was no correlation between the residuals, indicating that the regression model was appropriate. The F value was $p = 0.000$ to 60.833, indicating that the regression line is suitable for the model. Based on the t value, the independent variable with the highest explanatory power was found to be the program that included the provision and education of the smart plug and switch ($t = 6.715$, $p = 0.000$). Providing appropriate equipment to participants and educating them regarding energy-saving had the most significant impact on power usage. Results also showed that the number of smart plug and switch installations had a high impact on energy use reduction ($t = 3.072$, $p = 0.003$). These results suggest that merely using smart plugs and switches, which can cut off used power or standby power from the inside and outside of the dwelling, can provide an adequate level of power-saving. However, given that the control and real-time usage monitoring through related applications of the smart plug and switch, i.e., user-driven monitoring, contribute to power savings in use ($t = 2.355$, $p = 0.020$), the combined-comprehensive consideration and action of the physical–human factors to reduce power usage may increase the reduction of power consumption. However, the socio-economic characteristics mentioned in many studies were not significant in this regression model. Gender, age, income, and residential area were not relevant to the use of smart plugs and switches and reduced electricity use through related education.

The results of regression model II are shown in Tables 8 and 9. The model's dependent variable is the mean reduction rate of participants' power consumption over 15 months, and the independent variable is each type of household appliance connected to a smart plug and switch. Based on the adjusted R^2 value, the research model showed 59.9% and a Durbin–Watson value of 1.765, suggesting with a relatively high explanation power that it contributed quantitatively to reduced power consumption by appliances connected to smart plugs and switches. When considering independent variables that affect the dependent variables based on the t-values, TV sets, set-top boxes, and AV equipment that were mostly located in the living room had the highest effect, even though only 41 households had installed them ($t = 4.729$, $p = 0.000$). Smart plugs of seasonal home appliances installed in general electric heaters ($t = 3.668$, $p = 0.001$) and air conditioners ($t = 3.578$, $p = 0.001$) were found to have affected power usage reduction. Given that during the experiment, living room appliances such as set-top boxes consume the most standby power in Korean households [48], the actual power cut-off through smart plugs has a positive effect on energy reduction. In addition, cut-off of idle power and the power control of household appliances such as air purifiers, humidifiers, and dehumidifiers, which have been increasing recently, were shown to affect reduction of power use. Washing machines and dryers were also found to have affected the dependent variables; participants received education regarding efficient use and power reduction with these appliances. However, desktop computers and lighting devices using smart switches, known to use relatively high standby power, had no significant impact on the dependent variables, though they were installed in 70 and 69 households respectively, which are relatively high rates.

Table 8. Result of Regression Model II.

Variables (Std. Values)	Std. Error	β	t-Value	p-Value [1]	Tolerance
(constant)	0.218	–	16.194	0.000 ***	–
Lighting Fixtures	0.238	0.115	1.858	0.066	0.848
Washing and Drying Machine	0.249	0.194	2.994	0.003 **	0.774
Air Conditioner	0.246	0.234	3.668	0.000 ***	0.793
Living Appliances	0.230	0.192	3.213	0.002 **	0.908
Electric Heater	0.237	0.220	3.578	0.001 **	0.852
Kitchen Appliances	0.248	0.183	2.846	0.005 **	0.778
Desktop (or Laptop)	0.256	0.096	1.440	0.153	0.731
TV, AV devices, Set-top box,	0.239	0.293	4.729	0.000 ***	0.841
Etc.	0.263	0.242	3.543	0.001 **	0.693

[1] ** $p < 0.01$, *** $p < 0.001$.

Table 9. Summary of Regression Model II.

R	R^2	Adjusted R^2	F-Statistics	p-Value	Durbin-Watson
0.793	0.628	0.599	21.582	0.000 ***	1.765

*** $p < 0.01$.

4. Conclusions and Discussion

This study examined the possibility of reducing electricity use in homes through users' voluntary education and use of IoT-based smart plugs and switches to leap into smart cities and energy prosumers. Unlike previous studies that present the results of an energy data analysis at the urban level, or analyze the results of collective education for students or practitioners, actual members of households voluntarily installed smart plugs and switches, and monitored and quantitatively verified the trends in energy-reduction effects through training. Satisfaction with this program, which includes supplied devices and education, had the most significant impact on the power savings of each household; the number of applications designed for real-time power usage monitoring using applications, for which training was provided, had a positive effect on power savings at a relatively high rate. It is therefore expected that these strategies can impact energy savings in efficient homes if the program is implemented in collective residential areas such as apartment complexes with relatively clear management entities.

These results can be used as a basis for future education regarding how equipment such as smart switches and plugs can be used. Differences in the environmental, political, and economic size and characteristics of countries and cities will need to be considered. However, this study showed through a quantitative analysis that power management for TVs, set-top boxes, AV devices and air-conditioning and heating appliances, which are frequently used and have a relatively high standby power in the home, affects energy reduction over a long period of time. These generalized outcomes could lead to in-depth training and voluntary participation for individuals in related training, which could ultimately lead to user-led energy reduction.

However, there are practical limitations to the methods proposed in this study. The average energy reduction for the 125 samples used in this study was around 7.62 kWh. Based on the average household usage of about 200 kWh, the average monthly cost savings was about 1.2 USD based on the calculation method for the target's electricity costs. This means that consumers need years of use to recover their initial cost of investment. Thus, although it can have a relatively positive effect in new housing and housing complexes where various home IoT facilities are installed and provided during the construction phase, willingness to purchase related equipment may be reduced in existing housing and residential complexes; this will reduce effectiveness. Thus, in addition to tax reductions on the purchase of low-power home appliances when making similar plans, the relevant agencies, local governments, or the central government need to secure budget funds to introduce related equipment and education when implementing various residential projects. It is also necessary to provide a foundation for enhancing and implementing energy saving awareness without discrimination. Although many studies examined whether socio-economic characteristics of households affected energy savings, the results of this study model showed that such characteristics were not significant for the dependent variables. Given that there are differences in related programs for each country, city, and region, and that the study was limited to reducing power usage for those who participated in the program instead of samples based on population, further research will need to distinguish between the implications preceding studies and the similarities and differences of this study's topics.

Nevertheless, this study is meaningful in that it produced relevant results through long-term monitoring from a practical point of view of energy saving. It is also meaningful in that it provided a direction for energy saving from the perspective of future policy, education, citizens during the monitoring, and housing, a residential space for citizens.

Funding: This research received no external funding.

Acknowledgments: I thank the participant of this program who willingly participated whole process and provided their valuable research materials. I am also thankful to the anonymous reviewers for their useful suggestions.

Conflicts of Interest: The authors declare no conflict of interest.

References

1. U.S. Energy Information Administration. *Annual Energy Outlook 2011: With Projections to 2035*; EIA: Washington, DC, USA, 2011.
2. Pipattanasomporn, M.; Kuzlu, M.; Rahman, S. An algorithm for intelligent home energy management and demand response analysis. *IEEE Trans. Smart Grid* **2012**, *3*, 2166–2173. [CrossRef]
3. Anvari-Moghaddam, A.; Monsef, H.; Rahimi-Kian, A. Optimal smart home energy management considering energy saving and a comfortable lifestyle. *IEEE Trans. Smart Grid* **2014**, *6*, 324–332. [CrossRef]
4. Mohsenian-Rad, A.H.; Leon-Garcia, A. Optimal residential load control with price prediction in real-time electricity pricing environments. *IEEE Trans. Smart Grid* **2010**, *1*, 120–133. [CrossRef]
5. Pedrasa, M.A.A.; Spooner, T.D.; MacGill, I.F. Coordinated scheduling of residential distributed energy resources to optimize smart home energy services. *IEEE Trans. Smart Grid* **2010**, *1*, 134–143. [CrossRef]
6. Jiang, L.; Liu, D.Y.; Yang, B. *Smart Home Research, Proceedings of the 2004 International Conference on Machine Learning and Cybernetics, Shanghai, China, 26–29 August 2004*; IEEE Cat. No. 04EX826; IEEE: Piscataway, NY, USA, 2004; pp. 659–663.
7. Ricquebourg, V.; Menga, D.; Durand, D.; Marhic, B.; Delahoche, L.; Loge, C. The Smart Home Concept: Our Immediate Future. In Proceedings of the 2006 1st IEEE International Conference on E-learning in Industrial Electronics, Hammamet, Tunisia, 18–20 December 2006; pp. 23–28.
8. Shon, T.; Koo, B.; Choi, H.; Park, Y. Security architecture for IEEE 802.15. 4-based wireless sensor network. In Proceedings of the 2009 4th International Symposium on Wireless Pervasive Computing (IEEE), Melbourne, VIC, Australia, 11–13 February 2009; pp. 1–5.
9. Wang, C.; Liu, X.; Wang, H. A framework of intelligent agent based middleware for context aware computing. In Proceedings of the 2009 Fifth International Conference on Natural Computation (IEEE), Tianjin, China, 14–16 August 2009; pp. 107–110.
10. Mavroudi, A.; Divitini, M.; Gianni, F.; Mora, S.; Kvittem, D.R. Designing IoT applications in lower secondary schools. In Proceedings of the 2018 IEEE Global Engineering Education Conference (EDUCON), Tenerife, Spain, 17–20 April 2018; pp. 1120–1126.

11. Lee, G.M.; Crespi, N.; Choi, J.K.; Boussard, M. Internet of things. In *Evolution of Telecommunication Services*; Springer: Berlin/Heidelberg, Germany, 2013; pp. 257–282.
12. Global Smart Plug Market—Drivers and Forecast from Technavio. Available online: https://www.businesswire.com/news/home/20170228005699/en/Global-Smart-Plug-Market---Drivers-Forecast (accessed on 15 June 2020).
13. Musleh, A.S.; Debouza, M.; Farook, M. Design and implementation of smart plug: An Internet of Things (IoT) approach. In Proceedings of the 2017 International Conference on Electrical and Computing Technologies and Applications (ICECTA), (IEEE), Ras Al Khaimah, UAE, 21–23 November 2017; pp. 1–4.
14. Will a Smart Plug Pay For Itself? Available online: https://www.howtogeek.com/424558/will-a-smart-plug-pay-for-itself/ (accessed on 15 June 2020).
15. Hori, S.; Kondo, K.; Nogata, D.; Ben, H. The determinants of household energy-saving behavior: Survey and comparison in five major Asian cities. *Energy Policy* **2013**, *52*, 354–362. [CrossRef]
16. Ouyang, J.; Hokao, K. Energy-saving potential by improving occupants' behavior in urban residential sector in Hangzhou City, China. *Energy Build.* **2009**, *41*, 711–720. [CrossRef]
17. Alexandru, A.; Jitaru, E. *Education for Energy Saving in the House, Proceedings of the WSEAS International Conference on Energy Planning, Energy Saving, Environmental Education (EPESE'07), Arcachon, France, 14–16 October 2007*; WSEAS Press: Athens, Greece, 2007; pp. 84–89.
18. Ntona, E.; Arabatzis, G.; Kyriakopoulos, G.L. Energy saving: Views and attitudes of students in secondary education. *Renew. Sustain. Energy Rev.* **2015**, *46*, 1–15. [CrossRef]
19. Mohsenian-Rad, A.H.; Wong, V.W.; Jatskevich, J.; Schober, R.; Leon-Garcia, A. Autonomous demand-side management based on game-theoretic energy consumption scheduling for the future smart grid. *IEEE Trans. Smart Grid* **2010**, *1*, 320–331. [CrossRef]
20. Du, P.; Lu, N. Appliance commitment for household load scheduling. *IEEE Trans. Smart Grid* **2011**, *2*, 411–419. [CrossRef]
21. Böhringer, C.; Rutherford, T.F. Combining bottom-up and top-down. *Energy Econ.* **2008**, *30*, 574–596. [CrossRef]
22. Parag, Y.; Sovacool, B.K. Electricity market design for the prosumer era. *Nat. Energy* **2016**, *1*, 1–6. [CrossRef]
23. Yue, T.; Long, R.; Chen, H. Factors influencing energy-saving behavior of urban households in Jiangsu Province. *Energy Policy* **2013**, *62*, 665–675. [CrossRef]
24. Han, D.M.; Lim, J.H. Design and implementation of smart home energy management systems based on zigbee. *IEEE Trans. Consum. Electron.* **2010**, *56*, 1417–1425. [CrossRef]
25. Carroll, J.; Lyons, S.; Denny, E. Reducing household electricity demand through smart metering: The role of improved information about energy saving. *Energy Econ.* **2014**, *45*, 234–243. [CrossRef]
26. Gram-Hanssen, K. Standby consumption in households analyzed with a practice theory approach. *J. Ind. Ecol.* **2010**, *14*, 150–165. [CrossRef]
27. Faruqui, A.; Sergici, S.; Sharif, A. The impact of informational feedback on energy consumption—A survey of the experimental evidence. *Energy* **2010**, *35*, 1598–1608. [CrossRef]
28. Arasteh, H.; Hosseinnezhad, V.; Loia, V.; Tommasetti, A.; Troisi, O.; Shafie-khah, M.; Siano, P. IoT-based smart cities: A survey. In Proceedings of the 2016 IEEE 16th International Conference on Environment and Electrical Engineering (EEEIC), Florence, Italy, 7–10 June 2016; pp. 1–6.
29. Arshad, R.; Zahoor, S.; Shah, M.A.; Wahid, A.; Yu, H. Green IoT: An investigation on energy saving practices for 2020 and beyond. *IEEE Access* **2017**, *5*, 15667–15681. [CrossRef]
30. Tsai, K.L.; Leu, F.Y.; You, I. Residence energy control system based on wireless smart socket and Iot. *IEEE Access* **2016**, *4*, 2885–2894. [CrossRef]
31. Lutui, P.R.; Cusack, B.; Maeakafa, G. Energy efficiency for IoT devices in home environments. In Proceedings of the 2018 IEEE International Conference on Environmental Engineering (EE), Milan, Italy, 12–14 March 2018; pp. 1–6.
32. Ghazal, M.; Akmal, M.; Iyanna, S.; Ghoudi, K. Smart plugs: Perceived usefulness and satisfaction: Evidence from United Arab Emirates. *Renew. Sustain. Energy Rev.* **2016**, *55*, 1248–1259. [CrossRef]
33. Tanujaya, R.R.; Lee, C.Y.; Woo, J.; Huh, S.Y.; Lee, M.K. Quantifying Public Preferences for Community-Based Renewable Energy Projects in South Korea. *Energies* **2020**, *13*, 2384. [CrossRef]
34. Buylova, A.; Steel, B.S.; Simon, C.A. Public perceptions of energy scarcity and support for new energy technologies: A western US case study. *Energies* **2020**, *13*, 238. [CrossRef]

35. Streimikiene, D.; Balezentis, T. Willingness to Pay for Renovation of Multi-Flat Buildings and to Share the Costs of Renovation. *Energies* **2020**, *13*, 2721. [CrossRef]
36. Druică, E.; Goschin, Z.; Ianole-Călin, R. Energy Poverty and Life Satisfaction: Structural Mechanisms and Their Implications. *Energies* **2019**, *12*, 3988. [CrossRef]
37. Besley, J.C. The state of public opinion research on attitudes and understanding of science and technology. *Bull. Sci. Technol. Soc.* **2013**, *33*, 12–20. [CrossRef]
38. Gao, L.; Wang, S.; Li, J.; Li, H. Application of the extended theory of planned behavior to understand individual's energy saving behavior in workplaces. *Resour. Conserv. Recycl.* **2017**, *127*, 107–113. [CrossRef]
39. Aktamis, H. Determining energy saving behavior and energy awareness of secondary school students according to socio-demographic characteristics. *Educ. Res. Rev.* **2011**, *6*, 243–250.
40. Latre, S.; Leroux, P.; Coenen, T.; Braem, B.; Ballon, P.; Demeester, P. City of Things: An Integrated and Multi-Technology Testbed for IoT Smart City Experiments. In Proceedings of the 2016 IEEE International Smart Cities Conference (ISC2), Trento, Italy, 12–15 September 2016; pp. 1–8.
41. Yan, Z.; Zhang, P.; Vasilakos, A.V. A survey on trust management for Internet of Things. *J. Netw. Comput. Appl.* **2014**, *42*, 120–134. [CrossRef]
42. Cheng, H.C.; Liao, W.W. Establishing a lifelong learning environment using IOT and learning analytics. In Proceedings of the 2012 14th International Conference on Advanced Communication Technology (ICACT) (IEEE), Pyeongchang, Korea, 19–22 February 2012; pp. 1178–1183.
43. Lee, W.J. Satisfiers and dissatisfiers of smart IoT service and customer attitude. *Adv. Sci. Technol. Lett.* **2016**, *126*, 124–127.
44. Chatterjee, S.; Kar, A.K.; Gupta, M. Success of IoT in smart cities of India: An empirical analysis. *Gov. Inf. Q.* **2018**, *35*, 349–361. [CrossRef]
45. Martinsson, J.; Lundqvist, L.J.; Sundström, A. Energy saving in Swedish households. The (relative) importance of environmental attitudes. *Energy Policy* **2011**, *39*, 5182–5191. [CrossRef]
46. Wyatt, P. A dwelling-level investigation into the physical and socio-economic drivers of domestic energy consumption in England. *Energy Policy* **2013**, *60*, 540–549. [CrossRef]
47. Seoul Statistics Service. Available online: http://data.seoul.go.kr/dataService/boardList.do#submenu1 (accessed on 25 March 2020).
48. Blog of Ministry of Trade, Industry and Energy. Available online: https://blog.naver.com/mocienews/221436213752 (accessed on 1 July 2020).

energies

MDPI

Article

Safe and Secure Control of Swarms of Vehicles by Small-World Theory

Nicola Roveri *, Antonio Carcaterra, Leonardo Molinari and Gianluca Pepe

Department of Mechanical and Aerospace Engineering, Sapienza University of Rome, 00184 Rome, Italy; antonio.carcaterra@uniroma1.it (A.C.); leonardo.molinari17@gmail.com (L.M.); gianluca.pepe@uniroma1.it (G.P.)

* Correspondence: nicola.roveri@uniroma1.it

Received: 16 January 2020; Accepted: 22 February 2020; Published: 26 February 2020

Abstract: The present paper investigates a new paradigm to control a swarm of moving individual vehicles, based on the introduction of a few random long-range communications in a queue dominated by short-range car-following dynamics. The theoretical approach adapts the small-world theory, originally proposed in social sciences, to the investigation of these networks. It is shown that the controlled system exhibits properties of higher synchronization and robustness with respect to communication failures. The considered application to a vehicle swarm shows how safety and security of the related traffic dynamics are strongly increased, diminishing the collision probability even in the presence of a hacker attack to some connectivity channels.

Keywords: autonomous vehicle; platoon control; swarm behavior; small world theory; complex networks

1. Introduction

The fast development of new technologies in vehicle control and vehicle-to-vehicle (V2V) connectivity by advanced onboard sensors, presents the opportunity to optimize the collective behavior of a population of vehicles, improving the safety and also the security of the people and of the good mobility [1–5]. Safety relies on diminishing the probability of accidents because the vehicle-to-vehicle connectivity permits to transport the information on a long-range scale, with a better synchronization of the vehicle's motion [6–10]. On the other hand, one of the critical elements of this extended connectivity is that communication can be delayed [3,11], accidentally disturbed or broken, or purposely disturbed interfering with some communication channels, distorting or canceling some transmitted information by a hacker attack [12]. This second element requires the collective behavior of the traffic is robust with respect to the insertion or removal of connections [4,9]. Both the safety and security needs represent big challenges [4,10,13–15] in the autonomous vehicle technology development, and government and regulatory institutions, as well as large private companies, have these actions in their strategic agenda.

The present paper considers a new paradigm for V2V [16] connectivity, based on the addition of a few random long-range connections, to take advantage of the big opportunity the emerging new technologies offer. These additional communication channels are superimposed to the car-following mechanism, considered in more conventional traffic dynamics [17–23]. The original contribution of this investigation can be summarized as follows: (i) it is shown how few long-range connections are highly beneficial for the safety of the traffic system, diminishing the probability of collisions, offering a higher level of traffic flow synchronization, (ii) this safety benefit can be made robust with respect to the failure of some communication channels, the system exhibiting an intrinsic security if simple connectivity paradigms are followed, (iii) the method is based on re-adapting the Small-World theory, originally introduced in social sciences [24–26] to a network of cooperating vehicles. These results have an important practical value. In fact, the proposed V2V telecommunication network is of moderate complexity, including only a small number of interconnected vehicles, that produce even better results than a more ramified and capillary connectivity, reducing the cost of communication and the chance of failure. Moreover, because of the randomicity of the connections, the system connectivity can be easily reconfigured, randomly in space and periodically in time, making it secured. An additional argument to support the utility of this analysis is that V2V communications will be an evolutive process in which, in the beginning, only a small fraction of vehicles will be autonomous, and mixed autonomous and manned vehicles with long- and short-range communications, respectively, will coexist together for the times to come.

The theoretical background for this investigation views the traffic dynamics as a complex lattice the nodes of which are occupied by vehicles, subjected to both a convective and oscillatory motion. In the actual single-autonomous-vehicle, the controller takes data from the surrounding environment through on board-sensors (vision, acoustic and radar-based) that imitate the human perception. The same happens in large groups of animals, like a swarm of birds or a shoal of fishes [27–31]. Those animals, in general, have a limited perception of the surroundings since they move in highly populated groups. Each individual of the swarm perceives only a few similar about him and no chances to elaborate any navigation information of other elements of the swarm at a far distance. Nevertheless, despite the local nature of these decisions, under certain conditions, the swarm seems to act as a single body, showing a swarm intelligence [27,30], i.e., the emergence of an organized global behavior is observed.

In a new generation of autonomous vehicles, the increasing possibilities of vehicle-to-vehicle (V2V) and vehicles-to-infrastructure (V2I) [32] communications permit the autopilot to use information coming from remote parts of the swarm, i.e., received from similar individuals at long distance. This can potentially increase the collective intelligence of the system reducing the occurrence of collisions.

For example, along a queue on a highway, normally every vehicle adjusts its motion evaluating the distance and the velocity of the vehicle ahead. This phenomenon, based on a short-range interaction, has been widely investigated in the case of human drivers and is called car-following [17,18]. However, a more realistic scenario should include vehicles of different nature, as for instance autonomous vehicles and human-driven vehicles [33–35]. To guarantee an efficient integration and coordination, control algorithms are often designed so to mimic the human-like driving behavior [36,37], and, given the always-increasing number of vehicles, a collective intelligence, interweaving autonomous and non-autonomous vehicles at short and long-distance, becomes crucial to mitigate phenomena known as traffic waves [38–40]. Nevertheless, collective motions might be dominated by instability, and suitable control analytical methods must be provided to achieve safe and efficient global traffic motion [41,42].

The V2V communication systems allow some autonomous vehicles to act like no human driver could, using signals from some cooperating vehicles at a far distance in the queue. Is it possible to introduce a connectivity widening to induce an enhanced swarm intelligence? One of the main focus of this paper is to answer this question. In the following, it is proven that revisiting the small-world theory, few random long-range interactions between vehicles produce a strong motion synchronization along the queue, which is shown to reduce the probability of accidents.

2. Brief Resume of Small-World Theory Relevant for Swarm of Vehicles

The term small-world has been introduced by Stanley Milgram in 1967 in his seminal work on social networks, entitled "The Small-World Problem" [24]. In that article, Milgram introduces the small-world problem analyzing the intermediate acquaintance chain that connects two randomly chosen people in the world. It could happen that two people don't know each other directly, but they share a mutual acquaintance that is a person who knows both of them. In this case, only one acquaintance link connects the two people. Milgram's formulates the problem in this form: "Given any two people in the world, person X and person Z, how many intermediate acquaintance links are needed before X and Z are connected?" The experiment that Milgram conducted, brought to assert that, taken randomly two people in the United States, these are separated by a chain of relationships involving six acquaintance links. For this reason, his study is often called six degrees of separation.

Although the problem has been initially studied in the field of social networks, later it has been reconsidered in several other scientific fields approaching a strict definition of a small world model, still studied as a branch of the theory of graphs. This model, defined by Watts and Strogatz [26], provides a robust mathematical basis to the Milgram experiment. Besides the mathematical foundation, in this article an innovative significant concept is introduced, crucial for the present investigation. In the ordinary theory of graph, the connection topology is assumed to be either completely regular or completely random. In [26], the authors fill the gap, considering regular networks rewired to introduce some amount of disorder. They show these structures exhibit small characteristic path lengths, like random graphs, but with a modest addition or random connections. The dynamic of such small-world is shown to be characterized by high propagation speed and strong synchronization, a key effect in the present paper.

Technically, Watts and Strogatz consider a network made of vertices and unoriented links, initially characterized by a strongly regular pattern (in our case the short-range interactions of a car-following dynamics) that is randomly modified until a completely random pattern is obtained. The process is pictorially described in Figure 1.

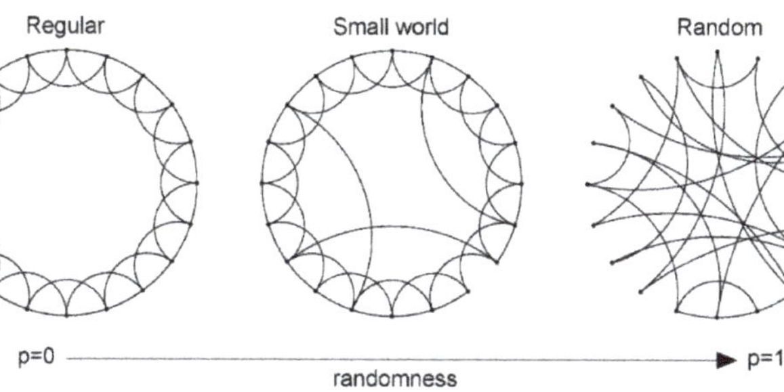

Figure 1. Example of the small-world model.

In Figure 1, p is the probability that in the network one of the initial *regular* connection has been changed to connect distant vertices (in our case, for the traffic network, the chance of including in communication between distant vehicles). This probability can be increased from $p = 0$, representing the regular connection architecture, up to $p = 1$, the case of a completely random graph.

The effects of this progressive modification are studied using two parameters:

- $C(p)$ Cluster coefficient, it quantifies the amount of local interaction, among local groups of close by nodes, that in our case is a measure of the car-following interaction in the swarm. To define the clustering coefficient, suppose that a vertex v has k_v neighbors: at most $k_v(k_v - 1)/2$ edges can exist between them and this happens when every neighbor of v is connected to every other neighbour of v. Let C_v denote the fraction of these allowable edges that actually exist and define $C(p)$ as the average of C_v overall v.
- $L(p)$: Characteristic length of the paths, is the characteristic separation that is present between two nodes of the graph, however, they are chosen. L is the number of edges in the shortest path between two vertices, averaged over all pairs of vertices. This parameter provides an indirect measure of the collective behavior of the network, that in our case is related to the synchronization effects of the swarm of vehicles.

Figure 2 shows how the global feature $L(p)$ is very sensitive to p, while the local feature $C(p)$ is insensitive to p variations, at least in the region of small p. One conclusion of great importance, and very attractive for vehicular traffic, is that the collective behavior of a network can be activated by insemination of a moderate perturbation of random connections, permitting the local information to travel much faster than possible in regular local connectivity architectures.

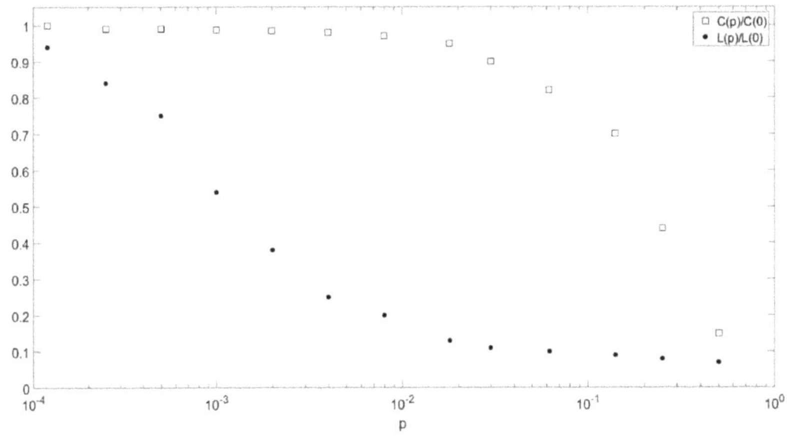

Figure 2. Path length $L(p)$ and clustering coefficient $C(p)$ for the family of randomly connected graph of Figure 1.

The basic idea of this paper is this effect can be used in the development of new connectivity architectures to be used among a swarm of autonomous vehicles. Namely:

(a) The swarm dynamics are initially based on short-range regular interactions taking place between individuals. This effect is related to the natural imitation of the human behavior that controls the motion of his own vehicle elaborating the information of the surrounding vehicles, perceived by proximity sensors.

(b) The synchronization of the vehicles of the swarm is a collective behavior that reduces the probability of collisions. It is related to the speed at which the information and the mechanical reactions propagate along with the swarm. Therefore, one winning strategy is based on promoting the collective effects in the swarm. This result can be achieved by introducing V2V long-range communications.

(c) The intriguing result of the small-world theory is represented by the fact this activation is possible using a moderate number of random long-range connectivity, meaning the perturbation to be introduced with respect to more classical traffic management is low-cost and technically feasible by a small number of hyper-connected cars in the swarm.

(d) An additional added value of this control strategy is related to the random nature of the additional connectivity. Since the driving parameter to activate the collective response is $L(p)$, controlled by p, it means if the connections are individually altered but leaving p constant, the level of collective behavior remains constant. This clarifies that in a connected swarm, we have the chance of modifying the connectivity architecture still preserving the synchronization. This is the root of the robustness of such architecture control that claims for better security of such systems. In fact, the connectivity architecture can be periodically or continuously modified, making useless to attack deliberately specific connections. Moreover, since there is not a central control unit, the system uses a distributed control logic the is made by connections of the individual. The intelligent behavior is because of the sparse random long-range connections, that is less prone to interference.

3. Complex Network of Vehicular Traffic

Car-following models are designed for a one-dimensional path. The leader has an assigned velocity law that represents the forcing term to the system. Every vehicle considers the vehicle in front as its leader and adjusts its acceleration accordingly: each follower may consider, for instance, the distance and the difference of velocity (given a characteristic delay) with regard to its leader, then computes its own acceleration. Therefore, any car-following model can be schematized as an oriented graph, where the kinematic information regarding the first vehicle is progressively transmitted downstream, to the other ones, while no information can travel upstream, from the following to the vehicle ahead.

As can be seen in Figure 3, it is possible to consider the flow of information originated by the leader, to study its propagation along the line of vehicles. Considering the notation for the vehicles expressed in the figure, vehicle n = 1th is the leader, n = 2th is the first of the followers and so on. The state of the leader (position and velocity) is continuously monitored by the vehicle 2nd, which suitably adapts its state, then the reaction of vehicle 2nd is evaluated by vehicle 3rd so that variations of the leader's motion are transmitted along the line of vehicles.

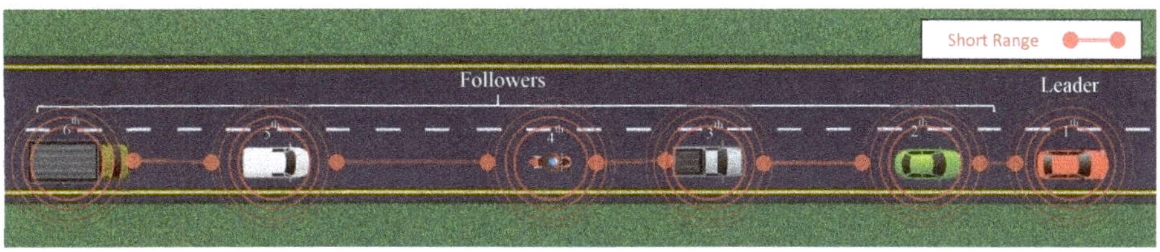

Figure 3. Graph of the standard car-following model.

It is now introduced the distance, i.e., the number of steps needed so that the information emanated from the leader reaches a certain vehicle n along the line. For instance, vehicle 2 receives the information directly from the leader so its distance can be considered equal to one, vehicle 3 is one step behind in the line, so the leader's input is indirectly received after two steps, and the vehicle n is at a distance $D_n = n - 1$ from the leader. The average distance from the leader is now introduced, as follows:

$$\overline{D} = \frac{\sum_{n=2}^{N} D_n}{N-1} \tag{1}$$

where D_n is the distance between the vehicle n to the leader. Equation (1) evaluates the mean distance from the leader for a line of vehicles and represents a global information regarding how fast the leader state is known to the other vehicles.

In the standard car-following model shown in Figure 3, Equation (1) becomes a simple sum of powers evaluated with the Faulhaber's formula:

$$\overline{D} = \frac{\sum_{n=2}^{N}(n-1)}{N-1} = \frac{(N-1)N}{2(N-1)} = \frac{N}{2} \tag{2}$$

The result carried out by Equation (2) is rather intuitive, indicating that the average distance from the leader linearly increases with the total number N of vehicles within the queue.

At this point, the standard car following model is modified, introducing a few random connections, to study if and how the findings of the small-world theory modify the average distance from the leader.

With reference to Figure 4, given a few, randomly chosen, vehicles, each one is connected not only with the vehicle ahead but also with a vehicle s-steps ahead (for the sake of simplicity, it will be briefly referred as the distant vehicle in the following), beings a random positive integer as well. The information regarding the states of the two vehicles is weighted so that the bi-connected vehicle adapts its state accordingly. The information continues to travel only in one direction, but random links between distant vehicles are added to the standard connection between neighbors.

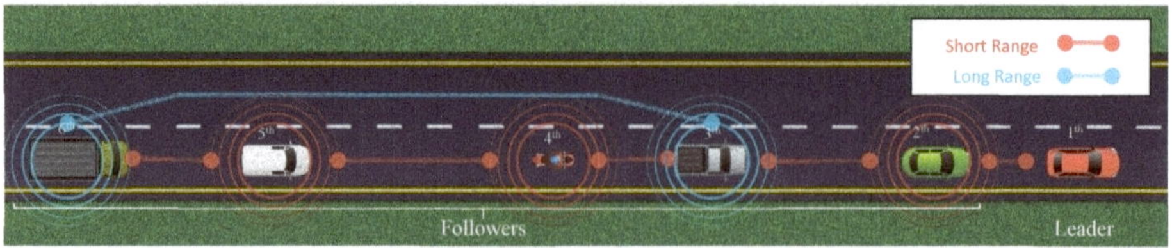

Figure 4. Graph of the small-world model.

For the modified model, the average distance (1) depends on the random, long-range connections that are introduced. For example as in Figure 4, if the vehicle 6th receives long-range information from the third vehicle, its distance from the leader can be calculated in two different ways:

- Minimum distance from the leader D_m: it considers the shortest path that the leader's information covers. For the vehicle 6th, the input arrives faster from vehicle 3rd than 5th, then the distance of vehicle 6th is considered as if it would follow the vehicle 3rd only, so $D_6 = 3$.
- Weighed distance from the leader D_w: the information deriving from the vehicle 3rd is joined to the one coming from the vehicle 5th. Set a and b the weight assigned to the vehicle ahead and the one of the distant vehicle, respectively, where $a + b = 1$, the distance from the leader is weighed accordingly. In the example examined, the distance from the leader is equal to $D_6 = a(D_3 + 1) + b(D_5 + 1)$.

By using these two indicators, a preliminary yet important understanding of how, random, small-world like, connections modifies the average distance, is obtained. To this aim, the modified model is numerically implemented in Matlab© (2019b, MathWorks company, Natick, MA, USA), which depends only on N, the total number of vehicles, and P, the density of long-range interactions. If, for instance, $N = 100$ and $P = 0.2$, then the generator randomly chooses 20 vehicles and for each of them, randomly assigns the distant vehicle connection with which communicate. Each vehicle can communicate with any of the vehicles ahead, so the vehicle n can have a long-range connection with vehicles between $[2; n - 2]$, the leader is indeed excluded and $n - 1$ is the standard connection between n and the vehicle ahead, which is always present, as shown in Figure 4.

Following the previous definitions, the distances $D_m(N, P)$ and $D_w(N, P)$ are numerically evaluated: in order to limit the dependence of the result from a specific set of random connections, for each selected value of the density P, a hundred different trials are computed, i.e., 100 different randomly connected systems are considered, then, the mean value of each distance is computed, i.e., $\overline{D}_m(N, P) = \sum_{k=1}^{100} \frac{D_{mk}(N,P)}{100}$, and P is increased by 0.5%. The analysis continues for each value of P studied. The distances are then normalized with the mean distance value for the standard car-following model, i.e., $\frac{N}{2}$ from Equation (2).

Figure 5 shows the results of the numerical analysis, for which comments follow below:

(1) The overall behavior of the system, regarding the transmission of the information, changes significantly with the introduction of P: the two distances have a strong decaying trend with respect to P.
(2) Even for low values of P, the distance from the leader diminishes rapidly, regardless of the formula used to compute it. For instance, considering $P = 0.1$, e.g., the random connections are only 10% in respect to the standard model, the Minimum normalized distance is roughly 0.06 and the Weighed normalized distance is 0.14, which mean the distances for the small world model are much shorter than the mean distance for the standard model, being 6% and 14%, respectively.

(3) The local connections remain almost unchanged since the information from the vehicle in front is still used from each vehicle even if few of them give it less importance, as imposed by the weight coefficient previously introduced.

The results above are rather general since they are obtained without studying the controller but just analyzing the information flow. These findings allow to conjecture that, for the modified models, a control imparted to the leader can be efficiently transmitted to the rest of the vehicles, or, in other words, the state of the whole system can be easily controlled, acting on the leader only, if a small-world network of following vehicles is employed.

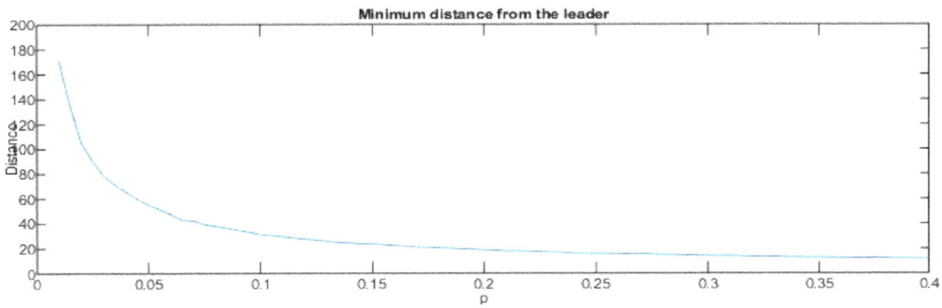

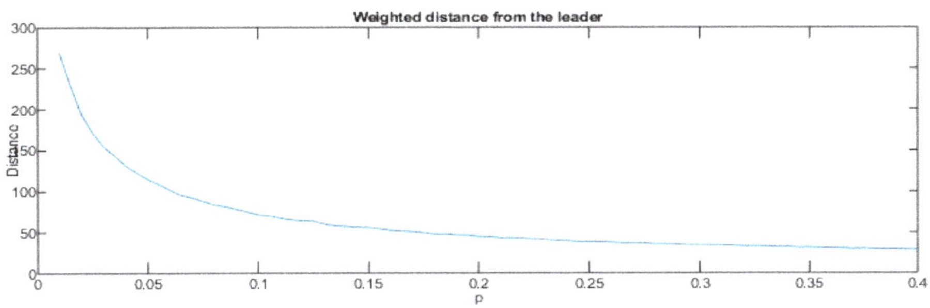

Figure 5. Starting from the plot on top, Minimum normalized distance and Weighted normalized distance of the small-world model in function of p.

3.1. Mathematical Introduction to the Standard Car Following Model

With reference to Figure 3, for the standard model each following vehicle, from 2 to N, modifies its state only in function of the state of the vehicle in front.

The information needed to implement a control law is the position and velocity, which can be measured somehow for the vehicle ahead, and with a characteristic delay time τ which can't be neglected. Then the equations that govern the dynamics of the standard model are now presented and commented:

$$\ddot{x}_n(t) = \beta_n(t)\left(\dot{x}_{n-1}(t-\tau) - \dot{x}_n(t-\tau)\right),\ n = 2,3,\ldots,N \tag{3}$$

with:

$$\beta_n(t) = \frac{\alpha\left(\dot{x}_n(t)\right)^m}{\left(x_{n-1}(t-\tau) - x_n(t-\tau)\right)^l},\ \ \alpha > 0,\ \ m,\ l \geq 0 \tag{4}$$

where x is the position of the vehicle with respect to a fixed reference frame, t is the time variable and τ is the characteristic delay of the model, also called the reaction time. Equation (3) shows that the acceleration of each vehicle linearly depends on:

The relative velocity with respect to the vehicle ahead, delayed by the reaction time.

A coefficient β, which measures the sensitivity to the velocity of the vehicle, evaluated at the time t since the velocity can be measured onboard with a neglectable delay, with respect to the relative, delayed distance.

α is a positive coefficient of proportionality, m, and l are nonnegative constants if $m = l = 0$ the model is linear.

Equation (3) is rewritten introducing relative variables, i.e., relative distance and relative velocity. The relative state variables of the system are defined as follows:

$$\begin{gathered} y_n(t) + b = x_{n-1}(t) - x_n(t) \\ \dot{y}_n(t) = v_n(t) = \dot{x}_{n-1}(t) - \dot{x}_n(t) \end{gathered} \tag{5}$$

where b is the distance between the vehicles in stationary conditions. Substituting the (5) in the (3) the following is obtained:

$$\ddot{x}_n(t) = \beta_n(t)(v_n(t-\tau)), \quad n = 2, 3, \ldots, N \tag{6}$$

where

$$\beta_n(t) = \frac{\alpha\left(\dot{x}_n(t)\right)^m}{(y_n(t-\tau)+b)^l} \tag{7}$$

In order to remove the dependence from $\dot{x}_n(t)$, Equation (7) is written in terms of relative variables:

$$\begin{gathered} v_2 + v_3 = \dot{x}_1 - \dot{x}_2 + \dot{x}_2 - \dot{x}_3 = \dot{x}_1 - \dot{x}_3 \\ v_2 + v_3 + v_4 = \dot{x}_1 - \dot{x}_2 + \dot{x}_2 - \dot{x}_3 + \dot{x}_3 - \dot{x}_4 = \dot{x}_1 - \dot{x}_4 \\ \vdots \\ \sum_{j=2}^{n} v_j = \dot{x}_1 - \dot{x}_n \rightarrow \dot{x}_n = \dot{x}_1 - \sum_{j=2}^{n} v_j \end{gathered} \tag{8}$$

where, for the sake of notation, the time dependence (t) has been omitted. Using the (8) in the (7) β_n is written as:

$$\beta_n(t) = \frac{\alpha\left(\dot{x}_1(t) - \Sigma_{j=2}^{n} v_j(t)\right)^m}{(y_n(t-\tau)+b)^l}, \quad n = 2,\ 3,\ N,\ \alpha > 0,\ m,\ n \geq 0 \tag{9}$$

Taking the derivative for the second of the (5):

$$\dot{v}_n(t) = \ddot{x}_{n-1}(t) - \ddot{x}_n(t) \tag{10}$$

then substituting Equation (6) in the right-hand side (rhs) variables of Equation (10) holds:

$$\dot{v}_n(t) = \beta_{n-1}(t)v_{n-1}(t-\tau) - \beta_n(t)v_n(t-\tau), \quad n = 3, 4, \ldots, N \tag{11}$$

From the second of the (5) and the (7) the state equations are obtained:

$$\begin{cases} \dot{y}_n(t) = v_n(t) \quad n = 2,\ 3, \ldots,\ N \\ \dot{v}_2(t) = \ddot{x}_1(t) - \beta_2(t)v_2(t-\tau) \\ \dot{v}_n(t) = \beta_{n-1}(t)v_{n-1}(t-\tau) - \beta_n(t)v_n(t-\tau),\ n = 3 \cdots N \end{cases} \tag{12}$$

where β is given by Equation (9). The model expressed by Equation (12) can be modified to introduce the small-world dynamics.

3.2. Mathematical Introduction to the Small-World Model

The small-world model is now introduced, adding random connections on the rhs of Equation (12) as depicted in Figure 4. After some algebra, fully worked out in Appendix A, the state equations for the small world model are found:

$$\begin{cases} \dot{y}_n = v_n \;\; n = 2, 3, \ldots, N \\ \dot{v}_2(t) = \ddot{x}_1(t) - \beta_{2,1}(t) v_2(t-\tau) \\ \dot{v}_3 = \beta_{2,1} v_2(t-\tau) - \beta_{3,1} v_3(t-\tau) \\ \dot{v}_n = a_{n-1}\beta_{n-1,1} v_{n-1}(t-\tau) + b_{n-1}\beta_{n-1,s} \sum\limits_{j=1}^{s} v_{n-1-j+1}(t-\tau) + \\ -\left(a_n \beta_{n,1} v_n(t-\tau) + b_n \beta_{n,s} \sum\limits_{j=1}^{s} v_{n-j+1}(t-\tau) \right), n = 4, 5, \ldots, N \end{cases} \tag{13}$$

for the sake of notation, the time dependence (t) has been omitted, the new coefficients β are given by Equations (A8) and s is a positive random integer that ranges within [2, n – 2].

Comparing Equation (12) with Equation (13), the modified model is rather like the standard one, since, under the small world hypothesis only a few vehicles have also long-range connections. Both models, the standard and the small-world are delayed differential models that need the initial condition of the state, $y_i(0), v_i(0)$, and the input of the leader to be solved.

4. Numerical Analysis and Discussion

The responses of the standard model expressed by Equation (12) and the small-world model expressed by the Equation (13) are now studied, putting the systems out of their stationary equilibrium positions. In both models, the stationary state is given by a constant velocity for each vehicle and a constant distance between the vehicles, equal to the parameter "b" in Equations (12) and (13). Since the aim of this study is comparing the performance of two models, all the parameters will remain constant through the numerical solutions for both models. Therefore, the assigned parameters are: $\alpha = \tau = l = m = 1$ and $b = 40$ (equilibrium distance in meters); the equilibrium velocity is assigned at 10/s, this state is altered by varying the input i.e., the leader's velocity. Please note that the equilibrium distance of 40 m is quite large for normal driving conditions and in respect of the selected speed: this is a first attempt made in order to avoid the problem of string instability [35,36], which is actually under an in-depth investigation and will be the object of a forthcoming paper, which will show how the small world model is able to ameliorate the problem. Anyhow, a second experiment will be also provided at the end of the section, considering a much shorter distance, of 20 m, which is closer to the usual safety distance among adjacent vehicles.

Numerical solutions of Equations (12) and (13) are obtained using Matlab's DDE23 function to solve the delayed differential equations. To study and compare the numerical results, two different inputs are assigned to the leader:

A harmonic perturbation of the Leader's velocity:

$$\dot{x}_1 = 10 + 3\sin\left(\frac{\pi}{10}t\right)\frac{\mathrm{m}}{\mathrm{s}} \tag{14}$$

A constant deceleration of the leader

$$\begin{cases} \dot{x}_1 = 10 - 4t & for\; t \leq 2s \\ \dot{x}_1 = 2 & for\; t > 2s \end{cases} \tag{15}$$

The harmonic velocity, Equation (14), has a mean value equal to the equilibrium velocity, an amplitude that prevents negative velocities (the minimum velocity is 7 m/s) and a period of 20 s that is long enough to represent a slow variation of the velocity.

The fast deceleration (Equation (15)) instead, varies the velocity from 10 m/s to 2 m/s in just 2 s to represent a rather intense braking maneuver. These two cases allow analyzing the behavior of the vehicles both in normal and emergency conditions: from a traffic point of view, the main goal of a car-following model is that the queue of vehicles is able to replicate the input imparted by the leader, which is the vehicle in front of the queue so that the entire line of vehicles almost move as a rigid body. A sinusoidal function appears to be the natural input to analyze the capability of the model to synchronize its motion with that of the leader. The breaking maneuver allows studying how fast the queue of the vehicles is able to react to a sudden and unsteady input imparted by the leader.

4.1. Analysis of the Standard Model

The sinusoidal input stated in Equation (14), is considered for a standard model with 500 vehicles. Please note that the selection of a rather large number of cars is required in order to have a significant number of long-range connected vehicles along the queue. Indeed, the small-world theory has been originally introduced in social and complex networks, where the number of nodes is usually huge and the theory relies on a very small density of long-range connections. The velocity and relative distance from the vehicle ahead for the first 7 vehicles that follow the leader are plotted in Figure 6: these curves are characterized by sinusoidal trends, whose amplitudes diminish with the distance from the leader (considering the distance as the number of vehicles between the vehicle and the leader), and progressively become out of phase with the leader motion.

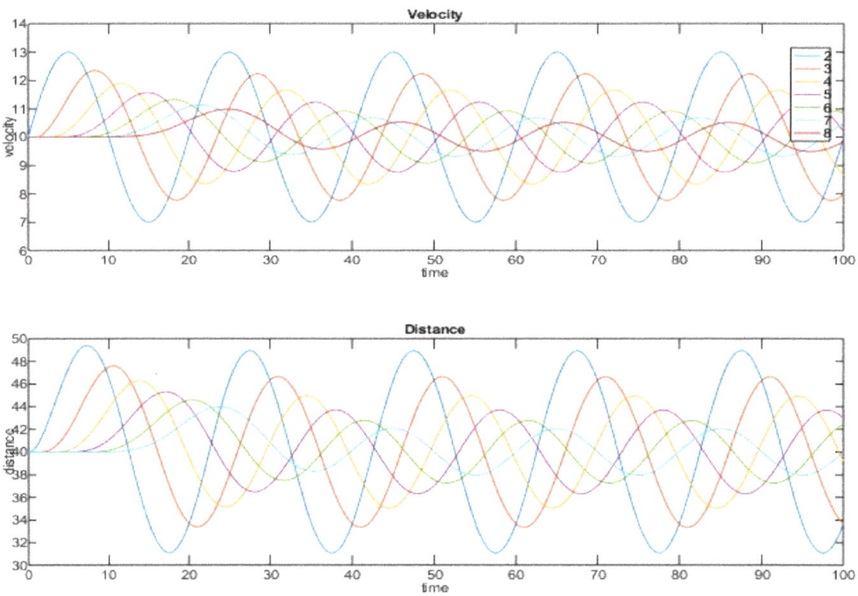

Figure 6. The subplot on (**top**) shows the velocity of the first 7 vehicles that follows the leader, which is subjected to the harmonic perturbation defined in Equation (14). Using the same symbols, the subplot on the (**bottom**) shows the relative distance between a vehicle and the one ahead.

The state of each vehicle only depends on the motion of the vehicle ahead and starts following its state after a time delay equal to the delay τ that appears in the differential equations Equation (12). This is clearly visible in Figure 6 where each curve moves from the initial value with a constant delay from the preceding. For each vehicle, the mean value of the velocity is always equal to the mean value of the leader's velocity, while the amplitude of oscillations decays rapidly from one vehicle to the next, because of the stability of the model as it is explained in Appendix B.

With reference to Equation (12) the coefficient β plays the role of damping parameter, which increases going down the line of vehicles. As shown in Figure 7, farther vehicles show indeed a transient behavior, characterized by a small oscillation, close to 1% of the leader's amplitude, followed by stationary oscillations with very small amplitudes. The first oscillation of velocity shown in Figure 7 is indeed characterized by a peak velocity, due to the impulsive application of the load.

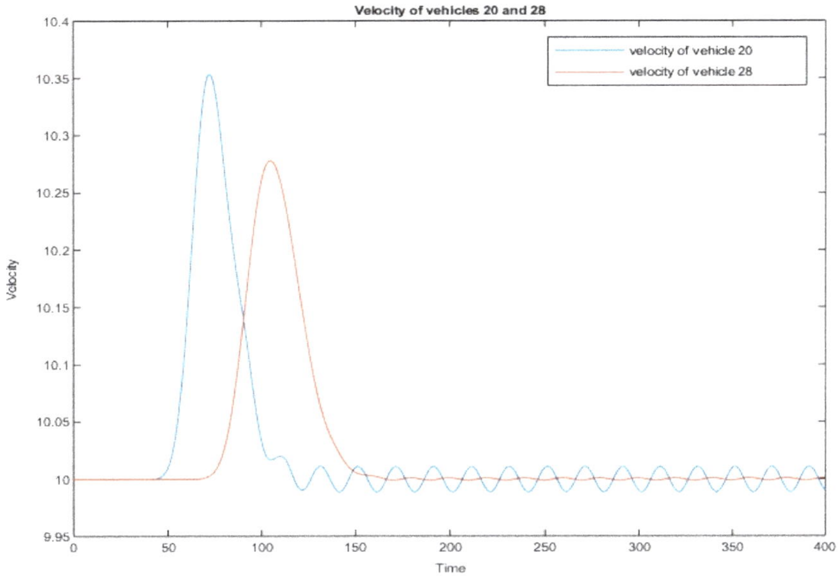

Figure 7. The velocity of vehicles 20 and 28 plotted versus time. The amplitude scale has been magnified in respect of Figure 6, to better appreciate the arrival (peak) of the harmonic disturbance imparted to the leader.

To analyze how the peak changes along the queue of vehicle, the maximum value of the velocity and the time at which takes place are shown in Figure 8 (the plot shows only the first 200 vehicles, over a total of 500, because the peak intensity quickly becomes negligible). The time at which the peak takes place linearly increases with the distance from the leader, the slope of the curve in Figure 8 is related to the characteristic time delay τ. The magnitude of the peak decays rapidly after the first vehicles, so that the harmonic perturbation emanated from the leader motion becomes, quickly, no longer detectable.

With reference to Figure 9, the motion of the barycenter of the whole queue of vehicles is almost constant, while the motion of the leader is oscillatory: as long as the number of vehicles increases, the oscillations of the barycenter becomes smaller and smaller until become negligible. For the standard model, the motion of the barycenter is related to leader motion only on average, i.e., same mean value, while it is not able to highlight the harmonic perturbation, i.e., the amplitude of oscillations is negligible.

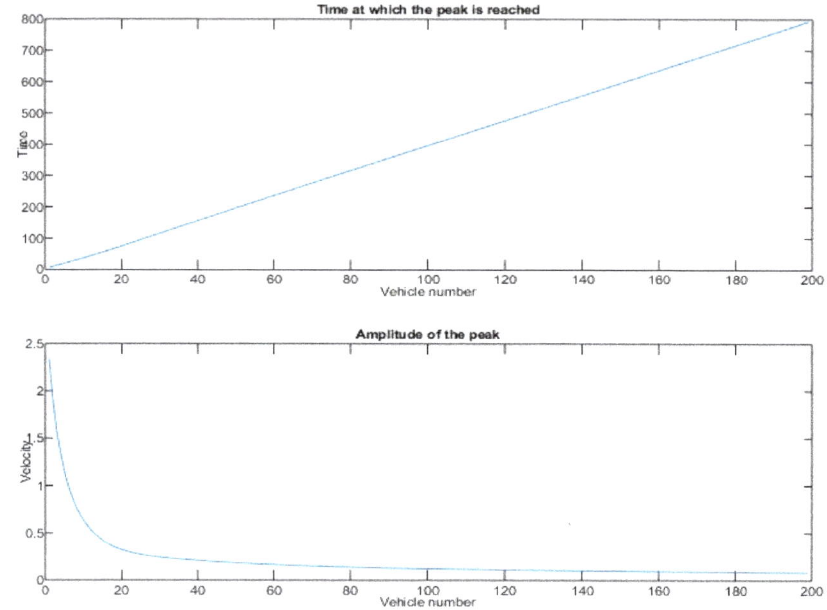

Figure 8. Time at which the harmonic disturbance imparted to the leader reaches each vehicle, on (**top**), and peak velocity, on the (**bottom**), versus the vehicle number.

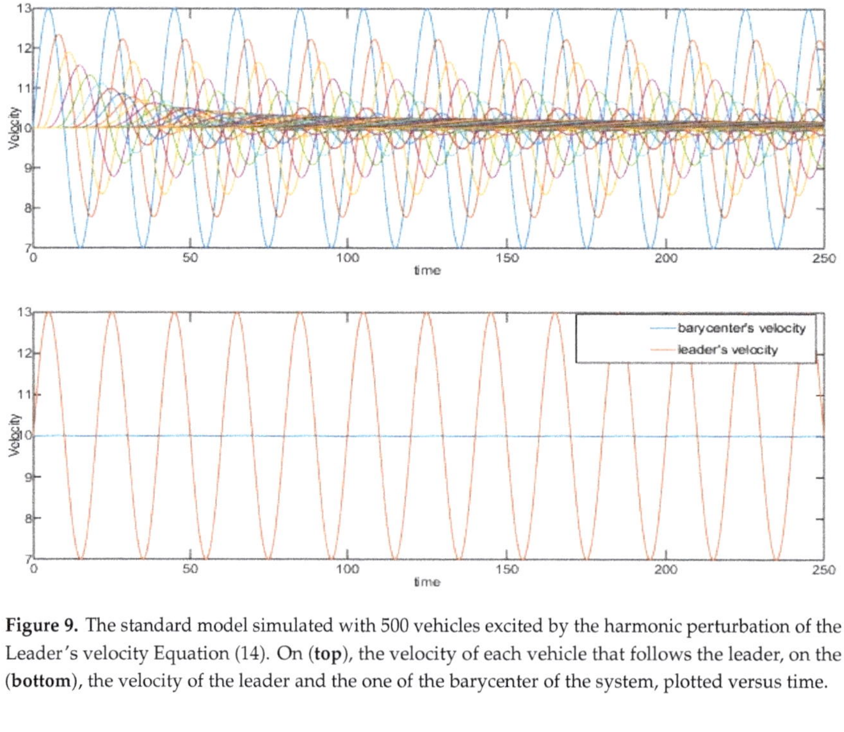

Figure 9. The standard model simulated with 500 vehicles excited by the harmonic perturbation of the Leader's velocity Equation (14). On (**top**), the velocity of each vehicle that follows the leader, on the (**bottom**), the velocity of the leader and the one of the barycenter of the system, plotted versus time.

Considering now the constant deceleration input imparted by the leader defined by Equation (15), for a set of 100 vehicles. As shown in Figure 10, the standard model responds to the braking maneuver in the most intuitive way: each vehicle starts breaking, following the vehicle ahead, after a time delay equal to the characteristic delay τ. The deceleration intensity decreases from one vehicle to the next since the stability of the model damps the original, transient, input emanated from the leader. As a result, the barycenter has a much smaller deceleration with respect to the one of the leader: the queue of 100 vehicles reaches indeed the leader's velocity after 450 s.

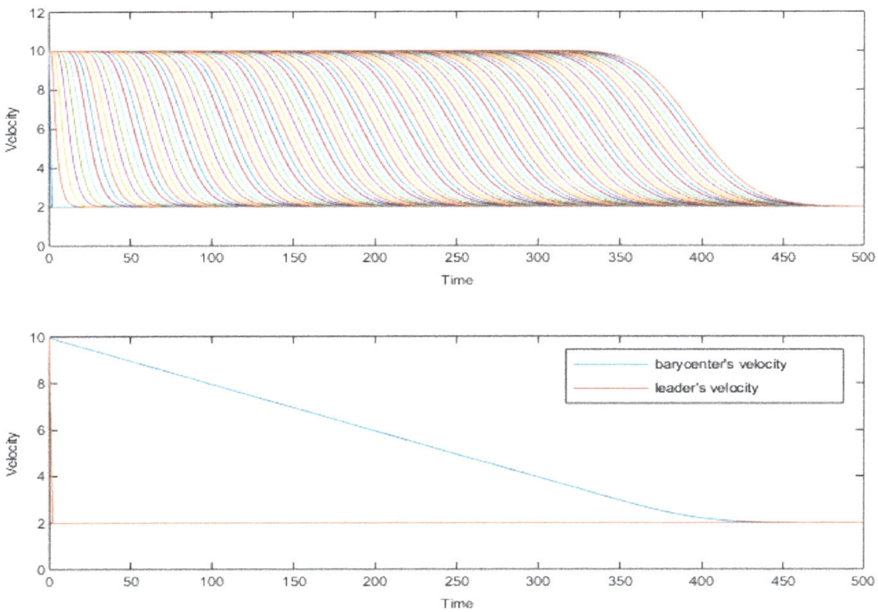

Figure 10. The standard model simulated with 100 vehicles excited by the braking maneuver imparted by the leader Equation (15). The velocity of each following vehicle, on (**top**), and velocity of the Leader and velocity of the barycenter, on the (**bottom**), plotted versus time.

4.2. Analysis of the Small-World Model

Even if the small-world model has a random component that makes each realization unique, it has been conjectured that the system response mainly depends on the density of random long-range connections. This paragraph shows how, even for low values of the density of long-range interactions, the platoon manages to reach a swarm behavior.

Keeping all the parameters unchanged with regard to the standard model ($\alpha = \tau = l = m = 1$ and $b = 40$) a density P = 5% of random long-range interactions is assigned to a platoon of 500 vehicles. Therefore, 475 vehicles will follow the same law as the ones in the standard model and the remaining 25, randomly selected along with the platoon, have two inputs: the one from the preceding vehicle and one from a distant vehicle ahead.

In Figure 11 the response of the modified platoon to the harmonic input defined in Equation (14) is represented. The first four vehicles after the leader show the same trend already encountered in Figure 9 for the standard model: this is due to the fact that, in this realization, the first long-range connection involves the fifth follower. Then the velocities of the following vehicles are condensed in a cluster, which is characterized by the same, smaller and roughly in phase, oscillations. Even if only a few vehicles have a long-range interaction, nearly all the vehicles in the platoon reach a nearly synchronized state, in less than 15 s after the application of the perturbation, this property will be further appreciated with the comparison between the barycenter's velocity for the standard and small-world model, carried out in Section 4.3 and shown in Figure 19.

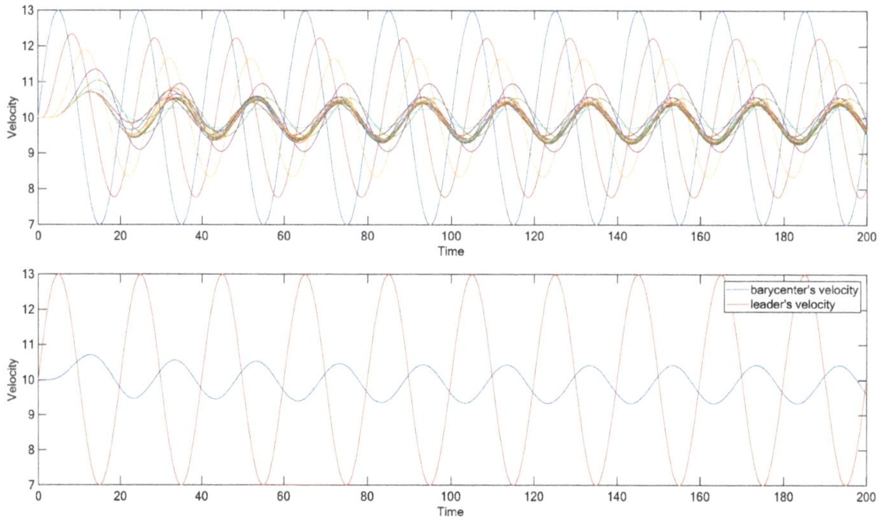

Figure 11. The small-world model simulated with 500 vehicles and with $p = 0.05$, excited by the harmonic perturbation of the Leader's velocity Equation (14). On (**top**), the velocity of each vehicle that follows the leader, on the (**bottom**), the velocity of the leader and the one of the barycenter of the system of 500 vehicles, plotted versus time.

The barycenter of the platoon is indeed clearly affected by the input of the leader, this shows that the small-world theory applied to a platoon of vehicles has the same effect as in any other field where it has been studied before: it allows the information to spread rapidly along with all the population. It should be noticed that the barycenter of the platoon in Figure 19 is roughly in the antiphase with respect to the leader motion.

The effect of the long-range interaction is even more evident for the transient input imparted by the leader, i.e., the deceleration introduced in Equation (15). To compare the standard and the small-world models, the platoon subject to the breaking is composed of the same number of vehicles, i.e., $N = 100$. In Figure 12 the 100 vehicles present 10% of random long-range connections that allow the platoon to follow the input of the leader after few seconds: the long-range interactions allow the farther vehicles to respond indeed to the leader's input after just a few seconds.

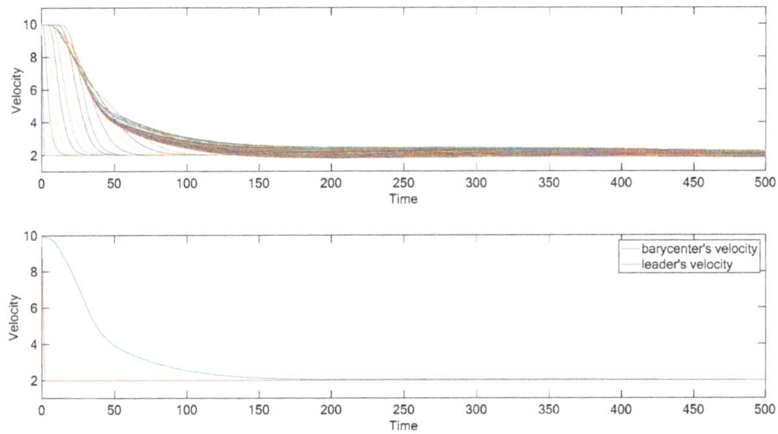

Figure 12. Small world model simulated with 100 vehicles and $p = 0.1$, excited by the braking maneuver imparted by the leader Equation (15). The velocity of each following vehicle, on (**top**), and velocity of the Leader and velocity of the barycenter, on the (**bottom**), plotted versus time.

Accordingly to the comments at the beginning of the section, the same experiments are repeated considering a much shorter distance, of 20 m, among adjacent vehicles, which is closer to the usual safety distance: results are shown in Figures 13–16. The figures show that, even in this case, the small world model still preserves the outlined features and outperforms the standard one. In case of the harmonic perturbation, the velocities of the majority of vehicles for the modified model are still condensed in a cluster, characterized by the same, small and in phase, oscillations, and the barycenter of the platoon is also affected by the input of the leader, as shown in Figure 15. The synchronization effect is observed also for the braking manoeuvre, since the majority of vehicles still respond to the leader's input after just a few seconds, as shown in Figure 15.

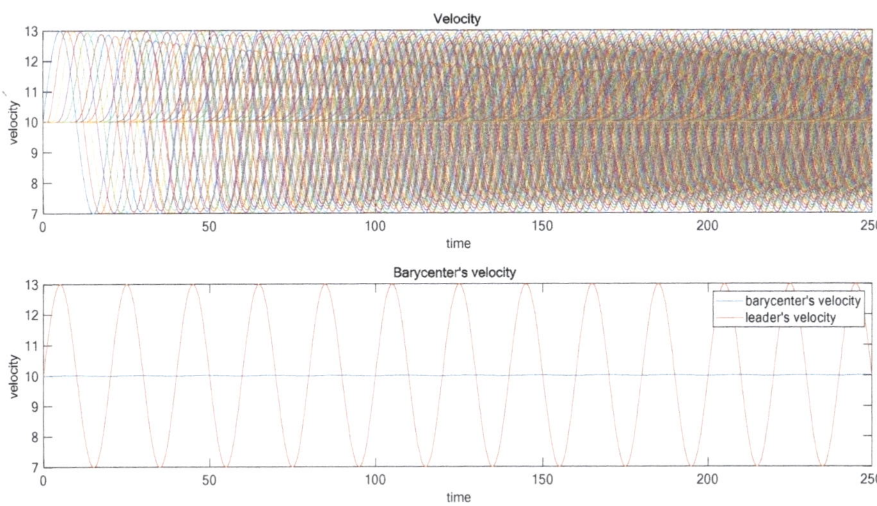

Figure 13. The standard model simulated with b = 20 m and excited by the harmonic perturbation of the Leader's velocity Equation (14); symbolism is the same in Figure 9.

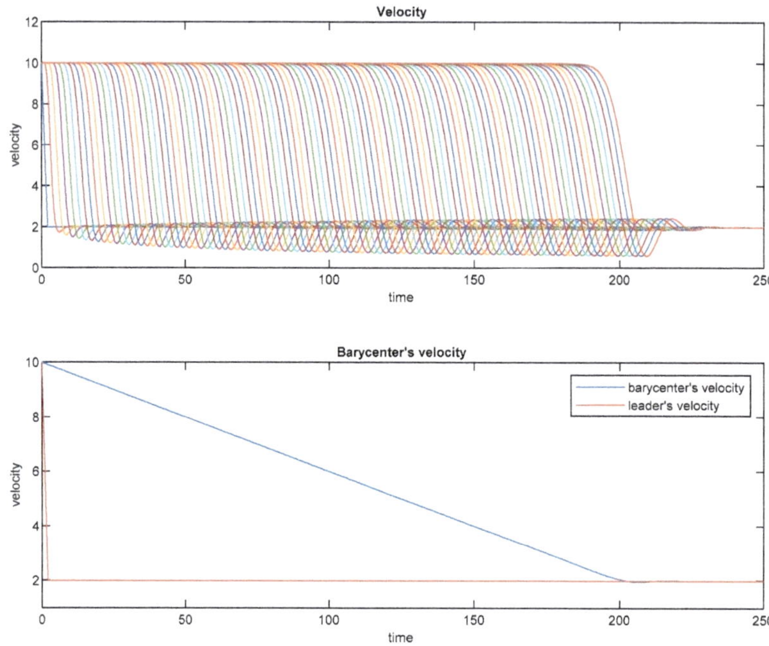

Figure 14. The standard model simulated with b = 20 m and excited by the braking maneuver imparted by the leader Equation (15); symbolism is the same in Figure 10.

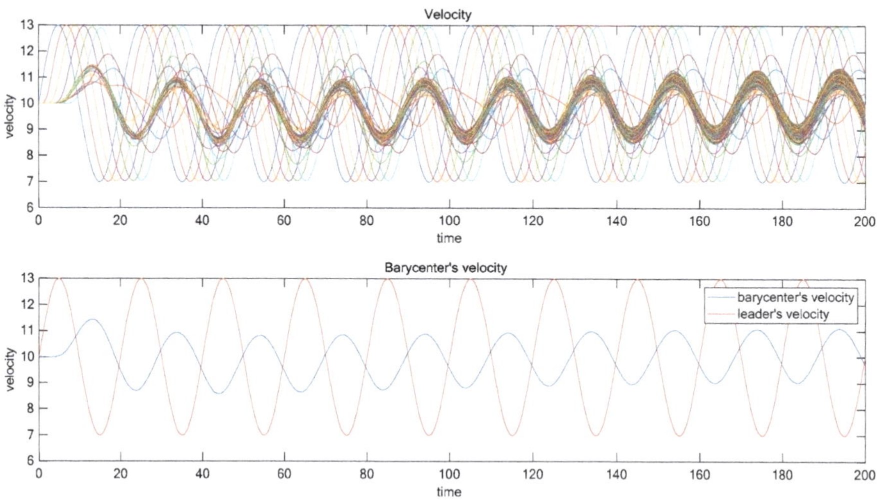

Figure 15. The small-world model simulated with b = 20 m and excited by the harmonic perturbation of the Leader's velocity Equation (14); symbolism is the same in Figure 11.

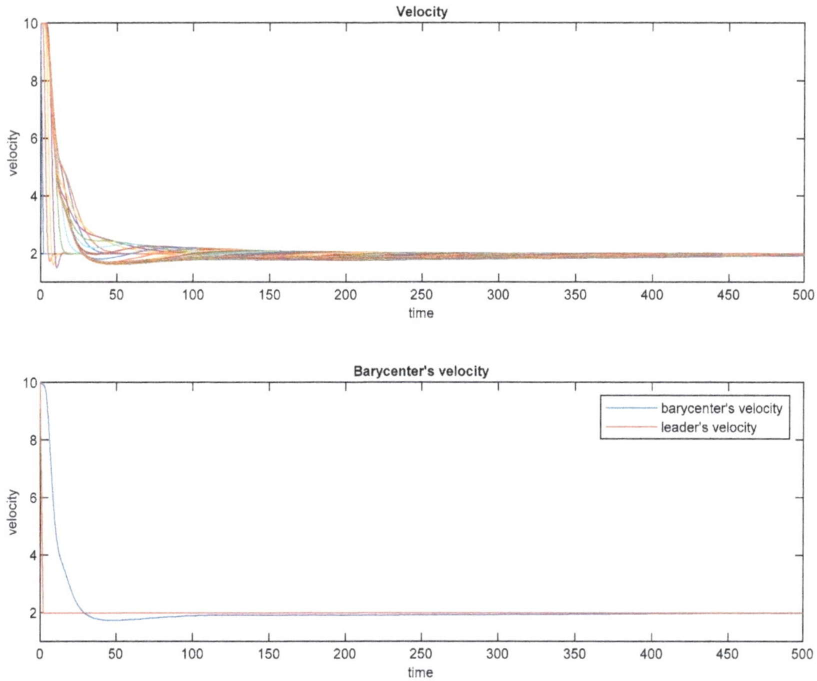

Figure 16. Small world model simulated with b = 20 m and excited by the braking maneuver imparted by the leader Equation (15); symbolism is the same in Figure 12.

4.3. Comparison between the Two Models

Figure 17 shows a comparison between the time evolution of the standard car-following and the small world models, considering the braking maneuver of the leader stated by Equation (15); please note that in the modified model the first vehicle that has a long-range connection is the 13th linked with vehicle 10th. As shown in Figure 17a, at the time 0 s both models have the same spatial configurations and the leader starts braking. As time unfolds, the small world model allows an earlier braking maneuver for the upcoming vehicles. In fact, a misalignment, between the standard and the modified model, is observed for the 14th vehicle, as shown in Figure 17b. The misalignment becomes larger and larger along with the queue of vehicles, as shown in Figure 17c, making the group velocity much shorter in respect of the one of the standard model.

Even if the two models are rather similar, the response of the platoon to the leader's input is substantially different. The leader's information spreads along with the platoon in two different ways as shown in Figure 18. Regarding the distance from the leader defined in Section 3, for the standard model the distance linearly increases with the vehicle number, because $D_n = n - 1$. For the small world models, the previous law holds between two consecutive long-range connections: the distance linearly increases with n up to the vehicle subjected to long-range connection, then the distance drop to a much lower value and then start to rise again, up to the next vehicle with the long-range connection. Regarding the average distance from the leader, the one of the small-world model presented in Figure 11 is roughly 38, which is much smaller than the one of the standard model, $D = N/2 = 250$.

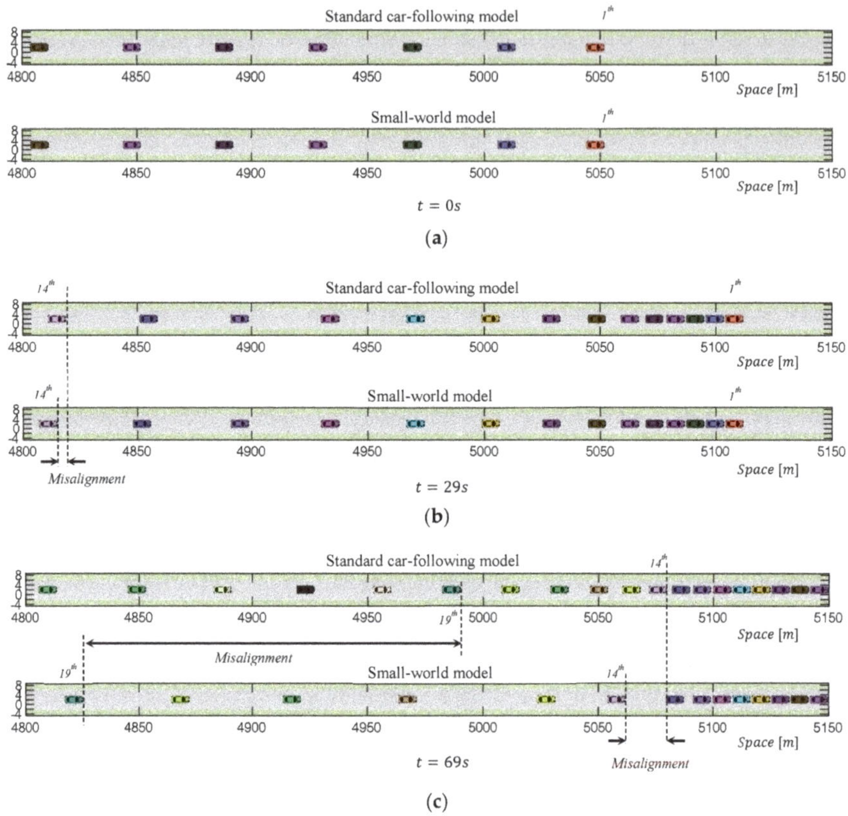

Figure 17. The standard car-following and small-world models are compared in the space domain: (**a**) the start alignment at $t = 0$ s; (**b**) the time frame at $t = 29$ s; (**c**) the time frame at $t = 69$ s.

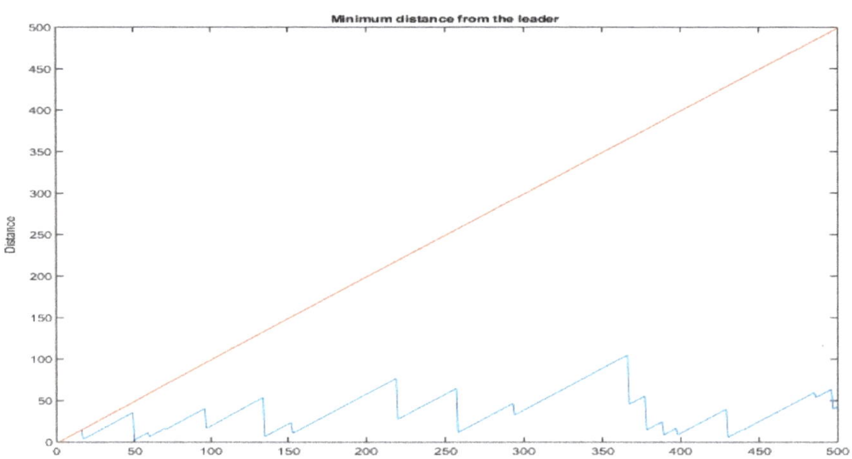

Figure 18. Minimum distance from the leader, for the standard and the small world model, plotted versus the vehicle number.

The greater ability of the small-world model to spread the leader's information gives to the platoon the ability to follow the leader's input way better than the standard model as shown in Figures 19 and 20. Figure 19 shows the comparison between the Barycenter's velocity for the standard and the small world models, subjected to the sinusoidal perturbation of the leader stated in Equation (14). For the standard model, the barycenter's velocity remains almost constant and does not highlight the harmonic perturbation imparted by the leader. For the small-world model, the amplitude of oscillations is much higher, as shown in Figure 11, many vehicles have synchronized oscillation, of the same amplitude: this swarming behavior has been obtained introducing only a 5% of long-range connections.

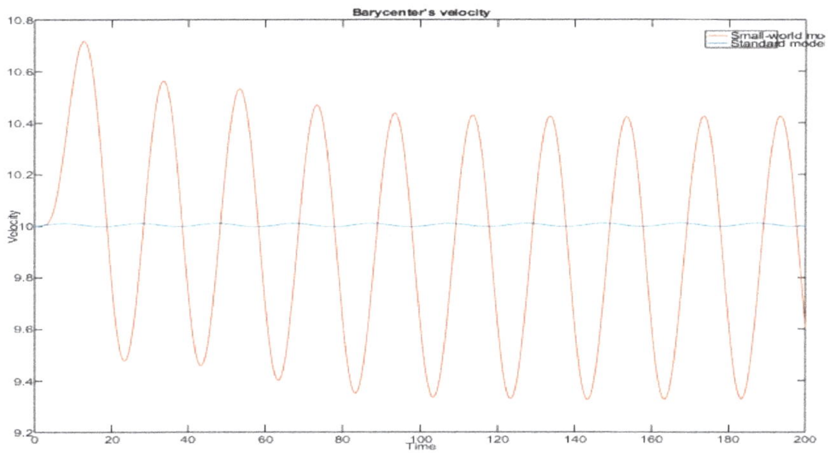

Figure 19. Comparison between the barycenter's velocity for the standard and small-world model, for a sinusoidal perturbation of the leader Equation (14), as in Figures 9 and 11.

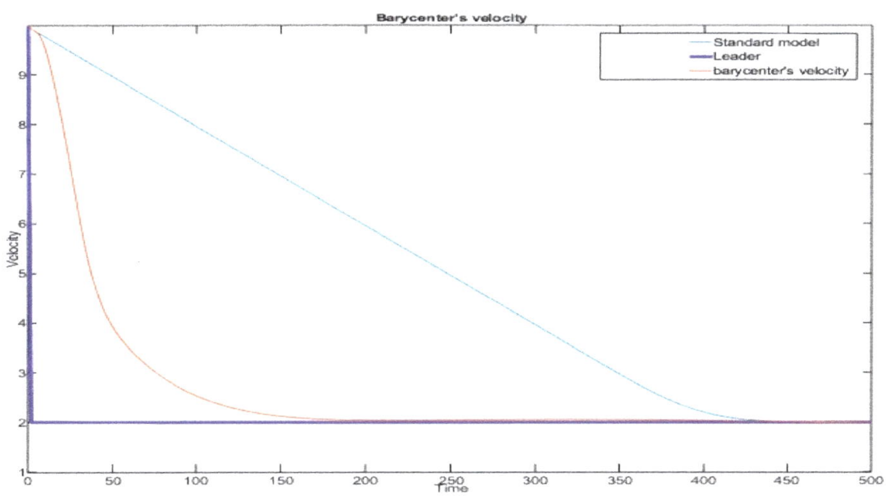

Figure 20. Comparison between the barycenter's velocity for the standard and small-world model, for the braking maneuver imparted to the leader Equation (15), as in Figures 10 and 12.

Regarding the braking maneuver introduced in Equation (15), the comparison in Figure 20 shows that the small-world model is much more reactive than the standard one. The random long-range interactions lead to a swarm behavior, which can hardly be obtained by a purposely designed controller, being able to follow the input given in a quarter of the time needed from the standard model. The small-world model is much more reactive and able to spread the leader's information: this is due to the different patterns of communication, which affects the stability of the model as well, as it is shown in Appendix B.

4.4. Influence of the Random Connections on the Small-World Model

Each random realization of a small world model gives to the swarm a different ability to follow the leader's input, with respect to the actual set of long-range connections. In order to evaluate the statistical validity of the properties highlighted in Sections 4.2 and 4.3 and to further understand how the density of long-range connections may alter the motion of the platoon, an ensemble of 100 trials is considered for each density value. In particular, for the sinusoidal input introduced in Equation (14), where the leader velocity is characterized by oscillations of amplitude 3 m/s, the numerical experiment is performed with $N = 500$ and P that ranges within the interval [1,10] % with a step of size 1%, once P is fixed 100 randomly connected small-world models are considered, then P is increased and the analysis is repeated.

For a certain P, the peak velocity of the barycenter changes with each samples of the statistical ensemble: the non-dimensional barycenter's velocity of the platoon, normalized with the amplitude of oscillations of the velocity of the leader, i.e., 3 m/s, is then used as an indicator of the system sensitivity to the actual placement of the long-range connections, results are shown in Figures 21 and 22. With reference to Figure 21, for low values of P, i.e., $P < 4\%$, each boxplot is small, the first, the second (median) and the third quartile are indeed packed and close to zero: most of the random samples have similar motions and the swarm behavior is not evident since the perturbation imparted by the leader has a low amplitude along the queue of vehicles. However, a few samples, i.e., the outliers marked with red crosses, show larger amplitudes, which means that even with very low values of P is possible, even if unlikely, to have a swarm behavior for a certain combination of long-range connections.

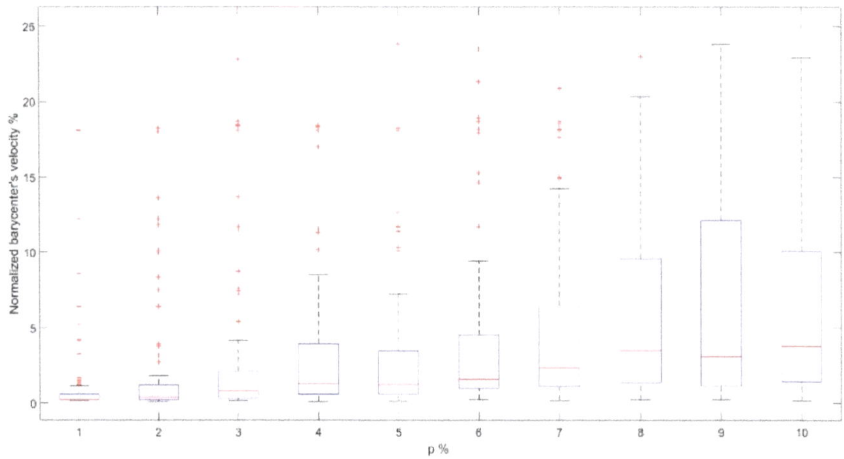

Figure 21. Boxplots of the statistical ensemble plotted versus the probability of long-range connections.

This can be better appreciated from Figure 22, where the number of samples whose nondimensional barycenter's velocity is greater than 10% (here briefly indicated with threshold level TL) is represented along the ordinate axes: up to $P = 4\%$ less than 8% of the samples go beyond TL.

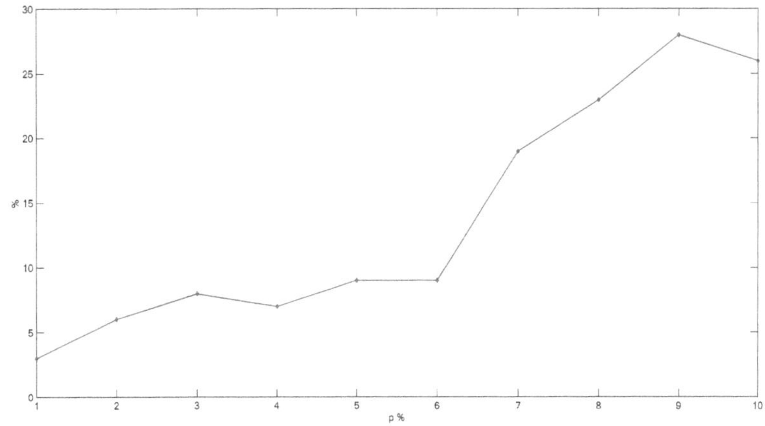

Figure 22. Percentage of samples with nondimensional barycenter's velocity greater than 10% plotted versus the probability of long-range connection.

For low to mid values of P, the median is very close to the first quartile, which means the samples are characterized by lower amplitudes. For the highest values of P, the median gets higher and further form the first quartile, so a collective behavior of the platoon is very likely to be observed among the vast majority of the samples. For values of P of the order of $6 \div 7\%$, the boxplots get indeed significantly larger, the quartiles not only get farther from one another, but also the median significantly increase its value, as shown in Figure 21: it means that the barycenter's velocity of most of the small world samples is able to follow the leader's input. Figure 22 shows that for $P > 6\%$ the number of samples that goes beyond TL suddenly grows increases from 10 to 20%.

4.5. Robustness of the Small-World Model

The robustness of the small world behavior can be analyzed evaluating the effects of a sudden interruption of a long-distance connection, for the model studied in Figure 11. In Figure 23 the vehicle 21 is subjected to a long-range interaction with vehicle 8 up to 100 s, then the connection is broken.

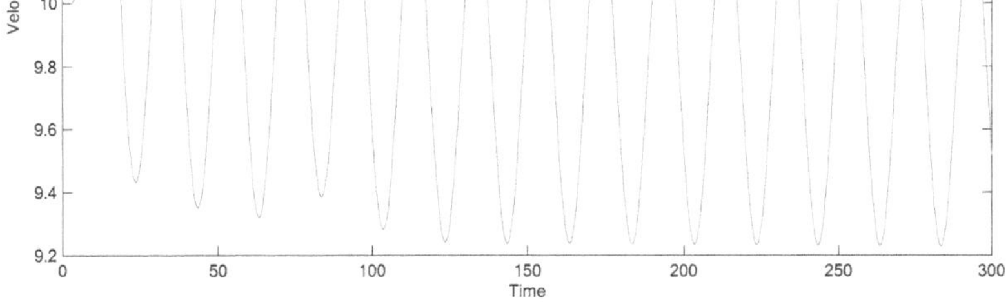

Figure 23. Small world model studied in Figure 11, the vehicle 21 breaks the long-range interaction after 100s.

Figure 23 shows that the velocity of the vehicle 21 undergoes only a small fluctuation, due to the connection interruption.

However, the whole system keeps on oscillating in the same fashion, as shown in Figure 24: the large cluster of vehicles that oscillates roughly in-phase and the barycenter's velocity are not appreciably affected by the connection broken.

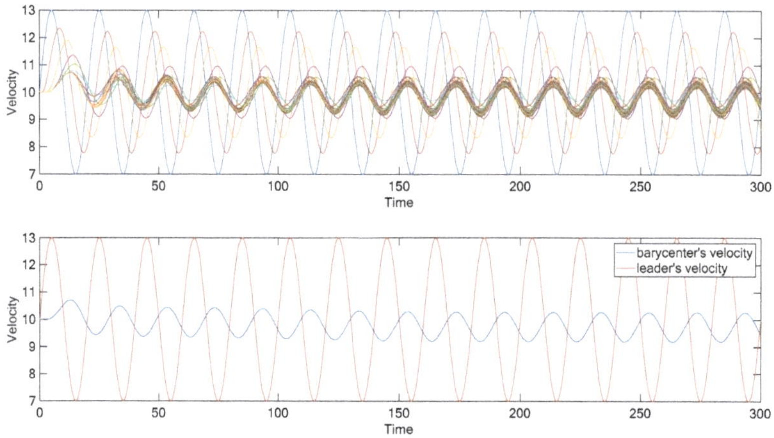

Figure 24. The small-world model studied in Figure 11 for a connection fail of vehicle 21 after 100 s. On top, the velocity of each vehicle that follows the leader, on the bottom, the velocity of the leader and the one of the barycenter of the system of 500 vehicles, plotted versus time.

5. Conclusions

In this paper, a car-following model has been examined. Due to the stability of the standard model, a large platoon of vehicles cannot be controlled simply altering the state of the leader vehicle and a sophisticated controller is generally required. In this regard, the small world theory has been introduced to analyze, if and how it is able to produce a collective behavior also in the case of vehicle platoons.

This work has shown that it is possible to deeply modify the behavior of a line of vehicles introducing a few random long-range connections: a swarm behavior is indeed achieved for a density of long-range interaction even smaller than 6%. The platoon exhibits a swarm behavior that makes the vehicles act, together, almost like a rigid body, able to quickly react to input commands imparted by the leader, as a sudden braking maneuver, i.e., the small-word model is able to reduce the breaking time of the platoon by a factor of 4, or to a cyclical perturbations of the velocity.

The properties of the small-world model have been also analyzed from a theoretical point of view, studying the stability properties of the standard and the small world linearized models. It was found that the small world model modifies the reactiveness and the stability of the original model: the eigenvalues corresponding to the vehicles subjected to the long-range interaction show smaller absolute values and the farther the long-range interaction the closer to zero the eigenvalue. The behavior of the platoon becomes indeed more reactive to the leader's input with respect to the standard model.

The paper represents, to the authors' knowledge, a first successful application of the theory, originally developed in the field of social networks, which has shown a rather effective way to regularize the traffic flow with a very small control effort: it was shown indeed that a swarm behavior can be achieved controlling only the leader vehicle if a few random, long-range connections, are considered. The modified models exhibit indeed properties of high synchronization and it is also robust to communication failures.

Thanks to these promising results, the authors are carrying out an effective control strategy, which relies on the small world theory and that requires a very small control effort. This will be the topic of a forthcoming paper, where the authors are investigating the main pros and cons provided by the small world model with respect to a queue of vehicles controlled via other algorithms already developed by the authors that use optimal control [15,43–46].

Author Contributions: Conceptualization, N.R.; methodology, N.R. and G.P.; software, L.M.; validation A.C.; formal analysis, G.P.; investigation, N.R.; data curation, L.M.; writing—original draft preparation, N.R.; writing—review and editing, N.R.; visualization, G.P. and L.M. All authors have read and agreed to the published version of the manuscript.

Funding: This research received no external funding.

Conflicts of Interest: The authors declare no conflict of interest.

Appendix A

The small-world model is now introduced, adding random connections on the rhs of Equation (12). The modified model can be expressed by the following:

$$\ddot{x}_n(t) = a_n\beta_{n,1}(t)\big(\dot{x}_{n-1}(t-\tau) - \dot{x}_n(t-\tau)\big) + b_n\beta_{n,s}(t)\big(\dot{x}_{n-s}(t-\tau) - \dot{x}_n(t-\tau)\big),\ n = 4,5,\ldots,N \tag{A1}$$

where s is a positive random integer that ranges within $[2,\ n-2]$, and

$$\begin{aligned} \beta_{n,1}(t) &= \frac{\alpha\big(\dot{x}_n(t)\big)^m}{(x_{n-1}(t-\tau)-x_n(t-\tau))^l} \\ \beta_{n,s}(t) &= \frac{\alpha\big(\dot{x}_n(t)\big)^m}{(x_{n-s}(t-\tau)-x_n(t-\tau))^l} \end{aligned} \tag{A2}$$

Equation (A1) shows that, when the vehicle n is subject to a long-range interaction, its acceleration depends on the state of the vehicle $n-s$ as well as on the vehicle $n-1$ weighted by coefficients a_n and b_n where $a_n + b_n = 1$. When the long-range connection is not present, $a_n = 1$ and $b_n = 0$ and the vehicle n follows the same low of the standard model.

Equation (A1) is rewritten using relative variables, previously defined:

$$\begin{aligned} y_n(t) + b &= x_{n-1}(t) - x_n(t) \\ \dot{y}_n(t) &= v_n(t) = \dot{x}_{n-1}(t) - \dot{x}_n(t) \end{aligned} \tag{A3}$$

At this point, using the property:

$$\begin{aligned} x_{n-s}(t) - x_n(t) &= -(x_n(t) - x_{n-s}(t)) \\ &= -(x_n(t) - x_{n-1}(t) + x_{n-1}(t) - x_{n-2}(t) + \ldots + x_{n-s+1}(t) - x_{n-s}(t)) \\ &= sb + \sum_{j=1}^{s} y_{n-j+1}(t) \end{aligned} \tag{A4}$$

it holds:

$$x_{n-s}(t) - x_n(t) = sb + \sum_{j=1}^{s} y_{n-j+1}(t) \tag{A5}$$

and taking the derivative, it holds:

$$\dot{x}_{n-s}(t) - \dot{x}_n(t) = \sum_{j=1}^{s} \dot{y}_{n-j+1}(t) \tag{A6}$$

Since $\dot{y}_n(t) = v_n(t)$, as stated by the second of the (A3), Equation (A5) becomes:

$$\dot{x}_{n-s}(t) - \dot{x}_n(t) = \sum_{j=1}^{s} v_{n-j+1}(t) \tag{A7}$$

Substituting the second of Equations (A3) and (A6) into Equation (A1), it holds:

$$\ddot{x}_n(t) = a_n\beta_{n,1}(t)v_n(t-\tau) + b_n\beta_{n,s}(t)\sum_{j=1}^{s} v_{n-j+1}(t-\tau),\ n = 2,3,\ldots,N \quad \text{(A8)}$$

where the coefficients given by Equations (A2) are expressed in relative terms:

$$\begin{aligned} \beta_{n,1}(t) &= \frac{\alpha\left(\dot{x}_1(t)-\sum_{j=2}^{n} v_j(t)\right)^m}{\left(y_n(t)(t-\tau)+b\right)^l} \\ \beta_{n,s}(t) &= \frac{\alpha\left(\dot{x}_1(t)-\sum_{j=2}^{n} v_j(t)\right)^m}{\left(sb+\sum_{j=1}^{s} y_{n-j+1}(t-\tau)\right)^l} \end{aligned} \quad \text{(A9)}$$

In analogy with the standard model, taking the derivative of the second of Equation (A3), it holds:

$$\dot{v}_n(t) = \ddot{x}_{n-1}(t) - \ddot{x}_n(t) \quad \text{(A10)}$$

Using Equation (A7) into the rhs of Equation (A9), it holds:

$$\begin{aligned} \dot{v}_n(t) = {} & a_{n-1}\beta_{n-1,1}(t)v_{n-1}(t-\tau) + b_{n-1}\beta_{n-1,s}(t)\sum_{j=1}^{s} v_{n-1-j+1}(t-\tau) \\ & -\left(a_n\beta_{n,1}(t)v_n(t-\tau) + b_n\beta_{n,s}(t)\sum_{j=1}^{s} v_{n-j+1}(t-\tau)\right),\quad n = 4,5,\ldots,N \end{aligned} \quad \text{(A11)}$$

Finally, combining the second Equations (A3) and (A10), the state equations for the small world model are found:

$$\begin{cases} \dot{y}_n(t) = v_n(t) \quad n = 2,\ 3, N \\ \dot{v}_2(t) = \ddot{x}_1(t) - \beta_{2,1}(t)v_2(t-\tau) \\ \dot{v}_3(t) = \beta_{2,1}(t)v_2(t-\tau) - \beta_{3,1}(t)v_3(t-\tau) \\ eq.\ (A10) \end{cases} \quad \text{(A12)}$$

where β are given by Equation (A8) and $s \in [2, n-2]$.

Comparing Equation (12) with Equation (13), the modified model is rather similar to the standard one, since, under the small world hypothesis only a few vehicles have also long-range connections. Both models are delayed differential models that need the initial condition of the state, $y_i(0), v_i(0)$, and the input to the leader to be solved.

The theory of delay differential equations [47] is not just a simple extension of the theory of ODE, even though there are similarities. The initial conditions for ordinary differential equations are constant values, while the initial conditions for delay differential equations are functions that depend on the delay time, which involves infinite numbers of initial values. The delay-differential equations explain the way the initial functions evolve as time unfolds, thus they are infinite dimensional systems. Generally, a delay-differential equation is difficult to solve analytically and, even if a closed-form analytical solution can be obtained, for example with Laplace transforms or other techniques [48], the structure of the solutions is so complicated that it is difficult to gain general conclusion about their behavior: it is often more easy/comprehensible the numerical solution of the problem.

Appendix B

The small-world model modifies the reactiveness and the stability of the original model. To understand why the modified model manages to follow the leader's input better than the standard model, the two models are linearized and compared in terms of their stability properties.

(a) Standard model linearization and stability

The linearization of the standard model is obtained using the first order term of the Taylor expansion around the equilibrium point. The equilibrium point is considered in stationary conditions, when all the velocities are equal to the (constant) velocity assigned to the leader and the distances are all equal to the parameter b in Equation (9).

Expressing these conditions in the relative variables introduced in Equation (5), the equilibrium results in:

$$\begin{aligned} y_n(t) &= x_{n-1}(t) - x_n(t) - b = 0 \\ \dot{y}_n(t) &= v_n(t) = \dot{x}_{n-1}(t) - \dot{x}_n(t) = 0 \end{aligned} \quad \text{(A13)}$$

Considering the state vector S and the equilibrium E, which, for the Equation (A11), consists of a null vector, the linearization is:

$$\dot{v}_n(S,t) = \dot{v}_n(E,t) + \nabla\dot{v}_n(E,t)\cdot(S-E) + o(\|S-E\|) \quad \text{(A14)}$$

and is obtained neglecting le last term. The resulting linearized model for the Equation (12) is

$$\dot{v}_n(t) = \frac{\alpha\dot{x}_1^m}{b^l}(v_{n-1}(t-\tau) - v_n(t-\tau)) \quad \text{(A15)}$$

The linearized model can be rewritten in the form:

$$\underline{\dot{V}}(t) = \underline{\underline{B}}\,\underline{V}(t-\tau) \quad \text{(A16)}$$

where $\underline{\underline{B}}$ is a lower triangular matrix. The elements along the diagonal of $\underline{\underline{B}}$ in Equation (A14) are negative and on the first under diagonal all the elements are positive, given the coefficients defined by Equation (A13), whose value is always equal to $\pm\frac{\alpha\dot{x}_1^m}{b^l}$.

The normal coordinates are then introduced in Equation (A14):

$$\underline{V}(t) = \underline{U}\,f(t) \quad \text{(A17)}$$

so:

$$\underline{U}\,\dot{f}(t) = \underline{\underline{B}}\,\underline{U}\,f(t-\tau) \quad \text{(A18)}$$

Considering:

$$\lambda = \frac{\dot{f}(t)}{f(t-\tau)} \quad \text{(A19)}$$

it is now possible to study the eigenvalues problem:

$$\left(\lambda\,\underline{\underline{I}} - \underline{\underline{B}}\right)\underline{U} = 0 \quad \text{(A20)}$$

From Equation (A18) the values of the eigenvalues λ and the eigenvectors $\underline{U}$ can be obtained: since the matrix $\underline{\underline{B}}$ is triangular, its eigenvalues are equal to the elements on the diagonal. Therefore, in this model the eigenvalues are all equal to $-\frac{\alpha\dot{x}_1^m}{b^l}$: since all the parameters are greater than zero, the system is stable.

The eigenvalues are dependent on the characteristic parameters of the model and on the leader's velocity. In particular, it is important to focus the attention on two aspects of these eigenvalues, to deduce a qualitative consideration: since the farther the negative eigenvalues are from zero, the more stable or less reactive the model is, an increase in leader's velocity decreases the reactiveness of the platoon. The same effect is obtained with the decrease of the denominator that can be caused by a reduction of b, i.e., the distance between the vehicles in stationary conditions, or a decrease of l.

The variations of disturbances along the line of cars are considered and referred to, as stability over cars: A line of cars is said to be stable over cars if small fluctuations in the movements of the first car are damped as they pass from one car to the next in the line. Otherwise, if the fluctuations are amplified, then the system is unstable.

The reactiveness and the stability over cars of the model are, then, in conflict. The more this model becomes stable, i.e., the amplitude of the negative eigenvalue increases, the less it becomes reactive.

(b) Small-world model linearization and stability

The linearization of the small-world model is computed using the first order term of the Taylor expansion around the equilibrium point. When a long-range interaction is not present, the linearized model is the same obtained in Equation (A13), while, when a long-range interaction is present, the constants s, an e bn in Equation (13) are different from zero.

Computing all the components of the Equation (A12), in this case the result, for the vehicle subject to a long-range interaction, is:

$$\begin{aligned}\dot{v}_n(t) = {} & b_{n-1}\frac{\alpha\dot{x}_1^m}{(sb)^l}v_{n-s}(t-\tau) + (b_{n-1}-b_n)\frac{\alpha\dot{x}_1^m}{(sb)^l}\sum_{i=n+1-s}^{n-2} v_i(t-\tau)\\ & +\left(a_{n-1}\frac{\alpha\dot{x}_1^m}{b^l} + (b_{n-1}-b_n)\frac{\alpha\dot{x}_1^m}{(sb)^l}\right)v_{n-1}(t-\tau)\\ & -\left(a_n\frac{\alpha\dot{x}_1^m}{b^l} + b_n\frac{\alpha\dot{x}_1^m}{(sb)^l}\right)v_n(t-\tau)\end{aligned} \tag{A21}$$

where the time dependence (t) has been omitted.

As well as in the standard model, the solution is dependent on $\dot{x}_1$, while in this case the vehicles subject to the long-range interactions have a random component given by s in Equation (A19). In analogy with the matrix representation in Equation (A14), since the matrix $\underline{\underline{B}}$ is still triangular, also for the small world model, the eigenvalues are equal to the elements on the diagonal, therefore the coefficient of the last term on the right hand side of Equation (A19) is the eigenvalue related to vehicle n

$$\lambda_n = -\left(a_n\frac{\alpha\dot{x}_1^m}{b^l} + b_n\frac{\alpha\dot{x}_1^m}{(sb)^l}\right) \tag{A22}$$

Since $a_n + b_n = 1$ and $s > 1$, the eigenvalue expressed in Equation (A20) has always a lower module if compared with the standard eigenvalue.

The eigenvalues corresponding to the vehicles subjected to the long-range interaction are negative, since the system remain stable, but show smaller (absolute) values in respect to the standard system. Since "s" in Equation (A19) is the number of vehicles between the vehicle n and the vehicle *n-s*, which is the one whose state is used by the controller n, the farther the long-range interaction, the closer to zero the eigenvalue. The stability over cars decreases, but it doesn't lead to unstable solutions. The behavior of the platoon is more reactive to the leader's input with respect to the standard model: the whole platoon is now able to follow the leader's velocity, in correspondence of perturbations from the equilibrium positions, in a short amount of time.

References

1. European Commission. *A European Strategy on Cooperative Intelligent Transport Systems, a Milestone Towards Cooperative, Connected and Automated Mobility*; EC: Brussels, Belgium, 2016.
2. European Vehicle Manufacturers Work Towards Bringing Vehicle-to-X Communication onto European Roads. 2015. Available online: https://www.car-2-car.org/index.php?id=214 (accessed on 1 September 2019).
3. Turri, V.; Besselink, B.; Johansson, K.H. Cooperative Look-Ahead Control for Fuel-Efficient and Safe Heavy-Duty Vehicle Platooning. *IEEE Trans. Control Syst. Technol.* **2017**, *25*, 12–28. [CrossRef]
4. Ghasemi, A.; Kazemi, R.; Azadi, S. Stable Decentralized Control of a Platoon of Vehicles With Heterogeneous Information Feedback. *IEEE Trans. Control Syst. Technol.* **2013**, *62*, 4299–4308. [CrossRef]
5. van Bart, A.; van Cornelie, J.G.D.; Visser, R. The Impact of Cooperative Adaptive Cruise Control on Traffic-Flow Characteristics. *IEEE Trans. Control Syst. Technol.* **2006**, *7*, 429–436.
6. Rajamani, R.; Tan, S.-H.; Law, B.K.; Zhang, W.-B. Demonstration of integrated longitudinal and lateral control for the operation of automated vehicles in platoons. *IEEE Trans. Control Syst. Technol.* **2000**, *8*, 695–708. [CrossRef]

7. Stotsky, A.; Chien, C.C.; Joannou, P. Robust Platoon-Stable Controller Design for Autonomous Intelligent Vehicles. *Math. Comput. Modell.* **1995**, 287–303. [CrossRef]
8. Zegers, J.C.; Semsar-Kazerooni, E.; Ploeg, J.; van de Wouw, N.; Nijmeijer, H. Consensus-based bi-directional CACC for vehicular platooning. In Proceedings of the 2016 American Control Conference (ACC), Boston, MA, USA, 6–8 July 2016.
9. Ploeg, J.; van de Wouw, N.; Nijmeijer, H. Lp String Stability of Cascaded Systems: Application to Vehicle Platooning. *IEEE Trans. Control Syst. Technol.* **2014**, *22*, 786–793. [CrossRef]
10. Fusco, M.; Semsar-Kazerooni, E.; Ploeg, J.; van de Wouw, N. Vehicular platooning: Multi-Layer Consensus Seeking. In Proceedings of the 2016 IEEE Intelligent Vehicles Symposium (IV), Gothenburg, Sweden, 19–22 June 2016; pp. 382–387.
11. Fernandes, P.; Nunes, U. Platooning With IVC-Enabled Autonomous Vehicles: Strategies to Mitigate Communication Delays, Improve Safety and Traffic Flow. *IEEE Trans. Intell. Trans. Syst.* **2012**, *13*, 91–106. [CrossRef]
12. Harris, M. *Researcher Hacks Self-Driving Car Sensors*; IEEE Spectrum: Bethesda, MD, USA, 2015.
13. Antonelli, D.; Nesi, L.; Pepe, G.; Carcaterra, A. Mechatronic Control of the Car Response Based on VFC. In Proceedings of the International Conference on Noise and Vibration Engineering (ISMA2018), Leuven, Belgium, 19 November 2018.
14. Laurenza, M.; Pepe, G.; Antonelli, D.; Carcaterra, A. Car collision avoidance with velocity obstacle approach: Evaluation of the reliability and performace of the collision avoidance maneuver. In Proceedings of the 5th International Forum on Research and Technologies for Society and Industry (RTSI), Florence, Italy, 9–12 September 2019.
15. Pepe, G.; Laurenza, M.; Antonelli, D.; Carcaterra, A. A new optimal control of obstacle avoidance for safer autonomous driving. In Proceedings of the 2019 AEIT International Conference of Electrical and Electronic Technologies for Automotive, AEIT AUTOMOTIVE 2019, Torino, Italy, 2–4 July 2019.
16. Bergenhem, C.; Hedin, E.; Skarin, D. Vehicle-to-Vehicle Communication for a Platooning System. *Proced. Soc. Behav. Sci.* **2012**, *48*, 1222–1233. [CrossRef]
17. Elefteriadou, L. *An Introduction to Traffic Flow Theory*; Springer: New York, NY, USA, 2014; Volume 84.
18. Wilhelm, W.; Schmidt, J.W. Review of Car-Following Theory. *Trans. Eng. J. ASCE* **1973**, *99*, 923–933.
19. Ran, B.; Huang, W.; Leight, S. Some solution strategies for automated highway exit bottleneck problems. *Trans. Res. Part C Emerg. Technol.* **1996**, *4*, 167–179. [CrossRef]
20. Ozaki, H. Reaction and anticipation in the car-following behavior. In Proceedings of the 13th International Symposium on Traffic and Transportation Theory, Lyon, France, 24–26 July 1993; pp. 349–366.
21. Ferrari, P. The instability of motorway traffic. *Trans. Res. Part B Methodol.* **1994**, *28*, 175–186. [CrossRef]
22. Gazis, D.C.; Herman, R.; Rothery, R.W. Nonlinear Follow-the-Leader Models of Traffic Flow. *Oper. Res.* **1961**, *9*, 545–567. [CrossRef]
23. Disbro, J.E.; Frame, M. *Traffic Flow Theory and Chaotic Behavior*; National Academy of Sciences: New York, NY, USA, 1989.
24. Milgram, S. The Small World Problem. *Psychol. Today* **1967**, *1*, 61–67.
25. Barrat, A.; Barthélemy, M.; Vespignani, A. *Dynamical Processes on Complex Networks*; Cambridge University Press: Cambridge, UK, 2008.
26. Watts, D.J.; Strogatz, S.H. Collective dynamics of 'small-world' networks. *Nature* **1998**, *393*, 440–442. [CrossRef]
27. Dorigo, M.; Birattari, M.; Li, X.; López-Ibáñez, M.; Ohkura, K.; Pinciroli, C.; Stützle, T. *Swarm Intelligence*; Springer: Berlin/Heidelberg, Germany, 2014.
28. Schöll, E.; Klapp, S.H.; Hövel, P. *Control of Self-Organizing Nonlinear Systems*; Springer: Berlin/Heidelberg, Germany, 2016.
29. Gross, T.; Sayama, H. *Adaptive Networks: Theory, Models and Applications*; Springer Science & Business Media: Berlin/Heidelberg, Germany, 2009.
30. Bonabeau, E.; Dorigo, M.; Theraulaz, G. *Swarm Intelligence: From Natural to Artificial Systems*; OUP: Oxford, UK, 1999.
31. Arenas, A.; Diaz-Guilera, A.; Kurths, J.; Moreno, Y.; Zhou, C. Synchronization in complex networks. *Phys. Rep.* **2008**, *469*, 93–153. [CrossRef]

32. Gerla, M.; Lee, E.; Pau, G.; Lee, U. Internet of vehicles: From intelligent grid to autonomous cars and vehicular clouds. In Proceedings of the 2014 IEEE World Forum on Internet of Things (WF-IoT), Seoul, Korea, 6–8 March 2014.
33. Ge, J.I.; Avedisov, S.S.; He, C.R.; Qin, W.B.; Sadeghpour, M.; Orosz, G. Experimental validation of connected automated vehicle design among human-driven vehicles. *Trans. Res. Part C Emerg. Technol.* **2018**, *91*, 335–352. [CrossRef]
34. Orosz, G. Connected automated vehicle design among human-driven vehicles. *IFAC-PapersOnLine* **2019**, *51*, 403–406. [CrossRef]
35. Hajdu, D.; Ge, J.I.; Insperger, T.; Orosz, G. Robust Design of Connected Cruise Control Among Human-Driven Vehicles. *IEEE Trans. Intell. Trans. Syst.* **2020**, *21*, 749–761. [CrossRef]
36. Qin, W.B.; Orosz, G. Experimental Validation of String Stability for Connected Vehicles Subject to Information Delay. *IEEE Trans. Control Syst. Technol.* **2019**, 1–15. [CrossRef]
37. Ge, J.I.; Orosz, G. Connected cruise control among human-driven vehicles: Experiment-based parameter estimation and optimal control design. *Trans. Res. Part C Emerg. Technol.* **2018**, *95*, 445–459. [CrossRef]
38. Stern, R.E.; Cui, S.; Monache, M.L.D.; Bhadani, R.; Bunting, M.; Churchill, M.; Hamilton, N.; Haulcy, R.; Pohlmann, H.; Wu, F.; et al. Dissipation of stop-and-go waves via control of autonomous vehicles: Field experiments. *Trans. Res. Part C Emerg. Technol.* **2018**, *89*, 205–221. [CrossRef]
39. Li, X.; Ghiasi, A.; Xu, Z.; Qu, X. A piecewise trajectory optimization model for connected automated vehicles: Exact optimization algorithm and queue propagation analysis. *Trans. Res. Part B Methodol.* **2018**, *118*, 429–456. [CrossRef]
40. Tilg, G.; Yang, K.; Menendez, M. Evaluating the effects of automated vehicle technology on the capacity of freeway weaving sections. *Trans. Res. Part C Methodol.* **2018**, *96*, 3–21. [CrossRef]
41. Feng, S.; Zhang, Y.; EbenLi, S.; Cao, Z.; Liu, H.X.; Li, L. String stability for vehicular platoon control: Definitions and analysis methods. *Annu. Rev. Control* **2019**, *47*, 81–97. [CrossRef]
42. Zhou, Y.; Ahn, S.; Wang, M.; Hoogendoorn, S. Stabilizing mixed vehicular platoons with connected automated vehicles: An H-infinity approach. *Trans. Res. Part B Methodol.* **2019**. [CrossRef]
43. Paifelman, E.; Pepe, G.; Carcaterra, A. An optimal indirect control of underwater vehicle. *Int. J. Control* **2019**, 1–15. [CrossRef]
44. Pepe, G.; Antonelli, D.; Nesi, L.; Carcaterra, A. FLOP: Feedback Local Optimality Control of the Inverse Pendulum Oscillations. In Proceedings of the International Conference on Noise and Vibration Engineering (ISMA2018), Leuven, Belgium, 19 November 2018.
45. Antonelli, D.; Nesi, L.; Pepe, G.; Carcaterra, A. A novel approach in Optimal trajectory identification for Autonomous driving in racetrack. In Proceedings of the 2019 18th European Control Conference (ECC), Naples, Italy, 25–28 June 2019.
46. Pensalfini, S.; Coppo, F.; Mezzani, F.; Pepe, G.; Carcaterra, A. Optimal control theory based design of elasto-magnetic metamaterial. *Proced. Eng.* **2017**, *199*, 1761–1766. [CrossRef]
47. Györi, I.; Ladas, G. *Oscillation Theory of Delay Differential Equations*; Clarendon Press: Oxford, UK, 1991.
48. Driver, R.D. *Ordinary and Delay Differential Equations*; Springer: Berlin/Heidelberg, Germany; New York, NY, USA, 1977.

energies

MDPI

Article

iABACUS: A Wi-Fi-Based Automatic Bus Passenger Counting System

Michele Nitti [1,2,*], Francesca Pinna [1], Lucia Pintor [1], Virginia Pilloni [1,2,*] and Benedetto Barabino [3,*]

1 Department of Electrical and Electronic Engineering (DIEE), University of Cagliari, 09123 Cagliari, Italy; f.pinna3000@gmail.com (F.P.); luciapintor90@gmail.com (L.P.)

2 National Telecommunication Inter University Consortium (CNIT), Research Unit of Cagliari, 09123 Cagliari, Italy

3 Department of Civil Engineering, Architecture, Land, Environment and Mathematics (DICATAM), University of Brescia, 25123 Brescia, Italy

* Correspondence: michele.nitti@unica.it (M.N.); virginia.pilloni@unica.it (V.P.); benedetto.barabino@unibs.it (B.B.)

Received: 21 February 2020; Accepted: 15 March 2020; Published: 19 March 2020

Abstract: Since the early stages of the Internet-of-Things (IoT), one of the application scenarios that have been affected the most by this new paradigm is mobility. Smart Cities have greatly benefited from the awareness of some people's habits to develop efficient mobility services. In particular, knowing how people use public transportation services and move throughout urban infrastructure is crucial in several areas, among which the most prominent are tourism and transportation. Indeed, especially for Public Transportation Companies (PTCs), long- and short-term planning of the transit network requires having a thorough knowledge of the flows of passengers in and out vehicles. Thanks to the ubiquitous presence of Internet connections, this knowledge can be easily enabled by sensors deployed on board of public transport vehicles. In this paper, a Wi-Fi-based Automatic Bus pAssenger CoUnting System, named iABACUS, is presented. The objective of iABACUS is to observe and analyze urban mobility by tracking passengers throughout their journey on public transportation vehicles, without the need for them to take any action. Test results proves that iABACUS efficiently detects the number of devices with an active Wi-Fi interface, with an accuracy of 100% in the static case and almost 94% in the dynamic case. In the latter case, there is a random error that only appears when two bus stops are very close to each other.

Keywords: smart phones; IEEE 802.11 standards; automatic passenger counting; mobile devices; cloud computing

1. Introduction

Urban mobility and flow of people throughout urban infrastructures have a great impact on several areas, such as tourism and transportation [1,2]. In particular, being able to accurately count the number of passengers represents one of the most relevant components of the transit service, since it provides a key measure of the effectiveness of Public Transportation Companies (PTCs) and is pivotal for the efficient planning of the transit network, both in the long and short term [3,4]. Indeed, long-term planning of routes and related timetables is enabled through the analysis of origin–destination matrices, which provide information on commuting flows. Moreover, such matrices give indications regarding the congested time and routes, which simplify short term planning strategies as well, e.g., through the re-assignment of buses to a specific route. Thus, long- and short-term planning contributes to the efficient use of resources and guarantees that buses run where and when passengers need them [5].

To obtain information regarding passenger volumes, PTCs have typically employed traditional mechanisms, going from non-automatic human visual counting, to Automatic Passenger Counting (APC) methods based on various data acquisition technologies (e.g., mat sensors [6], infrared sensors [7], video cameras [8]). These systems require to be installed on vehicles, and are usually quite expensive. With the advent of the Internet of Things (IoT), APC Systems (APCS) has seen a huge boost in the development of new methods to "observe" urban mobility, particularly in recent years. Nevertheless, the great success of recent APCS is primarily due to the advent of portable and mobile devices such as tablets, smartphones and smartwatches, which give new opportunities to collect detailed passengers' data and to track their movements throughout the cities [9,10]. However, so far, the research in this direction has been mainly focused on determining the travel mode from data collected by various sensors, such as accelerometers and Global Positioning System/Geographycal Information System (GPS/GIS) [11,12]. Furthermore, most of these approaches require that passengers have a specific smartphone app installed on their devices. Therefore, results severely depend on the willingness of passengers to participate. This is not compatible with APCS, which rely on crowd counting principles.

Only recently, Wi-Fi has been used to count the number of active interfaces detected nearby Wi-Fi Access Points (APs) [13]. These systems are typically based on the identification of the Media Access Control (MAC) address that is associated with a Wi-Fi interface. Examples of such systems are represented by the Amsterdam Airport Schiphol [14] and Transport for London (TfL) stations [15]. However, they can effectively work only for obsolete operating systems (earlier than Android 5.0, iOS 7 and Windows Phone 8). In fact, with the aim to protect their users' privacy against device tracking, Google, Apple and Microsoft introduced software randomization of MAC addresses [16]: the claimed MAC address is randomly generated by a software, and it periodically changes.The impact of MAC address randomization on mobile device tracking is studied in [17], where the authors analyze the performance of different radomization techniques implemented for several different commercial off-the-shelf operating systems. Therefore, tracking devices is becoming not feasible anymore, as it is confirmed by [15], where TfL claims that they were unable to construct journeys when MAC addresses were randomized.

In this paper, we present iABACUS (Wi-Fi-based Automatic Bus pAssenger CoUnting System), the first tool to tackle the issue of MAC randomization, by introducing a system that tracks passengers's devices from the moment they board a vehicle to the moment they alight. The goal of the paper is to show how it is possible to leverage the IoT to accurately count the number of devices, which can be considered equal to the number of passengers on the bus. Someone could argue that there is a mismatch between the number of passengers and the number of devices owned. While this is indeed true, the penetration of these devices heavily depends on the considered country: in countries with emerging economy, the percentage of people that do not own a smartphone or do not have an active Wi-Fi interface is higher w.r.t countries with advanced economy where users can own more than one connected device, such as a smartphone and a tablet, and then they are counted several times [18]. To adjust the output of the proposed iABACUS algorithm it is then necessary to calibrate it based on the considered scenario and then introduce a scaling factor; however, these tests are out of scope of this paper. The provided contribution is threefold:

- Since it tracks active Wi-Fi interfaces, iABACUS does not require that passengers take any action, which is a great advantage as compared to most emerging APCS. Moreover, since iABACUS counts the number of active Wi-Fi interfaces, it is not required that passengers install anything on their smartphone, nor do they have to connect to an AP;
- iABACUS is based on a de-randomization mechanism, which overcomes the issue of not being able to attribute two or more random MAC addresses to the same device. Furthermore, since the original MAC address is kept unknown, the identity of passengers cannot be inferred, and their privacy is preserved;

- Not only does iABACUS count passengers's devices, but it also tracks them throughout their journey on public transportation vehicles, by providing when they board or alight from the bus. Therefore, its functionality is not limited to passenger counting: it enables urban mobility observation and analysis, which provides a great support to short- and long-term PTC planning.

The remainder of the paper is structured as follows. Section 2 introduces the background on APC technologies. In Section 3 the model of iABACUS is thoroughly described. In Section 4 experiments are presented and discussed to assess the overall system performance and accuracy. Finally, conclusions are drawn in Section 5, as well as future directions.

2. State of the Art

Generally speaking, APCSs are realized through electronic devices used on transit vehicles. They mainly and accurately record raw boarding and alighting passengers' data. The research on APCSs and related analysis is quite extensive and one can distinguish between traditional APCSs and emerging ones.

2.1. Traditional Automatic Passenger Counting Systems

Traditional APCSs may be classified according to indirect or direct measures of passengers. In the case of an *indirect* measure, passengers may be estimated by weighing all on board passengers by load sensors on the ground or on the suspensions or on the braking system (e.g., [6,19]). This indirect APC technology provides the total weight of passengers on board but does not offer data on the flow of passengers.

In the case of a *direct* measure, passengers are estimated by counting people when they board or alight from the bus. Three main systems are presented, which include mat sensors infrared, and video image sensors (e.g., [6–8,20–28]). Usually, mat sensors measure the weight of a single person on the two steps of each door of the bus. These mats are activated when a minimum design weight is applied to the treadle (e.g., 15 kg—[6]). Infrared sensors measure the number of passengers by light beams, which are interrupted when they board or alight from the bus (e.g., [6,7,20–22]). The infrared system is the most used for this kind of application and easily found in commerce. Moreover, some tests showed that this technology is able to count both the total number of passengers and the maximum passenger load with high accuracy (e.g., [7,29]). However, despite this technology dominates in public transportation, the performance might get worse as the number of people grows [8,24]. Video image sensors measure passengers using cameras in the bus, which are able to recognize the passenger flow in both directions (e.g., [8,25–28]). These systems rely on dynamic image sequence processing to automatically count boarding and alighting passengers in a bus. They use several algorithms to (i) detect motion, (ii) estimate its direction, and (iii) validate the existence of a moving passenger. Each traditional APC technology is summarized in Table 1, which reports the main problem and a possible solution as well as pros and cons. Table 1 is self-explanatory. Nonetheless, all traditional technologies provide a measure of boarding and alighting passengers; therefore, the origin and destination of passengers need to be inferred only. Moreover, even if direct APCSs seem to work better than indirect ones, all technologies present their own drawbacks. The main drawbacks are the capital cost, due to the need to install more than one sensor per door, and the maintenance costs.

Table 1. Studies on traditional APC technologies (the list of references is representative, but not comprehensive).

Sources	APCs Technology	Description	Problem	Solution	Pros	Cons
[6,19]	Load sensors	The passenger is counted indirectly by devices on the ground, suspensions and/or breaking system of the vehicle	- Only Boarding and Alighting measures - Total weight of passengers on vehicles	- Passenger flows may be inferred - Origin and destination might be inferred	- Fast estimation of passenger volumes - Indirect technology to count passengers - No information on the flow of passengers - Physically integrated in the vehicle - Variability of the dynamic load on the shock absorbers	
[6,20,21]	Pressure sensitive and multi-switch treadle mats sensors	The passenger is counted directly when s/he boards/alights on the two steps of each door of the bus	- Only Boarding and Alighting measures	- Origin and destination might be inferred	- Direct technology to count passengers - High accuracy in the measurement	- Old technology - Not applicable in the case of low floor buses - Physically integrated on the vehicle - Need of slower passenger flows, possibly in a unique row - Capital and maintenance costs (e.g., mechanical parts in movement more sensitive to dirt and environment conditions)
[6,7,20–23,28]	Infrared sensors	The passenger is directly counted when s/he interrupts light beams during boarding/alighting operations	- Only Boarding and Alighting measures	- Origin and destination might be inferred	- Direct technology to count passengers - Most used and easily found in commerce - Installation on any kind of vehicle - High accuracy in the measurement	- The need to install more than one sensor per door - Low performance during congestion - Capital and maintenance costs (e.g., daily cleaning of the sensors is recommended)
[8,24–27]	Video Image sensors	The passenger is directly counted when cameras recognize his/her movement during boarding/alighting operations	- Only Boarding and Alighting measures	- Origin and destination might be inferred	- Direct technology to count passengers - Installation on any kind of vehicle - Allows to detect forms, sizes	- The need to install more than one sensor per door - Difficulties in case of strong variations of illumination and poor visibility - Capital and maintenance costs

2.2. Emerging Automatic Passenger Counting Systems

Emerging APCSs systems are quite recent and they are increasing rapidly with the development of ubiquitous Internet connection and IoT. Moreover, they outperform traditional systems in terms of capital and maintenance costs as bus operators do not have to install sensors or devices that have to be "physically integrated" on the bus. Using these systems, a passenger may be detected in an indirect way by counting the device carried out. It could be argued that some passengers might carry more than one device, thus overestimating the total passenger count. However, this is not a strong limitation of these systems. Indeed, some adjustment factors may be calibrated to improve the accuracy of scaling for inferred disaggregated boarding and alighting passengers.

The literature can be classified into three different systems: (1) large-scale cell phone data based on call details records; (2) apps installed into smartphones, and (3) Wi-Fi technology. Large-scale cell phone systems collect data when the device is connected to the cellular network. This may include a call both made or received, a short message that is sent or received, and/or when the user is connected to the internet (e.g., to browse the web). According to this technology, [30] presented a method to estimate at aggregate level passenger demand for public transportation services. More precisely, they showed how to extract significant origins and destinations of inhabitants to infer origin–destination matrices.

Smartphone app systems collect data by tracking the individual location of vehicles [31] or passengers [32] with high frequency or providing information on onboard passengers [33], also taking into account combined types of transportation [34]. For instance, [32] proposed a system of integrated methods to reconstruct and track the use of bus transit by passengers at a disaggregate level. These methods were based on the matching between location data from a smartphone app and automatic vehicle location data as in [35]. Participatory sensing is used in [34], where passengers are tracked throughout their journey using different transportation choices. Although smartphone app systems provide high-granularity data on the passenger trajectory, passengers need to install an app and give the consent to have their location tracked during their movements. As a result, the passenger engagement is a critical factor in encouraging the adoption and active participation in these activities. Moreover, once attracted, the motivation to sustain participation may be reliant upon factors such as simplicity and system design, feedback, and provision of incentives (e.g., [36,37]). Thus, if there are no benefits for passengers, the amount of data collected is quite scarce, thus inferring the total number of passengers is quite problematic, unless other systems are integrated (e.g., automatic fare collection).

Wi-Fi systems represent the newer technologies to collect data once a device have an active Wi-Fi interface, independently from the fact that its owner is connected to a network or not [38]. Indeed, these systems are based on the device discovery procedure that allows devices to discover other devices by acquiring their MAC address [39]. Recent studies were carried out in these years. For example, references [13,40] evaluated if a mobile AP installed on the bus helps detecting the relative position of a user device. This is to identify if a user device is inside or outside the bus. They showed that, unlike the received signal strength indication, bus speed may be a good indicator to whether a connection is established inside or outside the bus. Santos et al. [41] presented a multisource sensing infrastructure (i.e., Porto Living Lab) based on the IoT technology. It is conceived to detect on city-scale four phenomena, i.e., weather, environment, public transportation, and people flows. This infrastructure helps estimating the aggregate passengers' flow on buses using Wi-Fi connections and buses' on-board units.

It is noteworthy that passengers might have turned off the Wi-Fi, as it consumes battery [42]. Thus, Wi-Fi may result in an underestimation of passenger volumes. However, this is not a relevant problem as: (1) a proper scaling factor may be calibrated for each city in which this method will be applied. By a proper survey on a statistical sample of bus passenger is quite simple to calibrate a scaling factor characterized by a well-established level of confidence and margin of error, in order to estimate with statistical accuracy what is the percentage of people connected to a Wi-Fi network. For instance, in Dordrecht (the Netherlands), the number of people connected to Wi-Fi network ranges from 31% to 49%, thus a first factor may be applied to adjust passenger volumes [43]; (2) the Wi-Fi APs are increasing rapidly and many buses are expected to be equipped with them in the future.

Each emerging APC technology is summarized in Table 2, which is organized as Table 1. Although a passenger may be detected in an indirect way by counting the device carried out, Table 2 shows that all emerging technologies may help estimate the origin and destination of passengers. It could be argued that some passengers might carry more than one device, thus overestimating the total passenger count. However, this is not a strong limitation of these technologies. Indeed, some adjustment factors may be calibrated to improve the accuracy of scaling for inferred disaggregated boarding and alighting passengers.

Anyway, Wi-Fi systems seem to outperform large-scale cell phone and smartphone app systems as they depend neither on telco operators nor on passenger consents. Moreover, they enable to track passengers in an anonymous way. Hence, this system is expected to provide interesting results and will be adopted in this paper.

2.3. Gaps in the Literature

Overall, there are no doubts that all these studies provided interesting and captivating results for both research and practical applications. They provided evidences that different APCSs may be adopted for measuring passenger volumes.

However, by analyzing the literature on the adoption of these technologies, we have highlighted some gaps.

First, in traditional APCSs passengers cannot be tracked on their origin and destination bus stops, but only an aggregate estimation of passengers at each bus stop may be performed.

Second, emerging APCSs based on large-scale cell phone and/or smartphone apps may help to count boarding and alighting and estimate the origin and destination of passengers. However, collaboration with telco operators and/or consent of passengers are required steps to provide this estimation.

Therefore, it may be crucial to shed light on the estimation of the number of passengers, as well as on the estimation of origin destination matrices, using Wi-Fi systems that are based on the identification of the MAC address associated with devices' Wi-Fi interface. However, MAC address randomization procedures, recently introduced on mobile devices by operating system providers, make passenger tracking particularly difficult. For this reason, a de-randomization mechanism is introduced in iABACUS to overcome this issue. With reference to the approaches proposed in the literature, iABACUS provides multiple advantages: it enables anonymous counting of passengers; passengers are not required to take any action; flows of passengers are observed and analyzed, therefore contributing to short- and long-term urban mobility planning.

Table 2. Studies on emerging APC technologies (the list of references is representative, but not comprehensive).

Sources	APCs Technology	Description	Problem	Solution	Pros	Cons
[30]	Large-scale cell phone	The passenger is counted when his/her device is connected to the cellular network	Passenger may not carry the device	Origin and Destination can be estimated	- No action is required to passengers	- Indirect technology to count passengers - Need scaling factors to adjust data - Collaboration with Telco operators is required
[31–37]	Smartphone app	The passenger is counted when s/he uses the app	Passenger may not carry the device or may not installed the app	Origin and Destination can be estimated	- High granularity of trajectory data - Travel behavior	- Indirect technology to count passengers - Need scaling factors to adjust data - Privacy issues - Involvement of passengers - Not anonymous tracking - Benefits are required to support the participation
[13,38–41]	Wi-Fi System	The passenger is counted when s/he has active Wi-Fi	Passenger may not carry the phone and/or not have active the Wi-Fi	Origin and Destination can be estimated	- No action is required to passenger - Anonymous tracking - Low cost device for both capital and maintenance	- Indirect technology to count passengers - Need scaling factors to adjust data - Battery consumption

3. System Description

iABACUS is based on the detection of Wi-Fi signatures coming from any device, such as mobile phones, tablets and so on, with an active Wi-Fi interface. As shown in Figure 1, an on-board unit is installed on the bus. Thanks to the presence of a sniffer, it is in charge of collecting MAC addresses from the devices on board, store data and provide a first elaboration of the collected data (the de-randomization of the MAC addresses). Data are then transferred to the Cloud, either through a mobile connection on the run or through a Wi-Fi connection at the bus station, where they are further analyzed to count the actual number of devices on the bus.

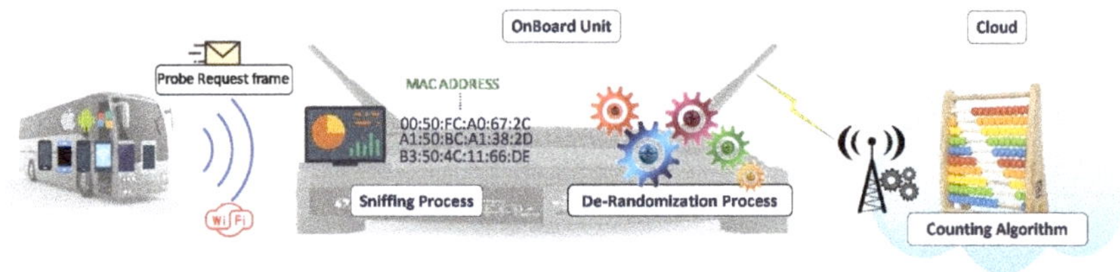

Figure 1. iABACUS.

Every time a Wi-Fi device has to deliver a message, it needs to know which AP to access. This information is provided to the devices by the concept of association, which is a necessary, but not sufficient, operation to support the connectivity of the devices. Indeed, before a device is allowed to send a data message via an AP, it must first become associated with the AP. To this, the device sends Probe Request frames, i.e., messages broadcast periodically from any active Wi-Fi interface to detect nearby APs.

The core of the proposed system is a Wi-Fi sniffer that collects and analyzes the Probe Request frames; these frames include information that can be associated univocally to the device that sent them, thus enabling its identification and counting. Indeed, as shown in Figure 2 (Note that the purpose of Figure 2 is not to describe the fields of a Probe Request frame, but just to show its structure. The interested reader is referred to [44] for further details about the standard.), one of the fields of the Probe Request frame is the MAC address of the source (SA), which is univocally assigned by the manufacturer to the Wi-Fi interface. Therefore, common Wi-Fi-based APCSs are able to find the number of different devices located close to the sniffer by counting the number of different MAC addresses identified in the received Probe Request frames.

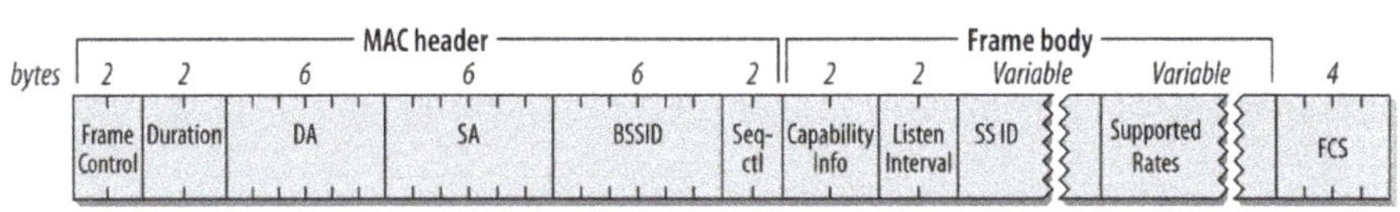

Figure 2. Wi-Fi Probe Request frame [44].

Nevertheless, current Wi-Fi-based APCS based on MAC identification can effectively work only for obsolete operating systems (earlier than Android 5.0, iOS 7 and Windows Phone 8). In fact, with the aim to protect their users' privacy against device tracking, the main mobile operating system providers have introduced software randomization of MAC addresses [16]: the MAC address included in Probe Request frames is not the real MAC address associated to the Wi-Fi interface anymore, but it rather is randomly generated and it periodically changes.

The characteristic that differentiates random MAC addresses from non-random ones is the fact that the first 6 octets of non-random MAC addresses are characterized by the Organizationally Unique Identifier (OUI), which is assigned by the Institute of Electrical and Electronics Engineers (IEEE) and identifies the manufacturers of existing network cards. Accordingly, the first 6 octets of the MAC address included in a Probe Request frame are compared with the list of all the OUIs that are associated to existing manufacturers of network cards. As depicted in Figure 1, if the MAC address is identified as a random MAC address, i.e., its first 6 octets do not correspond to any existing OUI, the *de-randomization algorithm* is run right after the sniffing process, before the counting algorithm. Otherwise, no de-randomization is needed, and the Probe Request can be passed directly to the counting algorithm. The main goal of the counting algorithm is to understand which of the Probe Requests received come from devices that are actually on the bus or are received due to cars or people moving near it.

In the following, the de-randomization process and the counting algorithm will be described in details.

3.1. De-Randomization Algorithm

The randomization of the MAC address introduced by operating system providers has allowed to hide the real MAC address of the network cards in the Probe Request frames of the device from which they are sent. In the Probe Request frames the real MAC addresses are replaced by random MAC addresses that are changed several times over a limited period of time. The change does not take place at regular intervals or according to predefined timing, but depending on the use of the device. Therefore the MAC address contained in the Probe Request frames is no longer sufficient to count the devices as it previously happened.

In this Section we introduce the de-randomization algorithm, which has the purpose to understand which MAC addresses are more likely to be brought back to the same device. Indeed, some parameters included in Probe Request frames can be exploited to estimate with sufficient reliability which frames containing different random MAC addresses may have been sent by the same device.

In particular, some fields of Probe Request frames remain constant even with randomized MAC addresses, as highlighted by the red square in Figure 3, also called tagged parameters [44].

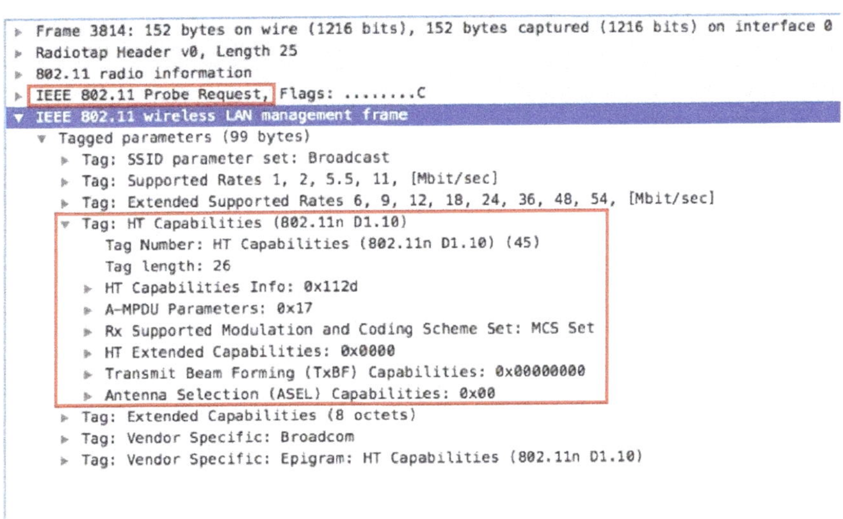

Figure 3. Tag Section Probe Request Frame.

This information is the same in all the Probe Request frames sent by the same device, even if the MAC address is randomized. However, this information identifies a particular family of devices but not the single device. In order for two frames containing different random MAC addresses to be traceable to the same device, the fact that these information are the same must therefore be considered as the first condition, necessary but not sufficient.

The de-randomization algorithm therefore needs to exploit other parameters, whose changes provide important information. Let us consider two MAC addresses received from the sniffer, namely MAC_i and MAC_j with MAC_i received before MAC_j. Accordingly, in order for the two MAC addresses to be traced back to the same source device, the instant of time when MAC_i was received, namely its timestamp, has to be lower than MAC_j's timestamp. Therefore, calling tg_i and tg_j the tagged parameters of respectively MAC_i and MAC_j, t_i^l the last timestamp associated to MAC_i and t_j^f the first timestamp associated to MAC_j (both expressed in seconds), the de-randomization algorithm starts if, and only if, both the following conditions are met:

$$\Gamma_{ij} = \begin{cases} tg_i \equiv tg_j & \text{(1a)} \\ t_i^l < t_j^f & \text{(1b)} \end{cases}$$

where the first condition of Equation (1) selects only devices that have the same transmitting characteristics, i.e., same tagged parameters, whilst the second condition of Equation (1) ensures that the last frame with MAC_i was sent before the first frame with MAC_j.

As stated above, the de-randomization algorithm assesses the probability that two random MAC addresses correspond to the same device. To compute this probability, we define a *score* for each couple of random MAC addresses identified. The score is calculated using the timestamp and another relevant parameter included in the Probe Request: the frame sequence number. The sequence number is a 12-bit code that progressively increases with each frame and is contained in the Sequence control (Seq-ctl in Figure 2). Its value varies from 0 to 4095 and once this maximum value is reached the numbering starts again from 0. Successive frames have increasing sequence numbers, even if the random MAC address changes.

The score is in inverse proportion to the difference in time and the difference in the sequence numbers between received frames with different MAC addresses. The difference in time is expressed as as:

$$\Delta T_{ij} = t_j^f - t_i^l \tag{2}$$

which expresses the continuity in the arrival of the frames, even for randomized addresses, and then guarantees that not too much time has passed since the change of address. The difference in the sequence numbers has a similar goal, i.e., to check the continuity in the received frames even if the MAC address is changed due to the randomization process. However, we need to take into account that the sequence number assumes values between 0 and 4095. Defining s_i^l the sequence number of the last frame corresponding to MAC_i and s_j^f the sequence number of the first frame corresponding to MAC_j, the resulting formula is the following:

$$\Delta S_{ij} = \begin{cases} s_j^f - s_i^l & \text{for } s_i^l < s_j^f \\ 4095 - (s_i^l - s_j^f) & \text{for } s_i^l > s_j^f \end{cases} \tag{3}$$

The score used in the algorithm is then defined as:

$$Score_{ij} = \begin{cases} \frac{1}{\Delta T_{ij}} \cdot \frac{1}{\Delta S_{ij}} & \text{for } \Gamma_{ij} \equiv true \\ 0 & \text{otherwise} \end{cases} \tag{4}$$

The score assumes values higher than 0 only if Equation (1) holds, i.e., if j and i have the same tagged parameters and they are received one after the other. For values higher than 0, the greater the value, the greater the probability that two MAC addresses are attributable to the same source device.

Once the score has been calculated for all the couples of random MAC addresses identified, the algorithm has the task of creating lists of MAC addresses traceable to the same source device. The process to create lists, depicted in the flowchart of Figure 4, is now explained. Whenever a frame with a new random MAC address MAC_j is received, the score is computed between MAC_j and all the other random MAC addresses that are already in a list. If no lists have been created yet, no scores will be computed; if all the scores are equal to 0, MAC_j certainly belongs to a new device. In both these case, a new list with MAC_j as the only element is created. Otherwise, the address MAC_m with the highest score with MAC_j, i.e., with the highest probability to belong to the same device, is identified. Considering the list L_x where MAC_m is located, if MAC_m is the last element of the list, MAC_j is appended to the list. If there is another MAC address MAC_n that follows MAC_m in list L_x, the $Score_{mj}$ has to be compared to $Score_{mn}$: if the first is lower than the latter, it means that it is more likely that MAC_m and MAC_n belong to the same device, with respect to MAC_m and MAC_j. Therefore, the process to find the right list for MAC_j starts again ignoring m. If, on the other hand, $Score_{mj}$ is higher than $Score_{mn}$, the probability that MAC_m and MAC_j belong to the same device is higher than the probability that MAC_m and MAC_n do. Accordingly, MAC_j is put in L_x right after MAC_m. Furthermore, the following MAC addresses in list L_x are not sure to belong to the same device anymore. Hence, the process is repeated again for MAC_n and all its following MAC addresses in L_x.

The procedure therefore assumes a recursive form. At the end, all the lists are those with greater probability that the MAC addresses present therein are traceable to the same device. Accordingly, all the frames belonging to the same list are tagged with the same MAC address.

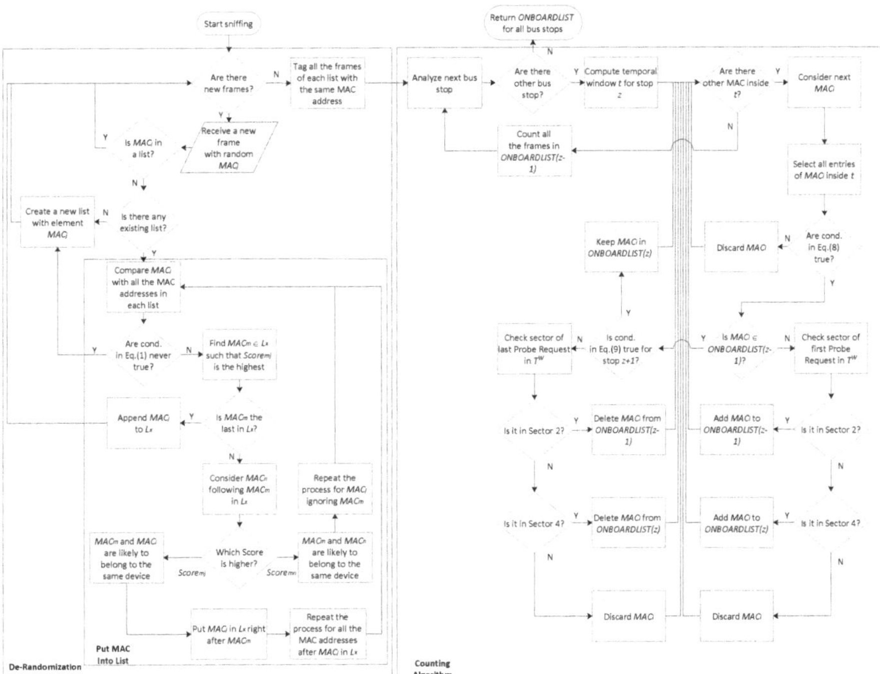

Figure 4. Flow chart of the system.

3.2. Passenger Counting Algorithm

Once the de-randomization process is solved, we need to analyze the resulting data, which represent the univocal MAC addresses of the sensed devices, in order to count the number of people on the bus.

With respect to a static situation, e.g., when counting people in a room, there are several issues that needs to be considered in order to accurately count only the people on the bus.

- the distance between two consecutive bus stops can be highly variable, from hundredth of meters to one-two kilometers;
- the Probe Requests are not sent regularly;
- the time the bus spends at each stop is variable and there may be stops where the bus does not stop;
- the sniffer can sense devices that are not on the bus, but are walking in the footpath, waiting at the bus stop or in the car near to the bus;

Let us consider the scenario depicted in Figure 5; our goal is to count the number of people on the bus and to know how many people board and alight from the bus during the last stop. To this, we define the set of bus stops $\mathcal{Z} = \{1, ..., z, ...Z\}$, so that for each stop z, we can define the time of arrival at the bus stop t_z^a and the time of departure from it t_z^d; then, we call *time spent*, the time interval the bus spends at each stop, which can be identified as follows:

$$\Delta T_z^S = t_z^d - t_z^a \tag{5}$$

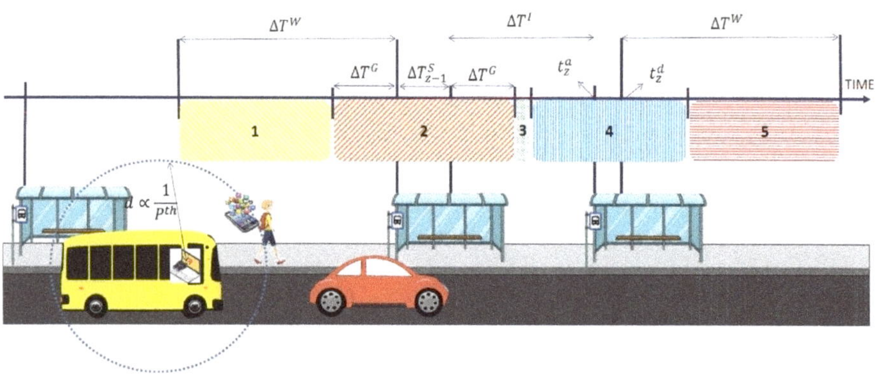

Figure 5. Counting problem and parameters.

We can also identify the *running time* as the time the bus needs in order to run from one stop to the next one as follows:

$$\Delta T_{z,z-1}^I = t_z^a - t_{z-1}^d \tag{6}$$

In order to overcome the issues mentioned earlier, we need to filter all the entries in our database. To address the variability in the reception frequency of the Probe Requests, for every stop z we need to examine a temporal window so to consider for the counting algorithm only the Probe Requests received with an instant of time included in the interval:

$$t \in [t_{z-1}^a - \Delta T^W, t_z^d + \Delta T^W] \tag{7}$$

where ΔT^W is a fixed parameter, called *watch time*.

After the temporal filtering, we need to understand if the remaining Probe Requests are transmitted by devices on the bus. To this, we applied a series of checks to all the entries in the filtered window. A first check is carried out to count the number of frames for each MAC address, namely N_i^{frame}, in order to identify devices encountered only for brief moments during the movement of the bus, such as the smartphone of a pedestrian or of a driver: if the number of frames captured is below a given threshold x, then the requests of these devices have been probably captured only a few times by the sniffer and then should be discarded.

Another check is implemented to evaluate the received power from the Probe Requests: received power is inversely proportional to the distance at which the transmitting device is located with respect to the sniffer, so it is possible to discard all the Probe Requests from devices which are too far and then are likely to be outside the bus, such as the case of a car that is moving in front of the bus as shown in Figure 5. In particular, we consider that the average power of the Probe Requests from MAC_i received by the sniffer must be higher than a certain threshold P^{th}, i.e., that $P_i^{avg} > P^{th}$; doing so, it is possible to discard even the cases of a car in a traffic jam which constantly moves away (lower power) and gets together again with the bus (higher power).

Finally, due to how Probe Requests are sent, i.e., through a train of near-time requests, we decided to consider another parameter, namely the permanence of a device on the bus ΔT^P, calculated as the difference between the last and the first frame received in the temporal window, namely $t_i^{l^*}$ and $t_i^{f^*}$, to assess if the requests belong to different trains of requests.

Summarizing, the entries of the generic MAC_i address inside a temporal window, as defined by Equation (7), are considered for the counting algorithm if they satisfy the following conditions:

$$\begin{cases} N_i^{frame} > x \\ P_i^{avg} > P^{th} \\ t_i^{l^*} - t_i^{f^*} > \Delta T^P \end{cases} \tag{8}$$

The number of unique MAC addresses left after the filtering steps gives a rough estimate of the number of devices on the bus. However, it is still important to assess at which stop the device boards and alights. This information is really useful to construct origin–destination matrices, but also to assess that the first or the last occurrence of a MAC address is not too far from any stops, thus indicating another vehicle travelling close to the bus with a similar speed, rather than a passenger boarding or alighting from the bus in motion.

To this, we consider that a device boarded or alighted from the bus only if the first or last Probe Request was received in a time frame of the bus stop, respectively. This is necessary because it is not possible to know in advance when the Probe Requests will be transmitted, so we must guarantee a *guard time*, both before arriving at the bus stop, $\Delta T_z^{G_b}$, and after leaving it, $\Delta T_z^{G_l}$, for those unlucky cases in which the first or the last Probe Request is sent just before reaching the bus stop or shortly after leaving it.

For every bus stop with a $\Delta T^S > 0$, there will be a time frame of $\Delta T_z^{G_b} + \Delta T_z^{G_l} + \Delta T_z^S$ available to receive the first or last Probe Request from a device. In general, the guard time can be considered constant for all the stops both before and after each stop, so we can indicate it as ΔT^G; however, if the running time between two stops $\Delta T_{z,z-1}^I$ is too small, i.e., less than $\Delta T_z^{G_b} + \Delta T_{z-1}^{G_l} = 2\Delta T^G$, then the guard time must be modified accordingly as $\Delta T_z^{G_b} = \Delta T_{z-1}^{G_l} = 0.5\Delta T_{z,z-1}^I$.

We finally need to understand, for each temporal window and for each device, if the device was actually on board and if it boarded or alighted during the considered timeframe. As shown in Figure 5, when analyzing a single timeframe, we can individuate different sectors:

- Sector 1, between the start of the temporal window and the start of the guard time for stop $z-1$.
- Sector 2, which analyzes the stop $z-1$, considering both the guard times, before and after the stop, and the time spent.
- Sector 3 examines the time between two stops.
- Sector 4, which takes into account the event for stop z.
- Sector 5, between the end of the guard time for stop z and the temporal window.

For each device, the counting algorithm first checks if the device was already considered on board. If it is the first time the sniffer has received requests from the device, the algorithm has to understand at which stop the device boarded: as we said earlier, a device can only board if the first Probe Request, $t_i^{f^*}$, has been received when the bus is near a bus stop. The device will be counted as boarding at stop $z-1$ if the first Probe Request is received in Sector 2 or at stop z if it is received in Sector 4. In all the other cases, i.e., when the Probe Request is received inside Sector 1, 3 and 5, all the entries for the device are considered as spurious and then discarded.

An on board device is considered as such for the whole temporal window; in this situation the condition on the number of frames for the next temporal window, described in Equation (8), is updates as follows:

$$N_i^{frame} + 1 > x \tag{9}$$

to take into account that the device is already on board. However, if the device does not pass the filtering step in the next temporal window, i.e., for stop $z+1$, then the algorithm goes back to the previous temporal window to check what happened by analyzing the last Probe Request received, $t_i^{l^*}$. Again, a device will be counted as alighting from the bus at stop $z-1$ or z only if the last Probe Request is received in Sector 2 or 4 respectively, otherwise all the entries for the device are discarded.

Noteworthy, every time the bus reaches a stop z, the algorithm is able to count the number of passengers related to the previous stop, by checking the next temporal window.

Figure 4 illustrates all the steps of the overall process. It starts with the sniffing of the Probe Requests and ends when all the bus stops have been analyzed to count the passengers.

4. Experiments

In this Section, the performance of iABACUS is evaluated. The setup of the experiments consists of a Wireshark sniffer, set in Monitor Mode, with a filter that picks up only Probe Request frames with broadcast destination address, i.e., ff:ff:ff:ff:ff:ff. The captured frames are sent to a MySQL database where they are collected and stored. To this, a connection is established by a Lua script, which forwards the captured frames to the database. The database is then queried thanks to a PHP application.

In the following, the experiments' results based on this setup will be shown. The first experiments, shown in Section 4.1, were made to test the de-randomization algorithm in a static scenario. Later, Section 4.2 presents the tests made to count passengers in a dynamic scenario.

4.1. Accuracy Evaluation for the De-Randomization Algorithm

The first tests were carried out in a static scenario, specifically to assess the accuracy of the de-randomization algorithm. Test results rely on the number of devices identified after the de-randomization process, i.e., the number of lists that are produced as output of the De-Randomization Algorithm described in Section 3.1

The experiments were run in a university room in a time window of 15 minutes. The following information was collected from the device owners inside the room: number of devices with active Wi-Fi interface; real MAC addresses of devices; brand, operating system type and version of devices.

The first parameter that was considered in the experiments is the transmission frequency. Indeed, according to the standard, the typical transmission frequency bands for Wi-Fi are 2.4 and 5 GHz. Nevertheless, while all the detected devices transmitted at 2.4 GHz, this was not always verified at 5 GHz. Furthermore, the 2.4 GHz Probe Request frames are the most complete and information-rich. Therefore, the 2.4 GHz was chosen as receiving frequency for the Wireshark sniffer. The possibility of gathering information on both frequencies was also assessed. However, since a sniffer can capture data only at one frequency at a time, this would require two sniffers placed in the same point. Furthermore, the process of combining the capture of the two sniffers and analyzing many more frames would significantly increase the computational load of the system, while not providing significant further information. For these reasons, the reception frequency was set to 2.4 GHz.

In order to set the most appropriate power threshold P^{th}, which limits the considered devices to those that are inside the room, the average power of all the Probe Request frames received, shown in Figure 6, was evaluated. As expected, the highest average power was measured for the devices that were closer to the sniffer. In addition, it can be noted that there is a distinct division between received power values higher than −51 dBm and lower than −61 dBm. Along with the distance, this latter higher attenuation can be ascribed to the presence of walls. Accordingly, the power threshold was set to −55 dBm. The parameters that were used for this test are summarized in Table 3.

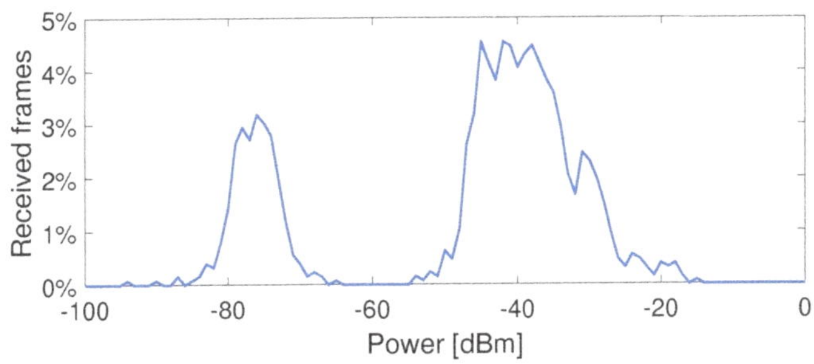

Figure 6. Percentage of received Probe Request frames with respect to their receiving power.

Table 3. Parameters of the experiments for the evaluation of the accuracy performance of the de-randomization algorithm.

Parameter	Value
Considered time window	15 min
Number of devices with active Wi-Fi-in the room	21
Frequency	2.4 GHz
Received power threshold	−55 dBm

The final results of iABACUS tested in this static case are shown in Table 4, where the number of MAC addresses counted by the proposed algorithm is compared with the number of MAC addresses counted by common Wi-Fi-based APCSs, i.e., systems that do not take into account the randomization of the MAC address and simply count the number of unique MAC addresses identified. As can be seen from the data in the table, iABACUS ensures that a much more accurate result is obtained, compared to common Wi-Fi-based APCS that does not take into account the randomization of MAC addresses. In particular, the system returned the exact number of devices present in the room (i.e., 21), while a traditional procedure would have recorded a significantly higher number of devices (i.e., 37). Since common Wi-Fi-based APCSs can only understand if a MAC address is random, and not if two different random MAC addresses can be associated to the same device, they associate a device to each different random MAC address received. Therefore, their error on the number of devices that implement the MAC address randomization is tied by the observation time window, that in this case was equal to 15 minutes: the longer the time window, the higher the number of times a device changes its MAC address and then the higher the number of devices counted.

Table 4. Results for the accuracy evaluation of the de-randomization algorithm.

Description	Value
Devices in the room	21
Devices counted by iABACUS	21
Devices counted by common Wi-Fi-based APCSs	37
Devices implementing MAC address randomization techniques counted by iABACUS	3
Devices implementing MAC address randomization techniques counted by common Wi-Fi-based APCSs	19

4.2. Passenger Counting Experiments

To evaluate the feasibility and performance of the proposed counting algorithm, a real dynamic environment has been reproduced. The main characteristic of this dynamic scenario is the possibility for people to move in and out of the sniffer caption's range due to the movement of the sniffer itself. This is the case of a sniffer inside a bus where people board and alight during bus stops.

The experiments were performed in the city of Cagliari, Italy. To limit their randomness, we made use of a car to simulate an existing bus path both in the bus stops' positions and for their time spent. The car's path is shown in Figure 7, which also highlights the 13 stops considered. The number of Wi-Fi devices involved is limited to 8, with three of them implementing the randomization of the MAC address. The events of people boarding and alighting from the bus were simulated by switching on and off the Wi-Fi interface of the devices respectively. The plan of the experiments is shown in Table 5, which represents for each device, labelled from device A to device H, the origin' and destination' stop; Table 5 also shows the results of the three tests performed, that will be discussed below, highlighting in pale blue color the cells where something wrong happened. All the experiments were carried out before or during the lunch break, in moderate traffic conditions. The path is around 2.4 km long and has three traffic lights: the car needed around 15 min to travel it, which is an average time also for a passenger travelling on the bus.

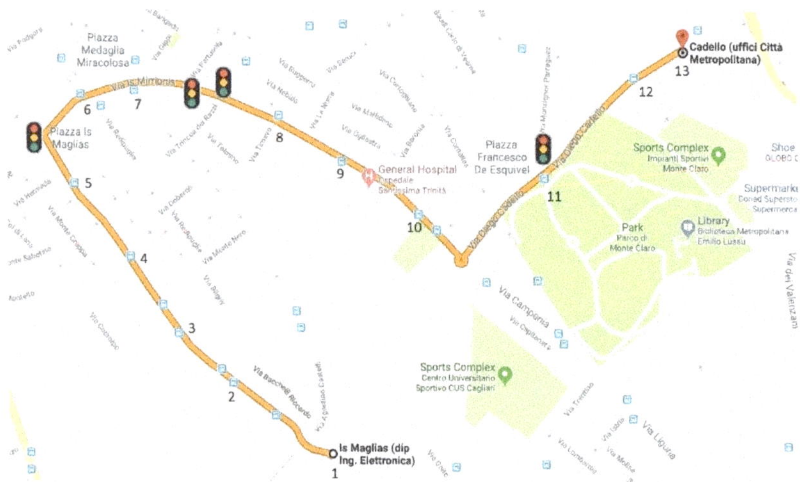

Figure 7. Path of the vehicle for the experiments.

Table 5. Origin and Destination matrix.

Device	Planned Experiment		First Test		Second Test		Third Test	
	O	D	O	D	O	D	O	D
A	1	2	NA	NA	1	2	1	2
B	1	6	1	6	1	6	1	6
C	1	9	1	8	1 5	3 9	1	9
D	2	12	4	12	2	12	3	12
E	6	8	6	8	6	8	6	8
F	6	9	6	9	6	9	6	9
G	8	13	8	13	8	13	8	13
H	9	10	NA	NA	9	10	9	10
X			3	5				

For the sniffer, the same setup described in Section 4 has been implemented on a Raspberry Pi 2 with Raspbian Stretch as Operating System to improve the portability of the solution. A USB Wi-Fi dongle is used as antenna to enable the Raspberry to collect the Probe Requests. The car's path has been monitored with a GPS module (GPS GY-NEO-6M v2), sending its position (latitude and longitude) to the Raspberry every 5 s. The GPS information is essential to obtain the time of arrival t_z^a and the time of departure t_z^d from each bus stop, as well as the running time $\Delta T^I_{z,z-1}$, which are key parameters for our counting algorithm. Moreover, the GPS enables the spatial mapping with the Probe Requests received by the Raspberry by comparing their timestamp.

Table 6 summarizes the value of the parameters considered during the performed tests. We have chosen reasonable values for the parameters in the first test; we then analyze the results of each test, comparing them with the planned experiment in order to tune them in the subsequent tests and correct the errors of the algorithm.

Table 6. Setting Parameters.

Parameter	Test 1	Test 2	Test 3
ΔT^W (minutes)	2	4	4
N^{frame}	5	3	1
P^{th} (dBm)	−55	−65	−65
ΔT^P (minutes)	1	1	1
ΔT_z^G (seconds)	10	20	20

The results of the first test show that devices A and H are missing, device D boards at stop 4 instead of stop 2 while device C alights at stop 8 instead of stop 9; moreover, we individuated a new device, labelled as X, that was not planned in the experiment, boarding at stop 3 and alighting at stop 5. We then analyzed step by step what happened during the trip:

- during the temporal filtering less than 5 occurrences of the requests for devices A and H were found, so since they were on board for only one stop, they were discarded from the algorithm; during our experiments, devices usually send a train of probe requests, i.e., a group of requests, every 40 s, but devices in energy-saving mode can send these requests with a lower frequency. By studying the capture from the sniffer, we notices that the frequency of the probe requests from devices A and H was really low, around 2–3 min, meaning that the devices were in energy saving mode.
- We have entries of device D from the temporal window of bus stop 2, unfortunately none of those entries were in the right Sector in order to count the device as on board, until the car arrived at the bus stop 4.
- Regarding device C, when analyzing bus stop 9, we found too few entries (only 3), so we checked the previous bus stop, i.e., bus stop 8, and found that the device last probe request was in Sector 4, so the algorithm signed the device as alighting at stop 8.
- Finally, between stop 3 and stop 5 we got stuck in a little traffic jam, so we needed a lot of time to travel this road section and our algorithm was able to accumulate a lot of entries from another device, maybe a smartphone from a car travelling behind or in front of us, that was then counted as a passenger on board.

For the second test, we increased the temporal window from 2 to 4 min and reduced the required number of frame from 5 to 3, in order to enable even the devices in energy save mode to be counted correctly; moreover, we also increased the guard time ΔT_z^G, so that it was more likely to find a device in the right bus stop, both boarding or alighting. We also decide to set a more conservative threshold for the received power, to avoid counting devices from outside the car.

From the analysis of the second test, we found only one error: device C alights at bus stop 3 but boards again at stop 5 to finish its trip correctly at stop 9. By checking the requests sent by the device, we noticed that they were fairly regular (about one train of requests every 42 s), but for some reason, around stop 3 and 5, the sniffer has missed some of them, resulting in the strange behaviour detected. However, we noticed that all the previous error were corrected, so we decided only to further reduce the number of frames required for a MAC address to be considered by the counting algorithm.

From the third test, we noticed that the problem regarding device C was solved and that no external devices were counted, even with a number of frames required equals to 1: this is due to the ability of the algorithm to filter out most of the probe requests thanks to the power threshold. However, we detected that device D was counted again as boarding at a different bus stop, as it happened during the first test, but not in the second one; this happens since stop 2 and 3 are really close, so that there is no Sector 3 among them and the two guard times from stop 2 and stop 3 overlap. Device D is then counted as boarding based on the first probe request received: in the unlucky case, that the probe request is delayed to stop 3, this will result as an error, that can not be avoided.

From these tests, we can infer simple rules to set the algorithm's parameters. The main goal is not to discard any device on the bus, even if they are in energy saving mode: to this, the temporal window should be quite large to be sure to detect at least one Probe Request, while the number of frames lose significance. Devices outside of the bus need then to be discarded through the use of the other parameters, such as the power threshold. However, these rules can only provide a first hint on the setting, since the algorithm needs to be adjusted based on the particular route and city under consideration.

5. Conclusions and Future Works

This paper proposed iABACUS, a novel automatic passenger counting system based on IoT for public transportation, which infers the number of people on-board based on Wi-Fi probe requests received by a sniffer installed on the bus. Due to the randomization of the MAC addresses, introduced with the latest mobile operating systems, current APCSs based on Wi-Fi are now obsolete. For this reason, iABACUS includes a de-randomization algorithm in order to understand which MAC addresses are more likely to be brought back to the same device; tests of the proposed algorithm on a total of 21 devices show that it is able to successfully recognize all the randomized MAC address. Moreover, our tests also highlight that the randomization of the MAC addresses is a serious problem, since even the presence of only three devices can lead to a count six times higher in a 15 min time interval.

We then investigated the counting algorithm by means of real experiments, simulating the bus behaviour with a car and considering eight passengers, boarding and alighting in different bus stops. Experiments are performed in stress conditions, in order to underline any problem with the algorithm. The results are quite good, as the counting algorithm, when set correctly, is able to count all the passengers; however, due to the low frequency with which devices send their probe requests, there may be random errors concerning the correct boarding and alighting from the bus of passengers.

As future works, we plan to consider how real use-case conditions, such as attenuation of the received Wi-Fi signal due to bus size or absorption by passenger bodies (especially for crowded buses), interference, and non-line-of-sight, affect iABACUS's performance. We will study how many sniffers are required and their optimal position, in order to discern correctly the power of all the devices on-board w.r.t. the ones outside: using more than one sniffer, maybe set to accumulate requests in a different frequency channel, and aggregating their results, could prove to be a better approach. Finally, we plan to study the random error related to the proximity of two bus stops in order to find a relation with the probability of its occurrence. Moreover, we plan to study the systematic error due to the non-connected users in our system; to this, we aim to compare the results of the proposed algorithm with the computed number of passengers estimated by the local bus transport in the city of Cagliari and propose an adjusting factor to be applied to our system. There are certainly differences among cities in the number of non-connected users, so we expect that a similar methodology should be carried out in every city.

Author Contributions: Conceptualization, all; methodology, all; software, F.P. and L.P.; validation, M.N., F.P., L.P. and V.P.; formal analysis, all; investigation, M.N., F.P., L.P. and V.P.; resources, M.N., V.P. and B.B.; data curation, M.N., F.P., L.P., V.P.; writing—original draft preparation, all; writing—review and editing, M.N., V.P. and B.B.; visualization, all; supervision, M.N., V.P. and B.B. All authors have read and agreed to the published version of the manuscript.

Funding: This work was partially supported by the Italian Ministry of University and Research (MIUR), within the Smart Cities framework (Project CagliariPort2020, ID: SCN_00281 and Project Cagliari2020, ID: PON04a2_00381), and by the Italian Ministry of Economic Development (MiSE, Project INSIEME, HORIZON 2020, PON 2014/2020 POS. 395).

Conflicts of Interest: The authors declare no conflict of interest.

References

1. Gong, H.; Chen, C.; Bialostozky, E.; Lawson, C.T. A GPS/GIS method for travel mode detection in New York City. *Comput. Environ. Urban Syst.* **2012**, *36*, 131–139. [CrossRef]
2. Kaiser, M.S.; Lwin, K.T.; Mahmud, M.; Hajializadeh, D.; Chaipimonplin, T.; Sarhan, A.; Hossain, M.A. Advances in Crowd Analysis for Urban Applications Through Urban Event Detection. *Proc. IEEE Trans. Intell. Transp. Syst.* **2017**, *19*, 3092–3112. [CrossRef]
3. Zhang, F.; Jin, B.; Wang, Z.; Liu, H.; Hu, J.; Zhang, L. On geocasting over urban bus-based networks by mining trajectories. *IEEE Trans. Intell. Transp. Syst.* **2016**, *17*, 1734–1747. [CrossRef]
4. Zhang, J.; Shen, D.; Tu, L.; Zhang, F.; Xu, C.; Wang, Y.; Tian, C.; Li, X.; Huang, B.; Li, Z. A real-time passenger flow estimation and prediction method for urban bus transit systems. *IEEE Trans. Intell. Transp. Syst.* **2017**, *18*, 3168–3178. [CrossRef]
5. Barabino, B.; Deiana, E.; Mozzoni, S. The quality of public transport service: The 13816 standard and a methodological approach to an Italian case. *Ing. Ferrov.* **2013**, *68*, 475–499.
6. Pinna, I.; Dalla Chiara, B.; Deflorio, F. Automatic passenger counting and vehicle load monitoring. *Ing. Ferrov.* **2010**, *65*, 101–138.
7. Barabino, B.; Di Francesco, M.; Mozzoni, S. An offline framework for handling automatic passenger counting raw data. *IEEE Trans. Intell. Transp. Syst.* **2014**, *15*, 2443–2456. [CrossRef]
8. Chen, C.H.; Chang, Y.C.; Chen, T.Y.; Wang, D.J. People counting system for getting in/out of a bus based on video processing. In Proceedings of the Eighth International Conference on Intelligent Systems Design and Applications, 2008, ISDA'08, Kaohsiung, Taiwan, 26–28 November 2008; Volume 3; pp. 565–569.
9. Gonzalez, M.C.; Hidalgo, C.A.; Barabasi, A.L. Understanding individual human mobility patterns. *Nature* **2008**, *453*, 779. [CrossRef]
10. Sohn, K.; Kim, D. Dynamic origin–destination flow estimation using cellular communication system. *IEEE Trans. Veh. Technol.* **2008**, *57*, 2703–2713. [CrossRef]
11. Wang, Y.; Chen, Y.J.; Yang, J.; Gruteser, M.; Martin, R.P.; Liu, H.; Liu, L.; Karatas, C. Determining driver phone use by exploiting smartphone integrated sensors. *IEEE Trans. Mob. Comput.* **2015**, *15*, 1965–1981. [CrossRef]
12. Pompei, F. Geolocation for LPT: Use of geolocation technologies for performance improvement and test of Local Public Transport. In Proceedings of the 2017 IEEE International Conference on Environment and Electrical Engineering and 2017 IEEE Industrial and Commercial Power Systems Europe (EEEIC/I&CPS Europe), Milan, Italy, 1–5 July 2017; pp. 1–5.
13. Santos, P.M.; Kholkine, L.; Cardote, A.; Aguiar, A. Context classifier for position-based user association control in vehicular hotspots. *Comput. Commun.* **2018**, *121*, 71–82. [CrossRef]
14. Amsterdam Airport Schiphol. Royal Schiphol Group Privacy Statement. Available online: https://www.schiphol.nl/en/privacy-policy/. (accessed on 12 February 2019).
15. Transport for London. Review of the TfL WiFi Pilot. Available online: http://content.tfl.gov.uk/review-tfl-wifi-pilot.pdf. (accessed on 12 February 2019).
16. Myrvoll, T.A.; Håkegård, J.E.; Matsui, T.; Septier, F. Counting public transport passenger using WiFi signatures of mobile devices. In Proceedings of the 2017 IEEE 20th International Conference on Intelligent Transportation Systems (ITSC), Yokohama, Japan, 16–19 October 2017; pp. 1–6.

17. Martin, J.; Mayberry, T.; Donahue, C.; Foppe, L.; Brown, L.; Riggins, C.; Rye, E.C.; Brown, D. A study of MAC address randomization in mobile devices and when it fails. *Proc. Priv. Enhancing Technol.* **2017**, *2017*, 365–383. [CrossRef]
18. Smartphone Ownership Is Growing Rapidly around the World, but Not Always Equally. Available online: https://pewrsr.ch/2Ngqr32. (accessed on 10 October 2019).
19. Nielsen, B.F.; Frølich, L.; Nielsen, O.A.; Filges, D. Estimating passenger numbers in trains using existing weighing capabilities. *Transp. A Transp. Sci.* **2014**, *10*, 502–517. [CrossRef]
20. Attanucci, J.; Vozzolo, D. *Assessment of Operational Effectiveness, Accuracy, and Costs of Automatic Passenger Counters*; Number HS-037 821; Transportation Research Board: Washington, DC, USA, 1983.
21. Furth, P.G.; Hemily, B.; Muller, T.; Strathman, J.G. Uses of archived AVL-APC data to improve transit performance and management: Review and potential. *TCRP Web Doc.* **2003**, *23*, 1–58.
22. Boyle, D.K. *Passenger Counting Systems*; Number 77; Transportation Research Board: Washington, DC, USA, 2008.
23. Kovács, R.; Nádai, L.; Horváth, G. Concept validation of an automatic passenger counting system for trams. In Proceedings of the 5th IEEE International Symposium on Applied Computational Intelligence and Informatics, SACI'09, Timisoara, Romania, 28–29 May 2009; pp. 211–216.
24. Choi, J.W.; Quan, X.; Cho, S.H. Bi-directional passing people counting system based on IR-UWB radar sensors. *IEEE Internet Things J.* **2018**, *5*, 512–522. [CrossRef]
25. Bartolini, F.; Cappellini, V.; Mecocci, A. Counting people getting in and out of a bus by real-time image-sequence processing. *Image Vis. Comput.* **1994**, *12*, 36–41. [CrossRef]
26. Harasse, S.; Bonnaud, L.; Desvignes, M. Finding people in video streams by statistical modeling. In *International Conference on Pattern Recognition and Image Analysis*; Springer: Berlin, Germany, 2005; pp. 608–617.
27. De Potter, P.; Kypraios, I.; Verstockt, S.; Poppe, C.; Van de Walle, R. Automatic Passengers Counting In Public Rail Transport Using Wavelets. *Automatika* **2012**, *53*, 321–334. [CrossRef]
28. Xiang-Yang, S.; Hao-Wei, W. Study on Method of Multi-feature Reduction Based on Rough Set in Passenger Counting. In Proceedings of the 2016 IEEE International Conference on Industrial Informatics-Computing Technology, Intelligent Technology, Industrial Information Integration (ICIICII), Wuhan, China, 3–4 December 2016; pp. 307–310.
29. Olivo, A.; Maternini, G.; Barabino, B. Empirical Study on the Accuracy and Precision of Automatic Passenger Counting in European Bus Services. *Open Transp. J.* **2019**, *13*, 250–260. [CrossRef]
30. Demissie, M.G.; Phithakkitnukoon, S.; Sukhvibul, T.; Antunes, F.; Gomes, R.; Bento, C. Inferring passenger travel demand to improve urban mobility in developing countries using cell phone data: A case study of Senegal. *IEEE Trans. Intell. Transp. Syst.* **2016**, *17*, 2466–2478. [CrossRef]
31. Gao, R.; Zhao, M.; Ye, T.; Ye, F.; Wang, Y.; Luo, G. Smartphone-based real time vehicle tracking in indoor parking structures. *IEEE Trans. Mob. Comput.* **2017**, *16*, 2023–2036. [CrossRef]
32. Carrel, A.; Lau, P.S.; Mishalani, R.G.; Sengupta, R.; Walker, J.L. Quantifying transit travel experiences from the users' perspective with high-resolution smartphone and vehicle location data: Methodologies, validation, and example analyses. *Transp. Res. Part C Emerg. Technol.* **2015**, *58*, 224–239. [CrossRef]
33. Chaudhary, M.; Bansal, A.; Bansal, D.; Raman, B.; Ramakrishnan, K.; Aggarwal, N. Finding occupancy in buses using crowdsourced data from smartphones. In Proceedings of the 17th International Conference on Distributed Computing and Networking, Singapore, 4–7 January 2016; p. 35.
34. Lu, Y.; Misra, A.; Wu, H. Smartphone sensing meets transport data: A collaborative framework for transportation service analytics. *IEEE Trans. Mob. Comput.* **2017**, *17*, 945–960. [CrossRef]
35. Lee, U.; Magistretti, E.; Gerla, M.; Bellavista, P.; Corradi, A. Dissemination and harvesting of urban data using vehicular sensing platforms. *IEEE Trans. Veh. Technol.* **2009**, *58*, 882–901.
36. Corsar, D.; Cottrill, C.; Beecroft, M.; Nelson, J.D.; Papangelis, K.; Edwards, P.; Velaga, N.; Sripada, S. Build an app and they will come? Lessons learnt from trialling the GetThereBus app in rural communities. *IET Intell. Transp. Syst.* **2017**, *12*, 194–201. [CrossRef]
37. Susilo, Y.O.; Liotopoulos, F.K. Measuring Door-to-Door Journey Travel Satisfaction with a Mobile Phone App. In *Quality of Life and Daily Travel*; Springer: Berlin, Germany, 2018; pp. 119–138.
38. Liu, H.; Yang, J.; Sidhom, S.; Wang, Y.; Chen, Y.; Ye, F. Accurate WiFi based localization for smartphones using peer assistance. *IEEE Trans. Mob. Comput.* **2013**, *13*, 2199–2214. [CrossRef]

39. Casetti, C.E.; Chiasserini, C.F.; Duan, Y.; Giaccone, P.; Manriquez, A.P. Data connectivity and smart group formation in Wi-Fi Direct multi-group networks. *IEEE Trans. Netw. Serv. Manag.* **2017**, *15*, 245–259. [CrossRef]
40. Kholkine, L.; Santos, P.M.; Cardote, A.; Aguiar, A. Detecting relative position of user devices and mobile access points. In Proceedings of the 2016 IEEE Vehicular Networking Conference (VNC), Columbus, OH, USA, 8–10 December 2016; pp. 1–8.
41. Santos, P.M.; Rodrigues, J.G.; Cruz, S.B.; Lourenço, T.; d'Orey, P.M.; Luis, Y.; Rocha, C.; Sousa, S.; Crisóstomo, S.; Queirós, C.; et al. PortoLivingLab: An IoT-Based Sensing Platform for Smart Cities. *IEEE Internet Things J.* **2018**, *5*, 523–532. [CrossRef]
42. Tu, L.; Wang, S.; Zhang, D.; Zhang, F.; He, T. ViFi-MobiScanner: Observe Human Mobility via Vehicular Internet Service. *IEEE Trans. Intell. Transp. Syst.* **2019**, pp. 1–13. [CrossRef]
43. Kyritsis, D. The Identification of Road Modality and Occupancy Patterns by Wi-Fi Monitoring Sensors as a Way to Support the "Smart Cities" Concept: Application at the City Centre of Dordrecht. Master's Thesis, TU Delft, Delft, The Netherlands, 2017.
44. IEEE Computer Society LAN/MAN Standards Committee. *IEEE Standard for Information Technology-Telecommunications and Information Exchange between Systems-Local and Metropolitan Area Networks-Specific Requirements Part 11: Wireless LAN Medium Access Control (MAC) and Physical Layer (PHY) Specifications*; IEEE Std 802.11; IEEE: New York, NY, USA, 2007.

energies

MDPI

Article

Efficiency of Telematics Systems in Management of Operational Activities in Road Transport Enterprises

Ryszard K. Miler [1], Marcin J. Kisielewski [1], Anna Brzozowska [2] and Antonina Kalinichenko [3,4,*]

1 Management and Finance Institute, WSB University in Gdańsk, Al. Grunwaldzka 238 A, 80-266 Gdańsk, Poland; rmiler@wsb.gda.pl (R.K.M.); mkisielewski@wsb.gda.pl (M.J.K.)
2 Faculty of Management, Business Informatics and Ecosystems, Czestochowa University of Technology, ul. J.H. Dąbrowskiego 69, 42-201 Częstochowa, Poland; annabrzozowskapcz@gmail.com
3 Institute of Environmental Engineering and Biotechnology, University of Opole, Kominka Street, 6, 45-032 Opole, Poland
4 Department of Information System and Technology, Poltava State Agrarian Academy, Skovorody 1/3, 36003 Poltava, Ukraine
* Correspondence: akalinichenko@uni.opole.pl; Tel.: +48-787-321-587

Received: 29 July 2020; Accepted: 15 September 2020; Published: 18 September 2020

Abstract: Implemented in road transport enterprises (RTEs) on a large scale, telematics systems are dedicated both to the particular aspects of their operation and to the integrated fields of the total operational functioning of such entities. Hence, a research problem can be defined as the identification of their efficiency levels in the context of operational activities undertaken by RTEs (including more holistic effects, e.g., lowering fuel/energy consumption and negative environmental impacts). Current research studies refer to the efficiency of some particular modules, but there have not been any publications focused on describing the efficiency of telematics systems in a more integrated (holistic) way, due to the lack of a universal tool that could be applied to provide this type of measurement. In this paper, an attempt at filling the identified cognitive gap is presented through empirical research analysing the original matrix developed by the authors that refers to the efficiency rates of organisational activities undertaken by RTEs. The purpose of this paper is to present a tool that has been designed to provide a holistic evaluation of efficiency of telematics systems in RTE operational management. The results are presented in a form of an individual (ontogenetic) matrix of the analysed companies, for which a determinant was calculated with the use of Sarrus' rule. Obtained in such a way, the set of values identified for the determinants of the subsequent ontogenetic matrices came as an arithmetic progression that characterised the scope and the level of the influence exerted by the implemented IT (information technology) systems on the organisational efficiency of operational activities undertaken by the analysed RTEs. We present a hypothesis stating that the originally developed matrix can be viewed as a reliable tool used for comparative analysis in the field of efficiency of telematics systems in RTEs, and this hypothesis was positively verified during the research. The obtained results prove the significant potential for the wide application of the discussed matrix, which can be used as a universal tool for the analysis and comparison of efficiency indicated by the integrated IT systems in the operational activities undertaken by RTEs.

Keywords: road transport enterprises (RTEs); telematics systems; operational activities of RTEs; a universal matrix of efficiency rates

1. Introduction

Undoubtedly, problems related to the organisational behaviour of road transport enterprises (RTEs) come as a group of highly dynamic issues because RTEs face numerous functional and operational problems in the contemporary transport market [1], along with challenges related to competitiveness [2]

(including economical approach) and the pursuit of process optimisation [3] (including fuel/energy efficiency). Hence, interest in all types of solutions that can be helpful in facing decision-making challenges in the field of management has been growing [4–6], and more efficient and more effective operation strategies [7], especially in the field of operational activities, have been researched [8,9]. A remedy to the above-mentioned problems may be found in the application of managerial information technology (IT) systems [7] (more broadly understood as telematics systems) [10,11], the functionalities of which are dedicated to the current handling of transport processes (including vehicle mileage record systems, users, contractor costs, generated costs, and energy/fuel consumption) [12]. Through the use of IT systems, there is also a possibility to optimise transport routes [13–15] and to utilise work factors (such as vehicle fleet [16] load capacity and drivers' working time) [17], especially in cases where there are no regular routes due to constant changes with customer orders [18,19].

When considered in this way, the features of RTE management with the use of IT systems are the reason for a continuous discussion between the practitioners (managerial staff) and theoreticians of management about their significance, roles, and efficiency in the process of making decisions aimed at the improvement of the efficiency of the operational functioning of enterprises [20–22] in regard to economy, energy, and the environment.

The above-mentioned findings motivated the authors to undertake the research presented in this paper. The research was intended to provide an unambiguous answer to the following question: how can one indicate a correlation between the implementation of telematics systems (as tools applied by RTEs to support their operational management) and the efficiency of these entities in the competitive market while considering the scarcity of generally available theoretical analysis and the lack of a synthetic tool (rate) for indicating and measuring the expected correlation?

An in-depth analysis of specialist literature [23–25] indicated that despite some publications on the improvement in efficiency observed at various organisations, the significance and role of IT systems in increasing organisational efficiency have been relatively scarcely discussed [26]. Most frequently, such considerations have been side-effects resulting from major discussions on other problems [3,27–29].

Moreover, in light of the research on RTEs, current publications on the operational efficiency of RTEs have mostly presented the results of research based on an analysis of efficiency referring to individual factors, such as the efficiency of drivers' working time management [30], the efficiency of using the mileage (using the loading capacity) [31], and the efficiency of forwarders' work [32,33].

A scarce number of publications have indicated that there is a relationship between the implementation of IT applications (dedicated to either functional or operational fields) and an increase in the efficiency observed in the particular operational field, e.g., the implementation of electronic tachographs and the more efficient use of drivers' working time [34,35]. However, there have not been publications focused on the comprehensive influence of the integrated implementation of IT systems (more than two modules) on the organisational efficiency of RTEs. Additionally, no universal tool that could be treated as an objective measure of the influence or correlation between the implementation of IT systems and changes (expected increase) observed in the operational efficiency of RTEs has been designed.

Furthermore, despite the fact that there are some considerable knowledge deficiencies in reference to the relationships between IT systems and the organisational efficiency [36] of RTEs, some considerable investments into IT systems are made every year.

Therefore, the statement that problems related to the efficiency of RTE functioning with the use of IT systems are still serious challenges to scientific research is a well-justified one. The results of empirical research on the above-mentioned problems are presented in this paper, and these can fill the cognitive and pragmatic (instrumental) gaps observed in the discussed field.

The main objective of the research was to indicate the relation between the application of IT systems to support management (or, in a broader sense, telematics systems) and the organisational efficiency of operational activities [37] undertaken by RTEs.

The pragmatic aim of the paper is to indicate the usefulness of a matrix that has been originally developed to provide a holistic evaluation of the above-mentioned relationship, regarding the operational efficiency of RTEs. The authors formulated a general hypothesis stating that the originally developed matrix can be viewed as a reliable tool for comparative analysis in the field of the efficiency of telematics systems in RTEs. The scope of the research included RTEs (more specifically, the field of their operational activities supported by the applied IT systems).

The research was divided into three stages. During the first stage, based the information obtained with the use of the CATI (computer-assisted telephone interviewing) method [38], the key fields of operational activities undertaken by RTEs, which are the most sensitive to the implementation of telematics systems, were identified. As a result, an original matrix of the organisational efficiency rates of operational activities was developed. The above-mentioned rates were determined by the implementation of IT systems in RTEs. During the second stage, eight RTEs were identified with the use of qualitative methods, such as the individual PAPI (paper and pen personal interviews carried out with the members of the managerial staff) and questionnaire method [39,40]. The selected RTEs met the criteria adopted in the research. Additionally, an empirical analysis of the basic rates and measures of the operational activity evaluation was conducted. As a result, a graphical presentation of the outcomes recorded before and after the implementation of IT systems is provided here. During the third stage of research, an evaluation of the usability of the suggested matrix for the holistic and comparative assessment of the influence exerted by telematics systems on the efficiency of operational processes of RTEs was conducted.

Considering the results obtained during the research, the authors hope that, in the theoretical aspect, the paper will become a starting point for further research into the positive correlation between the implementation of IT systems within the pragmatics of RTE operation and the increase in operational efficiency of particular RTEs. Considering economic practice, the ontogenetic matrix may become a universal tool to support managerial decisions made at the RTE operational level (after evaluation that will be provided during further research and a pilot programme).

2. Efficiency in Transport Services (With Use of Managerial IT Systems)

Efficiency is treated as a category superior to notions such as capability, profitability, productivity, effectiveness, agility, and rationality [41]. When considering such an approach, efficiency can be understood as the relationship between effort and effect [42], the capability for agile adjustments to changes, a criterion related to the abilities of an organisation to achieve its aim [43] and to implement its strategies [44], and a tool applied to measure effective operation and efficacy [45].

In the theory of management, the notion of organisational efficiency [46] is defined as the ability of a system for current and strategic adjustment to changes in the environment, as well as the productive utilisation of resources (including fuel/energy) in order to achieve the assumed structure of targets [47]. The components of organisational efficiency are managerial [48] and technical efficiency [49]. Technical efficiency refers to the efficiency of technology and to the scale of its application.

The axiomatic approach to the transport market makes it possible to prove that a fundamental unit responsible for organising transport operations is an RTE that undertakes its operational activities (transport processes and services) [50,51] and evaluates them from the perspective of its organisational efficiency.

The operational activities of RTEs involve activities that are strictly related to the provisions of transport services, the main element of which is the implementation of transport production processes [52], which involve the use and productivity of human resources, as well as technical and other materials such as fuel/energy.

Hence, the fundamental efficiency requirement for RTEs and the effectiveness of the transport process itself is the efficiency of transport processes [53], which can be indicated with the use of various rates. The rates that are most frequently used in the evaluation of operational efficiency of RTEs include those of the technical readiness of the vehicle fleet, vehicle fleet exploitation, roadworthy vehicle exploitation, the average daily time of vehicle operation (the number of vehicle working hours per day), using the vehicle fleet operating time, the average loading capacity of vehicles in operation, the average operating speed, using the mileage, and using vehicle loading capacity or the rate of fuel consumption [54,55].

Managerial efficiency mainly involves evaluating the levels and efficiency of decisions made by the managerial staff that directly affect the operational efficiency of a particular RTE [56,57] in regard to using human resources, knowledge, and know-how, as well as the technical and operational exploitation of transport means. Consequently, the operational efficiency of an RTE affects the way in which it is perceived by its potential customers and its competitive position [58,59]. At this point, we should refer to such evaluation rates as the manager's decision-making competences and decision area, the time of the decision-making process, the scope of instrumental/systemic support (e.g., IT technologies), and the correctness of decisions (based on the evaluation systems and theory of decisions) [50].

Within the research aims of this paper, the most interesting concept is technical efficiency, which involves the efficiency of technology and its applied scale. The authors believe that the above-mentioned concept refers not only to strictly technical problems [60]—such as rates of vehicle quality, the technical condition of the vehicle fleet, operation issues, service life, vehicle technical inspections, vehicle inspection management, and compatibility with innovative transport technologies (e.g., bimodal transport) [61,62], including the effect of the scale of the applied technologies—but also to technology that supports the first component of operational efficiency, namely the managerial efficiency of RTEs. IT technologies that have been implemented (telematics systems of road transport) and the scale of their application (the integration level of the fields covered by the operational activities of RTEs), which determine the increase in the efficiency of decisions made in a particular company, directly refer to the notion of the technical efficiency of RTEs, although they are applied as technical support provided to the processes of managerial decision-making (in the field of managerial efficiency) [2,63,64].

Taking the current high complexity of transport processes [65–68] into account, it can be said that the efficient functioning of RTEs depends on the use of IT systems, including integrated IT systems (IITSs). For the requirements of the research, a taxonomy of IT systems is provided in Figure 1, with the consideration of the following division:

(a) Telematics systems—intelligent systems of road transport (ITS—intelligent transport systems) [69].
(b) Systems of transport, fleet, and transport operation management (TMS/FMS—transport/fleet management systems).

In order to achieve the main objective of the research and to properly verify the assumed hypothesis (general and partial, detailed for each stage of research), the field of the research was identified in detail, along with the required tools and expected results. The field of research includes road (freight) transport enterprises; the pragmatics of organisational efficiency, methods, and tools applied to increase the organisational efficiency in these enterprises in the field; and their operational activities, with consideration of the following parameters:

(a) The Polish road transport enterprises met the following parameters:

- They had over 9 employees classified to the H section of Transport and Warehousing, according to the Polish Classification of Business Activities (due to the assumption that a very small RTE might not implement IT services for drivers' working time management or may not implement any IT solutions).
- They had a fleet of more than 6 vehicles during the first stage of research (due to the assumption that a very small RTE might not implement IT services for fleet and order management).
- They had a fleet of more than 50 vehicles during the second stage of research (due to the assumption that the effect of the scale of IITS implementation will be assured).
- They had implemented IT/IITS systems.

- They provide transport services to the countries in Western Europe (Germany-bound as the most frequent direction) during the second stage of research (due to the assumption that the effect of the scale of IT systems for drivers' working time management implementation will be assured).

(b) The scope of the efficiency category: organisational efficiency (economic efficiency as a complementary field).

(c) Temporal: the years 2011–2016.

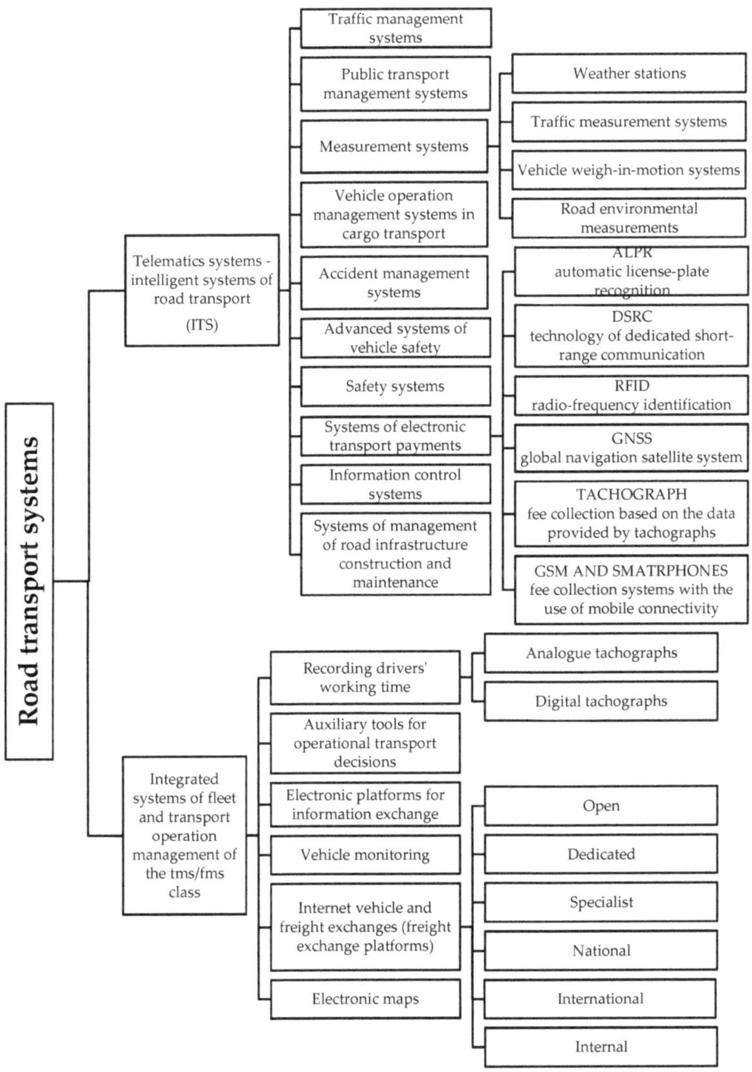

Figure 1. IT (information technology) systems supporting operation of road transport.

The main aim during the first stage of the empirical research was the identification of areas related to operational activities undertaken by RTEs that are the most sensitive to support provided by IT processes/systems (implemented in order to improve organisational efficiency). During this stage of research, the method of random choice was applied with the use of a random number generator. In reference to the terminology applied by the Central Statistical Office [70], the enterprises were selected in accordance with the following rules:

(a) The parameters of the analysed population were that the Polish RTEs had to have more than 9 employees and a fleet consisting of more than 6 vehicles (as presented in Table 1)—according to the data of 2014, the population included 3969 entities.
(b) For statistical research, we identified a population consisting of 500 RTEs at a confidence interval of 95%, a size of 0.5, and a maximum error at the level of 5% (the Infobrokering database system was used [71]).

Table 1. Enterprises (employing more than 9 drivers) operating in the sector of commercial road transport, as organised by the number of owned trucks/lorries and road tractor units.

Total	Enterprises with the Number of Lorries and Road Tractors					
	6 or Less	6–9	10–19	20–49	50–99	100 or More
			2014			
3969	276	673	1839	910	188	83
			2015			
3989	277	582	1860	981	191	98

Source: Transport drogowy w Polsce w latach 2014 i 2015, GUS, on-line version: http://stat.gov.pl/obszary-tematyczne/transport-i-lacznosc/transport/transport-drogowy-w-polsce-w-latach-2014-i-2015,6,4.html.

There were 356 answers provided via the questionnaire survey carried out with the CATI method; therefore, according to the sampling calculations, the condition of a representative sample was met (minimum 350 respondents) for the adopted assumptions.

The obtained results made it possible to identify the areas of operational activities undertaken by RTEs that are considered by the managerial staff of these entities as the most sensitive and the most important in terms of the support provided by IT systems to improve organisational efficiency. The results obtained during this stage of research are presented in Figure 2.

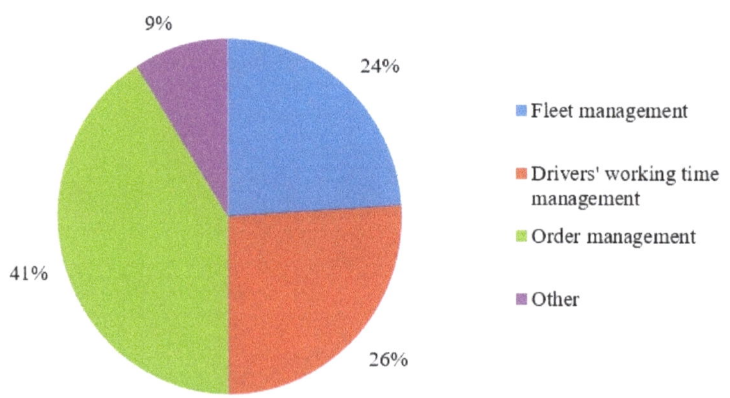

Figure 2. The fields of operational activities undertaken by the analysed enterprises that prove to be the most sensitive in terms of the support provided by IT systems.

While searching for an answer to the question how—in a holistic aspect (minimum 60% of operational activities)—an integrated IT system can affect organisational efficiency [72,73], an original matrix of rates was developed, as there have not been any other universal tools available so far.

The matrix refers to three key fields of operational activities undertaken by RTEs (fleet management [74], drivers' working time management [75], and transport order management). For each of the above-mentioned fields, three representative rates were identified, and these were included in the nine-element matrix (Figure 3). All rates included in matrix were chosen as being the most capable rates in terms of possible influence on RTE's operational efficiency, as well as being the most popular rates used in operational evaluation in RTEs by management (the most accessible rates calculated by all RTEs in the same way—convenient for any type of comparison).

		KEY FIELDS		
		Fleet management	Drivers' work management	Transport order management
KEY RATES AND MEASURES	Implemented in the integrated systems at the general level	Rate of technical readiness (A_t)	Using the driver's working time (T_{jk})	Number of daily orders per a forwarder (L_{zs})
	Implemented in the integrated systems at the operational level	Rate of using roadworthy vehicles (A_{gt})	Number of kilometres covered by one driver (K_{jp}) and by a team of drivers (K_{jz})	Number of kilometres per one order (L_{kmz})
	Implemented in the integrated systems at the specific level and/or in the dedicated systems	Rate of using the mileage of the vehicle fleet (B)	Vehicle loading time (T_{za}) and vehicle unloading time (T_{ro})	Average lead time (T_{pz})

Figure 3. Matrix of rates indicating the organisational efficiency of the operational activities undertaken by road transport enterprises (RTEs), as determined by the implementation of IT systems.

In the light of the above-mentioned findings, it was possible to formulate the main hypothesis of the stage 1 that the comprehensive use (over 60%) of IT systems in road transport enterprises improves the efficiency of their functioning by increase in their organisational efficiency in the field of their operational activities. The matrix of rates presented in the study can inform one regarding the scale of the influence exerted by IT systems on the efficiency of RTEs.

In conclusion, at this stage of research, an original matrix of the organisational efficiency rates of operational activities was developed as a necessary tool for research stage II.

3. The Course and the Results of the Research—Stage II (Rate Analysis)

The development of the rate matrix allowed the authors to reach the next stage of the empirical research aimed at the acquisition of more specific information related to the organisational efficiency of operational activities undertaken by RTEs in reference to the selected technical, technological, and economic parameters with the use of the individual questionnaire interview method (PAPI). The application of the PAPI method allowed the authors to collect quantitative and qualitative information about the analysed entities, and this information was indispensable for further research.

The main aim of the second stage of the empirical research was the evaluation of interdependencies and conditions related to the influence exerted by IT/IITS systems on the organisational efficiency of operational activities undertaken by RTEs. Based on the database used in the previous stage of research, a non-probability selection of the RTEs was conducted. From twenty-six selected RTEs (fulfilling research stage II conditions), only eight enterprises accepted an invitation to the next stage of research (for the requirements of the research and in order to anonymise as a prerequisite for participation in further research, they are labelled as A, B, C, D, E, F, G, and H), and they became the subjects of the focus research in the second part. The selected RTEs:

- Declared the implementation of the IT/IITS systems in the identified fields in 2013 (considering the fact that the implementation of the IT systems took place in various months, the data of 2013 were not included in the research because of their low usability for the assumed research aims), with a possibility of accessing the quantitative data of 2011–2012, which was the period before the implementation the discussed systems, and 2014–2016, which was the period after the implementation of these systems.
- Provided services of international (Germany-bound) transport; as such, the limitation to only one direction of transport services was justified when considering the possibility of comparing operational parameters observed in the same conditions—the choice of that destination came from the fact that almost every Polish RTE providing international transport services offers their services in that direction.

Based on the data obtained during the empirical research, a calculation of the selected organisational efficiency rates for RTE management before and after the implementation of IT/IITS systems was provided.

In the field of fleet management, the following rates were analysed:

(a) A_t—rate of technical readiness.
(b) A_{gt}—rate of using roadworthy vehicles.
(c) B—rate of using the mileage of the vehicle fleet.

The value of the technical readiness rate is highly significant for the operation of a road transport enterprise when considering its fundamental task, namely providing transport services. As seen in Figure 4, before the implementation of IT systems, the value of the discussed rate reported by four enterprises, A, B, D, and G, indicated a decreasing tendency, C and H indicated a slight increase, and E and F indicated a growing tendency.

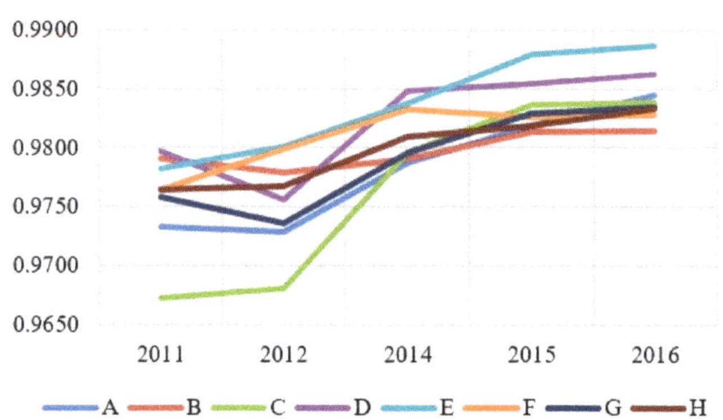

Figure 4. The average value of the technical readiness rate for the analysed road freight transport enterprises in the years 2011–2016.

The next rate, which defines the level of using roadworthy vehicles, was calculated based on the obtained data (Figure 5).

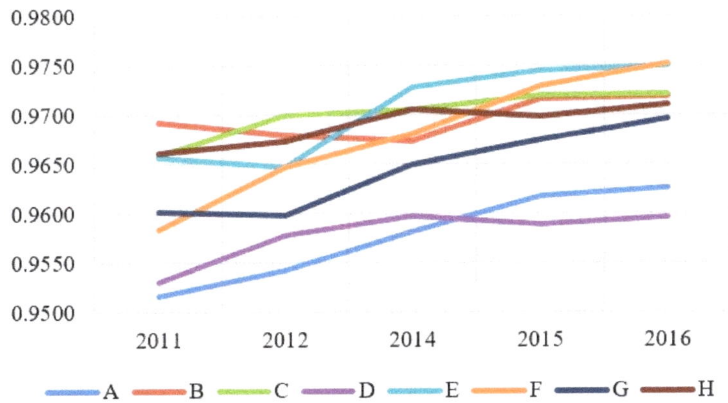

Figure 5. The average value of the using roadworthy vehicles rate for the analysed road freight transport enterprises in the years 2011–2016.

The using the mileage of the vehicle fleet rate is the relationship between the mileage covered by a vehicle with the cargo and the total mileage. However, the discussed distance must be augmented by a distance covered by the vehicle while driving from its depot base to the cargo loading point and a distance covered after the delivery of the cargo or a distance covered to reach another cargo loading point. It is possible to decrease the number of kilometres covered during empty running by the proper organisation of routes. This can be achieved with the use of relevant functions offered by IT systems. The values of the discussed rate in the analysed enterprises during the analysed period of time are presented in Figure 6.

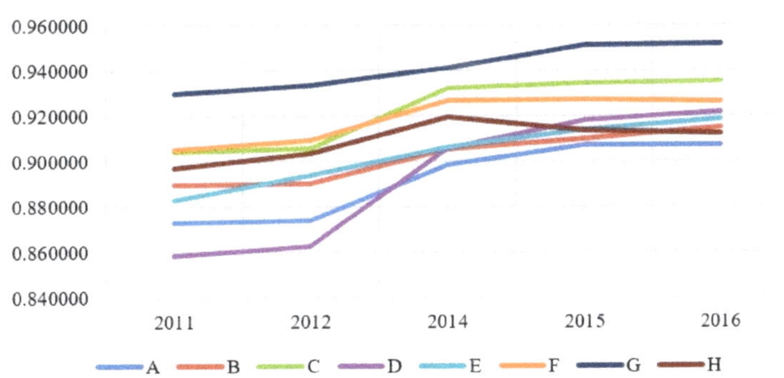

Figure 6. The average value of the using the mileage of the vehicle fleet rate for the analysed road freight transport enterprises in the years 2011–2016.

The next field that was analysed refers to the management of drivers' work. There are three representative rates in the discussed area:

(a) T_{jk}—using the driver's working time.
(b) K_{jz}—the number of kilometres covered by a team of drivers (two drivers).
(c) T_{zr}—the total time of vehicle loading and unloading.

After analysing the using the driver's working time rate (Figure 7), it was possible to state that its value has increased (by 5% on average) since IT system implementation in all the analysed cases.

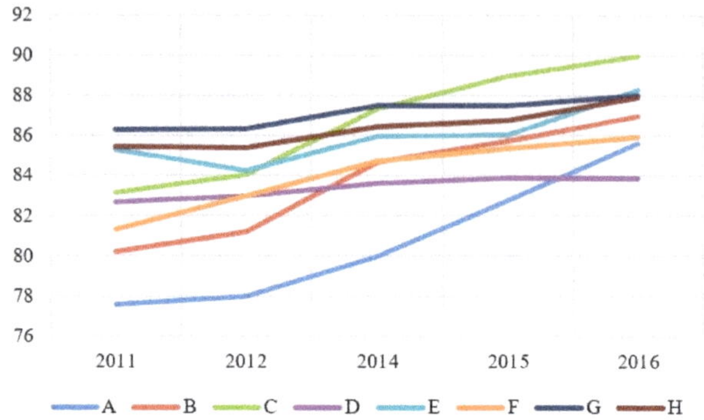

Figure 7. The average value of the using the driver's working time rate (%) for the analysed road freight transport enterprises in the years 2011–2016.

The next rate, which refers to the number of kilometres covered by a team of drivers (Figure 8), was increased by, on average, 15%.

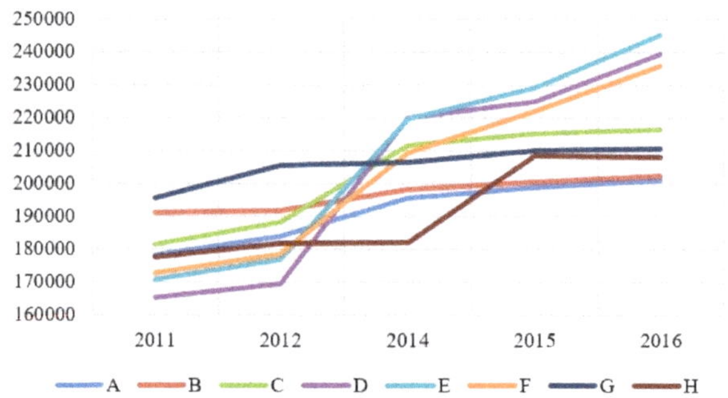

Figure 8. The average value of the annual number of kilometres covered by a team of drivers' rate (km) in Germany-bound transport for the analysed road freight transport enterprises in the years 2011–2016.

Between 2012 and 2016, the cargo loading time in the analysed enterprises was shortened by an average of 71 min. This indicated an average decrease of approximately 30% (Figure 9).

Figure 9. The average value of the vehicle loading time rate (minutes) in the analysed road freight transport enterprises in the years 2011–2016.

The next rate is related to the time of unloading a vehicle. It indicated a tendency similar to the previously discussed rate, namely that it has been considerably decreasing since 2014. In the time period between 2012 and 2016 in the analysed enterprises, the unloading time was shortened by an average of 55 min, or 30% (Figure 10).

Figure 10. The average value of the vehicle unloading time rate (minutes) in the analysed road freight transport enterprises in the years 2011–2016.

The last analysed field at this stage of research was transport order management. The following rates were analysed:

(a) L_{zs}—the number of daily orders per a forwarder.
(b) L_{kmz}—the number of kilometres per one order.
(c) T_{pz}—the average lead time.

In order to increase the scope of its operational activities, an enterprise needs a proper number of transport orders. Each forwarder must gain a certain number of orders to provide profitability to their enterprise. The rate of the average number of transport orders per a forwarder in the Germany-bound transport (Figure 11) increased by three transport orders, which is by 27%, on average in all the analysed enterprises since the implementation of the IT systems.

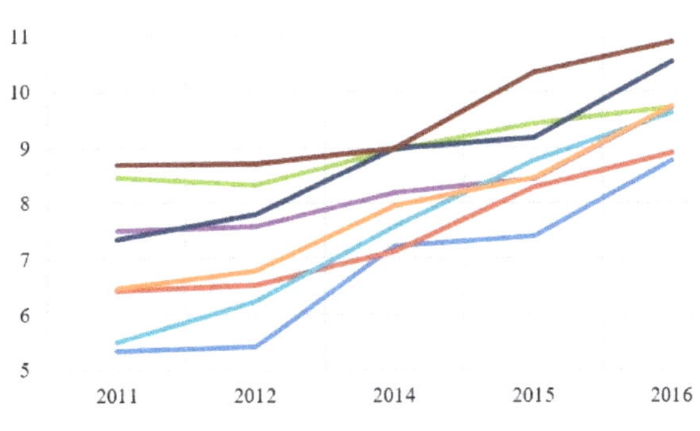

Figure 11. The average number of transport orders per day, received by a forwarder in the Germany-bound transport in the analysed road freight transport enterprises in the years 2011–2016.

The analysis of the number of kilometres per one transport order (Figure 12) indicated an average increase of 13% in the discussed enterprises.

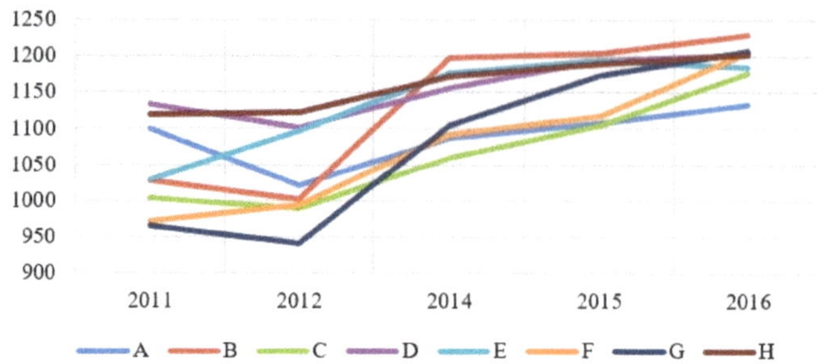

Figure 12. The average value of the rate of number of kilometres per one transport order in the Germany-bound transport in the analysed road freight transport enterprises in the years 2011–2016.

The time of making a decision about fulfilling an order in the analysed enterprises between 2012 and 2016 indicated an average decrease of 40% (Figure 13).

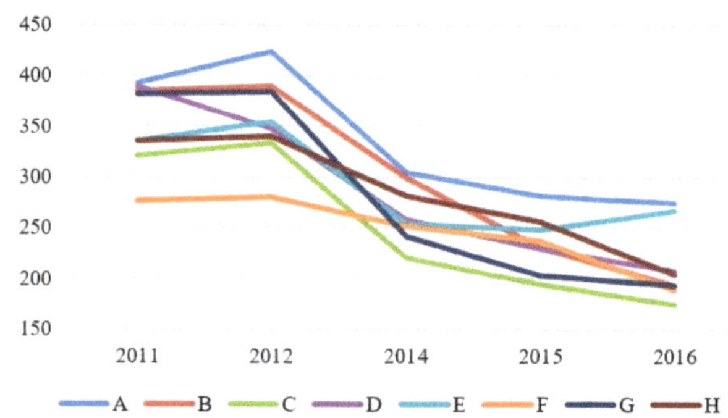

Figure 13. The average value of the rate indicating the time (minutes) required by a forwarder to make a decision about fulfilling an order in the Germany-bound transport in the analysed road freight transport enterprises in the years 2011–2016.

The results obtained during the first and the second stages of the research allowed the authors to provide an analysis with the use of an original matrix of rates that indicates the organisational efficiency of operational activities undertaken by RTEs, as determined by the implementation of IT systems. The matrix was applied as a universal tool to provide holistic and comparative analysis.

4. The Course and the Results of the Research—Stage III (Comparative Analysis with the Use of the Matrix)

During this stage of research, the results of the calculations provided during the second stage were collated in the form of an individual (ontogenetic) matrix for each of the analysed enterprises; the elements of the matrix are the calculated values of the parameters (rates and measures), and the values of the time rates (decrease in the time of the particular operations) are presented as absolute values. The main aim of this stage was to find an aggregated parameter of the comparative evaluation, which was a determinant of the individual rate matrix for each of the analysed enterprises (calculated with the use of Sarrus' rule [76]). The algorithm of the research process is presented in Figure 14.

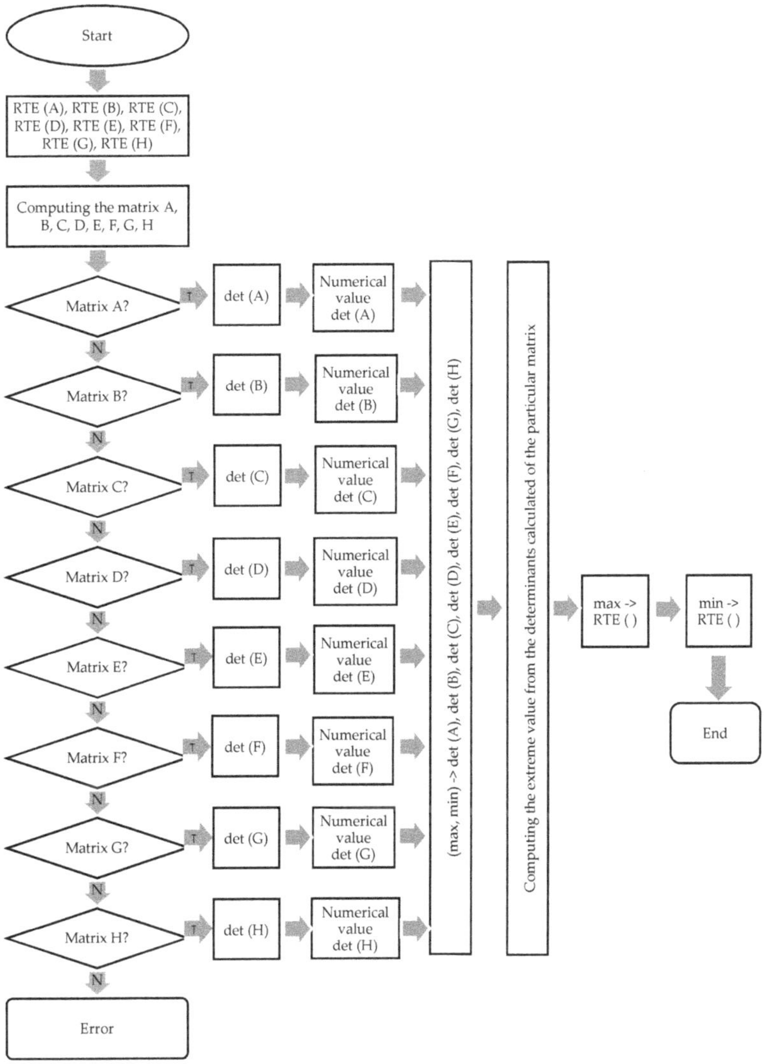

Figure 14. The algorithm for determining the researched RTEs with the highest and the lowest numerical values of the matrix determinants.

The research procedure for the development of individual (ontogenetic) matrices for each of the analysed RTEs was as follows:

- For each of the analysed enterprises, an ontogenetic matrix was developed in the following form:

$$A = \begin{bmatrix} A_t & T_{jk} & L_{zs} \\ A_{gt} & K_{jz} & L_{kmz} \\ B & T_{zr} & T_{pz} \end{bmatrix} \quad (1)$$

where A_t is the rate of technical readiness, A_{gt} is the rate of using roadworthy vehicles, B is the rate of using the mileage of the vehicle fleet, T_{jk} indicates using the driver's working time, K_{jz} is the number of kilometres covered by a team of drivers, T_{zr} is the time required for loading and unloading a vehicle, L_{zs} is the number of daily transport orders per a forwarder, L_{kmz} is the number of kilometres per one transport order, and T_{pz} is the average lead time.

- A matrix for RTE A in accordance with the values of rates and measures calculated in the previous stage of research was as follows:

$$\det(A)\begin{vmatrix} 0.0115 & 8 & 3,35 \\ 0.0084 & 17100 & 112 \\ 0.033426 & 42 & 149 \end{vmatrix} \tag{2}$$

- Calculation of the value of the determinant for the RTE A with the use of Sarrus' rule is:

$$\det(A) = 27{,}353.064366$$

- The above-mentioned steps were repeated to develop ontogenetic matrices for RTEs B–H, as follows:

$$\det(B) = \begin{vmatrix} 0.035 & 6 & 2,39 \\ 0.0039 & 10596 & 227 \\ 0.024460 & 52 & 197 \end{vmatrix} \tag{3}$$

$$\det(C) = \begin{vmatrix} 0.018 & 6 & 1,41 \\ 0.0022 & 28488 & 188 \\ 0.029678 & 72 & 159 \end{vmatrix} \tag{4}$$

$$\det(D) = \begin{vmatrix} 0.0106 & 1 & 2,14 \\ 0.0018 & 69900 & 102 \\ 0.058992 & 67 & 140 \end{vmatrix} \tag{5}$$

$$\det(E) = \begin{vmatrix} 0.0084 & 4 & 3,4 \\ 0.0103 & 68556 & 89 \\ 0.024543 & 71 & 87 \end{vmatrix} \tag{6}$$

$$\det(F) = \begin{vmatrix} 0.0029 & 3 & 2,97 \\ 0.0104 & 57444 & 211 \\ 0.016589 & 74 & 91 \end{vmatrix} \tag{7}$$

$$\det(G) = \begin{vmatrix} 0.0098 & 2 & 2,75 \\ 0.0098 & 4956 & 267 \\ 0.018282 & 60 & 191 \end{vmatrix} \tag{8}$$

$$\det(H) = \begin{vmatrix} 0.0066 & 3 & 2,18 \\ 0.0037 & 26676 & 81 \\ 0.008630 & 70 & 136 \end{vmatrix} \tag{9}$$

Finally, the determinants of the ontogenetic matrices for the analysed RTEs B-H, provided with the use of Sarrus' rule, have been calculated as follows:

- The value of the determinant for RTE B was 72,056.0336096.
- The value of the determinant for RTE C was 70,176.29874576.
- The value of the determinant for RTE D was 94,840.805556.
- The value of the determinant for RTE E was 45,526.1574408.
- The value of the determinant for RTE F was 12,293.91095648.
- The value of the determinant for RTE G was 8878.11541.
- The value of the determinant for RTE H was 23,406.2414516.

The obtained numerical values of the matrix determinants for the particular analysed RTEs (A–H) is presented in Figure 15 as an increasingly ordered sequence.

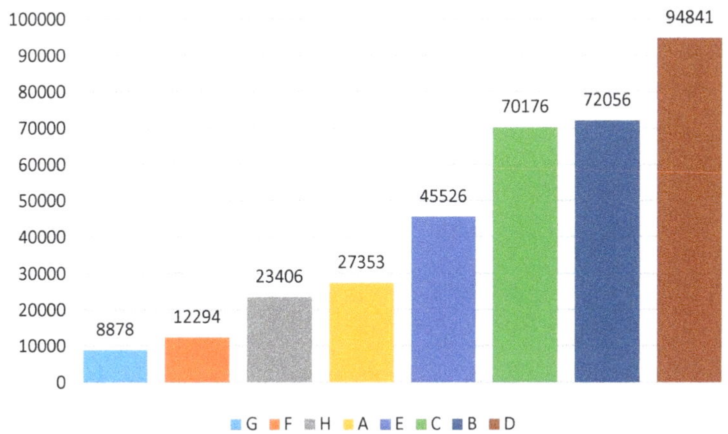

Figure 15. Increasingly ordered values of the ontogenetic matrix determinants calculated for the particular analysed RTEs (A–H).

The median [77,78] of the ontogenetic matrix determinants calculated for the analysed RTEs was 36,439.61. The determinants calculated for RTEs D, B, C, and E were above the median value, and the ontogenetic matrix determinants calculated for RTEs A, H, F, and G were below that value. This explicitly indicated that the level of the positive influence exerted by the implemented IT systems on the organisational efficiency of operational activities undertaken by the analysed enterprises was not the same.

The result achieved by RTE D (the highest level of the matrix determinant) was determined by:

- The highest increase in covered kilometres (which resulted in the broadest scope of the potential of positive influence exerted by the telematics systems in the field of mileage optimisation).
- The highest mileage covered by teams of drivers (two drivers in a team, which resulted in the largest scope of optimisation resulting from the systems implemented in the field of drivers' working time record and settlement or the possibility to form teams).
- A high increase in the number of days when the vehicles are roadworthy (resulting from the use of relevant systems applied to manage maintenance breaks and obligatory inspections; consequently, the possibility to accept more orders/to cover more kilometres was achieved).

The result achieved by RTE G (the lowest level of the matrix determinant) was determined by:

- The lowest mileage covered by teams of drivers (two drivers in a team, which resulted in the narrowest scope of optimisation resulting from the systems implemented in the field of drivers' working time record and settlement or the possibility to form teams).
- A very low level of using drivers' working time (which resulted in a narrow scope of system optimisation in the field of drivers' working time record and settlement, resulting in a decreased number of transport orders and covered kilometres).
- A low increase in the covered kilometres (which resulted in a low scope of the potential of positive influence exerted by the telematics systems implemented in the field of mileage optimisation).

5. Conclusions and Recapitulation

The theoretical considerations and the results of the empirical research presented above allowed the authors to positively verify all assumed research hypotheses.

According to the research findings, the originally developed matrix of rates indicating the organisational efficiency of operational activities, which were determined by the implementation of IT systems in RTEs, met the criteria defining the universal nature of the evaluation and also made it possible to provide a holistic assessment of the analysed RTEs.

The authors are aware of the fact that the operational efficiency of RTEs is affected by a number of factors (including those of an external nature that are independent of the RTEs themselves) [16,17]. However, during research, it was proved that understanding operational efficiency as a combination of managerial efficiency and technical efficiency (including the application of IT technologies to support the management of RTEs) and the purposeful selection of the most universal rates (in the opinion of managerial staff)—which comprise nine elements of the matrix—can adequately document the relationships between the selected components of the matrix, the integrated level of implementation of managerial telematics systems, and the increase of effectiveness in RTEs. In each analysed RTE, the matrix rate indicating organisational efficiency was found to significantly increase after the implementation of integrated IT systems. This proved the hypothesis that there is a correlation between those two elements.

Furthermore, the original method developed to assess the level and the scope of influence exerted by the implemented IT systems on the organisational efficiency of the operational activities undertaken by RTEs (a set of values of the ontogenetic matrix determinants), along with the analytical and algorithmic schemes of the procedures presented in the paper, is flexible enough to be applied to the requirements of comparative analysis in all types of RTEs.

The results of the research explicitly confirm the fact that the implementation of IT systems (especially when considered at the level of IITS, in a holistic way) can improve managerial efficiency in the field of operational activities undertaken by RTEs, and the fact that it can considerably and positively affect the results of operational activities, apart from the limits related to some restraining mechanisms. However, constant software updating—along with the regular replacement of hardware, permanent training provided to the staff working with the systems, and following customer demand and the needs of the evolving market—should also be remembered.

As far as the theoretical aspects are concerned, the authors believe that the results of their research present potential for further theoretical considerations regarding the influence exerted by IT/IITS on the organisational efficiency of RTEs. There is also some potential for theoretical considerations regarding relationships observed between the three variables of the selection of the matrix rates for the evaluation of the implementation results, the integrated level of implementing IT systems, and the accuracy and credibility of the evaluation provided in this way. From a practical point of view, the authors believe that the matrix presented in the paper has considerable potential to be broadly applied as a universal tool to examine and to compare the efficiency of integrated telematics systems in the operational activities undertaken by RTEs. In this way, the discussed matrix fills in a pragmatic gap in this field.

Considering the above-mentioned statements and the results of the previous available research (based on partial fields and selected systemic aspects), the authors believe that the main result of the research was the confirmed existence of a direct correlation between the implementation of integrated telematics systems and an increase in the organisational efficiency of RTEs. The authors believe that this paper also provides managerial staff with an innovative tool to measure the results of IT implementation in a more precise way based on comparable and generally available criteria (with a possibility of comparing particular RTEs). It can undoubtedly bring highly positive results for the process of managerial decision-making related to the purchase and implementation of integrated telematics systems for RTEs, as for the evaluation of managerial efficiency (through managerial decisions). The research of this paper will be continued because it is necessary to verify the matrix in a larger population, to verify a different configuration of the matrix elements, and to develop an application for algebraic operations. Further research will be also necessary to more precisely prove the scale of the influence exerted by IT systems, assuming interactions with other factors in determining efficiency of

RTEs (and how to measure it) and also assuming no interactions with other factors (with all conditions being considered equal).

The authors also believe that the more common and frequent (e.g., quarterly) calculation of matrix values can provide information from a dynamic perspective that should become a significant supporting element in the evaluation of managerial decision results. The fact that numerous matrix elements (rates) are treated as information of commercial sensitivity and the fact that managerial staff members may be unwilling to share data referring to the operational fields of their companies, even though such data are indispensable for calculation and comparison, come as serious limitations to the matrix discussed in the paper.

Considering direct conclusions drawn from the presented research, it is also possible to draw much more general (holistic) conclusions. The use of telematics systems in RTEs provides the opportunity to enrich management in the fields of economy, fuel/energy consumption, human resource productivity, the enlargement of vehicle/truck exploitation periods, and corporate environmental and social responsibility. Moreover, the results of the research may encourage RTEs to improve their technological potential (investment into telematics systems and support for managerial decisions), which should be translated into higher levels of technical/technological competence, higher levels of efficient management, and, eventually, increases in organisational efficiency/effectiveness. In the time of dynamic changes in the market, high uncertainty, the stochasticity of social and economic phenomena (e.g., the COVID-19 pandemic), stronger competitiveness in transport, low freight margins, possible changes in the role and significance of transport in logistic supply chains (in the oncoming post-COVID era), imperatively green and pro-ecological transport, and many other unpredictable factors, the value of winning of some market advantage by RTEs in the form of the organisational efficiency generated by new IT and telematics technologies cannot be underestimated.

Author Contributions: Data curation, M.J.K. and R.K.M.; formal analysis, M.J.K. and R.K.M.; methodology, R.K.M., A.K. and A.B.; project administration, M.J.K.; resources, R.K.M. and M.J.K.; writing—original draft, R.K.M. and M.J.K.; writing—review and editing, A.B. and A.K. All authors have read and agreed to the published version of the manuscript.

Funding: This research received no external funding.

Conflicts of Interest: The authors declare no conflict of interest.

Nomenclature

A_{gt}	the rate of using roadworthy vehicles
A_t	the rate of technical readiness
B	the rate of using the mileage of the vehicle fleet
det (A)	matrix A
K_{jz}	the number of kilometres covered by a team of drivers
L_{kmz}	number of kilometres per one transport order
L_{zs}	the number of daily transport orders per a forwarder
RTEs	road transport enterprises
T_{jk}	the driver's working time
T_{pz}	the average lead time
T_{zr}	the time required for loading and unloading a vehicle

References

1. Verdonck, L.; Caris, A.; Ramaekers, K.; Janssens, G. Collaborative Logistics from the Perspective of Road Transportation Companies. *Transp. Rev.* **2013**, *33*, 700–719. [CrossRef]
2. Radović, D.; Stević, Ž.; Pamučar, D.; Zavadskas, E.K.; Badi, I.; Antuchevičiene, J.; Turskis, Z. Measuring Performance in Transportation Companies in Developing Countries: A Novel Rough ARAS Model. *Symmetry* **2018**, *10*, 434. [CrossRef]
3. Hernandez, A.; Bernardo, M.; Cruz, C. Relating open innovation, innovation and management systems integration. *Ind. Manag. Data Syst.* **2016**, *116*, 1540–1556. [CrossRef]

4. Chatterjee, K. Modelling the impacts of transport telematics: Current limitations and future developments. *Transp. Rev.* **1999**, *19*, 57–80. [CrossRef]
5. Rosaldo, J.F.; Liu, R.R. An Agent-Based Approach to Assess Drivers' Interaction with Pre-Trip Information Systems. *J. Intellig. Transp. Syst.* **2005**, *9*, 1–10. [CrossRef]
6. Sussman, J.S. *Perspectives on Intelligent Transportation Systems (ITS)*; Springer: Boston, MA, USA, 2005; ISBN 978-0-387-23260-7.
7. Crainic, T.G.; Gendreau, M.; Potvin, J.Y. Intelligent freight-transportation systems: Assessment and the contribution of operations research. *Transp. Res. Part C Emerg. Technol.* **2009**, *17*, 541–557. [CrossRef]
8. Schlote, A.; Faizrahnemoon, M.; Radusch, I.; Crisostomi, E.; Häusler, F.; Shorten, R. A framework for real-time emissions trading in large-scale vehicle fleets. *IET Intell. Transp. Syst.* **2015**, *9*, 275–284. [CrossRef]
9. Joh, C.-H.; Lee, B.; Bin, M.; Arentze, T.; Timmermans, H. Exploring the use of travel information–identifying contextual market segmentation in Seoul, Korea. *J. Transp. Geogr.* **2011**, *19*, 1245–1251. [CrossRef]
10. Giannopoulos, G.A.; McDonald, M. Developments in transport telematics applications in Japan: Traffic management, freight and public transport. *Transp. Rev.* **1997**, *17*, 37–59. [CrossRef]
11. Xu, Y. Development of Transport Telematics in Europe. *GeoInformatica* **2000**, *4*, 179–200. [CrossRef]
12. Janecki, R.; Krawiec, S.; Sierpiński, G. Telematics and the transportation system's value. *IFAC Proc. Vol.* **2010**, *43*, 43–49. [CrossRef]
13. Taniguchi, E.; Yamada, T. Reliable Vehicle Routing and Scheduling with Time Windows Towards City Logistics. In *The Network Reliability of Transport*; Bell, M.G.H., Iida, Y., Eds.; Emerald Group Publishing Limited: Pergamon, Turkey, 2003; pp. 301–322. [CrossRef]
14. Spiekerman, K.; Neubauer, J. *European Accessibility and Peripherality: Concepts, Models and Indicators*; Nordregio Working Paper: Stockholm, Sweden, 2002; pp. 9–11. ISSN 1403-2511. Available online: https://www.diva-portal.org/smash/get/diva2:700463/FULLTEXT01.pdf (accessed on 28 July 2020).
15. Ye, N.; Wang, Z.; Zhang, Y.; Wang, R. A method of vehicle route prediction based on social network analysis. *J. Sens.* **2015**, *15*, 210298. [CrossRef]
16. Ning, Y.; Zhong-Qin, W.; Ru-Chuan, W.; Abdullah, A.H. Design of accurate vehicle location system using RFID. *Elektron. Elektrotechnika* **2013**, *19*, 105–110. [CrossRef]
17. Bäumler, I.; Kotzab, H. Intelligent Transport Systems for Road Freight Transport—An Overview. In *Dynamics in Logistics. Lecture Notes in Logistics*; Freitag, M., Kotzab, H., Pannek, J., Eds.; Springer: Berlin/Heidelberg, Germany, 2017. [CrossRef]
18. Coronado Mondragon, A.E.; Lalwani, C.S.; Coronado Mondragon, E.S.; Coronado Mondragon, C.E. Facilitating multimodal logistics and enabling information systems connectivity through wireless vehicular networks. *Eur. Transp. Res. Rev.* **2009**, *122*, 229–240. [CrossRef]
19. Quinet, E.; Vickerman, R. *Principles of Transport Economics*; Edward Elgar Publishing Ltd.: Cheltenham, UK, 2008; pp. 350–353, ISBN 9781840648652.
20. Giannopoulos, G.A. Towards a European ITS for freight transport and logistics: Results of current EU funded research and prospects for the future. *Eur. Transp. Res. Rev.* **2009**, *1*, 147–161. [CrossRef]
21. Kapsalis, V.; Fidas, C.; Hadellis, L.; Karavasilis, C.; Galetakis, M.; Katsenos, C. A Networking Platform for Real-Time Monitoring and Rule-Based Control of Transport Fleets and Transferred Goods. In Proceedings of the 13th International IEEE Conference on Intelligent Transportation Systems—(ITSC 2010), Funchal, Portugal, 19–22 September 2010; pp. 295–300. [CrossRef]
22. Kortüm, W.; Goodall, R.M.; Hedrick, J.K. Mechatronics in ground transportation-current trends and future possibilities. *Annu. Rev. Control* **1998**, *22*, 133–144. [CrossRef]
23. Gainnopouls, G.A. The application of information and communication technologies in transport. *Eur. J. Oper. Res.* **2004**, *152*, 302–320. [CrossRef]
24. Fouchal, H.; Bourdy, E.; Wilhelm, G.; Ayaida, M. A framework for validation of cooperative intelligent transport systems. In Proceedings of the 2016 IEEE Global Communications Conference (GLOBECOM), Washington, DC, USA, 4–8 December 2016; pp. 1–6. [CrossRef]
25. Fouchal, H.; Wilhelm, G.; Bourdy, E.; Wilhelm, G.; Ayaida, M. A testing framework for intelligent transport systems. In Proceedings of the 2016 IEEE Symposium on Computers and Communication (ISCC), Messina, Italy, 27–30 June 2016; pp. 180–184. [CrossRef]
26. Kolosz, B.; Grant-Muller, S. Sustainability assessment approaches for intelligent transport systems: The state of the art. *IET Intell. Transp. Syst.* **2016**, *10*, 287–297. [CrossRef]

27. Domingues, P.; Sampaio, P.; Arezes, P. Integrated management systems assessment: A maturity model proposal. *J. Clean. Prod.* **2016**, *124*, 164–174. [CrossRef]
28. Trierweiller, A.; Bornia, A.; Gisi, M.; Spenassato, D.; Severo-Peixe, B.; Rotta, M. An exploratory survey on the topic integrated management systems. *Braz. J. Oper. Prod. Manag.* **2016**, *13*, 184–193. [CrossRef]
29. Abad, J.; Dalmau, I.; Vilajosana, J. Taxonomic proposal for integration levels of management systems based on empirical evidence and derived corporate benefits. *J. Clean. Prod.* **2014**, *78*, 164–173. [CrossRef]
30. Rincon-Garcia, N.; Waterson, B.; Cherrett, T.J.; Salazar-Arrietac, F. A metaheuristic for the time-dependent vehicle routing problem considering driving hours regulations–Anapplication in city Logistics. *Transp. Res. Part A* **2020**, *137*, 429–446. [CrossRef]
31. Carlan, V.; Sys, C.; Vanelslander, T. Innovation in Road Freight Transport: Quantifying the Environmental Performance of Operational Cost-Reducing Practices. *Sustainability* **2019**, *11*, 2212. [CrossRef]
32. Gregson, N. Logistics at Work: Trucks, Containers and the Friction of Circulation in the UK. *Mobilities* **2017**, *12*, 343–364. [CrossRef]
33. Huang, S.T.; Bulut, E.; Duru, O. Service quality evaluation of international freight forwarders: An empirical research in East Asia. *J. Shipp. Trade* **2019**, *4*, 14. [CrossRef]
34. Baldini, G.; Sportiello, L.; Chiaramello, M.; Mahieu, V. Regulated applications for the road transportation infrastructure: The case study of the smart tachograph in the European Union. *Int. J. Crit. Infrast. Prot.* **2018**, *21*, 3–21. [CrossRef]
35. Borio, D.; Cano, E.; Baldini, G. Speed Consistency in the Smart Tachograph. *Sensors* **2018**, *18*, 1583. [CrossRef]
36. Steers, R.N. Problems in the measurement of organizational effectiveness. *Adm. Sci.* **1975**, *20*, 546–548. [CrossRef]
37. Farrell, M.J. The Measurement of Productive Efficiency. *J. R. Stat. Soc. Ser. A* **1957**, *120*, 253–290. [CrossRef]
38. Corkrey, R.; Parkinson, L.A. Comparison of four computer-based telephone interviewing methods: Getting answers to sensitive questions. *Behav. Res. Methods Instrum. Comput.* **2002**, *34*, 354–363. [CrossRef]
39. Bush, S.S.; Prather, L. Do electronic devices in face-to-face interviews change survey behavior? Evidence from a developing country. *Res. Politics* **2019**, *6*, 1–7. [CrossRef]
40. Vandenplas, C.; Beullens, K.; Loosveldt, G. Linking interview speed and interviewer effects on target variables in face-to-face surveys. *Surv. Res. Methods* **2019**, *13*, 249–265. [CrossRef]
41. Quinn, R.E.; Rohrbaugh, J. A Spatial Model of Effectiveness Criteria: Towards a Competing Values Approach to Organizational Analysis. *Manag. Sci.* **1983**, *29*, 273–393. [CrossRef]
42. Cameron, K.S. Effectiveness as Paradox: Consensus and Conflict in Conceptions of Organizational Effectiveness. *Manag. Sci.* **1986**, *32*, 513–644. [CrossRef]
43. McShane, S.L.; Von Glinow, M.A. *Organizational Behaviour*; Irwin/McGraw-Hill: Boston, MA, USA, 2000; ISBN1 1259562794, ISBN2 9781259562792.
44. Aktaş, E.; Çiçek, I.; Kıyak, M. The Effect of Organizational Culture on Organizational Efficiency: The Moderating Role of Organizational Environment and CEO Values. *Procedia-Soc. Behav. Sci.* **2011**, *24*, 1560–1573. [CrossRef]
45. Cameron, K.; Organizational Effectiveness. Wiley Encyclopedia of Management. 2015. Available online: https://doi.org/10.1002/9781118785317.weom110202 (accessed on 28 July 2020).
46. Venkataiah, P. Models of organizational effectiveness. *Osman. J. Manag.* **2006**, *2*, 1–7. Available online: http://ou-mba.ac.in/i/3.pdf (accessed on 28 July 2020).
47. Cameron, S.; Whetten, K.; David, A. Organizational effectiveness: A comparison of multiple Models. *J. Policy Anal. Manag.* **1983**, *3*, 477–478. [CrossRef]
48. Ostroff, C.; Schmitt, N. Configurations of organizational effectiveness and efficiency. *Acad. Manag. J.* **1993**, *36*, 1345–1361. [CrossRef]
49. Russell, R. Measures of technical efficiency. *J. Econ. Theory* **1985**, *35*, 109–126. [CrossRef]
50. Nyulásziová, M.; Palová, D. Implementing a decision support system in the transport process management of a small Slovak transport company. *J. Entrep. Manag. Innov.* **2020**, *16*, 75–105. [CrossRef]
51. Krasnyanskiy, M.; Penshin, N. Quality criteria when assessing competitiveness in road transport services. *Transp. Probl.* **2016**, *11*, 15–20. [CrossRef]
52. Grantner, J.; Bazuin, B.; Al-Shawawreh, J.; Castanier, M.P.; Hussain, S. Condition based maintenance for light trucks. In Proceedings of the 2010 IEEE International Conference on Systems, Man and Cybernetics, Istanbul, Turkey, 10–13 October 2010; pp. 336–342. [CrossRef]

53. Kozhukhovskaya, L.; Baskov, V.; Ignatov, A. Modular Management of Indicators of Efficiency and Safety of Transportation Processes. *Transp. Res. Procedia* **2017**, *20*, 361–366. [CrossRef]
54. Özen, M.; Fayyaz, M.; Tüydeş Yaman, H. Factors Affecting the Capacity Utilization of Road Freight Transport in Turkey. *Tek. Dergi* **2020**, *31*, 9813–9832. [CrossRef]
55. Kovács, G.Y. Development of performance evaluation software for road freight transport activity. *Pol. J. Manag. Stud.* **2017**, *15*, 121–134. [CrossRef]
56. Čejka, J.; Telecký, M.; Kolář, J. Appropriate Strategies of Transport Companies for More Efficient Management with the Aim of their Further Assessment Using the Operations Research Methods. *Naše More* **2016**, *63*, 98–101. [CrossRef]
57. Dubrovsky, V.; Yaroshevich, N.; Kuzmin, E. Transactional approach in assessment of operational performance of companies in transport infrastructure. *J. Ind. Eng. Manag.* **2016**, *9*, 389–412. [CrossRef]
58. Nakano, H. A Study on the Features of the Evolution Processes and Business Models of Global Enterprises in the Transport Sector. *Transp. Res. Procedia* **2017**, *25*, 3769–3788. [CrossRef]
59. Durišová, M.; Tokarcíková, E.; Virlanuta, F.O.; Chodasová, Z. The Corporate Performance Measurement and Its Importance for the Pricing in a Transport Enterprise. *Sustainability* **2019**, *11*, 6164. [CrossRef]
60. Tingting, Y.; Xuefeng, G.; Yuehui, Q.; Weiran, X.; Huayi, W. Efficiency Evaluation of Urban Road Transport and Land Use in Hunan Province of China Based on Hybrid Data Envelopment Analysis (DEA) Models. *Sustainability* **2019**, *11*, 3826. [CrossRef]
61. Kos, S.; Vukić, L.; Brčić, D. Comparison of External Costs in Multimodal Container Transport Chain. *Promet-Traffic Transp.* **2017**, *29*, 243–252. [CrossRef]
62. Guo, R.Y.; Szeto, W.Y.; Long, J. Trial-and-error operation schemes for bimodal transport systems, Transportation Research Part B. *Methodological* **2020**, *131*, 106–123. [CrossRef]
63. Bokovets, V.V.; Zamkova, N.L.; Makhnachova, N.M. Assessment of the Effectiveness of Enterprise Management Components in Modern Conditions. *Sci. Bull. Polissia* **2017**, *4*, 27–32. [CrossRef]
64. Kalinichenko, A.; Havrysh, V. Environmentally friendly fuel usage: Economic margin of feasibility. *Ecol. Chem. Eng. S* **2019**, *26*, 241–254. [CrossRef]
65. Malekian, R.; Kavishe, A.; Maharaj, B.T.; Gupta, P.; Singh, G.; Waschefort, H. Smart vehicle navigation system using hidden markov model and RFID technology. *Wirel. Pers. Commun.* **2016**, *90*, 1717–1742. [CrossRef]
66. Hannan, M.; Habib, S.; Javadi, M.; Samad, S.; Muad, A.; Hussain, A. Inter-vehicle wireless communications technologies, issues and challenges. *Inf. Technol. J.* **2013**, *12*, 558–568. [CrossRef]
67. Muerza, V.; Larrodé, E.; Moreno-Jiménez, J.M. Identification and selection of ICTs for freight transport in product service supply chain diversification. *Ind. Manag. Data Syst.* **2017**, *117*, 1469–1484. [CrossRef]
68. Rémy, G.; Mehar, S.; Sophy, T.; Senouci, S.; Jan, F.; Gourhant, Y. Green fleet management architecture: Application to economic itinerary planning. In Proceedings of the 2012 IEEE Globecom Workshops, Anaheim, CA, USA, 3–7 December 2012; pp. 369–373. [CrossRef]
69. McDonald, M.; Keller, H.; Klijnhout, J.; Mauro, V. *Intelligent Transport Systems in Europe: Opportunities for Future Research*; World Scientific Publishing Co. Pte. Ltd.: Singapore, 2006; ISBN 978-981-270-082-7.
70. Transport Drogowy w Polsce w Latach 2014 i 2015. GUS. Available online: http://stat.gov.pl/obszary-tematyczne/transport-i-lacznosc/transport/transport-drogowy-w-polsce-w-latach-2014-i-2015,6,4.html (accessed on 4 September 2019).
71. InfoBrokering. Bazy Danych Polskich Firm od InfoBrokering–Jakosc Zbudowana na Doswiadczeniu. Available online: https://www.infobrokering.com.pl/baza-firm-polskich (accessed on 14 October 2019).
72. Jimenez, G.; Novoa, L.; Ramos, L.; Martinez, J.; Alvarino, C. Diagnosis of initial conditions for the implementation of the integrated management system in the companies of the land cargo transportation in the City of Barranquilla (Colombia). In *HCI, CCIS*; Stephanidis, C., Ed.; Springer: Cham, Germany, 2018; Volume 852, pp. 282–289. [CrossRef]
73. Kopishynska, O.; Utkin, Y.; Kalinichenko, A.; Jelonek, D. Efficacy of the cloud computing technology in the management of communication and business processes of the companies. *Pol. J. Manag. Stud.* **2016**, *14*, 104–114. [CrossRef]
74. Stefansson, G.; Lumsden, K. Performance issues of Smart Transportation Management systems. *Int. J. Prod. Perform. Manag.* **2009**, *58*, 55–70. [CrossRef]
75. Hallé, S.; Chaib-draa, B. A collaborative driving system based on multiagent modelling and simulations. *J. Transp. Res. C Emerg. Technol.* **2005**, *13*, 320–345. [CrossRef]

76. Sobamowo, M.G. On the Extension of Sarrus' Rule to n × n (n > 3) Matrices: Development of New Method for the Computation of the Determinant of 4 × 4 Matrix. *Int. J. Eng. Math.* **2016**, 9382739. [CrossRef]
77. Brownrigg, D.R.K. The weighted median filter. *Commun. ACM* **1984**, *27*, 807–818. [CrossRef]
78. Tukey, J.W. *Exploratory Data Analysis*; Pearson: Lebanon, Indiana, USA, 1977; ISBN1 978-0201076165, ISBN2 0201076160.

energies

MDPI

Article

Improved Particle Swarm Optimization for Sea Surface Temperature Prediction

Qi He [1], Cheng Zha [1], Wei Song [1,*], Zengzhou Hao [2], Yanling Du [1], Antonio Liotta [3] and Cristian Perra [4]

1 Department of Information Technology, Shanghai Ocean University, Shanghai 201306, China; qihe@shou.edu.cn (Q.H.); m170550834@st.shou.edu.cn (C.Z.); yldu@shou.edu.cn (Y.D.)
2 State Key Laboratory of Satellite Ocean Environment Dynamics, Second Institute of Oceanography, Ministry of Natural Resources, Hangzhou 310012, China; hzyx80@sio.org.cn
3 School of Computing, Edinburgh Napier University, Edinburgh EH10 5DT, UK; a.Liotta@napier.ac.uk
4 Department of Electrical and Electronic Engineering, University of Cagliari, Via Marengo, 2, 09100 Cagliari, Italy; cperra@ieee.org
* Correspondence: wsong@shou.edu.cn

Received: 25 February 2020; Accepted: 13 March 2020; Published: 15 March 2020

Abstract: The Sea Surface Temperature (SST) is one of the key factors affecting ocean climate change. Hence, Sea Surface Temperature Prediction (SSTP) is of great significance to the study of navigation and meteorology. However, SST data is well-known to suffer from high levels of redundant information, which makes it very difficult to realize accurate predictions, for instance when using time-series regression. This paper constructs a simple yet effective SSTP model, dubbed DSL (given its origination from methods known as DTW, SVM and LSPSO). DSL is based on time-series similarity measure, multiple pattern learning and parameter optimization. It consists of three parts: (1) using Dynamic Time Warping (DTW) to mine the similarities in historical SST series; (2) training a Support Vector Machine (SVM) using the top-k similar patterns, deriving a robust SSTP model that offers a 5-day prediction window based on multiple SST input sequences; and (3) developing an improved Particle Swarm Optimization (PSO) method, dubbed LSPSO, which uses a local search strategy to achieve the combined requirement of prediction accuracy and efficiency. Our method strives for optimal model parameters (pattern length and interval step) and is suited for long-term series, leading to significant improvements in SST trend predictions. Our experimental validation shows a 16.7% reduction in prediction error, at a 76% gain in operating efficiency. We also achieve a significant improvement in prediction accuracy of non-stationary SST time series, compared to DTW, SVM, DS (i.e., DTW + SVM), and a recent deep learning method dubbed Long-Short Term Memory (LSTM).

Keywords: sea surface temperature; sea surface temperature prediction; similarity measure; support vector machine; particle swarm optimization; local search

1. Introduction

Sea Surface Temperature (SST) is one of the crucial factors affecting the ocean climate. The occurrence of events such as El Niño, storm surges and red tides are closely related to SST. Therefore, in recent years, Sea Surface Temperature Prediction (SSTP) has attracted more and more attention in various marine-related fields such as marine meteorology, navigation, marine disaster prevention and mitigation, and marine fisheries. So far, researchers across the world have proposed many methods to predict SST, which can be divided into three categories: statistical forecasting [1], numerical forecasting [2], and empirical forecasting [3].

SST data are long-term data sequences and typically involve large data volumes. Many scholars regard SSTP as a time-series regression problem, and derive prediction models by fitting the curve of historical data [4]. However, SST data is well-known to suffer from high levels of redundant information, which makes it very difficult to realize accurate predictions through time-series regression, nor to capture the complex dynamics of SST trends. In 2013, Lins et al., provided a prominent study showing how classic Support Vector Machine (SVM) could be used to predict SST in terms of raw data, slope and curvature [5]. Yet, they showed that SVM exhibited very similar performance when using raw data and curvature features, and was more effective than using slope. Better prediction accuracy was achieved with the more recent Long Short-Term Memory (LSTM) method [6]. This type of neural network has been widely used in diverse areas, thanks to its suitability for processing time series. In fact, Zhang et al. [7] were the first to use LSTM to predict SST. First, the SST sequence features were learned by LSTM layers. Then, a fully-connected layer was used to map the output of the LSTM layers to the final prediction result. Both SVM and LSTM tackle prediction accuracy. Yet, neither method is sufficiently efficient. Even more critically, marine operational forecasting requires significant improvements in SST trend predictions. Typically, empirical methods are used, of which the Analog Complexing (AC) algorithm [8] is a representative example. Nevertheless, AC has not shown sufficient accuracy in practical applications [9].

In this paper, our aim is to propose a new SSTP method, which not only can predict SST rapidly and accurately, but also can better model the SST trend. The main idea of our method is to use similarity measure to mine historical SSTs in order to extract sequences having similar trends, and then feed these into a suitable prediction method to infer future SSTs.

When using this method to perform SSTP, we need to consider two important issues. First, time-series mining is critically sensitive to the accuracy of the similarity measure method; thus, any unexpected error in the mined data will dominate the prediction result. We solve this problem by a combination of similarity measure and multiple time-series regression.

Another criticality is introduced by the choice the regression model parameters, which have a dramatic impact on the performance of the prediction model. We tackled this issue as a bi-objective optimization problem. In fact, our solution is dubbed DSL, since it is evolved from three building blocks: Dynamic Time Warping (DTW) [10], SVM [11], and Local-search enabled parameters optimization [12]. It consists of three parts:

1. Using Dynamic Time Warping (DTW) to mine the similarities in historical SST series. DTW has been chosen as the result of an experimental, comparative analysis of three representative time-series SST similarity methods. It led to the highest prediction accuracy, it better modeled the SST trends, and was found to be suitable to mining SST long-term time series.
2. Training a Support Vector Machine (SVM) using the top-k similar patterns, deriving a robust SSTP model that offers a 5-day prediction window based on multiple SST input sequences. Learning from multiple time-series sequences was instrumental to facilitating consistency enhancement and noise cancellation, thus achieving high prediction accuracy.
3. Developing an improved Particle Swarm Optimization (PSO) method, dubbed LSPSO, which uses a local search strategy to achieve the combined requirement of prediction accuracy and efficiency. We were striving for optimal model parameters, to pursue SST prediction efficiency, providing a new way for marine operational forecasting.

Overall, our work provides a new SSTP method that not only improves prediction accuracy and speed, but also better models the trend of SST changes. An important element was to find a suitable method to optimize the parameters of the combined DTW + SVM prediction method (dubbed DS hereafter). We achieved this goal by developing an improved Particle Swarm Optimization (PSO) method, which was responsible for a 16.7% improvement in prediction accuracy, in terms of reducing the Root Mean Square Error (RMSE), and for a 76% reduction in prediction time. We also achieved a significant improvement in prediction accuracy of non-stationary SST time series, compared to DTW, SVM, DS (i.e., DTW + SVM), and a recent deep learning method dubbed LSTM (Long-Short Term Memory).

The rest of this paper is organized as follows. In Section 2, we describe related work on SSTP, similarity measures, and optimization methods. The main idea of DSL is introduced in Section 3. Experimental results and performance evaluation are shown in Section 4. Section 5 presents our conclusions.

2. Related Work

SST data is typically long-time-sequence data, hence many researchers have regarded SSTP as a time-series regression problem, thus applying time-series prediction methods to SSTP. The traditional time-series prediction methods such as Autoregressive (AR) [13], Moving Average (MA) [14] and Autoregressive Moving Average (ARMA) [15] are linear. Yet, SST has non-stationary and nonlinear characteristics, thus these linear methods are not well-suited to the practical application of SSTP.

Therefore, researchers have proposed some nonlinear methods to predict SST. Li et al. [16] have used SVM to predict SST and achieved good results. AC is often used for hydrological forecasting, and can also be used for time-series prediction. It can better mine the hidden information in the sequence, yet the method is sensitive to factors such as Pattern Length (PL), Interval Step (IS), and similarity measure.

In the process of time-series data mining, the similarity measure is the basis of clustering, association rules and prediction. Euclidean distance [17] is a commonly used time-series similarity measure. It is simple to calculate, but can only process time series of equal length. It cannot handle the sequence stretching and bending on the time axis.

DTW is based on the idea of Dynamic Programming (DP) [18], which was originally applied in isolate word speech recognition [19]. It was then introduced into the study of similarity measures of time series by Berndt et al. [20], achieving good results. DTW overcomes the shortcomings of Euclidean distance. It will not only measure the similarity of time series of different lengths, but also supports the stretching and bending of sequences on the time axis. It has a good measurement accuracy and robustness, which is why it is widely used.

Cosine similarity [21] uses the cosine of the angles of two vectors in the vector space to measure the similarity between them. The closer the cosine value is to 1, the closer the angle is to 0, the more similar they are. Therefore, the method can better distinguish the difference in the direction of differentiation, and is not sensitive to the cosine value.

By comparing the evaluation index of the different similarity measures, one that is suitable for measuring the similarity of SST sequences can be selected, and the multi-objective optimization method can be used as a solution to select the appropriate values for PL and IS.

Many scholars have studied the theory and application of multi-optimization algorithms. Differential Evolution (DE) [22] is a simple but efficient parallel search algorithm proposed by R. Storn in 1997, whose principle is similar to genetic algorithms. Zitzler et al. [23] compared the Non-dominated Sorting Genetic Algorithm (NSGA) [24] with the Niche Pareto Genetic Algorithm (NPGA) [25], and the Vector-Evaluated Genetic Algorithm (VEGA) [26], finding that NSGA has the best performance. Yet, the computational complexity of NSGA is higher.

Aiming at the shortcomings of NSGA, Deb et al. [27] proposed the Elitist Non-dominated Sorting Genetic Algorithm (NSGA-II), which reduces the time complexity of Pareto dominating sorting [28]. It is suitable for dealing with low-dimensional, multi-objective optimization problems. However, when dealing with high-dimensional, multi-objective problems, the crowding distance [29] is not applicable in high-dimensional space, and the computational complexity is also high.

Inspired by the predatory behavior of natural bird populations, Kennedy [30] proposed Particle Swarm Optimization (PSO). PSO is an evolutionary computation method based on individual improvement, population cooperation and competition mechanisms. It has the important characteristics of being simple and easy to implement, which makes it suited to single-objective optimization. When optimizing multiple targets (as in MOPSO [31]), M. Reyes et al. proposed OMOPSO (Optimal Multi-Objective Particle Swarm Optimization) [32], which uses the crowding factor to select the leaders, based on Pareto dominance. Other researchers proposed hybrid optimization strategies that combine two or more heuristic optimization techniques [33,34], which could integrate the merits of different algorithms or achieve a near optimal solution quickly.

Considering that DS will fall into local optimum when optimizing the SSTP model, we propose LSPSO (Particle Swarm Optimization algorithm combined with Local Search strategy), which uses the Pareto dominance relationship to measure the advantages and disadvantages of the solution, and uses local search for non-dominated solution sets to enhance the local search ability. LSPSO has a strong ability to explore and can approach the true Pareto frontier.

3. The Proposed DSL Method

Our method involves three main algorithms, DTW, SVM, and LSPSO, and we have dubbed it DSL for short. Given an SST sequence $F = F_1, F_2, \ldots, F_{|F|}$ (whereby $|F|$ is the length of the sequence F), our aim is to predict the SST for the subsequent five days. Figure 1 shows the flow chart of our SSTP algorithm based on the similarity measure.

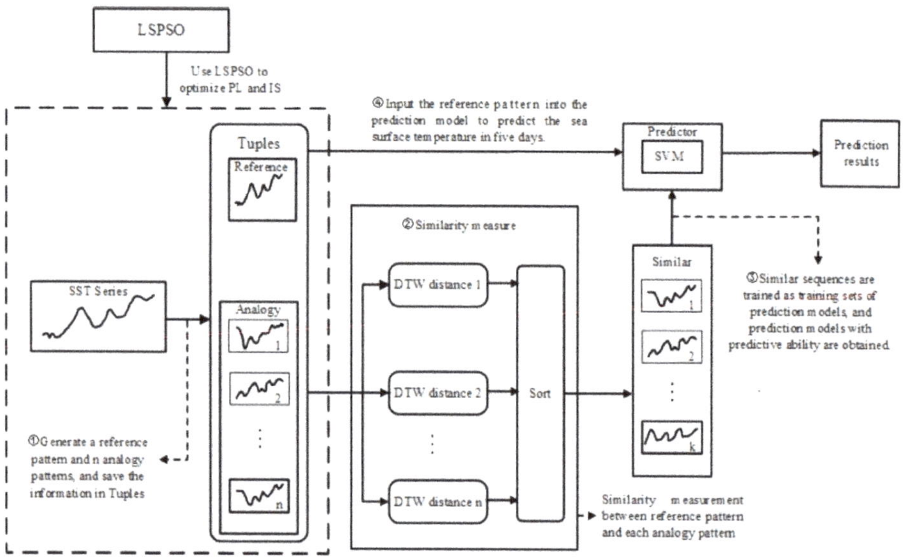

Figure 1. Flow chart of SSTP algorithm based on similarity measure.

The main steps of DSL are as follows: (1) Read the SST sequence F, generating a reference pattern and n analog patterns, and storing the information in the tuples T. (2) Calculate the DTW distances of the reference pattern and each analog pattern separately, and sort them in ascending order according to the DTW distance; the first k modes can be regarded as analog patterns similar to the reference pattern. (3) The obtained analog patterns are used as the input of the SVM model, and the corresponding SST of the following five days in the sequence F are used as the output of the SVM model; an SVM model with predictive ability is obtained through training. (4) The reference pattern is used as the input to the SVM model, generating a 5-day prediction of the SST. Materials and Methods should be described with sufficient details to allow others to replicate and build on published results. Please note that publication of your manuscript implicates that you must make all materials, data, computer code, and protocols associated with the publication available to readers. Please disclose at the submission stage any restrictions on the availability of materials or information. New methods and protocols should be described in detail while well-established methods can be briefly described and appropriately cited.

The reference pattern represents the current SST trend, while the analog pattern is the historical SST trend. In the process of generating the reference and the analog patterns, the PL and IS are uncertain; so before using DSL to predict SST, it is necessary to use LSPSO to optimize the model parameters and obtain the appropriate PL and IS.

3.1. Generation and Trend Prediction

In order to mine the similar trend to the current SST from the historical SST, we need to generate the reference and the analog patterns. The reference pattern represents the recent trend, whereas the analog pattern represents the historical trend. Based on the persistence and similarity of SST, we can calculate the similarity between reference pattern and analog pattern to mine the historical trends that are close to the current trends, and then use historical trends to predict current trends.

Algorithm 1 specifies the process of generating the reference and the analog patterns. Step 2 takes out the last five days of *F* for evaluating the prediction model; step 3 generates the reference pattern C; steps 4–9 generate the analog pattern A and save the reference and the analog patterns in the tuples T.

Algorithm 1. Generating the reference pattern C and the analog pattern A.

```
Input:
SST sequence: F; PL: m; IS: step;
Output:
      Tuples: T;
1:  T←Φ; A←Φ; C←Φ; D←Φ; Q←Φ;
2:  Take the last five days of F as the true value, and remove the last five days of F to obtain the sequence D;
3:  Take the SST of the last m days of D as the reference pattern C, and remove the SST of the m days to obtain
the sequence Q;
4:  t = 1; // t record the starting position of each analog pattern
5:  while ((t + m − 1)<(|F| − m − 5)) do
6:        Take the SST of the sequence Q from t to t + m - 1 as the analog pattern A;
7:        Save the generated reference and analog modes in the tuples T;
8:        t = t + step;
9:  End While
```

The similarity of the SST sequence is measured by the DTW distance, which can be performed in two steps: firstly, calculate the distance matrix by calculating the distance between the SST points of A and C; then, find an optimal path in the matching distance matrix. $C = c_1, c_2, \ldots, c_m$ and $A = a_1, a_2, \ldots, a_m$ define a matrix of m rows and m columns, with elements $d(a_i, c_i) = |a_i - c_i|$. W is the distance matrix between A and C, and can be expressed as per Equation (1).

$$W = \begin{bmatrix} d(a_1, c_1) & \cdots & d(a_1, c_m) \\ d(a_2, c_1) & \cdots & d(a_2, c_m) \\ \vdots & \ddots & \vdots \\ d(a_m, c_1) & \cdots & d(a_m, c_m) \end{bmatrix} \quad (1)$$

In the distance matrix W, the similarity relationship between A and C is represented by a set of consecutive matrix elements, which is referred to as a curved path L, $L = (w_1, w_2, \ldots w_K)$, where $m \leq K \leq 2m - 1$.

Path selection needs to meet the following constraints: (1) The start and end points of A and C are aligned, whereby the starting point of the path is (a_1, c_1), and the ending point is (a_m, c_m). (2) Any point on the path can only move along the adjacent elements of the matrix each time. That is, if $w_1 = (a_1, c_1)$ then next point $w_2 = (a, c)$ for the path, to satisfy $a - a_1 \leq 1$ and $c - c_1 \leq 1$. (3) Any point on the path can only move one way along the time axis each time. That is, if $w_1 = (a_1, c_1)$ then next point $w_2 = (a, c)$ for the path to satisfy $a - a_1 \geq 0$ and $c - c_1 \geq 1$. The shortest path distance that satisfies the above constraints is the DTW distance, which can be expressed by Equation (2).

$$DTW(\mathrm{A},\mathrm{C}) = \min\left\{\sum_{k=1}^{K} w_k\right\} \tag{2}$$

Through the above steps, the DTW distance between each analog pattern and the reference pattern can be calculated. The smaller the DTW distance, the more similar the two patterns are. So far, we have obtained analog patterns similar to reference patterns from F, showing how to use these analog patterns to predict the SST for the next five days. AC is calculating the average of the SST for the next five days that corresponds to these analog patterns in F as the SST prediction value for the next five days.

Yet, this method does not make full use of the information obtained through data mining, which leads to low prediction accuracy. In response to this problem, this paper uses these analog patterns to establish prediction modes. Because these analog patterns have a small sample size, they can only be modeled using small sample prediction methods to maximize information utilization. Although Back Propagation (BP) neural networks [35] can be used for small-sample prediction, model training is difficult and prone to failing during the training process. SVM is suitable for processing small samples and nonlinear problems, and has good robustness and high prediction accuracy. Considering that SST has nonlinear features, this paper combines SVM with DTW to construct a predictive model. In this paper, the analog patterns similar to the reference patterns are the input of the SVM, and the SST of the next five days corresponding to these analog patterns in F is used as the output of the SVM. Through training, we obtain a predictive SVM model and then use the reference model as the model input to get the SST for the next five days.

The training samples of the SVM are $(\mathrm{X}_i, \mathrm{Y}_i)$, $i = 1, 2, \ldots, k$, where $\mathrm{X}_i \in R^m$ is the input, $\mathrm{Y}_i \in R^n$ is the output, and m and n are the dimensions of the variable, and k is the number of samples. The goal of the SVM is to build the following regression function:

$$f(\mathrm{X}) = W\varphi(\mathrm{X}) + b = \begin{bmatrix} W_1\varphi(\mathrm{X}) + b_1 \\ \vdots \\ W_n\varphi(\mathrm{X}) + b_n \end{bmatrix} \tag{3}$$

$\varphi(\mathrm{X})$ can map data from the original space to the high-dimensional space, which can transform the nonlinear problem in the original space into a problem that can be solved in high-dimensional space. W and b are weights and offsets, respectively. To determine W and b we proceed with the minimization (4).

$$\min \frac{1}{2}\sum_{j=1}^{n} \|W\|^2 + \mathrm{C}\sum_{j=1}^{n}\sum_{i=1}^{k} L_j(f_j(\mathrm{X}_i), \mathrm{Y}_i^j) \tag{4}$$

where C is the balance factor,$L_j()$ is the loss function, and j is the dimension of the output variable. The loss function is usually defined as follows:

$$L_j(f_j(\mathrm{X}_i), \mathrm{Y}_i^j) = \begin{cases} 0, f_j(\mathrm{X}_i) - \mathrm{Y}_i^j < \varepsilon_j, \varepsilon_j > 0, \\ f_j(\mathrm{X}_i) - \mathrm{Y}_i^j - \varepsilon_j, otherwise. \end{cases} \tag{5}$$

The multivariate nonlinear regression SVM model can be finally expressed by the Lagrangian method as:

$$f_j(\mathrm{X}) = \sum_{i=1}^{k} (a_i^j - a_i^{j*})K(\mathrm{X}_i, \mathrm{X}) + b_j \tag{6}$$

where a_i^j, a_i^{j*} are Lagrangian multipliers, and $K(X_i, X)$ is a kernel function. RBF is the most widely used kernel function, which not only can implement nonlinear mapping, but also has fewer parameters. Therefore, this paper uses RBF as the kernel function of SVM. By training the above model, an SVM with predictive ability can be obtained, and then, inputting the reference pattern into the model, we can predict the SST for the next five days.

Although SSTP can be implemented in this way, the PL and IS may dramatically affect the accuracy and efficiency of the DS. In order to determine the appropriate PL and IS, this paper proposes LSPSO and applies it to model parameter optimization.

3.2. Parameter Optimization Using LSPSO

The basic operations of a standard PSO are particle speed and position updates:

$$v_{i,d}^{k+1} = wv_{i,d}^{k} + c_1 r_1 (p_{Pbest,d}^{k} - x_{i,d}^{k}) + c_2 r_2 (p_{Gbest,d}^{k} - x_{i,d}^{k}) \tag{7}$$

$$x_{i,d}^{k+1} = x_{i,d}^{k} + v_{i,d}^{k} \tag{8}$$

where w is the inertia weight, c_1, c_2 are learning factors, r_1, r_2 are mutually independent random numbers in the interval [0, 1], k is the number of evolutions, d is the search space dimension. Pbest is the optimal location for each particle search, and Gbest is the optimal location for the entire particle swarm search.

When PSO optimizes the parameters of the SST model, the search ability is weak in the late search, and it is easy to fall into local optimum, which is not conducive to optimize the model parameter. Therefore, we proposed LSPSO, which differs from the PSO as summarized below:

(1) PSO adopts a random method in population initialization, which leads to each particle appearing in a random distribution state in space, lacking the guidance of prior knowledge, which is not conducive to the particle close to the optimal solution. In this paper, the Beta distribution strategy is used to initialize the population, which is beneficial to the rapid formation of the surrounding situation by the particles to the optimal solution.

(2) The global and local search capabilities of PSO are mutually constrained and tend to fall into local optimum in the later stages of search. This paper uses local search strategy to enhance the local search ability of PSO, so that the improved PSO has independent global and local search capabilities.

(3) Particles tend to cross out of bounds during flight. Particles flying faster than the bound range will mutate; flying speeds that exceed half of the constraint range and below the constraint range decelerate to prevent particles from crossing the boundary.

When LSPSO optimizes the parameters of DS, it can be expressed as follows:

$$\begin{gathered} \min \text{RMSE} = G_1(PL, IS) \\ \min \text{RT} = G_2(PL, IS) \\ s.t \\ PS = \{x | 1 \le x \le 360, x \in Z\} \\ IS = \left\{y \middle| 1 \le y \le 30, y \in Z\right\} \end{gathered} \tag{9}$$

where RMSE gives the accuracy of the DS; RT is the running time of the DS; PL represents the pattern length; and IS represents the interval step. Figure 2 depicts the flow chart of LSPSO, including the following steps:

Step 1: Randomly initialize the population using the Beta strategy; each particle has two attributes: the pattern length and the interval step;

Step 2: Calculate the RMSE and RT of each particle, and then filter out the non-dominated solution according to the non-dominated relationship, and store it in the external population, S;

Step 3: Local search of the external population, S by local search, to enhance the local search ability of the particle, and obtain the population S';

Step 4: Control the population in population D by crowding distance;

Step 5: Update Pbest and Gbest;

Step 6: Update the velocity and position of the particle according to Formulas (7) and (8);

Step 7: Determine whether the termination condition is met. If it does, output the optimal solution set and select the appropriate parameters for predicting SST; otherwise, add G to 1, and return to step 2.

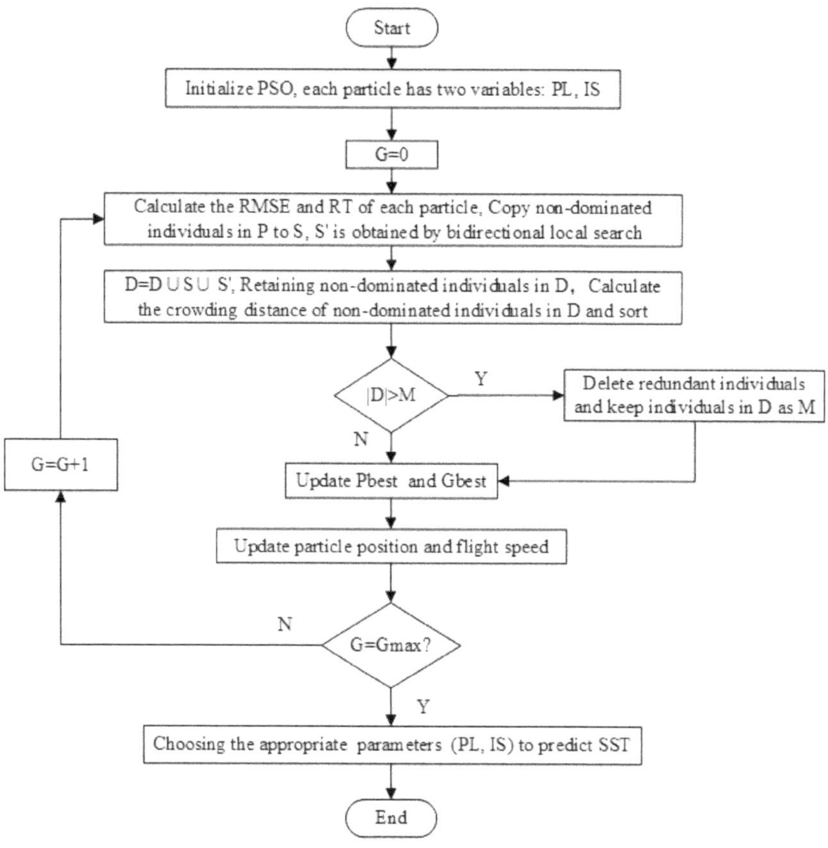

Figure 2. Flow chart of LSPSO.

The traditional particle swarm optimization algorithm typically uses random initialization to generate the initial solution. Due to the lack of guidance of prior information, it is not conducive to the initial particle to move closer to the optimal solution. In this paper, the Beta distribution initialization strategy is used to initialize the population, which is beneficial to the particle to form the enclosing situation to the optimal solution. The shape of the Beta function is a symmetric U-shape, and the candidate set is most likely to be located near the boundary of the search space. At this time, the global optimal solution is better enclosed within the initial particle population. The Beta distribution function is defined as:

$$\beta(x;m,n) = \frac{x^{m-1}(1-x)^{n-1}}{B(m,n)}, 0 < x < 1. \tag{10}$$

The denominator is defined as follows:

$$B(m,n) = \int_0^1 t^{m-1}(1-t)^{n-1}dt \tag{11}$$

Pareto dominance relationships are used in many optimization algorithms to measure the pros and cons of a solution, usually defined as follows:

Definition 1. *(non-dominated relationship). Let x, y be the two decision variables in equation (12), if x dominates y, then if and only if:*

$$\forall i \in \{1,2,\ldots,n\}, \exists j \in \{1,2,\ldots n\}; f_i(x) \le f_i(y) \cap f_j(x) < f_j(y). \tag{12}$$

Definition 2. *(non-dominated solution). Assuming that there is a solution set P, where individual q is not dominated by any other individual, then q is a non-dominated individual in P. The subset of non-dominated individuals of P is called the non-dominated set (NDset) of P.*

$$NDset = \{q|q \in P \text{ and does not exist } p \in P, \text{make } p \text{ dominate } q\}$$

Since only the non-dominated solution in the population P needs to be selected, in the process of comparing different particles, once the particles are dominated by other particles, they are not compared with other particles, which reduces the complexity of the algorithm. Through the non-dominated relationship, a non-dominated solution set S can be obtained. The solutions contained in the set are non-dominated solutions, and some better solutions can be found around the non- dominated solution.

The local search ability of the algorithm is enhanced by the local search strategy, which is beneficial to improve the local search ability of the algorithm. For the non-dominated solution set S, one of the individuals is $x_{i,k} = (x_{1,i,k}, x_{2,i,k}, \ldots x_{n,i,k})^T$, i represents the i-th individual in S, k is the number of evolutions, and n represents the dimension of the variable. The search for the m-th variable of the individual $x_{i,k}$ in two directions can be expressed below as $L_{m,i,t}$ and $R_{m,i,t}$.

$$L_{m,i,t} = x_{m,i,t} - c\times|P_{m,i,t} - G_{m,i,t}| \tag{13}$$

$$R_{m,i,t} = x_{m,i,t} + c\times|P_{m,i,t} - G_{m,i,t}| \tag{14}$$

where P and G are randomly selected from the non-dominated solution set S, and c is the interference coefficient. New individuals can be generated by Equations (13) and (14). Algorithm 2 expresses the complete process of local search.

The non-dominated solution set S obtained from each evolution, the local searched population S′ and the external archive D, are merged into a new external archive D. The non-dominated solutions in D are used as external archives to ensure the diversity of the population.

As the number of iterations increases, the size of the external archive will gradually increase too. It is unrealistic to include all non-dominated solutions in D. At this time, it is feasible to obtain a finite non-dominated subset that can represent the solution space. In order to obtain a more uniform non-dominated solution set, this paper uses the crowded distance to remove redundant individuals and controls the number of non-dominated solutions in D.

After updating Gbest and Pbest, we use Formulas (7) and (8) to update the velocity and position of the particle. Since the particles will cross the boundary when flying, the particles will be mutated for the particles whose particle velocity exceeds the constraint range. The formula is as follows:

$$M_{n,i,t} = E_{n,i,t} \times (1 - rand\,(-0.2, 0.2)) \tag{15}$$

where E is randomly selected from the external archive D, and M is the mutated particle. When the flying speed of the particle exceeds half of the constraint range, but is still below the constraint range, the particle velocity is set to half of the constraint range to prevent the particles from crossing the boundary. Through iteration, if the termination condition is met, the optimal solution set can be output finally, and a reasonable parameter combination is selected from the solution set as a parameter of the prediction model for predicting the SST for the next five days.

Algorithm 2. Local Search.

```
Input:
        The non-dominated solution set: S;
        Number of non-dominated solutions: |S|;
        The dimension of the search space: n;
Output:
        External population: S';
S'←Φ;
For I = 1 to |S| do
    Randomly select two individuals P and G from S
    For j = 1 to n do
        Generate individual L and R by Equations (13) and (14)
    End For
    Keep better individuals in both L and R in S'
End For
```

4. Results and Discussion

4.1. Experimental Environment and Data

The experimental environment of this paper has the following specifications: Windows 10 operating system; Intel Core i5 CPU; 2.6 GHz clock; and 8Gbyte RAM. The algorithms MODE, NSGA-II, and OMOPSO are implemented using Matlab, while the other components have been done in Python. The experimental data includes 9 SST sequences provided by the Second Institute of Oceanography, including SSTs from 1 January 2004 to 31 December 2016.

4.2. Evaluation Indicators and Test Functions

In this paper, we used three indicators to evaluate our model performance. We used Root Mean Square Error (RMSE) [36] to assess the overall accuracy of the SSTP models. To evaluate the convergence and uniformity of the parameter optimization algorithm, we used Generational Distance (GD) [37] and Spacing (SP) [38] as the indicators, and introduced three test functions. The details of these indicators and test functions are described in the following.

RMSE is a measure of the deviation between the observed value and the true value. The smaller the value, the more accurate the prediction. The formula of RMSE is as follows:

$$RMSE = \sqrt{\frac{\sum_{i=1}^{n} (Y\_real_i - Y\_pred_i)^2}{n}} \tag{16}$$

where Y_real is the true value, Y_pred is the predicted value, and n is the number of days to predict.

The Pareto optimal solution set obtained by the multi-objective optimization algorithm should maintain the convergence of the solution and the uniformity of the distribution. In order to evaluate the convergence and uniformity of the Pareto frontier obtained by the multi-objective optimization algorithm, to obtain as many Pareto optimal solutions as possible, and to approach the true Pareto frontier as best as possible, GD was used as the solution convergence performance evaluation.

The smaller the value of GD, the better the convergence of the solution set. Secondly, the Pareto optimal solution should be evenly distributed along the Pareto frontier as much as possible, and SP was used as the index of uniform distribution performance evaluation. The smaller the SP, the more uniform the solution set distribution. GD is defined as:

$$GD = \sqrt{\sum_{i=1}^{n} \frac{d_i^2}{n}} \tag{17}$$

where n is the optimal number of Pareto solutions, and d_i is the distance of the i-th Pareto optimal solution in the objective space from the nearest individual of the Pareto frontier. SP is defined as:

$$SP = \sqrt{\frac{1}{n-1} \sum_{i=1}^{n} (\bar{d} - d_i)^2} \tag{18}$$

where n is the optimal solution number of Pareto, d_i is the distance of the i-th Pareto optimal solution from other individuals in the objective space, and $\bar{d}$ is the average value of d_i.

LSPSO was applied to the bi-objective optimization of accuracy and efficiency of the SSTP. In order to verify the feasibility of LSPSO in the bi-objective optimization problem, three commonly used bi-objective test functions were selected for testing: BNH [39], SRN [40] and TNK [41]. We compared the GD and SP indicators of the optimal frontier obtained by MODE, NSGA-II, and OMOPSO, respectively. Table 1 lists the characteristics of these three test functions.

Table 1. Bi-objective optimization problems.

Function	Number of Variables	Number of Objectives	Analytical Pareto Frontier
BNH	2	2	Connected/Convex
SRN	2	2	Disconnected/Convex
TNK	2	2	Disconnected/Nonconvex

4.3. Analysis of Experimental Results

In this work we designed three sets of experiments: (1) to compare the advantages and disadvantages of the similarity measures in SSTP and choose the best method; (2) to verify the effectiveness of LSPSO by comparing it with the MODE, NSGA-II, and OMOPSO algorithms; (3) to compare the performance of DSL with other SSTP methods.

Experiment 1: Comparison of similarity measures. *Applying AC to SSTP requires choosing a suitable similarity measure method to measure the similarity of SST sequences. Therefore, we first computed the Euclidean distance, the cosine distance, and the DTW distance to measure the similarity of SSTs. Then, we chose the optimal similarity measure based on this principle: the better the similarity is measured, the smaller the SSTP error is. The error of SSTP was measured by Root Mean Square Error (RMSE).*

Table 2 shows the achieved RMSE in relation to SST predictions based, respectively, on Euclidean distance, Cosine distance, and DTW distance similarity measures. The first nine rows provide RMSE values for nine different SST sequences, with the average values given in the last row. When using the Euclidean distance to predict SST, the average RMSE is 0.4949, which is slightly higher than (at times comparable to) DTW, but much better than Cosine. The reason for this is that the number of days in which the SST changes regularly is not fixed, and the Euclidean distance does not support scaling. However, the DTW distance can better overcome this deficiency, so DTW can better reflect the similarity of SST changes.

Table 2. RMSE obtained by predicting SST with three similarity measures.

Station	Euclidean	Cosine	DTW
1	0.5289	1.3223	0.4866
2	0.5525	1.2907	0.5136
3	0.4813	0.8982	0.4606
4	0.4641	0.6669	0.4637
5	0.4788	1.4734	0.4544
6	0.5247	0.6666	0.4630
7	0.5364	0.5549	0.5204
8	0.4437	1.3070	0.4429
9	0.4439	1.4096	0.4161
Avg	0.4949	1.0655	0.4690

The average RMSE when using the cosine distance to predict SST is 1.0655. This is much higher than the other two indexes, leading to a much worse prediction ability. This happens because the cosine distance uses the cosine of the angle of the SST vector to measure its similarity, which only reflects the trend of SST changes, and is not sensitive to the value of the SST itself. The DTW and Euclidean distances are based on the SST values. In summary, DTW has the highest prediction accuracy among the three. Therefore, we use only DTW in the remaining experiments, below.

Experiment 2: Verification of LSPSO. *An LSPSO algorithm was proposed in this paper. To verify its effectiveness, three classical bi-objective test functions (BNH, SRN and TNK) were selected. GD and SP were used as evaluation indicators. The number of populations, N was set to 100, and the number of iterations G = 250 was compared with MODE, NSGA-II and OMOPSO, respectively. Each algorithm ran independently 30 times for each test function, computing the mean and variance of GD and SP values. Analysis of Variance (ANOVA) [42] was used to test the significant difference of the GD and SP indicators between LSPSO and the other models (MODE, NSGA-II, and OMOPSO). In general, P values smaller than 0.05 indicate that there is a significant difference.*

Figures 3–5 show the solution obtained by LSPSO for BNH, SRN and TNK functions and the true Pareto frontier. The red circle is the optimal solution obtained by LSPSO, and the black line is the true Pareto front. The optimal solution sets of LSPSO are convergent and evenly distributed on the true Pareto front.

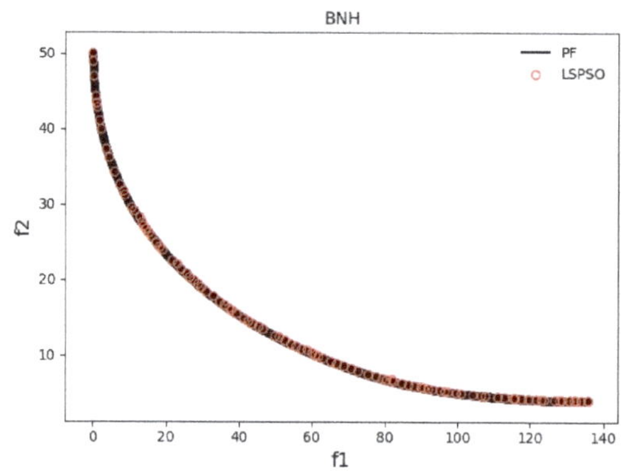

Figure 3. Pareto optimal solutions obtained by LSPSO for BNH.

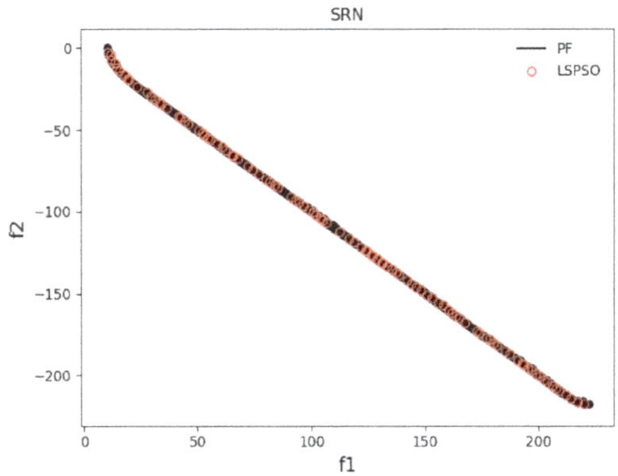

Figure 4. Pareto optimal solutions obtained by LSPSO for SRN.

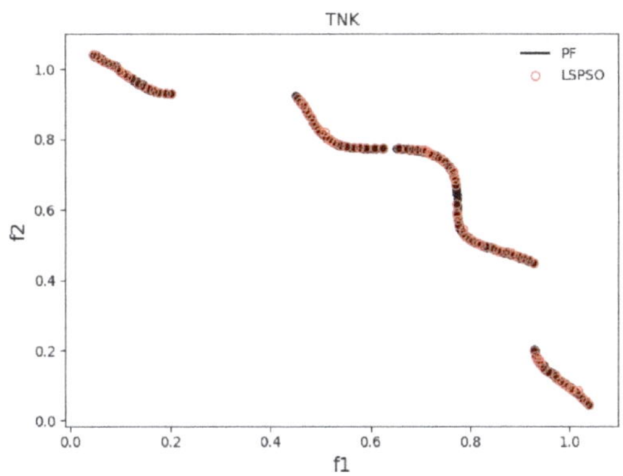

Figure 5. Pareto optimal solutions obtained by LSPSO for TNK.

Table 3 shows the GD and SP of MODE, NSGA-II, OMOPSO, and LSPSO in solving BNH, SRN, and TNK, respectively. Overall, our LSPSO is better than the other three methods according to the mean and std values of GD and SP.

Table 3. The GD and SP obtained by MODE, NSGA-II, OMOPSO and LSPSO in solving BNH, SRN and TNK.

Function		MODE		NSGA-II		OMOPSO		LSPSO	
		GD	SP	GD	SP	GD	SP	GD	SP
BNH	Mean	0.5498	1.9369	0.1386	1.1174	0.1405	0.9745	0.1350	0.7127
	Std	8.77×10^{-2}	0.5502	$1.56e \times 10^{-2}$	6.50×10^{-2}	1.46×10^{-2}	7.94×10^{-2}	1.36×10^{-2}	6.95×10^{-2}
SRN	Mean	2.8408	4.1774	0.6819	1.8991	0.5876	1.3269	0.4644	1.5033
	Std	0.4715	0.5601	9.02×10^{-2}	0.1593	6.58×10^{-2}	0.1020	3.77×10^{-2}	0.1149
TNK	Mean	4.96×10^{-3}	4.34×10^{-2}	2.45×10^{-3}	3.73×10^{-2}	2.68×10^{-3}	3.39×10^{-2}	2.41×10^{-3}	3.39×10^{-2}
	Std	6.91×10^{-4}	2.37×10^{-3}	2.24×10^{-4}	7.84×10^{-4}	2.44×10^{-4}	7.25×10^{-4}	1.46×10^{-4}	6.90×10^{-4}

When dealing with the BNH function, our LSPSO method outperforms the other three methods in the uniformity of the solution set, achieving a statistical significance of SP ($p < 0.001$). This can be observed from Figure 3, where the solution set of LSPSO is uniformly distributed on the Pareto front for BNH. The convergence of LSPSO is much better than that of MODE in term of GD ($p < 0.001$), but it is not significantly better than NSGA-II and OMOPSO ($p = 0.37$ and $p = 0.15$, respectively).

For the SRN function, the GD of LSPSO is significantly better than those on the other three methods ($p < 0.001$), indicating a good convergence. Regarding the uniformity of the solution set distribution (SP), although LSPSO is significantly better than MODE and NSGA-II ($p < 0.001$), it is worse than OMOPSO ($p < 0.001$). This maybe because the true Pareto frontier of the SRN function shows linearly distributed and global search has advantages when dealing with such problems, but when the true Pareto frontier of the test function is non-linear, our method can achieve much better results.

For the TNK function, the average and std of the GD and SP obtained by LSPSO are better than those on the other methods, indicating that LSPSO handles TNK well. According to the significance test, our LSPSO has similar performance with NSGA-II in terms of GD ($p = 0.48$), and with OMOPSO in terms of SP ($p = 0.95$).

In summary, LSPSO can better handle the bi-objective test functions, and provide effective support for the parameter optimization of DS. MODE, NSGA-II and OMPSO have strong global search capabilities and insufficient local search capabilities, so the solution sets obtained are not uniform and easily fall into a local optimum. The global and local search capabilities of PSO are mutually constrained and tend to fall into local optimum in the later stages of search. In this paper, a local search strategy is used to enhance the local search capability of the PSO, so that the improved PSO has independent global and local search capabilities. Therefore, the obtained solution set has better convergence and more uniform distribution.

Experiment 3: Comparison of DSL performance with other SSTP methods. *DSL is a combination of DTW + SVM (DS) and LSPSO. Define the space composed of the PL and the IS as the search space of the particle, and the prediction accuracy and efficiency set as the optimization objectives, LSPSO can compute the appropriate PL and IS that are set to the parameter of DS to predict SST. Here, we compare the performance of DSL with DTW, SVM, DS, LSTM in SSTP. In the situation of predicting 5-day SST with a given t-day of SST data, the DTW method is to find the most similar series of SST from historical data and take its following 5 days as the prediction. The SVM and LSTM both are trained by fitting nonlinear changes in SST. DS is to train a SVM by top-k similar series selected according to DTW.*

The comparison results are shown in Table 4. The average value of RMSE obtained by DS in predicting SST is 0.4468, which is lower than the average RMSE of the DTW. This indicates that the combination of DTW and SVM can effectively utilize the information after DTW mining. Secondly, the prediction results of the DS algorithm are better than those of SVM. This is because the SST sequence contains a lot of redundant information, which will interfere with the model during prediction, making the SVM prediction accuracy low.

Table 4. RMSE obtained when predicting SST by DTW, SVM, DS, LSTM and DSL.

Station	DTW	SVM	DS	LSTM	DSL
1	0.4866	0.7403	0.4640	0.5093	0.3255
2	0.5136	0.7117	0.4979	0.5075	0.4056
3	0.4606	0.7338	0.4387	0.4839	0.3334
4	0.4637	0.8383	0.4569	0.5458	0.3454
5	0.4544	0.6208	0.4259	0.4534	0.3862
6	0.4630	0.8694	0.4414	0.5710	0.4300
7	0.5204	0.8532	0.5121	0.5781	0.4816
8	0.4429	0.7840	0.3936	0.5154	0.2924
9	0.4161	0.7660	0.3911	0.5259	0.3534
Avg	0.4690	0.7686	0.4468	0.5211	0.3726

LSTM gains the average RMSE of 0.5211, which is worse than DTW, DS and DSL. Although LSTM is developed for dealing with long and short-term prediction problem, it does not work well in predicting SST. This is probably due to the non-stationarity of SST.

DSL obtains the optimal parameters of PL and IS by LSPSO, and uses them in DS to predict SST. Comparing the RMSE values between DS and DSL, it can be found that the performance of DSL prediction of SST is better than DS, indicating that the parameters of DS can be effectively optimized by LSPSO, which was responsible for a 16.7% improvement in prediction accuracy, in terms of reducing the RMSE. In summary, the overall effect of DSL prediction of SST is optimal, indicating that the method can effectively predict SST and verify the effectiveness of the proposed method.

We demonstrated the predicted results by different SSTP methods. Figure 6 shows a sample, randomly selected from the results. The black line represents the true values; the blue line represents the predicted results by using DTW; the red and yellow lines correspond to SVM and LSTM, respectively; the predicted results by using DSL are shown in green. It is clear that the results predicted by our method (green) are the closest ones to ground truth (black), including also the trend changes. On the other hand, the change trend of the other methods fluctuates significantly.

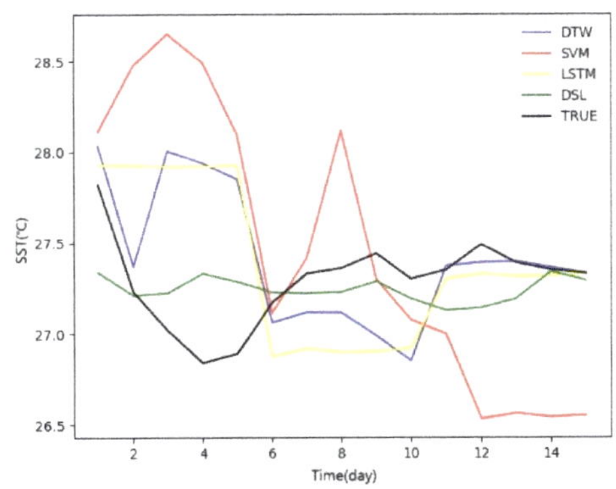

Figure 6. Comparison of SST prediction results by different methods, where black line is ground truth, green line is our model's prediction.

Finally, Figure 7 shows the operating efficiency of DS at predicting SST, before and after LSPSO optimization. RT represents the running time in seconds. The first nine columns provide RT for nine different SST sequences, with the average values given in the last column. The red histogram is the RT of the DS before optimization, and the blue histogram is the RT of DS after optimization. It can be clearly seen that, for each SST series, the RT before optimization is much longer, with an average 76% acceleration. Our results verify the effectiveness of the proposed method.

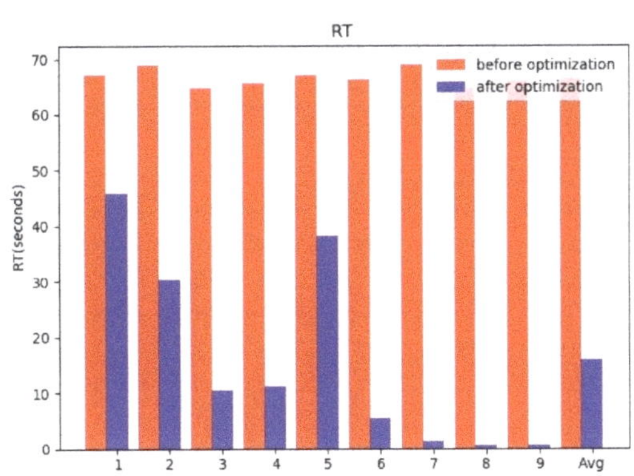

Figure 7. Comparison of RT before and after optimization.

5. Conclusions

This paper constructs a simple yet effective SSTP model, dubbed DSL, based on time-series similarity measure, multiple pattern learning and parameter optimization. We approach this complex issue through three stratagems: (1) Using Dynamic Time Warping (DTW) to mine the similarities in historical SST series. This can extract similar patterns of historical SST series simply and accurately. (2) Training an SVM using the top-k similar patterns, deriving a robust SSTP model that offers a 5-day prediction window based on multiple SST input sequences. Learning from multiple time-series sequences was instrumental to facilitating consistency enhancement and noise cancellation, thus achieving high prediction accuracy. (3) Developing an improved PSO method, dubbed LSPSO, to find the optimal parameters of the SSTP model, which uses a local search strategy to achieve the combined requirement of prediction accuracy and efficiency.

Our method strives for optimal model parameters (pattern length and interval step) and is suited for long-term series, leading also to significant improvements in SST trend predictions. The efficiency LSPSO was verified to have better convergence and more uniform distribution by comparing with different optimization methods such as MODE, NSGA-II and OMOPSO. With the LSPSO, the SST prediction using SVM and DTW achieved a 16.7% reduction in prediction error, at a 76% gain in operating efficiency. We also achieved a significant improvement in prediction accuracy of non-stationary SST time series compared to the more recent LSTM deep learning method. In general, our method provides a new way for marine operational forecasting.

In the work carried out so far, we have assumed that SST changes were affected only by internal factors. As a future extension, we shall strive to take into account also other external factors (such as air temperature, air pressure, etc.). This will require considering the correlations between SST and these factors. We plan to use association rules to analyze impact on SST, and then use multi-factor prediction models to predict SST, striving for more comprehensive and robust predictions.

Author Contributions: Conceptualization, Q.H. and W.S.; methodology, C.Z., W.S., C.P. and A.L.; software, C.Z.; investigation, Y.D.; resources, Z.H.; writing—original draft preparation, Q.H. and C.Z.; writing—review and editing, W.S., C.P. and A.L.; supervision, Q.H. and W.S. All authors have read and agreed to the published version of the manuscript.

Funding: The work is supported by the National Key R&D Program of China (2016YFC1401902), the National Natural Science Foundation of China (41671431), the Program for the Capacity Development for Shang Local Colleges (17050501900), the National Oceanic Administration Digital Ocean Science and Technology Key Laboratory Open Fund (B201801029) and the open fund of State Key Laboratory of Satellite Ocean Environment Dynamics, Second Institute of Oceanography, MNR (QNHX1913).

Acknowledgments: The sea surface temperature data were provided by State Key Laboratory of Satellite Ocean Environment Dynamics, the Second Institute of Oceanography, the Ministry of Natural Resources. The authors would also like to thank the anonymous reviewers for their very competent comments and helpful suggestions.

Conflicts of Interest: The authors declare that they have no known competing financial interests or personal relationships that could have appeared to influence the work reported in this paper.

References

1. Repelli, C.A.; Nobre, P. Statistical prediction of sea-surface temperature over the tropical Atlantic. *Int. J. Climatol.* **2014**, *24*, 45–55. [CrossRef]
2. Mendoza, V.M.; Villanueva, E.E.; Adem, J. Numerical experiments on the prediction of sea surface temperature anomalies in the Gulf of Mexico. *J. Mar. Syst.* **1997**, *13*, 83–99. [CrossRef]
3. De Elvira, A.R.; Bevia, M.O.; Narvaez, W.D.C. Empirical forecasts of tropical Atlantic sea surface temperature anomalies. *Q. J. R. Meteorol. Soc.* **2000**, *126*, 2199–2210. [CrossRef]
4. Wang, X.; Wu, J.; Liu, C. Exploring LSTM based recurrent neural network for failure time series prediction. *J. Beijing Univ. Aeronaut. Astronaut.* **2018**, *44*, 772–784.
5. Lins, I.D.; Araujo, M.; Moura, M.D.C.; Silva, M.; Droguett, E.L. Prediction of sea surface temperature in the tropical Atlantic by support vector machines. *Comput. Stat. Data Anal.* **2013**, *61*, 187–198. [CrossRef]

6. Zazo, R.; Nidadavolu, P.S.; Chen, N.; Gonzalez-Rodriguez, J.; Dehak, N. Age Estimation in Short Speech Utterances Based on LSTM Recurrent Neural Networks. *IEEE Access* **2018**, *6*, 22524–22530. [CrossRef]
7. Zhang, Q.; Wang, H.; Dong, J.; Zhong, G.; Sun, X. Prediction of Sea Surface Temperature Using Long Short-Term Memory. *IEEE Geosci. Remote. Sens. Lett.* **2017**, *14*, 1745–1749. [CrossRef]
8. Lorenz, E.N. Atmospheric Predictability as Revealed by Naturally Occurring Analogues. *J. Atmos. Sci.* **1969**, *26*, 636–646. [CrossRef]
9. He, C.; Zhu, B.; Zhang, M.; Zhuang, Y.; He, X.; Du, D. Customers' Risk Type Prediction Based on Analog Complexing. *Procedia Comput. Sci.* **2015**, *55*, 939–943. [CrossRef]
10. Baghi, A.G.; Ask, P.; Babić, A. A pattern recognition framework for detecting dynamic changes on cyclic time series. *Pattern Recognit.* **2015**, *48*, 696–708.
11. Zhou, J.; Yang, Y.; Ding, S.X.; Zi, Y.; Wei, M. A Fault Detection and Health Monitoring Scheme for Ship Propulsion Systems Using SVM Technique. *IEEE Access* **2018**, *6*, 16207–16215. [CrossRef]
12. Mavrovouniotis, M.; Muller, F.M.; Yang, S. Ant Colony Optimization with Local Search for Dynamic Traveling Salesman Problems. *IEEE Trans. Cybern.* **2017**, *47*, 1743–1756. [CrossRef] [PubMed]
13. Tao, Y.; He, Q.; Yao, Y. Solution for a time-series AR model based on robust TLS estimation. *Geomat. Nat. Hazards Risk* **2019**, *10*, 768–779. [CrossRef]
14. Akrami, S.A.; El-Shafie, A.; Naseri, M.; Santos, C.A.G. Rainfall data analyzing using moving average (MA) model and wavelet multi-resolution intelligent model for noise evaluation to improve the forecasting accuracy. *Neural Comput. Appl.* **2014**, *25*, 1853–1861. [CrossRef]
15. Baptista, M.; Sankararaman, S.; De Medeiros, I.P.; Nascimento, C.; Prendinger, H.; Henriques, E. Forecasting fault events for predictive maintenance using data-driven techniques and ARMA modeling. *Comput. Ind. Eng.* **2018**, *115*, 41–53. [CrossRef]
16. Li, Q.-J.; Zhao, Y.; Liao, H.-L.; Li, J.-K. Effective forecast of Northeast Pacific sea surface temperature based on a complementary ensemble empirical mode decomposition–support vector machine method. *Atmos. Ocean. Sci. Lett.* **2017**, *10*, 261–267. [CrossRef]
17. Elmore, K.L.; Richman, M. Euclidean Distance as a Similarity Metric for Principal Component Analysis. *Mon. Weather Rev.* **2001**, *129*, 540–549. [CrossRef]
18. Sun, T.; Liu, H.; Yu, H.; Chen, C.L.P. Degree-Pruning Dynamic Programming Approaches to Central Time Series Minimizing Dynamic Time Warping Distance. *IEEE Trans. Cybern.* **2017**, *47*, 1719–1729. [CrossRef]
19. Zhang, Y.; Li, X.S.; Song, Y. A Recognition Judgment Method of Isolated-Word Speech-Recognition. *Appl. Mech. Mater.* **2014**, *543*, 2337–2340. [CrossRef]
20. Berndt, D.J.; Clifford, J. Finding patterns in time series: A dynamic programming approach. In *Advances in Knowledge Discovery and Data Mining*; American Association for Artificial Intelligence: Menlo Park, CA, USA, 1996; pp. 229–248.
21. Eghbali, S.; Tahvildari, L. Fast Cosine Similarity Search in Binary Space with Angular Multi-Index Hashing. *IEEE Trans. Knowl. Data Eng.* **2019**, *31*, 329–342. [CrossRef]
22. Storn, R.; Price, K. Differential Evolution–A Simple and Efficient Heuristic for global Optimization over Continuous Spaces. *J. Glob. Optim.* **1997**, *11*, 341–359. [CrossRef]
23. Zitzler, E.; Thiele, L. Multi-objective optimization using evolutionary algorithms—A comparative case study. *Int. Conf. Parallel Probl. Solving Nat. (PPSN-V)* **1998**, *1498*, 292–301.
24. Sun, D.; Benekohal, R.F.; Waller, S.T. Multi-objective traffic signal timing optimization using non-dominated sorting genetic algorithm. In *International Conference on Genetic and Evolutionary Computation*; Springer: Berlin/Heidelberg, Germany, 2003; pp. 2420–2421.
25. Tongur, V.; Ülker, E. B-Spline Curve Knot Estimation by Using Niched Pareto Genetic Algorithm (NPGA). *Proc. Adapt. Learn. Optim.* **2015**, *5*, 305–316.
26. Schaffer, J.D. Multiple Objective Optimization with Vector Evaluated Genetic Algorithms. In Proceedings of the 1st International Conference on Genetic Algorithms, Pittsburgh, PA, USA, 24–26 July 1985; pp. 93–100.
27. Deb, K.; Agrawal, S.; Pratap, A. A Fast Elitist Non-dominated Sorting Genetic Algorithm for Multi-objective Optimization: NSGA-II. *Evol. Comput.* **2000**, *1917*, 849–858.
28. Zhong, X.; Yin, H.; He, Y. Joint Downlink Power and Time-Slot Allocation for Distributed Satellite Cluster Network Based on Pareto Optimization. *IEEE Access* **2017**, *1*, 99. [CrossRef]

29. Raquel, C.R.; Naval, P.C. An effective use of crowding distance in multiobjective particle swarm optimization. In Proceedings of the Genetic and Evolutionary Computation Conference, Washington, DC, USA, 25–29 June 2005; pp. 257–264.
30. Eberhart, R.; Kennedy, J. Particle swarm optimization. In Proceedings of the IEEE International Conference on Neural Networks, Perth, WA, Australia, 27 November–1 December 1995; pp. 1942–1948.
31. Coello, C.; Toscano-Pulido, G.; Lechuga, M. Handling multiple objectives with particle swarm optimization. *IEEE Trans. Evol. Comput.* **2004**, *8*, 256–279. [CrossRef]
32. Reyes, M.; Coello, C. Improving pso-based multiobjective optimization using crowding, mutation and ε-dominance. In *Evolutionary Multi-Criterion Optimization*; Springer: Berlin/Heidelberg, Germany, 2005; pp. 505–519.
33. Lin, J.T.; Chiu, C.-C. A hybrid particle swarm optimization with local search for stochastic resource allocation problem. *J. Intell. Manuf.* **2015**, *29*, 481–495. [CrossRef]
34. Mousa, A.; El-Shorbagy, M.; Abd-El-Wahed, W.; El-Shorbagy, M. Local search based hybrid particle swarm optimization algorithm for multiobjective optimization. *Swarm Evol. Comput.* **2012**, *3*, 1–14. [CrossRef]
35. Zhang, R.; Duan, Y.; Zhao, Y.; He, X. Temperature Compensation of Elasto-Magneto-Electric (EME) Sensors in Cable Force Monitoring Using BP Neural Network. *Sensors* **2018**, *18*, 2176. [CrossRef]
36. Ma, R.; Xu, W.; Liu, S.; Zhang, Y.; Xiong, J. Asymptotic mean and variance of Gini correlation under contaminated Gaussian model. *IEEE Access* **2016**, *4*, 8095–8104. [CrossRef]
37. Menchacamendez, A. *GDE-MOEA: A New MOEA Based on the Generational Distance Indicator and ε-Dominance*; CEC: Sendai, Japan, 2015; pp. 947–955.
38. Deb, K.; Mohan, M.; Mishra, S. Evaluating the ϵ-Dominance Based Multi-Objective Evolutionary Algorithm for a Quick Computation of Pareto-Optimal Solutions. *Evol. Comput.* **2005**, *13*, 501–525. [CrossRef] [PubMed]
39. Binh, T.T.; Korn, U. Scalar optimization with linear and nonlinear constraints using evolution strategies. In Proceedings of the International Conference on Computational Intelligence, Coast, Australia, 10–11 July 1997; pp. 381–392.
40. Srinivas, N.; Deb, K. Muiltiobjective optimization using non-dominated sorting in genetic algorithms. *Evol. Comput.* **1994**, *2*, 221–248. [CrossRef]
41. Tanaka, M.; Watanabe, H.; Furukawa, Y.; Tanino, T. GA-based decision support system for multicriteria optimization. In Proceedings of the IEEE International Conference on Systems, Columbia, BC, Canada, 22–25 October 1995; pp. 1556–1561.
42. Webster, R.; Lark, R.M. Analysis of variance in soil research: Let the analysis fit the design. *Eur. J. Soil Sci.* **2018**, *69*, 126–139. [CrossRef]

energies

MDPI

Article

An Alertness-Adjustable Cloud/Fog IoT Solution for Timely Environmental Monitoring Based on Wildfire Risk Forecasting

Athanasios Tsipis [1,*], Asterios Papamichail [1], Ioannis Angelis [1], George Koufoudakis [1] and Georgios Tsoumanis [2] and Konstantinos Oikonomou [1]

1 Department of Informatics, Ionian University, 49100 Corfu, Greece; aspapa@ionio.gr (A.P.); iangelis@ionio.gr (I.A.); gkoufoud@ionio.gr (G.K.); okon@ionio.gr (K.O.)

2 Department of Informatics and Telecommunications, University of Ioannina, 45110 Arta, Greece; gtsoum@uoi.gr

* Correspondence: atsipis@ionio.gr; Tel.: +30-26610-87734

Received:15 June 2020; Accepted:14 July 2020; Published: 17 July 2020

Abstract: Internet of Things (IoT) appliances, especially those realized through wireless sensor networks (WSNs), have been a dominant subject for heavy research in the environmental and agricultural sectors. To address the ever-increasing demands for real-time monitoring and sufficiently handle the growing volumes of raw data, the cloud/fog computing paradigm is deemed a highly promising solution. This paper presents a WSN-based IoT system that seamlessly integrates all aforementioned technologies, having at its core the cloud/fog hybrid network architecture. The system was intensively validated using a demo prototype in the Ionian University facilities, focusing on response time, an important metric of future smart applications. Further, the developed prototype is able to autonomously adjust its sensing behavior based on the criticality of the prevailing environmental conditions, regarding one of the most notable climate hazards, wildfires. Extensive experimentation verified its efficiency and reported on its alertness and highly conforming characteristics considering the use-case scenario of Corfu Island's 2019 fire risk severity. In all presented cases, it is shown that through fog leveraging it is feasible to contrive significant delay reduction, with high precision and throughput, whilst controlling the energy consumption levels. Finally, a user-driven web interface is highlighted to accompany the system; it is capable of augmenting the data curation and visualization, and offering real-time wildfire risk forecasting based on Chandler's burning index scoring.

Keywords: environmental monitoring; precision agriculture; Internet of Things; wireless sensor networks; cloud/fog computing; fire risk forecasting; Chandler burning index; wildfires

---

1. Introduction

There has been a recent spike of research activity in precision agriculture and environmental sustainability [1]. One popular direction, focuses on the advances of *cloud computing* enablers [2] and *wireless sensor networks* (WSNs) [3]. These systems embed pioneering wireless technologies, such as ZigBee [4,5], to monitor field conditions in secure and credible ways [6].

The current trend, however, especially when considering the wide proliferation of the *Internet-of-Things* (IoT) applications [7,8], and the opportunities that they bring [9], is to employ a *cloud/fog computing* environment [10–12], offering computing, networking, and storage support near the end user, minimizing response times and greatly improving operational capacity and scalability. Cloud servers (typically located in large data centers) are expected to have increased computational capabilities, whereas *fog devices* located in close proximity to the end users will obtain similar, though less-in-power attributes [13], and transmute them into suitable candidates for offloading cloud elastic resources and alleviating the communication overhead and traffic burden.

1.1. Challenges and Motivation

Despite the turn towards these information communication technologies (ICT), current practices focus heavily on standalone and trivial data logger systems, whose main tasks are data acquisition, regarding specific environmental facets, from spatially distributed sensors [14]. These initiatives, although useful, lack sophistication and do not take advantage of the full potential brought about by the assimilation of cloud/fog IoT platforms.

That being said, the design, implementation, and large-scale installation of a fully functional IoT platform will provide the involved parties with instant decision-making models, yielding an analytical understanding of natural systems. When combined with other visualization and knowledge tools, such as area maps or hazard scale metrics, they can become highly effective actuators for adaptive field administration, targeted interventions, and personalized notification procedures. Nevertheless, their programming is considered a devious process that requires expertise and extensive know-how, since in most cases it relies on custom-designed equipment and non-standard hardware solutions [15]. Therefore, the procurement of a complete design methodology across all system elements is imperative and the main motivation behind the current work, which attempts to address these issues with the proposition of a complete field-to-stakeholder IoT solution.

Meanwhile, the challenges that arise are numerous, especially when it comes to monitoring catastrophic events in nature [16], e.g., floods [17], earthquake activity [18], and volcanic eruptions [19]; or biological dangers to the plantation itself, e.g., vegetation deceases and pest infestations, which deeply distort the health of the farms, forests, and wild-lands, thereby requiring clear and fast tracking of the agents that cause them, and a continuous stream of data regarding the conditions that drive their spreading. For example, consider the work found in [20], which, much like the study presented here, reports on the development of a WSN-based IoT system, coupled with cloud/fog liturgies, to address time-sensitive agents that affect the health of olive groves.

Still, one of the major challenging environmental hazards to confront is the wildfire, mainly due to the large number of different variables affecting this complex phenomenon, which can wreak havoc on vast areas of land [21]. With that said, the main causal factors remain high temperature and low relative humidity, especially in prolonged drought during summer seasons, which is a common sight in regions such as the Mediterranean basin [22]. However, in conjunction with heterogeneous geographic and micro-climate habitats, such as the ones found in the island regions of Greece, their appearance becomes increasingly hard to predict. In such circumstances, in order to offer targeted fire suppression techniques, the WSNs must be deployed in various landscapes with potentially diversified prevailing conditions (e.g., under different altitudes, wind conditions, temperature/humidity levels, etc.), making the system complex and hard to manage. Consequently, the procurement of an ergonomic, adaptable, and easy-to-use IoT system, able to promptly manage data from various input sources, along with suitable visualization, forecasting, and decision-support functionality, is of the utmost importance for understanding habitat inter-dependencies and a key element of the proposed system in the current work.

Lastly, most existing systems for environmental monitoring or fire prediction/detection require significant computational resources and do not incorporate self-acting operational conformation towards achieving increased precision and energy conservative automation. To the best of the authors' knowledge, this is the first complete end-to-end IoT implementation that attempts to tackle these issues based on local decision-making processes at the fog layer.

1.2. Contribution

The fundamental premise of this paper lies in the presentation of a three-layered cloud/fog computing architecture, suitable for facilitating smart agricultural applications, especially those related to wildfire monitoring, and to propose a low-cost, WSN-based IoT system that seamlessly embeds the proposed logic. The presented architecture follows the main directions of the cloud/fog computing paradigm [10] based on the general framework found in [12]. In particular, Guardo et al. [12] revealed how the cloud/fog hybrid architecture can be effectively utilized in agriculture and they evaluated it using a prototype of ten nodes. In the current work, the basic principles are similar with an emphasis on the special case of wildfires; hence, a prototype system of 25 sensing nodes, forming six (6) distinct WSNs, each equipped with its own sink node, which is in turn connected to a fog device and then to the remote cloud server, is assumed for evaluation purposes.

The prototype has been developed in the facilities of the Ionian University and is able to autonomously adjust its operations to mirror the necessity for efficient wildfire alertness and lower power consumption. To demonstrate its potential, the prototype was put to the test in a controlled laboratory environment, with the most important evaluation metric being response time, which is considered vital for smart applications of the future. Experimental results showcase how such an architecture can indeed improve precision, by effectively reducing the average response time across all used platforms, with parallel energy efficiency, and high accuracy and throughput rates.

To further validate the criticality-adaptable behavior, the prototype was calibrated to deal with one of the most imminent threats to rural and outdoor environments, i.e., wildfires. Greece is known for suffering from heavy fire activity [23]. In fact, the annually burned area exceeds 245,000 acres of land, and so the experiment has great value for future farming in the area. As a case-study scenario, the Island of Corfu was considered for the firefighting period of 2019, to showcase the architecture's conforming character under different fire risk ignition severity degrees. Experimentation under these conditions showed great promise, highlighted the role of fog computing in dealing with such extreme phenomena, and verified the flexibility of the hardware components used.

Moreover, to better visualize and manage the systems' outputs, a user-driven graphical user interface (GUI) has also been developed to accompany the prototype system and assist the involved parties in the decision-making process, named the *Fog-assisted Environmental Monitoring System*, or F.E.MO.S. In short. That being said, F.E.MO.S. provides users with the ability to oversee the whole sensing process as it enfolds with suitable data curation and visualization. In fact, based on the two environmental sensed parameters (i.e., temperature and relative humidity), it is capable of objectively assessing the fire ignition risk, based on the popular "Chandler burning index" (CBI) [24], and generating an indication of the fire risk severity, subsequently allowing for targeted countermeasures that will mitigate the hazard. In this direction, automated notification alerts are generated to instantly mobilize the authorities to take appropriate mitigation actions.

The main contributing factors of the current paper can be summarized in the next bulletpoints.

1. A robust three-layered cloud/fog computing architecture for environmental monitoring, capable of dynamically conforming its sensing functionality to meet stringent latency requirements and the needs for energy conservation, and high accuracy and throughput.
2. A thorough presentation of its data flow and operations, starting from the initialization of the field WSNs and reaching up to the remote cloud infrastructure, in order to contextualize the steps undertaken from data acquisition to the creation of the appropriate response analysis.
3. The design, analysis, and development of a proof-of-concept prototype, mirroring the given architecture and utilizing state-of-art and low-cost hardware modules for transparent interactions.
4. Its performance evaluation primarily via the response time metric, which is crucial for time-sensitive agricultural applications of the future, especially those keeping track of wildfire activity.
5. The experimentation with real fire risk data considering the fire fighting season of 2019 for Corfu Island, which demonstrates how the considered approach can be effectively utilized to deal with such phenomena and showcases its alertness-adjustable character.
6. The implementation of an accompanying user-friendly web application to monitor the system's behavior and data curation and acquire real-time information relating to the monitored fields' health, including CBI-based fire risk severity forecasting along with the autonomous generation of appropriate notification alerts to actuate fast mobilization and countermeasures.

The aim here is on timely environmental monitoring, especially regarding wildfire ignition prediction and early detection; nonetheless, the novelty lies with the considered cloud/fog IoT solution that can be utilized for a wide range of time-sensitive agricultural applications with simple system modifications. Since its functionality is easily adjustable to network alterations or ecosystem changes, it can be easily customized and expanded to map its activity to the farmers' occasional needs and demands regarding other smart agriculture and forestry applications.

The remainder of the present paper is organized as follows. In Section 2 necessary background concepts are summarized. The proposed hybrid cloud/fog computing architecture, networking principles, and data flow methodology are underscored in Section 3. The evaluation process is presented in Section 4, while the appliance experiment for the case study of wildfires is showcased in Section 5. Finally, Section 6 concludes the paper and outlines directions for future work.

2. Literature Background

The current section presents related work regarding IoT systems and WSNs in the environmental/agricultural monitoring and fire prediction/detection/protection sectors.

2.1. Internet of Things and Wireless Sensor Networks

For both existing and upcoming applications, given the standardization process of the emerging Fifth-Generation (5G) of mobile communications [25], one of the major aspects that requires careful consideration is the support of multiple devices with parallel real-time processing of large volumes of data [26]. This is essential for the assimilation of IoT decision-making functionalities, with low energy consumption [27] and high accuracy and throughput [28].

An IoT system consists of smart devices that collect, transmit, and act on data they acquire from the environment, without the need for human intervention [15]. Each IoT device transmits its data either directly to the Internet or through a gateway, which are then gathered at a central station for further computation and analysis. However, the centralized paradigm does not always meet the storage and process requirements for the amount of data. This becomes abundantly obvious when considering the diversity of these devices and their lightweight and resource-constrained nature [29].

An alternative road refers to the benefits brought by the integration of *cloud computing* [30]. Moreover, the Internet necessities for low latency and high mobility push the cloud functionality to the edges of the IoT network [31], making way for *fog computing* [32], offering higher cognition and agility [33,34]. That being so, the IoT appliances have made a great leap in the direction of meeting high quality of service guarantees set by the upcoming 5G era, especially when combined with heterogeneous WSN systems [35,36].

WSNs generally consist of battery-powered sensor nodes that are spatially distributed in a wide area, capable of sensing environmental conditions, using powerful processors with low energy consumption [37], a subject of the utmost importance when realizing IoT platforms [38]. The data are often transmitted in a multi-hop manner towards a sink node, which can either store them locally or transmit them to a central location [39], e.g., a collection server.

Until recently, such systems faced many challenges, mainly due to the lack of wide-area connectivity and energy resources, and sometimes harsh environmental conditions. However, modern WSNs, though their inbuilt routing and relay capabilities [40], can quickly adjust to topology changes, allowing their large-scale deployment, even in areas where the battery replenishment may not always be feasible. A popular WSN configuration nowadays embeds Arduino boards [41] and utilizes the ZigBee module [5].

2.2. Related Research in the Agricultural/Environmental Monitoring Sector

Precision agriculture and smart farming [42] are considered two of the most rapidly evolving sciences of the twentieth century and major pillars for boosting productivity and economic growth [43]. Ergo, modern research turns to the previous ICT solutions and their seamless migration towards a multidisciplinary model [11] to support these sectors [28,44]. In fact, innovative enablers and wireless technologies (like SigFox [45], LoRa [46], NB-IoT [47], GSM-IoT [48], and ZigBee [49,50]) have empowered the involved stakeholders with the ability to experiment, manage, and record the dynamics of complex systems [51].

With that in mind, many systems, besides deceasing spread [52] or pest infestations [53], now tackle other aspects for climate protection, as identified in [54]. A recent appliance that meets these guidelines was described in [55]; the paper details a control system for monitoring field data originating from camera and sensor nodes deployed in crops. It then actuated the control devices adhering to threshold constraints relating to specific climate agents.

One of the major WSN obstacles refers to the energy needed to keep the network "alive". Many works attempt to address this barrier. Suárez-Albela et al. [56] identified in this regard the opportunities that arise from smart micro-controllers, such as Orange Pis. Meanwhile in [57], a green WSN node suitable for fog computing platforms named "FROG" was proposed, which introduces proactive power management tools for smart farming. In the present project, the WSN heavy demands on energy are counterbalanced with the use of Arduino boards that have been proven power-efficient (e.g., in [27]), along with ZigBee antennas, which yield significant energy gains [58].

Likewise, the utilization of Raspberry Pis has also revolutionized the data curation process. The work in [59] details a WSN, where information is collected by a Raspberry Pi acting as the base station. In the case of [59], however, the Raspberry Pi was used as a database and web server to manage the data. Similarly, in [60], the overseeing Raspberry Pi was responsible for data acquisition and analysis, while in [61] it created appropriate visualization. Contrarily, in the current study the Raspberry Pis are assigned the role of driving the data processing procedure through fog computing methods.

Diving deeper into the cloud/fog architecture, the works in [62] proposed a scalable fog network architecture to increase coverage and throughput. Emphasis was given to cross-layer channel access and routing, combining inputs generated in multiple networks. However, this work does not involve the use of open-access hardware and software utilities offered by Arduino and ZigBee, respectively, as in the current work. On the other hand, Bin Baharudin et al. [63] showcased the benefits of using Raspberry Pis as fog gateways in a three-layered IoT architecture, similarly to the one incorporated here, using ZigBee for communication. Clearly, ZigBee has been identified as a reliable and affordable standard for smart agriculture realization (e.g., [64]), thereby becoming the central field communication protocol here.

On a different path, the authors in [65] explored agricultural WSNs consisting entirely of Raspberry Pis. In their system, WSN administration was enabled using a GUI, developed in "MATLAB" and installed on the base station's board. Likewise, Zamora-Izquierdo et al. [66] developed an IoT platform for greenhouse automation that allows human operators to configure the individual system components through an "HTML5" interface. Understandably, GUIs are essential for an enhanced end-to-end IoT monitoring solution. This is clearly demonstrated in [14] and highly acknowledged in the current work, which also provides a user-driven GUI for delivering the system outputs in a user-friendly manner.

The majority of WSNs are deployed in uncontrolled areas, making them vulnerable to various types of attacks. Souissi et al. [67] try to tackle this issue by introducing trust in three different levels, namely, the data acquisition level (when the node takes a measurement), the network level (between the nodes of the network), and the data fusion level (during the aggregation and the processing of the measurements). Fortino et al. [68] performed a comparison of the existing architectures by modeling trust in IoT environments. Meanwhile, there is always a possibility that incoming packets may suffer from data distortions. To detect such occurrences, a field is usually used, called "checksum." Alternatively, Cao et al. [69] attempt to overcome this problem by deducing the measurements' correctness, based on predefined boundaries (e.g., for the temperature the boundary could be set in the $[-5, 40]$ °C). Another problem that WSN-based system operators have to consider is the false data detection. Casado-Vara et al. [70] offered a distributed algorithm that allows the collected temperature data to be self-corrected by the neighboring nodes' readings. In the current project, simple data validation is conducted in the considered fog computing network, where the fog devices evaluate the consistency of the received data packets.

Lastly, a special case of IoT refers to their assimilation for the detection and management of extreme events caused by climate change or other ecological agents. In general, this is a difficult ordeal due to the complex nature and conditions that lead up to their emergence; however, with WSN-based IoT infrastructures, new opportunities have come to light [71], enabling time-critical data curation, while achieving high semantic correlation and efficient risk forecasting [72]. To mention a few, consider the following for extreme weather estimation [73], air pollution detection [74], earthquake prediction [18], flood warning [75], landslide analysis [76], oceanic monitoring [77], etc. Unfortunately, one of the most sensitive and unpredictable hazardous phenomena is the wildfire, which leads to extensive catastrophes around the globe. The next subsections describe to relevant research into these events.

2.3. Related Research in the Wildfire Monitoring Sector

A plethora of systems have been developed to monitor the wildlands for fire threats. Nevertheless, the integration of WSNs for fire regime tracking has not yet been thoroughly explored, although there exists increasing research activity towards this direction (e.g., [78–80]), because compared to conventional methods, such as satellite imagery, which is affected by weather conditions (e.g., clouds), the amount of smoke, the image resolution, etc., WSNs offer faster detection [81].

According to Li et al. [82], these WSNs must possess four key aspects, namely, reactivity, reliability, robustness, and network lifespan elongation. Consequently, exertions are placed in realizing systems with these attributes. For instance, the research by [83] outlines a novel WSN monitoring methodology that adopts a maintenance process to detect temperature anomalies. On the other hand, [84] follows a contiguous approach to the one presented here, to adjust their prototype's sampling and reporting rates based on temperature fluctuations, while the work in [85] provided a fire detection WSN clustering solution with notable reduction in energy consumption. The current study adjusts the response time of the system by using the risk degree that Greece's General Secretariat for Civil Protection (GSCP) publishes every day for the fire fighting season. Additionally, the interval between measurement readings depends on the risk degree, resulting in energy conservation when the fire risk is low.

Obviously, energy conservation is important to prolonging the WSN's life. Having this in mind, the authors of [86] have developed an energy-efficient fire monitoring protocol, i.e., "EFMP", over clustered-based WSNs. Their results showed potential for overall energy consumption reduction, by forming a multi-layer cluster hierarchy depending on forest fire propagation. Despite its effectiveness, the EFMP introduces additional system complexity for dynamically computing the hierarchy. Ergo, alternative communication protocols have also been proposed. For instance, in [87,88] the authors have designed WSN systems, where the data transactions are conducted over the ZigBee protocol. Their experimentation results conduced that ZigBee is a powerful enabler for fire weather monitoring. Similarly, this study uses the ZigBee to realize the communication between the WSNs' nodes.

At the same time, many frameworks sanctioned by different emerging ICT have been also explored, e.g., [89,90], aiming to accomplish reliable field data dissemination, promptly predict wildfire ignitions, and autonomously launch avoidance responses. Similarly, Kaur and Sood [91] proposed a fog-assisted IoT framework for forecasting fire incidents. The framework, close to the current approach, comprises three layers, namely, the data accumulation, fog, and cloud layers. Experimentation exhibited high precision in assessing the susceptibility of the considered habitat towards wildfire spreading. Note that many frameworks, like the ones presented in [92,93], employ WSNs established by Arduino sensory nodes along with Raspberry Pi gateways, since their highly customizable and flexible modes allow for the fast adaptation to the ever-changing climate conditions that favor the fire regime. That being said, the presented study also employs a similar architecture, using Arduino-powered sensory nodes and Raspberry Pis as intermediary nodes between the WSNs and the remote cloud server.

Moving on, Roque and Padilla [94] developed a prototype using Arduino Uno that communicates through Sigfox. Their prototype is able to detect fires utilizing sensors for temperature and smoke/gas concentrations. Despite their positive performance in terms of response time, their solution is only able to detect fire. In contrast, the current approach is able to also predict fire ignitions, by utilizing temperature and humidity readings. The same applies when considering the case of the LoRaWAN prototype found in [95]. On the other hand, works like [93,96–98] take a different approach through neural networks, resulting in highly accurate fire models via pattern recognition, especially when considering long-term monitoring. These works highlight the efficiency of machine/deep learning approaches that are lightweight enough to run even on Raspberry Pis. Although the current work does not include such algorithms, it is understandable that they can easily be embedded in future work to increase precision and timely warnings. Other studies employ alternative means of detection, e.g., Khan et al. [99] used cameras, while Kalatzis et al. [100] used unmanned aerial vehicles. However, these are more expensive to deploy and on many occasions do not fare well under conditions of heavy rain, fog, snow, mist, etc., and so fail to promptly acknowledge fire incidents.

Whatever the case, the main aim in all the aforementioned works remains the early fire detection in order to launch appropriate remedy countermeasures (like the ones shown in [101,102]), and to execute evacuation procedures (e.g., [103]), while conserving precious energy resources [104] and reducing the end-to-end delay [105], as outlined by [106]. With that said, a close alternative approach to the one followed here is located in [107], which provides a similar IoT platform for the semantic correlation of the generated raw data and their interpretation in terms of imposed fire risk, based on the popular "fire weather index" (FWI) [108]. That particular index is very accurate; however, it requires costly hardware installations. Contrarily, the IoT solution presented here utilizes the CBI [24], which is also precise, but most importantly it relies solely on atmospheric agents, making it a suitable candidate for low-cost implementations. To better contextualize this claim, the following subsection enlists some of the most popular fire danger indexes (FDI) and their requirements.

2.4. Overview of Fire Danger Indexes

Fire outbreaks are affected by different factors related to various physical processes and events. To quantify the fire risk situation, different FDIs have been proposed that combine different quantity environment variables to compute the ignition risk [109].

The McAthur Mark 5 FDI, also known as the "forest fire danger index" (FFDI), is one of the oldest measures, dating back to the 1960s. It is mostly utilized in Australia and is characterized by five ratings, these being low, moderate, high, very high, and extreme. Noble et al. [110] expressed the FFDI as an equitation based on wind speed, relative humidity, temperature, and drought effects.

Contrarily, the FWI is a fire risk model issued by the Canadian Forestry Service in the 1970s [108]. It is affected by four meteorological parameters regarding the noon temperature, noon relative humidity, 24 h precipitation levels, and the maximum speed of the average wind. Its mathematical output procures a number ranging from 0 to 25, which is then suitably mapped to the very low, low, moderate, high, or extreme fire severity risk indications. Although it is considered a highly accurate metric, de Groot et al. [111] showed that it requires significant calibration for the classification thresholds to suit the local weather conditions appropriately.

Compared to FFDI and FWI, a simpler solution has been proposed by Sharples et al. [112], which is correspondingly called the "simple fire danger index" (F). F takes into account the wind velocity, fuel moisture content, temperature, and relative humidity to divide the fire danger into a five-level scale that ranges from low to extreme. Although simpler than the FFDI and FWI, it has generated mixed performance results that depend on the site of deployment [109].

The aforementioned FDIs are costly when observed from the perspective of equipment utilization. Moreover, they demand a priori the collection of large volumes of data, in order to produce accurate risk classification, especially when considering highly heterogeneous environments such as the Mediterranean Basin. The CBI, however, which was initially proposed by the Chandler et al. [24] in the 1980s, is solely based on weather conditions. Hence, it only requires the air temperature and relative humidity conditions to calculate the immediate fire risk. This makes CBI cost-efficient, since it does not postulate high equipment expenses for the collection and analysis of the field data, rendering it an ideal candidate for low-cost implementations and time-critical applications, as is the case here.

3. System Design and Configuration

Having discussed relevant literature, it is now feasible to venture forth and thoroughly present the proposed IoT solution. Although the scope here focuses on wildfire, the presented architecture is generic and can easily cope and comply with other types of timely environmental monitoring applications.

3.1. The Considered Cloud/Fog Computing Network Architecture

The particular cloud/fog hybrid architecture proposed here (i.e., Figure 1) follows a simple layering model to categorize the available services based on resource availability.

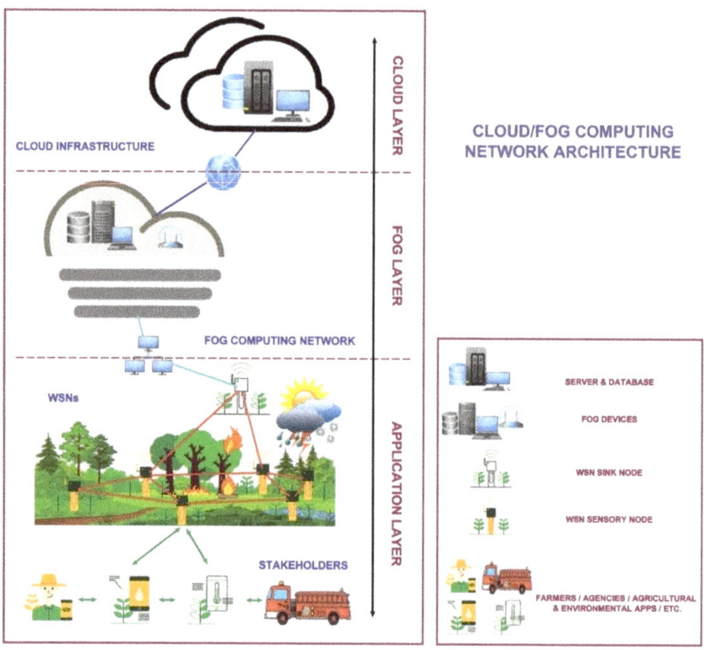

Figure 1. The considered three-layered cloud/fog computing IoT architecture for environmental monitoring, with an emphasis on fire detection applications.

Cloud computing providers commonly employ data centers considering various parameters, such as user proximity and energy consumption [113]. Thus the top layer, i.e., the *cloud layer*, usually includes a cloud infrastructure, formed by data centers offering amenities and resources, which are dynamically allocated based on the users' demands. These services may include, among other things, storage, networking, and server (computational power, rendering tools, etc.) privileges, as illustrated in Figure 1.

As expected, cloud data centers are usually situated in remote, safe locations and require expensive installations. To procure an affordable and scalable solution, the fog computing paradigm has been proposed. Fog devices, similarly to their cloud-only counterparts, can act as mini data centers that are generally cheaper and highly accessible, extending the service provisioning to the edges of the network [114], forming an intermediary layer, i.e., the *fog layer*. In this way, the computational burden is alleviated and resources are freed, avoiding traffic bottlenecks and increasing the cloud/fog system's overall capacity.

The *application layer*, located at the bottom layer, corresponds here to agricultural/environmental monitoring applications that have already been described in Section 2, including the ones that address the subject of fire handling events, running on the deployed WSNs. The last are comprised of devices called sensory nodes, capable of monitoring environmental agents, such as temperature and humidity, and a sink node, which is tasked with collecting all the field readings and reporting back to the higher layers for further computation. Note that end users also include the various stakeholders that manage the particular WSNs—farmers, firefighting services, researchers, environmental agencies, etc., possibly through various GUIs—different displays, smart or mobile devices, diversified platforms, etc., and are responsible for initiating appropriate countermeasures in cases of emergency (e.g., fire ignition) or high risk (e.g., when temperature and humidity favor the spreading of wildfire).

Clearly, the end users on most occasions can access fog devices in their near vicinity, without the necessity for establishing connections with the remote cloud servers. As a result, communication bandwidth is saved and the proposed architecture can trigger actions with lesser delay, reducing network congestion near the cloud servers if the need arises.

3.2. Hardware and Software Specifications

In the current subsection, a proposed customizable configuration comprising popular hardware micro-controllers for realizing the WSNs and a fog computing network is provided. These rely on market-based low-cost hardware solutions and established software operating systems and formats.

The WSNs are synthesized by sensory nodes, governed by the sink node, which is responsible for data collection. All nodes consist of Arduino micro-controller boards. Arduino [115] is an open-source electronics platform, popular for its hosting capacity and simple configuration, which can easily embed various components and modules, thereby extending its functionality and fast-tracking the creation of prototypes.

In detail, each sensory node is made of an Arduino Uno. To enable wireless connectivity, the board is enhanced by an Arduino wireless Secure Digital (SD) shield, with a Digi XBee ZigBee module [116]. The shield encapsulates three different protocol stacks: IEEE 802.15.4, DigiMesh, and ZigBee. The last one specifies a spacial carrier-sense multiple access with collision avoidance (CSMA/CA) protocol for creating wireless networks from small, low-power digital radio antennas [59]. Using wireless communications, although convenient, opens the door to potential security threats. Fortunately, the XBee makes securing the network a trivial task, since it uses encryption and a secret key, named the "network key", to ensure transmission protection and packet integrity [41].

Regarding the sink node, this differs because of its special properties regarding the collection and management of all the sensed data. For this reason, an Arduino Mega is utilized, which boasts greater memory capacity. The sink node is also enhanced with a wireless SD shield and a similar ZigBee radio module, allowing wireless communication with the Arduino Uno sensors of its assigned WSN.

Dwelling deeper into the WSN connection scheme, according to [117], three basic types of node roles are identified. The *coordinator* is a key structural component during the WSN initiation, tasked with setting up the conditions for its formation, including the selection of the operating channel, the assignment of a personal area network (PAN) identity (ID), and the establishment of a suitable routing traffic plan. There can only be a single coordinator in each WSN; hence, for the system at hand, the Arduino Mega is assigned this role during the field installation. In comparison, there can exist multiple *routers*, which are intermediate nodes with routing properties, tasked with relaying data from other nodes that cannot directly communicate with the coordinator due to long distances, i.e., the *end devices*. The routers and end devices, for the described system, are the Arduino Uno nodes.

The Arduino Mega is also utilized as a gateway for transmitting the field data to the overseeing fog device, through serial communication. With that said, the fog computing network is formed by Raspberry Pis, each responsible for receiving and analyzing readings from near WSNs and then relaying the data, if necessary, to the central cloud computing infrastructure using the local area network, in this case the Third-Generation (3G) of mobile communication connectivity. Unlike the Arduino devices, the Raspberry Pis have increased computational power that can be upgraded with cloud elastic resources, by offloading cloud demands in close proximity to the WSNs [118]. Additionally, they are able to seamlessly connect/interact with the former, thereby making them an ideal device for hosting fog-related processes.

Of course, there are many alternatives that can be employed to realize both the WSNs and the fog computing network. For instance, consider the Banana Pis or Orange Pis [119]. Although these are valid counter-proposals, the Arduino and Raspberry Pi devices were selected due to their extensive documentation, low-cost peripherals, highly configurable nature, and facile intercommunication. Similarly, the XBee modules, although they offer less reliability or range than other wireless antennas, such as the LoRa-based antennas [120], the former were chosen due to their affordable character, ease of programming, and low network and deployment cost, especially when considering large-scale, heterogeneous, and geographically distributed installation sites. In any case, the adoption of an alternative solution is strongly attainable since the elastic and extendable nature of the proposed system allows for such modifications with ease. That being said, for convenience and to better contextualize the economic aspects of the utilized hardware, Appendix A describes in more detail the specifications of the incorporated micro-controllers, whereas Appendix B compares wireless communication technologies.

3.3. Data Flow and Processing Methodology

The first phase of setting up the system, as depicted in Figure 2, involves the WSN initialization stage, where all nodes are deployed and then join and form the WSN [41]. This stage refers only to the WSN since it is assumed that the fog computing network and cloud computing infrastructure are already fired up and running, awaiting new raw data.

The first to power-on is the ZigBee coordinator, which in turn initializes the remaining WSN hardware. This is accomplished by initiating the protocol stack and performing an energy detection scan [4] to obtain a list of secure potential channels. Thereafter, it will continue with an active scan, where it chooses a free channel and enables the joining process, during which a routing plan is established. Moreover, due to WSNs being vulnerable to exogenous environmental factors, the clocks of the individual nodes experience de-synchronization. Hence, the proposed system encompasses a synchronization period during which the nodes calculate their time offset and synchronize their clocks accordingly, using as a reference point the coordinator's time.

Upon completion of the above phase, the various nodes start sensing environmental data. This occurs at regular intervals using their attached sensory modules. For the project at hand, which is targeted at fire alertness, the sensors track temperature and humidity, both variables critical to wildfire spreading. Conveniently, both come packed together in the DHT22 digital temperature/humidity sensor (also named AM2302 and depicted in Figure A1).

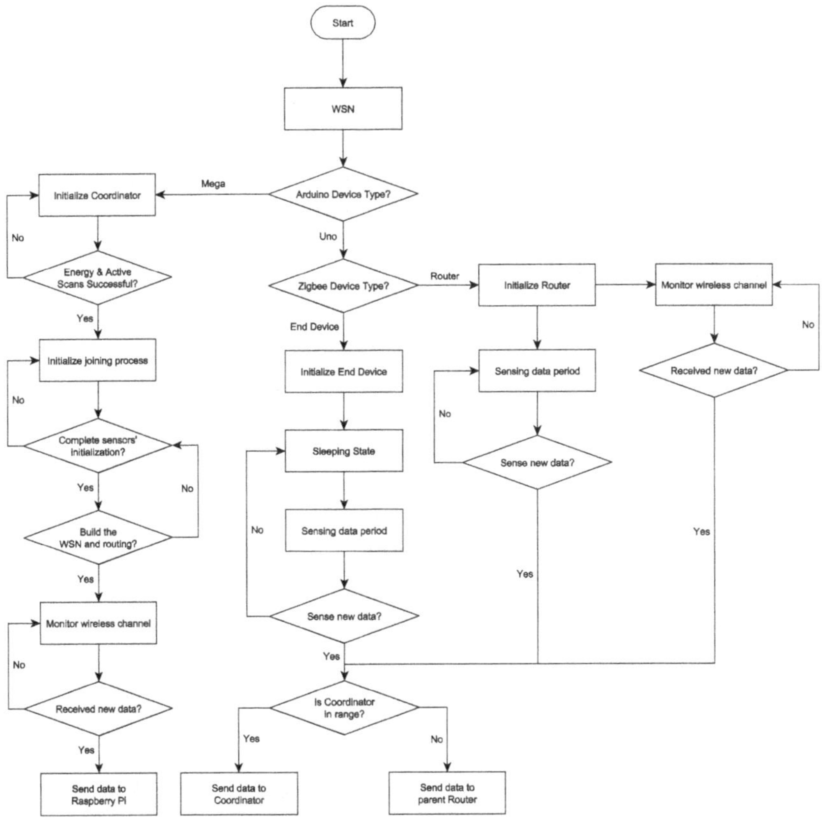

Figure 2. Flow chart of all processes taking place in the WSNs.

After a sensing period ends, the new data are transmitted towards the sink node over the established routing paths. If the nodes are too far away to convey the data directly by themselves (i.e., the coordinator is out of range), then they forward the data to their parent router. This check is repeated on all nodes along the given route until the data reach the coordinator. For this process to be reliable, the routers and the coordinator must continuously monitor the wireless channel for incoming signals.

Upon data reception, the Arduino Mega adopts a new role, forming a gateway for connecting the WSN with the fog computing network, comprised of several Raspberry Pis. For simplicity, these are connected with their assigned Arduino Mega devices through serial connections. The Raspberry Pis, as reported in Figure 3, listen to their serial channel for incoming data so as to initiate a connection. Note that the Raspberry Pis may potentially communicate with one or more Arduino Mega devices, collecting, in the second case, data from multiple WSNs.

The moment the Raspberry Pis receive a stream of data, they decide if they will perform the necessary calculations themselves or forward them to the central server infrastructure. This depends on the adopted implementation road, since the increased capabilities of the Raspberry Pis allow for such modifications in their configuration to support customized solutions. For the case study at hand, a simple distributed approach was considered; ergo, the fog devices' processing behavior was developed as follows. For every set of data packets received, a percentage of them would be computed on the spot by the fog computing network and the rest would be forwarded to the cloud computing infrastructure. In this way, the trafficking and computational load at the cloud servers would remain low and balanced by the fog devices, and so the system's overall response time would decrease.

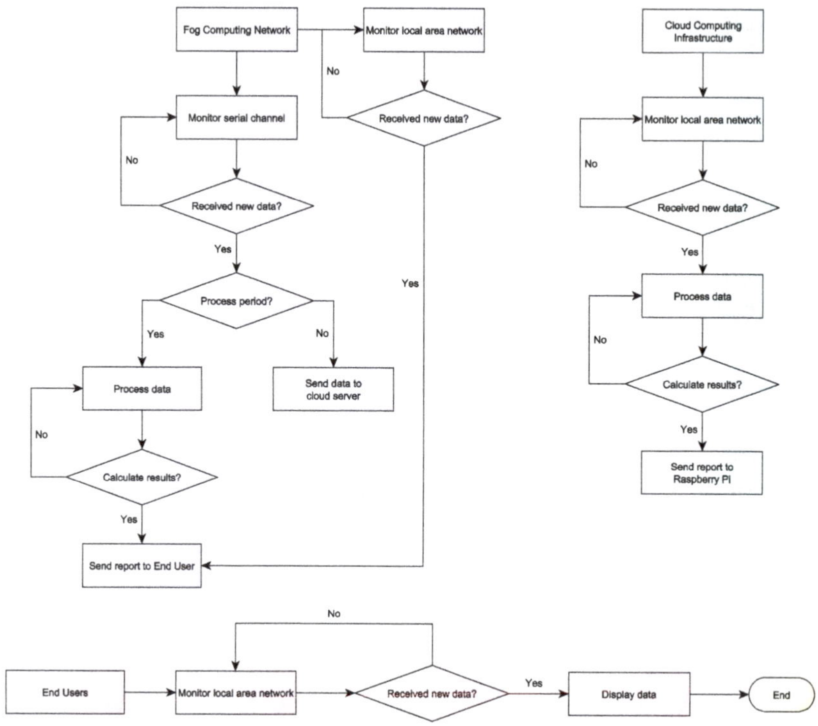

Figure 3. Flow chart of all processes occurring in the cloud/fog side of the proposed system.

The last can be further explained considering that, as the data packets are transmitted along the route, they are queued inside intermediate node buffers. The same occurs when the data reach the cloud server. However, if the data arriving rate exceeds the processing rate of the server or the data gets too many in number and too big in size, this will lead to the creation of network bottlenecks. In turn, the response time of the system might increase dramatically, especially in cases where there exist spikes and fluctuations in the sensed data, like in the cases of sudden fire spreading.

At any rate, when the sensed data are collected, computed, and analyzed, a system decision is generated and a response report is created. In the first case (where the computation takes place in the fog), the response is directly transmitted through the local area network to the end users' devices. Meanwhile, if the computation occurs in the cloud, the generated report travels backward towards the corresponding fog devices and then to the end users that administrate the corresponding WSNs.

4. Evaluation

The current section validates the effectiveness of the presented cloud/fog architecture and evaluates its adaptability in terms of reliable monitoring. To do so, an experimental prototype was designed in a closed control laboratory environment in the facilities of the Ionian University, located on Corfu Island.

4.1. Experimentation Setup

The designed experimental prototype comprised 25 Arduino Uno Rev 3 nodes, six (6) Arduino Mega 2560 nodes, and three (3) Raspberry Pi 3 Model B fog nodes (for more details about these models consult the Appendix A). In particular, six WSNs were considered, each containing a various number of Arduino Uno devices and an Arduino Mega coordinator/sink node, along with their arsenal of sensory and communication modules. Each sink node was linked with a serial cable to a Raspberry Pi. Following, the three fog devices communicated via 3G with the central cloud infrastructure, comprised of a virtual machine (VM) server with its accompanying database and storage, located in a different building of the Ionian University's facilities, as depicted in Figure 4. The exact setup of the WSNs in conjunction with the corresponding fog devices is outlined in Table 1. In fact, the number of sensory nodes varies among the WSNs in order to create diversified traffic load case scenarios (from low to high) at the respected fog devices and obtain a more objective assessment of their conforming capabilities.

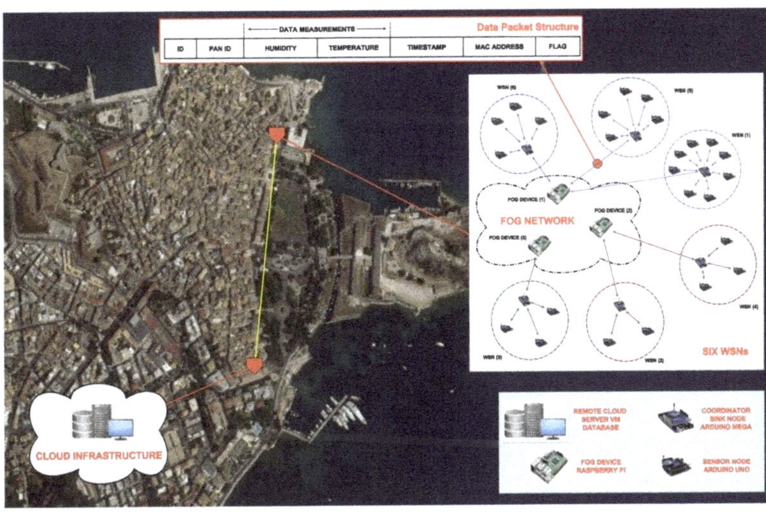

Figure 4. The experimental prototype system setup, programmed and installed in different locations of the Ionian University's facilities in Corfu Town.

Table 1. Deployment setup of the WSNs and fog network

WSN ID	Number of Sensory Nodes	Fog Device
One (1)	Eight (8)	One (1)
Two (2)	Two (2)	Two (2)
Three (3)	Three (3)	Three (3)
Four (4)	Three (3)	Two (2)
Five (5)	Five (5)	One (1)
Six (6)	Four (4)	One (1)

For simplicity, it was assumed that the WSNs are composed of their coordinator and all sensory nodes were assigned the role of routers in the constructed PANs, leading to more scalability. The sensed measurements along with the appropriate PAN ID were encapsulated inside data packets at regular intervals with a $+/-5$ s of random increment, in order to force diversity in the packet generation rate and minimize the appearance of packet collisions. In addition, the Arduino Uno timestamped each packet and encapsulated their own medium access control (MAC) address. Finally, the packets were marked with a unique ID and flagged based on the processing location (i.e., the fog or the cloud). The complete structure of a data packet is depicted in the upper part of Figure 4.

Each Arduino Mega stored the incoming information inside its SD memory card and then forwarded the data packet its assigned Raspberry Pi. The latter stochastically handled the incoming sensed data, based on a probability function, deciding whether to process the data locally or convey the data to the remote cloud server. More information regarding the server's resources can be found in Appendix A.

Both fog and cloud performed the same processing by extracting the sensed data and calculating the average values for each discrete variable. However, if the process took place directly at the fog network the data packet was flagged with a zero (0), whereas if it was carried out at the cloud, it was flagged with a one (1). Each data packet was processed once based on its unique identifier. Then the procured data (e.g., average temperature and humidity) were used to overwrite the corresponding fields in the data packet, while at the same time being stored in separate database logs and compatible formats for future reference.

At this point, it is important to state that data probity validation occurs for each data packet received to ensure that no data corruption exists and to exclude values that are miscalculated. Initially, the sensory and sink nodes' antennas perform a checksum on the data packets to verify their composition. Then, upon reception of a new data packet, the Raspberry Pis also check the structural integrity of the packet. First, based on the ID and PAN ID fields they cross-validate the origin of the data packet based on their assigned WSNs and previous observations. Then the measurement fields (i.e., the HUMIDITY and temperature) are assessed for their correctness, e.g., they must humidity be float numbers with a defined number of digits. Additionally, to verify that packets with information losses or distorted data are not taken into consideration, the fog devices inspect the format of all fields that must meet certain criteria; e.g., the ID and MAC ADDRESS fields must always process a specific sequence of characters. Data packets that infringe these conditions are automatically discarded. By embedding such mechanics, the fog devices are able to filter the data and offer basic data validation. Of course, there are other more sophisticated ways to verify the data integrity and offer reliability. For example, future work will include the cross-validation of the data based on pre-defined thresholds regarding the expected environmental and seasonal conditions in the monitored lands. For instance, during the summer pic a lower bound of 15 °C could be enforced and data packets that violate this constraint will be automatically dropped. Correspondingly, an upper bound of 50 °C could be placed and when violated the fog devices will assess the credibility of the reading based on the measurements previously obtained by the other sensory nodes in the immediate neighborhood. If multiple readings are received in sequence that record extreme heat, especially when originated in multiple sources, then the alarm will be raised appropriately.

4.2. Experimentation Results

For the presented prototype, response time is one of the most important performance metrics for timely monitoring and accurate correlation, and is computed as the *Round Trip Time* (*RTT*), i.e., the time that elapses between the moment a sensory node sends a new data packet and the moment it receives a reply from the cloud/fog system. In other words, the *RTT* is the sum of network and processing delays, which can be better observed in Figure 5, where the arrows depict the transmissions and the gray boxes illustrate the process time at each device. Note that Figure 5 represents the simplest scenario where the sensor is situated in the sink node's immediate neighborhood, i.e., one hop away.

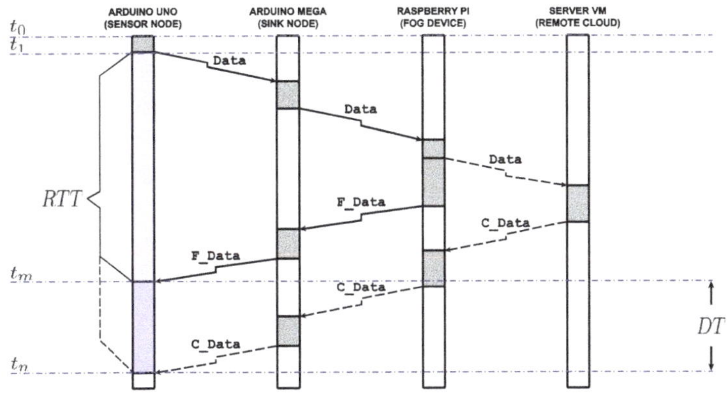

Figure 5. Calculation of the *RTT*, for both cloud and fog computing scenarios. The gray boxes symbolize the processing times along the route, whereas the arrows the transmissions that occur.

In detail, at time t_0 the Arduino Uno generates a data packet, hereafter called `Data`, containing the newly obtained measurements and their timestamp, which will be forwarded to the Arduino Mega at t_1, when the channel becomes available. Upon reception from the Arduino Mega, the `Data` undergoes some processing. When this is over, they are sent to the overseeing Raspberry Pi. Once there, the fog device decides with a probability P as to where the data process will eventuate, i.e, locally or remotely.

In the former case, the Raspberry Pi decodes the `Data`, processes the information, and extracts the average values, computing all relative information gathered up until that moment from the specific WSN. The average values are encapsulated into a new data packet, named `F_Data`, replacing the corresponding values of the received `Data`, but keeping intact the values of the ID, $TIMESTAMP$, and $MAC\_ADDRESS$ fields. Then the `F_Data` is flagged accordingly, by replacing the value of the $FLAG$ field, in order for the system to acknowledge the fog network as the processing location. Next, the `F_Data` begins its journey back towards the sensor node that generated the measurements. At the coordinator it stays a period equal to the time it takes for the Arduino Mega to read the $MAC\_ADDRESS$ and decide the route the `F_Data` must take to reach the respective Arduino Uno. Finally, it is forwarded to the last, which reads the $TIMESTAMP$ field and compares it to its current clock, calculating in this way the total RTT. Note again that in order for this procedure to be accurate there exists a simple clock synchronization method enforced at regular intervals.

In contrast, during the latter case, the Raspberry Pi conveys the data packet to the remote cloud server VM. The server performs similar actions to the ones described earlier. However, the information is now encoded into a data packet, named `C_Data`, and flagged with a value indicating the corresponding location, i.e., the cloud infrastructure. Similarly, the `C_Data` then travels backward the network route until it reaches the appropriate Arduino Uno, which, at that point calculates the RTT.

In Figure 5, the fog processing scenario is depicted through straight-arrow transmissions, whereas the cloud processing scenario is illustrated through dotted-arrow transmissions. Notice that the `F_Data` reaches the Arduino Uno in t_m, while the `C_Data` arrives at t_n, where $t_n = t_m + DT$. The DT is the difference between the two calculated RTTs, and is viewed as a measure estimate of the system's performance in terms of response time for multiple values of P.

Following the last assumption, Figure 6 encapsulates the results of the approach by presenting the average achieved RTT, as a function of P, for different experimental runs with varying time intervals regarding the data packet generation rate. The error bars demonstrate the upper and lower 95% confidence intervals. Clearly, for all depicted cases as P increases the RTT decreases. In fact, starting from the extreme scenario, where all processing occurs in the remote cloud server, i.e., for $P = 0$, the average $RTT \simeq 1160$ ms. Then, for every increment in P, more operations are performed in the fog network and so the RTT steadily drops, reaching its minimum value for the opposing extreme scenario, where all processing takes place directly in the fog network, i.e., for $P = 1$ the average $RTT \simeq 1080$ ms. Any spikes in the depicted plots are attributed to cloud/fog processing pulsations and delays posed by the ZigBee CSMA/CA protocol or other network intermediates during experimentation. What is more, no clear tendency can be derived by the abatement in the time interval period between sensory readings, eliciting that the fog network successfully coped with the incoming traffic in all circumstances.

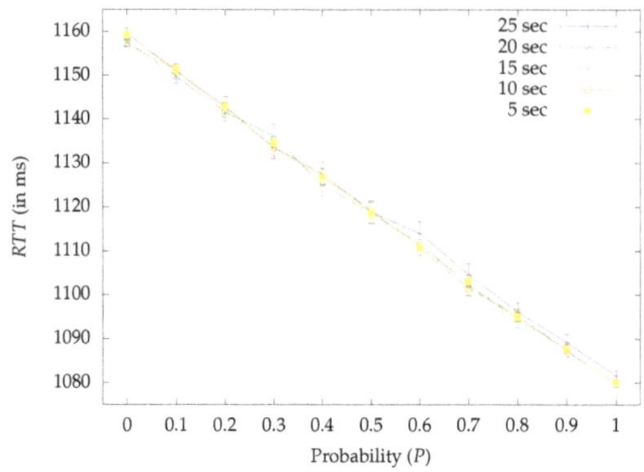

Figure 6. Experimental results of the system's total mean RTT as a function of the probability P for different data packet generation intervals, i.e., for 25 s, 20 s, 15 s, 10 s, and 5 s. The error bars represent the upper and lower 95% confidence intervals.

To further highlight the latency reduction behavior, Figure 7 decomposes each depicted case of Figure 6, by showcasing the mean RTT from the viewpoint of the involved fog devices. Even at a lower level of illustration the behavior still holds, proving once more the response time reduction across all fog devices as probability P is increased. To complete the system's RTT deconstruction Figure 8 offers a microscopic view of the system's RTT decaying behavior by encasing the same results from the scope of the deployed WSNs. In all six depicted cases the particular decreasing behavior is verified once more.

Finally, it is noteworthy that the number of WSNs and eventually of sensory nodes (according to Table 1) does not impact the responsiveness. Actually, all three Raspberry Pis successfully managed to handle the incoming traffic load, and thus it is further assumed that the RTT is affected by other processes running at the background during the experiment at the cloud or the fog devices.

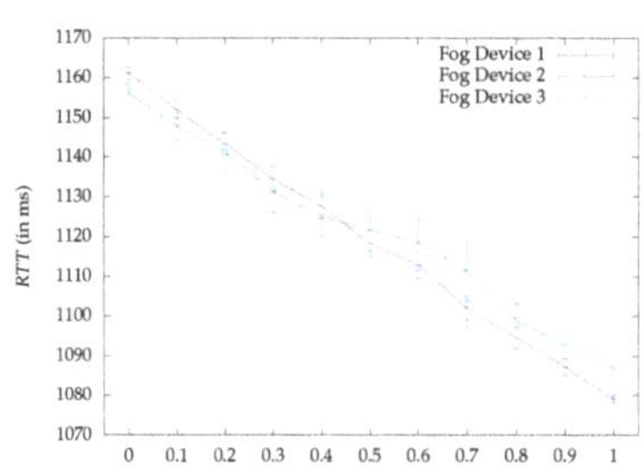

(a)

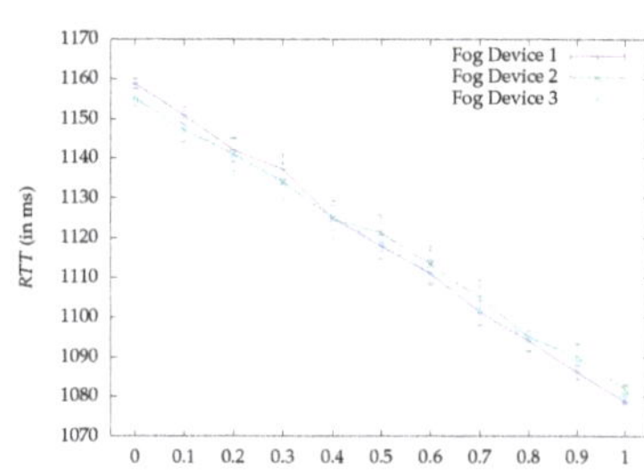

(b)

Figure 7. *Cont.*

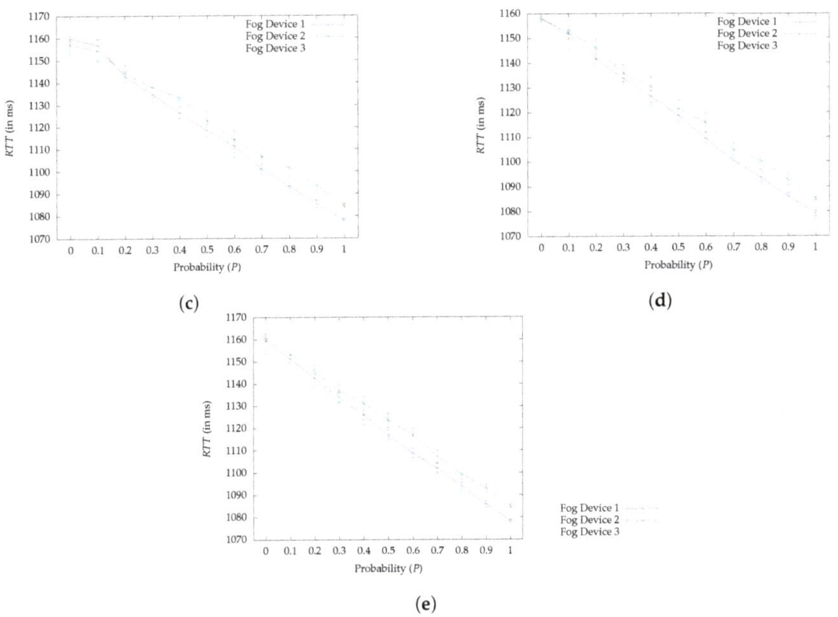

Figure 7. Experimental results of the system's total mean *RTT* value, through the viewpoint of the involved fog devices. The error bars correspond to the 95% upper and lower confidence intervals. (**a**) Interval: 25,000 (in ms). (**b**) Interval: 20,000 (in ms). (**c**) Interval: 15,000 (in ms). (**d**) Interval: 10,000 (in ms). (**e**) Interval: 5000 (in ms).

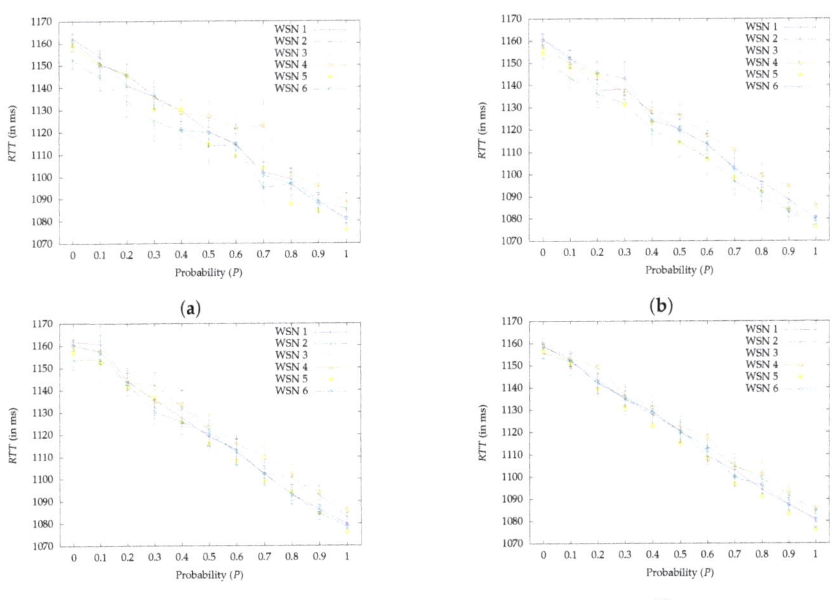

Figure 8. *Cont.*

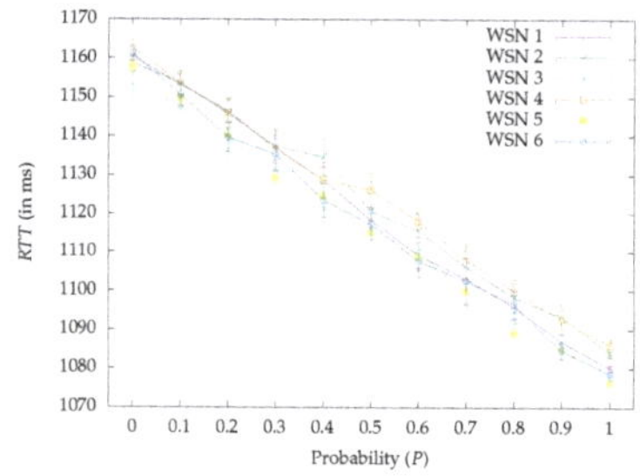

(e)

Figure 8. Experimental results regarding the mean *RTT*, through the scope of the deployed WSNs. The error bars refer to the upper and lower 95% confidence intervals. (**a**) Interval: 25,000 (in ms). (**b**) Interval: 20,000 (in ms). (**c**) Interval: 15,000 (in ms). (**d**) Interval: 10,000 (in ms). (**e**) Interval: 5000 (in ms).

5. System Conformation Based on Wildfire Risk Forecasting

To demonstrate the potential of the presented solution, which does not impose the irreducible network delays associated with conventional environmental IoT systems, whilst remaining energy efficient, the prototype is put against one of Greece's most hazardous states of emergency, i.e., the wildfires, to determine how well it can react and adapt its behavior to deal with a potential crisis. Additionally, a user-friendly designed GUI for data visualization and fire risk forecasting is presented.

5.1. The Case of Greece's Wildfires

According to Greece's Fire Brigade's (GFB) official statistics, 8006 forest fires were recorded during the year 2018 alone [121], while for 2019 the number is much higher, reaching 9502 fire fronts. The year 2007, however, was reportedly the worst of the last thirty years, since the country, during the summer, was hit by three consecutive heat waves (over 46 °C each), which coupled with strong winds and low relative humidity (around 9%), resulted in forest fires breaking out all over Greece. In fact, according to the European Space Agency (ESA), Greece has witnessed more wildfire activity during the summer of 2007 than other European countries have experienced over the last decade [122]. This led to 11,996 forest fires by the end of the year, burning over 675,000 acres of land, which was a European record for that period. To put this in more perspective, just on Corfu Island, an area of barely 236 square miles, in the year 2019 alone, 171 wildfires were recorded, burning around 137.5 acres of land, a great percentage of which consisted of farming lands and forests [121].

To deal with the fire peril and increase the degree of readiness, the GSCP during the firefighting period (from June the 1st to October the 31st), issues a daily map depicting the fire risk degree regarding all regions of Greece for the following day. In this way, the GSCP warns the corresponding authorities to prepare for the possibility of environment-threatening fire events in their respective regions and the citizens to stay on alert. The map is color-coded based on a five risk rate scale, starting from green, which indicates low risk, and up to red, which raises alarm to the highest possible level.

Having this information available is important because it allows for targeted adjustments to the proposed system's configuration, in order to maximize the effectiveness in detecting such catastrophic events. Consequently, for the purposes of the current experiment the cloud/fog system was programmed to autonomously adapt its behavior based on the risk degree scale regarding Corfu Island, where the experiment takes place. Figure 9 describes this procedure in depth.

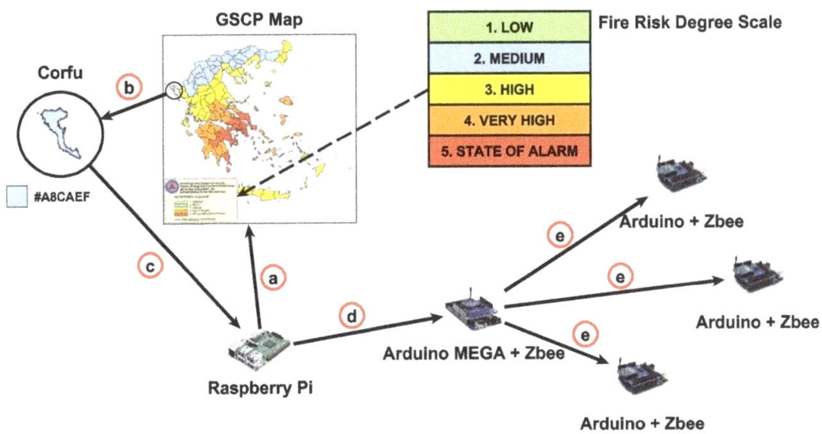

Figure 9. Steps for retrieving the risk degree and relaying the information to a WSN.

Once per day, during the early hours, when the map has been already published, the Raspberry Pis connect to the website of the GSCP (*step a* in Figure 9), which holds the information regarding the updated map [123]. Then they extract the specific risk degree regarding Corfu Island. To do so, the fog devices recognize the particular pixels that form Corfu on the map and retrieve their RGB color-code. For example, in Figure 9, where the fire risk for Corfu is set to "medium", the RGB color-code is determined as "#A8CAE" (i.e., *step b*). By reading the particular value, the Raspberry Pis immediately assess, according to the fire risk degree scale, the risk to be at level two (i.e., *step c*). After obtaining this value, they relay the information to their assigned Arduino Mega devices (i.e., *step d*), which in turn broadcast the information to their WSN, as illustrated in *step e* of Figure 9.

By the end of this procedure, all WSNs' nodes have received the fire risk degree and can modify their operation accordingly. Specifically, in order to be energy efficient, during the experiment, the sensors mapped their sampling behavior based on the risk degree that was fed into the WSNs by the fog devices. As such, when the sensory nodes received a degree of one (1), the interval period that intervened between two successive sensing periods was stretched to 25 s, with the goal of preserving battery energy and ensuring that communication bandwidth is not wasted, since there was no need for heavy monitoring at the time. For every degree that was added to the fire risk scale, however, the particular interval was autonomously reduced by 5 s. In this way, the generation rate of the packets was sped up as the risk got higher and continuous monitoring was required.

Table 2 captures the described behavior by presenting the interval period for every fire risk degree along with the probability *P* that drives the fog processing. In order to keep the main cloud infrastructure informed about the general ongoing situation in the fields, even in cases where the fire risk degree is five, a small percentage of the generated packages are still transmitted to the cloud server.

Table 2. System behavior under the different degrees of the Fire Risk.

Fire Risk Degree	Interval Period	Value of *P*
One (1)	25 s	5%
Two (2)	20 s	25%
Three (3)	15 s	50%
Four (4)	10 s	75%
Five (5)	5 s	95%

To test its performance the IoT system was instructed to retrieve each fire risk degree regarding Corfu, for the firefighting period of 2019. Ergo, 153 values corresponding to the five (5) months, were extracted and used as an input. The upper part of Figure 10 illustrates these values organized per month. Note that only once the risk achieved a degree of four (4), while no days existed that reached the value of five (5). Having these values stored, made it practicable to use them as system parameters. Thus, each day was represented with twenty (20) min of experimentation time.

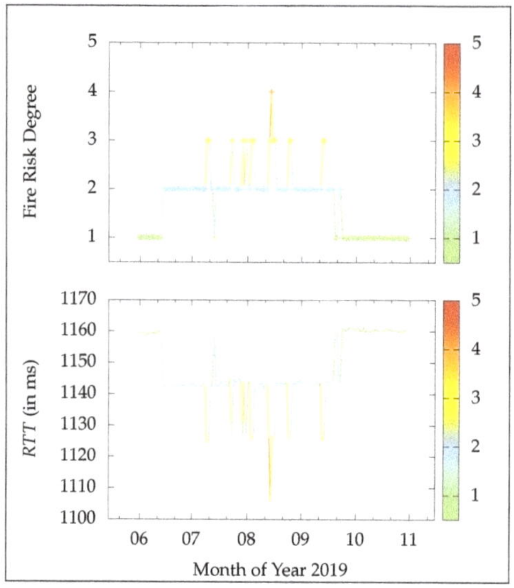

Figure 10. The fire risk degree (upper part) versus the average RTT in ms (bottom part), per day during the firefighting season of 2019.

The system's alertness encapsulates perfectly the change in the fire risk degrees as time goes on. Specifically, the system's obtained average RTT over all six deployed WSNs, is adjusted autonomously by pushing/pulling functionality towards/from the fog computing network, without human intervention. The result of this procedure is shown in the bottom part of Figure 10. In fact, the average RTT is reduced as the degrees climb the fire risk scale, enabling faster field-health analysis. In fact, it reaches its minimum value on the day that reported a fire risk degree of four (4), i.e., $RTT \leq 1110$ ms, during the pic of the firefighting season. In comparison, when the fire risk is one (1), most notably recorded during the beginning and ending of the firefighting season, the mean RTT is maximized, i.e., $RTT \simeq 1160$ ms, since most processing activity befalls the cloud server and there is no need for heavy field monitoring.

Regarding the traffic load, i.e., the data packet generation rate here, the system dynamically adapts its sampling periods, as validated in the upper part of Figure 11. Ergo, for low risk, the data packet generation rate is slowed down to save communication bandwidth and conserve precious energy, untimely elongating the WSNs' lifespan. The last is perfectly mirrored in the bottom part of Figure 12, where the energy consumption is visualized per day (i.e., for the 20 min experimental cycle here). This equates to the sum of the energy spent by all sensory modules, i.e., the temperature/humidity sensor to generate a new reading or not, plus the energy consumed by all antennas for their two alternate states, i.e., for transmission or idle respectively. Specifically, the DHT22 when in standby mode uses 50 uA while in reading mode 1.5 mA [124]. As for the XBee antenna, it consumes 31 mA in idle state and 120 mA in transmission state, respectively [125]. Likewise, as the degrees rise, the demand for higher precision impels the system to prioritize the need for continuous wildfire monitoring, thereby increasing the packet generation and transmission rates. In particular, for fire risk equal to one (1) the number of data packets per day is clustered around 900, translating to roughly 2.567 kJ of energy consumption. Contrarily, for the day reporting a degree of four (4), the packet number is launched to almost double and close to 1800, leading to increased energy consumption, which reports a value of around 2.572 kJ.

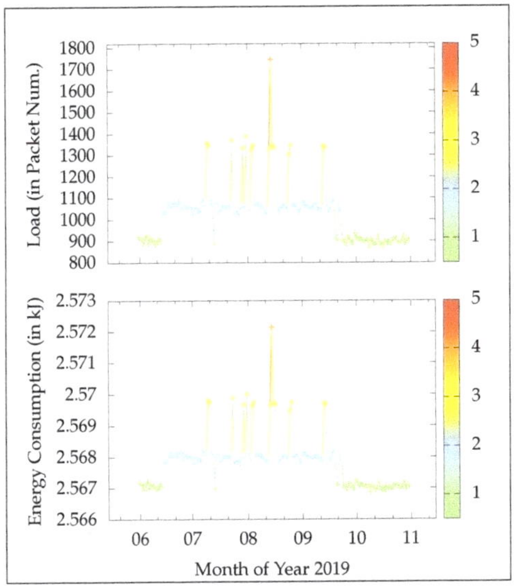

Figure 11. The traffic load, i.e., the number of transmitted packets (upper part), versus the corresponding energy consumption (bottom part) in kJ, per day of the firefighting season of 2019.

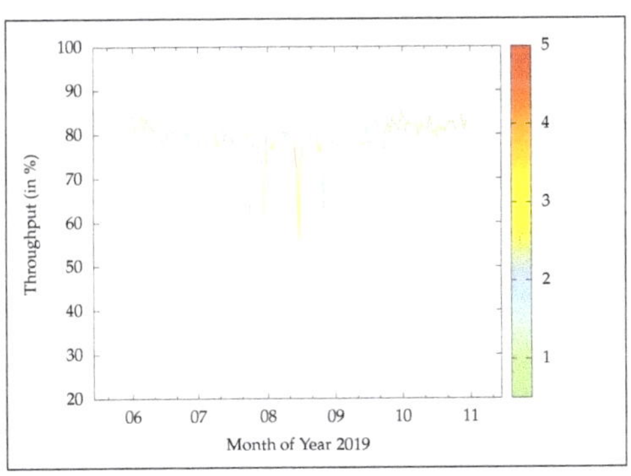

Figure 12. The achieved throughput in percent, per day during the firefighting season of 2019.

Although the aforementioned differences, especially in regards to energy consumption, at first glance might not seem crucial, it is understandable that under a real-world field deployment, where the number of sensory devices could grow to hundreds or even thousands of nodes and the time period will be expanded to real days, then the number of transmissions will greatly increase. However, the distributed fog processing methods adopted here will significantly reduce the *RTT* when compared to cloud-only approaches, especially considering the long distances that data packets will have to transverse to reach the remote cloud server from the field sites. In conjunction with multi-hop topologies and CSMA/CA re-transmissions and carrier sensing waiting times, it is obvious that the gap between the energy consumption during a low-risk day and an extreme-risk day will also be vastly larger. Moreover, in a real-world scenario, many of the WSNs' nodes will be assigned the role of end devices according to the ZigBee standard. Ergo, they will be permitted to enter "sleep" mode, by powering-off their antennas between transmissions to save additional battery usage. The network's lifetime can then be extended until the point of the first router or sink failure, which can jeopardize the WSN's connectivity, cutting access to the sink or fog network respectively. Nevertheless, this can also be addressed with energy-replenishment and harvesting tools, like solar-panel power-banks. Besides, ZigBee embeds re-routing methods that can promptly establish a new routing plan among the routers. With that said, it is possible to estimate the lifetime, by computing the average energy consumption per router/sink for the different fire risk degrees (using past observations), so as to predict the possible time of failure based on their residual battery energy from the previous day.

To verify that the rise in data packet traffic load does not sever the system's credibility, Figure 12 depicts the total throughput for each day of the experiment, based on the number of successfully retrieved data packet *IDs*. In most circumstances, even during heavy network traffic, the achievable throughput fluctuates around 80%. A few exceptions (i.e., throughput drops) are clearly attributed to the underlying CSMA/CA protocol. This finding is crucial because it enables the almost real-time forecasting/detection of a fire incident, with minimum loss of data, allowing for accurate monitoring and early notification of the authorities in case of fire ignition, in order to launch appropriate countermeasures that will mitigate the danger. Additionally, it is clear that the system can effectively cope with the computational burden, achieving high network performance, and retaining its accountability intact.

5.2. F.E.MO.S.: The Fog-Assisted Environmental Monitoring System

A key factor for truly providing an end-to-end IoT solution is the ability to visualize the data in a user-friendly manner through a proper GUI. Besides, as already mentioned the whole purpose of the considered monitoring cloud/fog system is to offer end users in the application layer the opportunity to quickly access and make sense of the field data, informing them about the ongoing status and alerting them regarding potential environmental hazards, like probable wildfire ignitions. The necessity becomes even more prevalent when the monitored lands are geographically scattered, with potentially diversified exposure to elements, altitudes, weather conditions, etc., and so the sensed data may vary substantially among the WSNs, resulting in complexities that jeopardize the decision making process.

To this end, a simple, yet accurate and extendable, web application has been developed for visualizing the streams of data arriving from either the cloud or the fog, suitably named the "**F**og-assisted **E**nvironmental **Mo**nitoring **S**ystem", or *F.E.MO.S.* in short. Both the server and the Raspberry Pis feed into F.E.MO.S. In real-time the captured environmental readings from the WSNs, and then F.E.MO.S. generates proper data-gram charts of their behavior in relation to time. Figure 13 depicts the web GUI dashboard of the F.E.MO.S. The dashboard is separated into panels, which are as follows.

Starting from the top central panel, namely, "Data Visualization", this is where the generated field data are plotted and custom-made figures are created regarding the chosen environmental parameters in relation to time. For the demo prototype system, the possible parameters to choose from are temperature (in °C) and relative humidity (in %). By default, F.E.MO.S. will plot the mean values of both, computed by the accumulated data from all installed WSN nodes, to procure an overall overview of the prevailing weather conditions in the deployment sites during the latest week, as is the case in Figure 13. Clearly, the generated plots are of high resolution and showcase the precision of the sensory nodes. Next to the bottom left corner of the panel, an indication of the current energy consumption (in kJ) is also provided to inform the stakeholders about the nodes' power utilization and help them in cases where energy replenishment is required. This value is updated on an hourly basis.

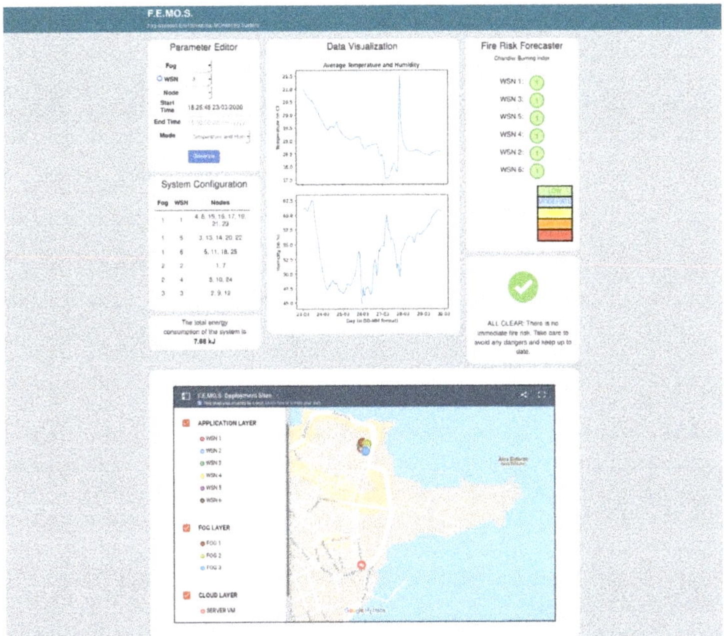

Figure 13. The web GUI dashboard of F.E.MO.S.

To customize the plots the left top panel, namely, "Parameter Editor", enables the users to configure the data visualization. In this respect, they may choose from a plethora of available options, to create targeted filtered queries and generate appropriate plots. The first option determines the type of system entity. There are mainly three types, these being the fog computing network, the deployed WSNs, or the sensor nodes themselves. However, in all three, the user is free to select either an averaged view of all sensory readings or further filter the data source. Ergo, he/she can designate which specific Raspberry Pi, WSN, or sensory node to explore respectively. A complete map of the system setup is provided, in the form of a Table, named "System Configuration", below the parameter editor, so as to assist the user in searching the appropriate device ID. Moreover, to access past measurement logs, a precise time period for data retrieval can be specified, in an hour/date format. Finally, a separate drop-down field is attributed to choosing the desired environmental parameter to populate the data visualization panel. In Figure 13, a simple user input example is shown, just before data generation.

While the aforementioned panels focused only on the presentation of the raw data, it was also considered imperative to augment F.E.MO.S. with cognition behavior, by offering semantic correlation of the readings in terms of wildfire forecasting, and thus highlight its potential in sensitive and timely decision-making procedures. To this end, the right top panel, suitably named the "Fire Risk Forecaster", presents the fire risk in each separate deployed WSN, using a color-coded, five-degree scale. To evaluate the risk, the CBI is utilized here [24], which is based solely on weather conditions. As such, CBI uses the air temperature and relative humidity to calculate in real-time a numerical index of the fire danger at the corresponding WSN sites, according to the following formula:

$$CBI = \frac{((110 - 1.373 \times RH) - 0.54 \times (10.20 - T)) \times (124 \times 10^{-0.0142 \times RH})}{60}, \quad (1)$$

where T is the current atmospheric temperature (in °C) and RH is the current relative humidity (in %). That number is then equated to the fire risk severity of either low, moderate, high, very high, or extreme, and mapped to the F.E.MO.S. color-coded scale mentioned earlier based on Table 3.

Table 3. F.E.MO.S. fire risk forecaster scale

Chandler Burning Index	Label & Color Code	Fire Risk Forecasting Rating
CBI $<$ 50	LOW (Green)	1
50 $\leq$ CBI $<$ 75	MODERATE (Blue)	2
75 $\leq$ CBI $<$ 90	HIGH (Yellow)	3
90 $\leq$ CBI $\leq$ 97.5	VERY HIGH (Orange)	4
CBI $>$ 97.5	EXTREME (Red)	5

Having this information available in real-time is exceptionally important, since it facilitates the dynamic conformation of the cloud/fog system to meet the requirements for greater or lesser environmental monitoring, as explained in Section 5.1. Moreover, it enables the early notification of the stakeholders and respective authorities for a possible fire threatening event at the monitored lands, allowing for timely interventions and targeted countermeasures (e.g., the launch of water-spraying mechanisms). Actually, to further increase mobilization, when the fire risk forecaster reports a fire risk rating equal or greater than three (3) at any given WSN (i.e., for CBI $\geq$ 75), F.E.MO.S. automatically generates and sends alerts to the registered email addresses of the responsible parties. An example of two such notifications is shown in Figure 14 for the case where the fire risk forecasting hits the severity score of four (4) and five (5), i.e., for "very high" and "extreme" danger respectively.

Similarly, F.E.MO.S. also provides in a separate panel the corresponding alert indication. In Figure 13, it is observable that all WSNs have a fire risk score of one (1), suitably highlighted with the color green. As a result, the indication is also a green check marker symbol, declaring that the ongoing situation in the WSN deployment sites is safe, and so there is no need for alarm. Contrarily, in Figure 15, an example of how a caution warning alert indication will look like in the case where the fire risk in the WSNs reaches the yellow indication, i.e., "high", is given, by zooming in on the fire risk forecaster and the corresponding notification panel of F.E.MO.S.

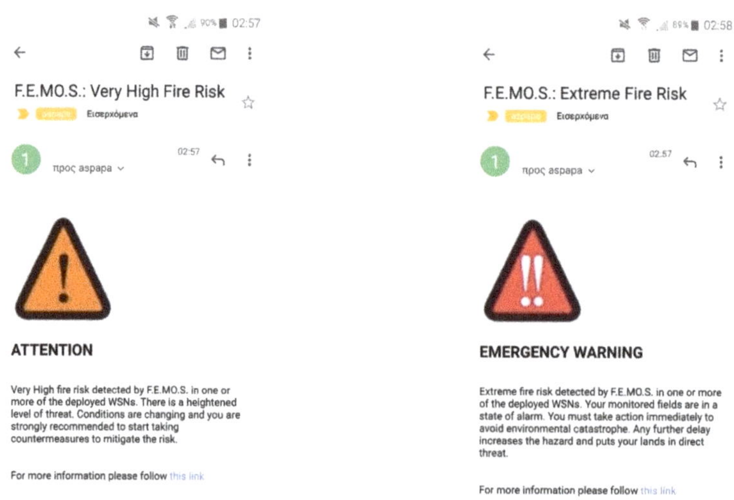

(**a**) Very high fire risk (**b**) Extreme fire risk

Figure 14. F.E.MO.S. automated notification alert messages for the cases of (**a**) very high and (**b**) extreme fire risk forecasting.

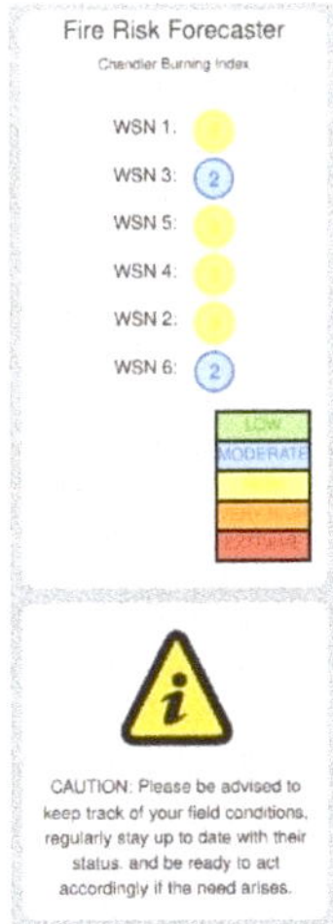

Figure 15. F.E.MO.S. indication regarding the case of a high fire risk scenario in one or more WSNs.

Lastly, for the users' convenience, F.E.MO.S. also provides a separate panel containing a map with all information regarding the marked deployment sites, organized in different layers that follow the considered three-layered system network architecture. The map is scrollable and flexible to allow users to explore the installation sites even when the system entities are situated in geographically different or remote locations, with long distances among them. An instance of the map panel, regarding the deployment sites of the prototype system, can be seen in the bottom part of Figure 13.

6. Conclusions and Future Directions

A hybrid, three-layered system architecture for smart and timely environmental monitoring, embedding affordable IoT and WSN appliances, while in its core following a cloud/fog computing approach, was presented here. The architecture was then extensively studied and its data flow functions were analyzed in depth, starting from the field nodes and up to the moment the information was delivered to the appropriate parties. Furthermore, the current work reported on the design and implementation of a demo prototype that can easily conform its functionality to address critical environmental challenges, based on the described architecture. To this end, a hardware/software solution was also proposed that is affordable and suitable for fast prototyping, while utilizing highly customizable components and controllers capable of networking, communication, measurement sensing, and processing.

Moreover, extensive experimentation was conducted, in controlled laboratory conditions in the facilities of the Ionian University, to evaluate the prototype's performance and highlight its alertness-conforming characteristics. A popular metric for precision environmental monitoring was considered, i.e, the response time. Initial experimentation to investigate the system's elements interactions and seamless adaptability captured the expected behavior and validated its effectiveness in dealing with time-sensitive agricultural and environmental applications. In all cases presented, the average response time was reduced as more operations took place in the intermediate fog computing network. That being said, for the two extreme cases, i.e., in the cloud-only processing and fog-only processing, the average RTT values were 1160 ms and 1080 ms respectively, indicating an overall 80 ms improvement across all system elements without exception.

To further highlight its performance, the developed prototype was put to the test under a real case scenario involving one of Greece's most notable environmental hazards, the wildfires. Again the results support the research claims, as it is shown that the system is able to adapt its operation and alertness, sufficiently addressing the problem, based on the five-degree fire risk scale retrieved from the GSCP.

Hence, it was able to reserve communication bandwidth (on average $\simeq$900 data packets) and decrease energy consumption, when the fire risk was low, all the while increasing its monitoring precision, effectively raising its sensory readings to almost the double (i.e., $\simeq$1800 data packets), when there existed high probability of fire ignition, achieving in most cases an average throughput over 80%.

To contextualize the system outputs and augment the data visualization process, a user-driven web-based GUI was also developed to accompany the prototype and allow users to proactively partake in the monitoring process. F.E.MO.S., as is its name, provides friendly panels to enable the filtering of the raw data and the flexible presentation of the monitored environmental conditions and system entities. Actually, it even goes a step further, by enhancing the decision-making process and offering real-time CBI-based fire risk forecasting, considering the prevailing weather conditions at each separate WSN deployment site. To increase alertness and actuate higher mobilization, targeted alert notifications are generated automatically in cases of emergency, while systematic logs are maintained to successfully correlate the data and allow the stakeholders to efficiently keep track of the status and health of the monitored lands and energy consumption levels of the system modules.

The paper also focused on the prototype's limitations. That being so, the biggest drawback of the approach is the lack of actual large-scale deployment in a real-world environment, which due to the poor network connectivity, power limitations, hostile environmental conditions, multi-hop routing, etc., will certainly affect the credibility and reliability of the sensor readings. However, the system in its current form can act as a proof-of-concept testbed, to fast test various other parameters or technologies prior to field deployment and investigate more deeply their implications, without endangering hardware/software elements. Moreover, it can significantly boost system debugging and offer insights regarding cases where device/module failures or data corruptions occur. The last is considered critical for the successful adoption of the approach; therefore, appropriate actions must also be undertaken to validate the accountability and credibility of the measured readings with the adoption of appropriate filtering and threshold methods. Other constraints, relating to automation, diversified network topologies, sensor clock synchronization, power utilization, and so on, are also taken into consideration during the prototype's design process, through proper software solutions and engineering decisions. The latter, although custom-made, remain generic and flexible, following established practices, and so can be easily modified to fit and include other appliances. Nevertheless, more research must be conducted towards the direction of securing these functionalities and augmenting their operations, even under extreme scenarios, e.g., during an actual fire outbreak. Another subject that raises attention relates to the optimal placement of the sensor nodes. For the case at hand, this was not necessary; however, during field deployment, this aspect will definitely go a long way towards alleviating possible WSN bottlenecks, routing issues, transmission collisions, and ultimately energy consumption. The same applies in the case where a proper subset of sensor nodes (backbone network) is discovered for each WSN, e.g., using a dominating set methodology, to boost information collection and dissemination. Obviously, tackling these challenges and limitations will greatly improve the system's overall performance, especially when considering that field deployment will involve a large number of sensory nodes.

With that being said, future guidelines will explore the system's standardization and its large-scale deployment in outdoor areas, where its activity will be extensively recorded and documented, as well as assessed in comparison with alternative techniques and systems. In fact, through mathematical formulation, it is feasible to identify additional crucial climate variables that lead to the formation of wildfire devastating events, in order to implement targeted field interventions. To this end, substitute fire burning indexes, such as the FWI, will be also researched, and their accuracies will be evaluated. Moreover, with a few tweaks in the code, the considered cloud/fog prototype system and F.E.MO.S. could be easily configured to support additional sensor modalities, and by extension alternative agricultural and environmental applications. In this respect, they will be enriched with extra monitoring tools (e.g., for determining smoke/gas emissions, soil moisture, wind velocity, pH levels, and rain intensity), which among other things, can also target or detect other environmentally hazardous events, for instance, earthquakes, rain floods, volcanic eruptions, tree deaths, pest infestations, etc. To manage the huge amount of raw data acquired by these additions, machine learning algorithms will be investigated, which coupled with the F.E.MO.S. forecasting metrics will lead to new prediction models, fostering high accuracy and enhanced decision-support mechanics. Furthermore, alternative wireless technologies (e.g., LoRa, NB-IoT, etc.) and hardware controllers (e.g., Orange Pis) will be tested to discover the optimum configuration. Great effort will be put toward providing a complete, low-cost IoT solution, reflecting the daily needs and accomplishing alignment with the expectations of future smart agriculture and environmental protection and preservation.

Author Contributions: Conceptualization, A.T. and A.P.; methodology, A.T., A.P., and I.A.; software, A.P. and G.K.; validation, K.O. and I.A.; formal analysis, A.T. and G.K.; investigation, G.K. and G.T.; resources, I.A. and G.T.; data curation, A.P.; writing—original draft preparation, A.T. and I.A.; writing—review and editing, A.P. and K.O.; visualization, A.T. and G.T.; supervision, A.T.; project administration, K.O. All authors have read and agreed to the published version of the manuscript.

Funding: This research received no external funding.

Acknowledgments: Acknowledgments This work was supported in part by project "A Pilot Wireless Sensor Networks System for Synchronized Monitoring of Climate and Soil Parameters in Olive Groves", (MIS 5007309) which is partially funded by European and National Greek Funds (ESPA) under the Regional Operational Programme "Ionian Islands 2014-2020".

Conflicts of Interest: The authors declare no conflict of interest.

Abbreviations

The following abbreviations are used in this manuscript:

3G	Third Generation of Wireless Mobile Telecommunications
5G	Fifth Generation of Wireless Mobile Telecommunications
CBI	Chandler Burning Index
CPU	Central Processing Unit
CSMA/CA	Carrier-Sense Multiple Access with Collision Avoidance
ESA	European Space Agency
F	Simple Fire Danger Index
FDI	Fire Danger Index
FFDI	Forest Fire Danger Index
F.E.MO.S.	Fog-Assisted Environmental Monitoring System
FWI	Fire Weather Index
GFB	Greece's Fire Brigade
GSCP	General Secretariat for Civil Protection
GUI	Graphical User Interface
ID	Identity
ICT	Information Communication Technologies
IoT	Internet of Things
MAC	Media Access Control
PAN	Personal Area Network
RAM	Random Access Memory
RTT	Round Trip Time
SD	Secure Digital
VM	Virtual Machine
WSN	Wireless Sensor Network

Appendix A. The Prototype's Utilized Hardware and Software Specifications

The utilized hardware micro-controllers for the design of the experimental cloud/fog IoT prototype include Arduino Uno, Arduino Mega, and Raspberry Pi devices. The current appendix encloses detailed information regarding technical specifications of these technologies, their selected models, and their accompanying modules, in addition to the resources utilized by the remote cloud server VM.

- *Arduino Uno*: In the current project implementation, the considered WSN sensors consist of an Arduino Uno Rev. 3, which is built on top of the Atmel ATmega328P micro-controller. This is in turn enhanced with a Digi XBee-PRO S2C ZigBee module [116] for wireless communication. The sensors were equipped with a DHT22 sensory module which is able to calculate the temperature in the scale of −40 °C to 80 °C, with a ±5 °C inaccuracy, and assess the humidity atmospheric levels in a scale of 0% to 100%, with an accuracy deviation between 2% and 5%.
- *Arduino Mega*: For the programming of the WSNs' sink nodes, an Arduino Mega 2560 micro-controller board was chosen, which is based on the ATmega2560. The sink nodes were augmented with communication capabilities using a wireless SD shield and a Digi XBee-PRO S2C module. They were also equipped with an SD memory card to save logs regarding the incoming readings. Moreover, they serially forwarded the data packets to their overseeing Raspberry Pi at a data rate of 115,200 bps.
- *Raspberry Pi Model B*: The fog devices composing the second hierarchy layer of the system's architecture, correspond to Raspberry Pis 3 Model B. This model was chosen due to its low-cost and low-power consumption attributes and its ability for wireless and serial connectivity. Essentially it is a small computer board that supports a number of different operating systems. For the purposes of current work, the Debian-based Linux operating system, named "Raspbian", was used.
- *Cloud Server VM*: The cloud server runs on a Unix-based VM, with a four-core central processing unit (CPU) and 4 GB of random access memory (RAM), which is part of the Ionian University's central cloud data center infrastructure, capable of high-speed computation and data transmission.

To put the aforementioned technologies in more perspective, Table A1 enlists technical specifications of the three micro-controller boards used, whereas Figure A1 depicts them after their assembly.

Table A1. Technical specifications of the devices used in this paper for the realization of the WSNs and fog computing network.

Specification	Arduino Uno Rev 3 [126]	Arduino Mega 2560 [127]	Raspberry Pi 3 Model B [128]
Microcontroller	ATmega328P	ATmega2560	Broadcom BCM2837 64 bit
Connectivity	-	-	Bluetooth 4.1 Classic/Low Energy, CSI, 10/100 Ethernet, 2.4 GHz 802.11b/g/n wireless
RAM	2 KB SRAM, 32 KB Flash Memory	8 KB SRAM, 256 KB Flash Memory	1GB LPDDR2 (900 MHz)
Pins	14 (of which 6 provide PWM output)	54 (of which 14 provide PWM output)	40-pin GPIO header
CPU	Intel Quark (x86) 16 MHz	Intel Quark (x86) 16 MHz	4 × ARM Cortex-A53, 1.2 GHz
GPU	-	-	Broadcom VideoCore IV @ 250 MHz
MSRP	≃20 €	≃35 €	≃40 €

Figure A1. From left to right, the utilized Arduino Uno with the attached Digi XBee-PRO S2C module and DHT22 Temperature and Humidity Sensor, the Arduino Mega with its Digi XBee-PRO S2C module and microSD card slot, and the Raspberry Pi 3 Model B enclosed in a protective case.

Appendix B. Comparison of Existing Wireless Technologies

The current appendix contains the comparisons of alternative wireless technologies in order to showcase the affordable character of the adopted approach in the current work and propose alternative solutions that can easily be incorporated in future system alterations. As such, Table A2 compares various communication technologies and their characteristics, including the utilized ZigBee.

Table A2. Wireless technologies comparison table [25,58,129–133].

Wireless Technology	Range	Security	Deployment Cost	Power Usage	Maximum Data Rate
Zigbee	≤100 m	LOW	LOW	LOW	250 Kbps
LoRa	≤20 Km	HIGH	LOW	LOW	50 Kbps
NB-IoT	≤10 Km	HIGH	HIGH	HIGH	200 Kbps
Sigfox	≤50 Km	HIGH	MEDIUM	MEDIUM	100 Bps
BLuetooth	≤50 m	HIGH	LOW	HIGH	2 Mbps
LTE	≤30 Km	HIGH	MEDIUM	MEDIUM	1 Mbps
Z-Wave	≤100 m	LOW	MEDIUM	LOW	100 Kbps
Weigtless	≤5 km	HIGH	LOW	MEDIUM	100 Kbps

References

1. Yost, M.; Sudduth, K.; Walthall, C.; Kitchen, N. Public-private collaboration toward research, education and innovation opportunities in precision agriculture. *Precis. Agric.* **2019**, *20*, 4–18. [CrossRef]
2. Mekala, M.S.; Viswanathan, P. A Survey: Smart agriculture IoT with cloud computing. In Proceedings of the IEEE International Conference on Microelectronic Devices, Circuits and Systems (ICMDCS), Vellore, India, 10–12 August 2017; pp. 1–7.
3. Ojha, T.; Misra, S.; Raghuwanshi, N.S. Wireless sensor networks for agriculture: The state-of-the-art in practice and future challenges. *Comput. Electron. Agric.* **2015**, *118*, 66–84. [CrossRef]
4. Baronti, P.; Pillai, P.; Chook, V.W.; Chessa, S.; Gotta, A.; Hu, Y.F. Wireless sensor networks: A survey on the state of the art and the 802.15. 4 and ZigBee standards. *Comput. Commun.* **2007**, *30*, 1655–1695. [CrossRef]
5. Kalaivani, T.; Allirani, A.; Priya, P. A survey on Zigbee based wireless sensor networks in agriculture. In Proceedings of the IEEE 3rd International Conference on Trendz in Information Sciences & Computing (TISC2011), Chennai, India, 8–9 December 2011; pp. 85–89.
6. Gupta, M.; Abdelsalam, M.; Khorsandroo, S.; Mittal, S. Security and privacy in smart farming: Challenges and opportunities. *IEEE Access* **2020**, *8*, 34564–34584. [CrossRef]
7. Lee, I.; Lee, K. The Internet of Things (IoT): Applications, investments, and challenges for enterprises. *Bus. Horizons* **2015**, *58*, 431–440. [CrossRef]
8. Chiang, M.; Zhang, T. Fog and IoT: An overview of research opportunities. *IEEE Internet Things J.* **2016**, *3*, 854–864. [CrossRef]
9. Popović, T.; Latinović, N.; Pešić, A.; Zečević, Ž.; Krstajić, B.; Djukanović, S. Architecting an IoT-enabled platform for precision agriculture and ecological monitoring: A case study. *Comput. Electron. Agric.* **2017**, *140*, 255–265. [CrossRef]
10. Bonomi, F.; Milito, R.; Zhu, J.; Addepalli, S. Fog computing and its role in the Internet of Things. In Proceedings of the First Edition of the MCC Workshop on Mobile Cloud Computing, Helsinki, Finland, 17 August 2012; ACM: New York, NY, USA, 2012; pp. 13–16.
11. Channe, H.; Kothari, S.; Kadam, D. Multidisciplinary model for smart agriculture using internet-of-things (IoT), sensors, cloud-computing, mobile-computing & big-data analysis. *Int. J. Comput. Technol. Appl.* **2015**, *6*, 374–382.
12. Guardo, E.; Di Stefano, A.; La Corte, A.; Sapienza, M.; Scatà, M. A Fog Computing-based IoT Framework for Precision Agriculture. *J. Internet Technol.* **2018**, *19*, 1401–1411.
13. Dastjerdi, A.V.; Gupta, H.; Calheiros, R.N.; Ghosh, S.K.; Buyya, R. Fog computing: Principles, architectures, and applications. In *Internet of Things*; Morgan Kaufmann, Elsevier: Amsterdam, The Netherlands, 2016; pp. 61–75.
14. Nundloll, V.; Porter, B.; Blair, G.S.; Emmett, B.; Cosby, J.; Jones, D.L.; Chadwick, D.; Winterbourn, B.; Beattie, P.; Dean, G.; et al. The design and deployment of an end-to-end IoT infrastructure for the natural environment. *Future Internet* **2019**, *11*, 129. [CrossRef]
15. Sethi, P.; Sarangi, S.R. Internet of things: Architectures, protocols, and applications. *J. Electr. Comput. Eng.* **2017**, *2017*. [CrossRef]

16. Ray, P.P.; Mukherjee, M.; Shu, L. Internet of things for disaster management: State-of-the-art and prospects. *IEEE Access* **2017**, *5*, 18818–18835. [CrossRef]
17. Visconti, P.; Primiceri, P.; Orlando, C. Solar powered wireless monitoring system of environmental conditions for early flood prediction or optimized irrigation in agriculture. *J. Eng. Appl. Sci.* **2016**, *11*, 4623–4632.
18. Alphonsa, A.; Ravi, G. Earthquake early warning system by IOT using Wireless sensor networks. In Proceedings of the IEEE International Conference on Wireless Communications, Signal Processing and Networking (WiSPNET), Chennai, India, 23–25 March 2016; pp. 1201–1205.
19. Awadallah, S.; Moure, D.; Torres-González, P. An Internet of Things (IoT) Application on Volcano Monitoring. *Sensors* **2019**, *19*, 4651. [CrossRef]
20. Tsipis, A.; Papamichail, A.; Koufoudakis, G.; Tsoumanis, G.; Polykalas, S.E.; Oikonomou, K. Latency-Adjustable Cloud/Fog Computing Architecture for Time-Sensitive Environmental Monitoring in Olive Groves. *AgriEngineering* **2020**, *2*, 175–205. [CrossRef]
21. Meyn, A.; White, P.S.; Buhk, C.; Jentsch, A. Environmental drivers of large, infrequent wildfires: the emerging conceptual model. *Prog. Phys. Geogr.* **2007**, *31*, 287–312. [CrossRef]
22. Pausas, J.G.; Llovet, J.; Rodrigo, A.; Vallejo, R. Are wildfires a disaster in the Mediterranean basin?–A review. *Int. J. Wildland Fire* **2009**, *17*, 713–723. [CrossRef]
23. Papadopoulos, A.; Paschalidou, A.; Kassomenos, P.; McGregor, G. Investigating the relationship of meteorological/climatological conditions and wildfires in Greece. *Theor. Appl. Climatol.* **2013**, *112*, 113–126. [CrossRef]
24. Chandler, C.; Cheney, P.; Thomas, P.; Trabaud, L.; Williams, D. Fire in forestry. In *Forest Fire Management and Organization*; John Wiley & Sons: New York, NY, USA, 1983; Volume 2.
25. Akpakwu, G.A.; Silva, B.J.; Hancke, G.P.; Abu-Mahfouz, A.M. A survey on 5G networks for the Internet of Things: Communication technologies and challenges. *IEEE Access* **2017**, *6*, 3619–3647. [CrossRef]
26. Li, S.; Da Xu, L.; Zhao, S. 5G Internet of Things: a survey. *J. Ind. Inf. Integr.* **2018**, *10*, 1–9. [CrossRef]
27. Jawad, H.M.; Nordin, R.; Gharghan, S.K.; Jawad, A.M.; Ismail, M. Energy-Efficient Wireless Sensor Networks for Precision Agriculture: A Review. *Sensors* **2017**, *17*. [CrossRef] [PubMed]
28. Tzounis, A.; Katsoulas, N.; Bartzanas, T.; Kittas, C. Internet of Things in agriculture, recent advances and future challenges. *Biosyst. Eng.* **2017**, *164*, 31–48. [CrossRef]
29. Yaqoob, I.; Ahmed, E.; Hashem, I.A.T.; Ahmed, A.I.A.; Gani, A.; Imran, M.; Guizani, M. Internet of things architecture: Recent advances, taxonomy, requirements, and open challenges. *IEEE Wirel. Commun.* **2017**, *24*, 10–16. [CrossRef]
30. Botta, A.; De Donato, W.; Persico, V.; Pescapé, A. Integration of cloud computing and internet of things: A survey. *Future Gener. Comput. Syst.* **2016**, *56*, 684–700. [CrossRef]
31. Yu, W.; Liang, F.; He, X.; Hatcher, W.G.; Lu, C.; Lin, J.; Yang, X. A survey on the edge computing for the Internet of Things. *IEEE Access* **2017**, *6*, 6900–6919. [CrossRef]
32. Puliafito, C.; Mingozzi, E.; Longo, F.; Puliafito, A.; Rana, O. Fog computing for the internet of things: a Survey. *ACM Trans. Internet Technol.* **2019**, *19*, 1–41. [CrossRef]
33. Cao, H.; Wachowicz, M.; Renso, C.; Carlini, E. Analytics everywhere: Generating insights from the internet of things. *IEEE Access* **2019**, *7*, 71749–71769. [CrossRef]
34. Bellavista, P.; Berrocal, J.; Corradi, A.; Das, S.K.; Foschini, L.; Zanni, A. A survey on fog computing for the Internet of Things. *Pervasive Mob. Comput.* **2019**, *52*, 71–99. [CrossRef]
35. Xu, L.; Collier, R.; O'Hare, G.M. A survey of clustering techniques in WSNs and consideration of the challenges of applying such to 5G IoT scenarios. *IEEE Internet Things J.* **2017**, *4*, 1229–1249. [CrossRef]
36. Qiu, T.; Chen, N.; Li, K.; Atiquzzaman, M.; Zhao, W. How can heterogeneous Internet of Things build our future: a survey. *IEEE Commun. Surv. Tutor.* **2018**, *20*, 2011–2027. [CrossRef]
37. Akyildiz, I.; Su, W.; Sankarasubramaniam, Y.; Cayirci, E. Wireless sensor networks: A survey. *Comput. Networks* **2002**, *38*, 393–422. [CrossRef]
38. Abbas, Z.; Yoon, W. A survey on energy conserving mechanisms for the internet of things: Wireless networking aspects. *Sensors* **2015**, *15*, 24818–24847. [CrossRef] [PubMed]
39. Akyildiz, I.F.; Su, W.; Sankarasubramaniam, Y.; Cayirci, E. A survey on sensor networks. *IEEE Commun. Mag.* **2002**, *40*, 102–114. [CrossRef]
40. Jindal, V. History and Architecture of Wireless Sensor Networks for Ubiquitous Computing. *History* **2018**, *7*, 214–217.
41. Kooijman, M. *Building Wireless Sensor Networks Using Arduino*; Packt Publishing Ltd.: Birmingham, UK, 2015.

42. Bacco, M.; Berton, A.; Ferro, E.; Gennaro, C.; Gotta, A.; Matteoli, S.; Paonessa, F.; Ruggeri, M.; Virone, G.; Zanella, A. Smart farming: Opportunities, challenges and technology enablers. In Proceedings of the IEEE IoT Vertical and Topical Summit on Agriculture-Tuscany (IOT Tuscany), Tuscany, Italy, 8–9 May 2018; pp. 1–6.
43. McConnell, M.D. Bridging the gap between conservation delivery and economics with precision agriculture. *Wildl. Soc. Bull.* **2019**, *43*, 391–397. [CrossRef]
44. Farooq, M.S.; Riaz, S.; Abid, A.; Abid, K.; Naeem, M.A. A Survey on the Role of IoT in Agriculture for the Implementation of Smart Farming. *IEEE Access* **2019**, *7*, 156237–156271. [CrossRef]
45. Joris, L.; Dupont, F.; Laurent, P.; Bellier, P.; Stoukatch, S.; Redouté, J.M. An Autonomous Sigfox Wireless Sensor Node for Environmental Monitoring. *IEEE Sens. Lett.* **2019**, *3*, 01–04. [CrossRef]
46. Botero-Valencia, J.; Castano-Londono, L.; Marquez-Viloria, D.; Rico-Garcia, M. Data reduction in a low-cost environmental monitoring system based on LoRa for WSN. *IEEE Internet Things J.* **2018**, *6*, 3024–3030. [CrossRef]
47. Yao, Z.; Bian, C. Smart Agriculture Information System Based on Cloud Computing and NB-IoT. In Proceedings of the 2019 International Conference on Computer Intelligent Systems and Network Remote Control (CISNRC 2019), Shanghai, China, 29–30 December 2019. [CrossRef]
48. Biswas, S. A remotely operated Soil Monitoring System: An Internet of Things (IoT) Application. *Int. J. Internet Things Web Serv.* **2018**, *3*, 32–38.
49. Jawad, H.M.; Jawad, A.M.; Nordin, R.; Gharghan, S.K.; Abdullah, N.F.; Ismail, M.; Abu-Al Shaeer, M.J. Accurate Empirical Path-loss Model Based on Particle Swarm Optimization for Wireless Sensor Networks in Smart Agriculture. *IEEE Sensors J.* **2019**. [CrossRef]
50. Li, N.; Xiao, Y.; Shen, L.; Xu, Z.; Li, B.; Yin, C. Smart Agriculture with an Automated IoT-Based Greenhouse System for Local Communities. *Adv. Internet Things* **2019**, *9*, 15. [CrossRef]
51. Kumar, S.A.; Ilango, P. The impact of wireless sensor network in the field of precision agriculture: A review. *Wirel. Pers. Commun.* **2018**, *98*, 685–698. [CrossRef]
52. Azfar, S.; Nadeem, A.; Alkhodre, A.; Ahsan, K.; Mehmood, N.; Alghmdi, T.; Alsaawy, Y. Monitoring, Detection and Control Techniques of Agriculture Pests and Diseases using Wireless Sensor Network: a Review. *Int. J. Adv. Comput. Sci. Appl.* **2018**, *9*, 424–433. [CrossRef]
53. Azfar, S.; Nadeem, A.; Basit, A. Pest detection and control techniques using wireless sensor network: A review. *J. Entomol. Zool. Stud.* **2015**, *3*, 92–99.
54. Grift, T. The first word: the farm of the future. *Resour. Mag.* **2011**, *18*, 1.
55. Chunduri, K.; Menaka, R. Agricultural Monitoring and Controlling System Using Wireless Sensor Network. In *Soft Computing and Signal Processing*; Springer: Berlin, Germany, 2019; pp. 47–56.
56. Suárez-Albela, M.; Fernández-Caramés, T.M.; Fraga-Lamas, P.; Castedo, L. A practical evaluation of a high-security energy-efficient gateway for IoT fog computing applications. *Sensors* **2017**, *17*, 1978. [CrossRef]
57. Castillo-Cara, M.; Huaranga-Junco, E.; Quispe-Montesinos, M.; Orozco-Barbosa, L.; Antúnez, E.A. FROG: a robust and green wireless sensor node for fog computing platforms. *J. Sensors* **2018**, *2018*. [CrossRef]
58. Hossein Motlagh, N.; Mohammadrezaei, M.; Hunt, J.; Zakeri, B. Internet of Things (IoT) and the energy sector. *Energies* **2020**, *13*, 494. [CrossRef]
59. Nikhade, S.G. Wireless sensor network system using Raspberry Pi and zigbee for environmental monitoring applications. In Proceedings of the IEEE International Conference on Smart Technologies and Management for Computing, Communication, Controls, Energy and Materials (ICSTM), Avadi, Chennai, India, 6–8 May 2015, pp. 376–381.
60. Flores, K.O.; Butaslac, I.M.; Gonzales, J.E.M.; Dumlao, S.M.G.; Reyes, R.S. Precision agriculture monitoring system using wireless sensor network and Raspberry Pi local server. In Proceedings of the IEEE Region 10 Conference (TENCON), Singapore, 22–25 November 2016; pp. 3018–3021.
61. Deshmukh, A.D.; Shinde, U.B. A low cost environment monitoring system using raspberry Pi and arduino with Zigbee. In Proceedings of the International Conference on Inventive Computation Technologies (ICICT), Tamilnandu, India, 26–27 August 2016; Volume 3, pp. 1–6.
62. Ahmed, N.; De, D.; Hussain, I. Internet of Things (IoT) for Smart Precision Agriculture and Farming in Rural Areas. *IEEE Internet Things J.* **2018**, *5*, 4890–4899. [CrossRef]

63. Bin Baharudin, A.M.; Saari, M.; Sillberg, P.; Rantanen, P.; Soini, J.; Jaakkola, H.; Yan, W. Portable fog gateways for resilient sensors data aggregation in internet-less environment. *Eng. J.* **2018**, *22*, 221–232. [CrossRef]
64. Keshtgari, M.; Deljoo, A. A wireless sensor network solution for precision agriculture based on zigbee technology. *Wirel. Sens. Netw.* **2012**. [CrossRef]
65. Cabaccan, C.N.; Cruz, F.R.G.; Agulto, I.C. Wireless sensor network for agricultural environment using raspberry pi based sensor nodes. In Proceedings of the IEEE 9th International Conference on Humanoid, Nanotechnology, Information Technology, Communication and Control, Environment and Management (HNICEM), Manila, Philippines, 1–3 December 2017; pp. 1–5.
66. Zamora-Izquierdo, M.A.; Santa, J.; Martínez, J.A.; Martínez, V.; Skarmeta, A.F. Smart farming IoT platform based on edge and cloud computing. *Biosyst. Eng.* **2019**, *177*, 4–17. [CrossRef]
67. Souissi, I.; Azzouna, N.B.; Said, L.B. A multi-level study of information trust models in WSN-assisted IoT. *Comput. Networks* **2019**, *151*, 12–30. [CrossRef]
68. Fortino, G.; Fotia, L.; Messina, F.; Rosaci, D.; Sarné, G.M. Trust and Reputation in the Internet of Things: State-of-the-Art and Research Challenges. *IEEE Access* **2020**, *8*, 60117–60125. [CrossRef]
69. Cao, X.; Chen, J.; Zhang, Y.; Sun, Y. Development of an integrated wireless sensor network micro-environmental monitoring system. *ISA Trans.* **2008**, *47*, 247–255. [CrossRef]
70. Casado-Vara, R.; Prieto-Castrillo, F.; Corchado, J.M. A game theory approach for cooperative control to improve data quality and false data detection in WSN. *Int. J. Robust Nonlinear Control* **2018**, *28*, 5087–5102. [CrossRef]
71. Adeel, A.; Gogate, M.; Farooq, S.; Ieracitano, C.; Dashtipour, K.; Larijani, H.; Hussain, A. A survey on the role of wireless sensor networks and IoT in disaster management. In *Geological Disaster Monitoring Based on Sensor Networks*; Springer: Berlin, Germany, 2019; pp. 57–66.
72. Poslad, S.; Middleton, S.E.; Chaves, F.; Tao, R.; Necmioglu, O.; Bügel, U. A semantic IoT early warning system for natural environment crisis management. *IEEE Trans. Emerg. Top. Comput.* **2015**, *3*, 246–257. [CrossRef]
73. Kodali, R.K.; Sahu, A. An IoT based weather information prototype using WeMos. In Proceedings of the IEEE 2nd International Conference on Contemporary Computing and Informatics (IC3I), Greater Noida, India , 14–17 December 2016; pp. 612–616.
74. Ayele, T.W.; Mehta, R. Air pollution monitoring and prediction using IoT. In Proceedings of the IEEE Second International Conference on Inventive Communication and Computational Technologies (ICICCT), Coimbatore, India, 20–21 April 2018; pp. 1741–1745.
75. Ghapar, A.A.; Yussof, S.; Bakar, A.A. Internet of Things (IoT) architecture for flood data management. *Int. J. Future Gener. Commun. Netw.* **2018**, *11*, 55–62. [CrossRef]
76. Abraham, M.T.; Satyam, N.; Pradhan, B.; Alamri, A.M. IoT-based geotechnical monitoring of unstable slopes for landslide early warning in the Darjeeling Himalayas. *Sensors* **2020**, *20*, 2611. [CrossRef]
77. Shaikh, S.F.; Hussain, M.M. Marine IoT: Non-invasive wearable multisensory platform for oceanic environment monitoring. In Proceedings of the IEEE 5th World Forum on Internet of Things (WF-IoT), Limerick, Ireland, 15–18 April 2019; pp. 309–312.
78. García, E.M.; Serna, M.Á.; Bermúdez, A.; Casado, R. Simulating a WSN-based wildfire fighting support system. In Proceedings of the IEEE International Symposium on Parallel and Distributed Processing with Applications, Sydney, Australia, 10–12 December 2008; pp. 896–902.
79. Kovács, Z.G.; Marosy, G.E.; Horváth, G. Case study of a simple, low power WSN implementation for forest monitoring. In Proceedings of the IEEE 12th Biennial Baltic Electronics Conference, Tallinn, Estonia, 4–6 October 2010; pp. 161–164.
80. Cantuña, J.G.; Bastidas, D.; Solórzano, S.; Clairand, J.M. Design and implementation of a Wireless Sensor Network to detect forest fires. In Proceedings of the IEEE Fourth international conference on eDemocracy & eGovernment (ICEDEG), Quito, Ecuador, 19–21 April 2017; pp. 15–21.
81. Yu, L.; Wang, N.; Meng, X. Real-time forest fire detection with wireless sensor networks. In Proceedings of the IEEE International Conference on Wireless Communications, Networking and Mobile Computing, Wuhan, China, 26 September 2005; Volume 2, pp. 1214–1217.
82. Li, Y.; Wang, Z.; Song, Y. Wireless sensor network design for wildfire monitoring. In Proceedings of the IEEE 6th World Congress on Intelligent Control and Automation, Dalian, China, 21–23 June 2006; Volume 1, pp. 109–113.

83. Díaz, S.E.; Pérez, J.C.; Mateos, A.C.; Marinescu, M.C.; Guerra, B.B. A novel methodology for the monitoring of the agricultural production process based on wireless sensor networks. *Comput. Electron. Agric.* **2011**, *76*, 252–265. [CrossRef]
84. Manolakos, E.S.; Logaras, E.; Paschos, F. Wireless sensor network application for fire hazard detection and monitoring. In Proceedings of the International Conference on Sensor Applications, Experimentation and Logistic, Athens, Greece, 25 September 2009; Springer: Berlin, Germany, 2009; pp. 1–15.
85. Liu, Y.; Liu, Y.; Xu, H.; Teo, K.L. Forest fire monitoring, detection and decision making systems by wireless sensor network. In Proceedings of the IEEE Chinese Control and Decision Conference (CCDC), Shenyang, China, 9–11 June 2018; pp. 5482–5486.
86. Ha, Y.g.; Kim, H.; Byun, Y.c. Energy-efficient fire monitoring over cluster-based wireless sensor networks. *Int. J. Distrib. Sens. Networks* **2012**, *8*, 460754. [CrossRef]
87. Zhang, J.; Li, W.; Han, N.; Kan, J. Forest fire detection system based on a ZigBee wireless sensor network. *Front. For. China* **2008**, *3*, 369–374. [CrossRef]
88. Jadhav, P.; Deshmukh, V. Forest fire monitoring system based on ZIG-BEE wireless sensor network. *Int. J. Emerg. Technol. Adv. Eng.* **2012**, *2*, 187–191.
89. Trivedi, K.; Srivastava, A.K. An energy efficient framework for detection and monitoring of forest fire using mobile agent in wireless sensor networks. In Proceedings of the IEEE International Conference on Computational Intelligence and Computing Research, Coimbatore, India, 18–20 December 2014; pp. 1–4.
90. Muhammad, K.; Ahmad, J.; Baik, S.W. Early fire detection using convolutional neural networks during surveillance for effective disaster management. *Neurocomputing* **2018**, *288*, 30–42. [CrossRef]
91. Kaur, H.; Sood, S.K. Fog-assisted IoT-enabled scalable network infrastructure for wildfire surveillance. *J. Netw. Comput. Appl.* **2019**, *144*, 171–183. [CrossRef]
92. Khalaf, O.I.; Abdulsahib, G.M.; Zghair, N.A.K. IOT fire detection system using sensor with Arduino. *AUS* **2019**, *26*, 74–78.
93. Jadon, A.; Omama, M.; Varshney, A.; Ansari, M.S.; Sharma, R. Firenet: A specialized lightweight fire & smoke detection model for real-time iot applications. *arXiv* **2019**, arXiv:1905.11922.
94. Roque, G.; Padilla, V.S. LPWAN Based IoT Surveillance System for Outdoor Fire Detection. *IEEE Access* **2020**, *8*, 114900–114909. [CrossRef]
95. Brito, T.; Pereira, A.I.; Lima, J.; Valente, A. Wireless Sensor Network for Ignitions Detection: an IoT approach. *Electronics* **2020**, *9*, 893. [CrossRef]
96. Khan, S.; Muhammad, K.; Mumtaz, S.; Baik, S.W.; de Albuquerque, V.H.C. Energy-efficient deep CNN for smoke detection in foggy IoT environment. *IEEE Internet Things J.* **2019**, *6*, 9237–9245. [CrossRef]
97. Muhammad, K.; Khan, S.; Elhoseny, M.; Ahmed, S.H.; Baik, S.W. Efficient fire detection for uncertain surveillance environment. *IEEE Trans. Ind. Inform.* **2019**, *15*, 3113–3122. [CrossRef]
98. Cui, F. Deployment and integration of smart sensors with IoT devices detecting fire disasters in huge forest environment. *Comput. Commun.* **2020**, *150*, 818–827. [CrossRef]
99. Khan, R.H.; Bhuiyan, Z.A.; Rahman, S.S.; Khondaker, S. A smart and cost-effective fire detection system for developing country: an IoT based approach. *Int. J. Inf. Eng. Electron. Bus.* **2019**, *11*, 16.
100. Kalatzis, N.; Avgeris, M.; Dechouniotis, D.; Papadakis-Vlachopapadopoulos, K.; Roussaki, I.; Papavassiliou, S. Edge computing in IoT ecosystems for UAV-enabled early fire detection. In Proceedings of the IEEE International Conference on Smart Computing (SMARTCOMP), Taormina, Italy, 18–20 June 2018; pp. 106–114.
101. Vimal, V.; Nigam, M.J. Forest Fire Prevention Using WSN Assisted IOT. *Int. J. of Eng. & Tech.* **2018**, *7*, 1317–1321.
102. Antunes, M.; Ferreira, L.M.; Viegas, C.; Coimbra, A.P.; de Almeida, A.T. Low-Cost System for Early Detection and Deployment of Countermeasures Against Wild Fires. In Proceedings of the IEEE 5th World Forum on Internet of Things (WF-IoT), Limerick, Ireland, 15–18 April 2019, pp. 418–423.
103. Jiang, H. Mobile Fire Evacuation System for Large Public Buildings Based on Artificial Intelligence and IoT. *IEEE Access* **2019**, *7*, 64101–64109. [CrossRef]
104. Xu, Y.H.; Sun, Q.Y.; Xiao, Y.T. An Environmentally Aware Scheme of Wireless Sensor Networks for Forest Fire Monitoring and Detection. *Future Internet* **2018**, *10*, 102. [CrossRef]
105. Lule, E.; Bulega, T.E. A scalable wireless sensor network (WSN) based architecture for fire disaster monitoring in the developing world. *Int. J. Comput. Netw. Inf. Secur.* **2015**, *7*, 40. [CrossRef]

106. Yang, Y.; Prasanna, R.; Yang, L.; May, A. Opportunities for WSN for facilitating fire emergency response. In Proceedings of the IEEE Fifth International Conference on Information and Automation for Sustainability, Colombo, Sri Lanka, 17–19 December 2010.
107. Kalatzis, N.; Routis, G.; Marinellis, Y.; Avgeris, M.; Roussaki, I.; Papavassiliou, S.; Anagnostou, M. Semantic interoperability for iot platforms in support of decision making: an experiment on early wildfire detection. *Sensors* **2019**, *19*, 528. [CrossRef]
108. Van Wagner, C.E. *Structure of the Canadian Forest fire Weather Index*; Environment Canada, Forestry Service: Ottawa, ON, USA, 1974; Volume 1333.
109. Hamadeh, N.; Karouni, A.; Daya, B.; Chauvet, P. Using correlative data analysis to develop weather index that estimates the risk of forest fires in Lebanon & Mediterranean: Assessment versus prevalent meteorological indices. *Case Stud. Fire Saf.* **2017**, *7*, 8–22.
110. Noble, I.; Gill, A.; Bary, G. McArthur's fire-danger meters expressed as equations. *Aust. J. Ecol.* **1980**, *5*, 201–203. [CrossRef]
111. de Groot, W.J.; Wang, Y. Calibrating the fine fuel moisture code for grass ignition potential in Sumatra, Indonesia. *Int. J. Wildland Fire* **2005**, *14*, 161–168. [CrossRef]
112. Sharples, J.; McRae, R.; Weber, R.; Gill, A.M. A simple index for assessing fire danger rating. *Environ. Model. Softw.* **2009**, *24*, 764–774. [CrossRef]
113. Agusti-Torra, A.; Raspall, F.; Remondo, D.; Rincón, D.; Giuliani, G. On the feasibility of collaborative green data center ecosystems. *Ad Hoc Networks* **2015**, *25*, 565–580. [CrossRef]
114. Hong, K.; Lillethun, D.; Ramachandran, U.; Ottenwälder, B.; Koldehofe, B. Mobile fog: A programming model for large-scale applications on the internet of things. In Proceedings of the Second ACM SIGCOMM Workshop on Mobile Cloud Computing, Hong Kong, China, 12 August 2013, pp. 15–20.
115. Banzi, M.; Shiloh, M. *Getting Started with Arduino: The Open Source Electronics Prototyping Platform*; Maker Media, Inc.: Sebastopol, CA, USA, 2014.
116. Faludi, R. *Building Wireless Sensor Networks: With ZigBee, XBee, Arduino, and Processing*; O'Reilly Media, Inc.: Sebastopol, CA, USA, 2010.
117. Farahani, S. *ZigBee Wireless Networks and Transceivers*; Newnes, Elsevier: Oxford, UK, 2011.
118. Osanaiye, O.; Chen, S.; Yan, Z.; Lu, R.; Choo, K.K.R.; Dlodlo, M. From cloud to fog computing: A review and a conceptual live VM migration framework. *IEEE Access* **2017**, *5*, 8284–8300. [CrossRef]
119. Ojo, M.O.; Giordano, S.; Procissi, G.; Seitanidis, I.N. A Review of Low-End, Middle-End, and High-End Iot Devices. *IEEE Access* **2018**, *6*, 70528–70554. [CrossRef]
120. Khutsoane, O.; Isong, B.; Abu-Mahfouz, A.M. IoT devices and applications based on LoRa/LoRaWAN. In Proceedings of the IEEE IECON 2017-43rd Annual Conference of the IEEE Industrial Electronics Society, Beijing, China, 29 October–1 November 2017; pp. 6107–6112.
121. GFB. Data Sets. 2020. Available online: https://www.fireservice.gr/en_US/synola-dedomenon (accessed on 27 May 2020).
122. ESA. Greece Suffers More Fires in 2007 than in Last Decade, Satellites Reveal. 2007. Available online: http://www.esa.int/esaCP/SEMMGZLPQ5F_index_0.html (accessed on 2 July 2020).
123. GSCP. Daily Fire Risk Map. 2019. Available online: https://www.civilprotection.gr/en/daily-fire-prediction-map (accessed on 7 December 2019).
124. Adafruit. Digital Relative Humidity and Temperature Sensor AM2302/DHT22. Available online: https://cdn-shop.adafruit.com/datasheets/Digital+humidity+and+temperature+sensor+AM2302.pdf (accessed on 3 July 2020).
125. Digi. XBee/XBee-PRO S2C Zigbee RF Module User Guide. Available online: https://tinyurl.com/y5posdyh (accessed on 3 July 2020).
126. Aduino. ARDUINO UNO REV3. 2020. Available online: https://store.arduino.cc/arduino-uno-rev3 (accessed on 4 July 2020).
127. Aduino. ARDUINO MEGA 2560 REV3. 2020. Available online: https://store.arduino.cc/arduino-mega-2560-rev3 (accessed on 4 July 2020).
128. Foundation, R.P. Raspberry Pi 3 Model B. 2020. Available online: https://www.raspberrypi.org/products/raspberry-pi-3-model-b/ (accessed on 4 July 2020).

129. Al-Sarawi, S.; Anbar, M.; Alieyan, K.; Alzubaidi, M. Internet of Things (IoT) communication protocols. In Proceedings of the IEEE 8th International Conference on Information Technology (ICIT), Jordan, 27–29 December 2017; pp. 685–690.
130. Glória, A.; Cercas, F.; Souto, N. Comparison of communication protocols for low cost Internet of Things devices. In Proceedings of the IEEE South Eastern European Design Automation, Computer Engineering, Computer Networks and Social Media Conference (SEEDA-CECNSM), Kastoria, Greece, 23–29 September 2017; pp. 1–6.
131. Yaqoob, I.; Hashem, I.A.T.; Mehmood, Y.; Gani, A.; Mokhtar, S.; Guizani, S. Enabling communication technologies for smart cities. *IEEE Commun. Mag.* **2017**, *55*, 112–120. [CrossRef]
132. Ali, A.I.; Partal, S.Z.; Kepke, S.; Partal, H.P. ZigBee and LoRa based Wireless Sensors for Smart Environment and IoT Applications. In Proceedings of the IEEE 1st Global Power, Energy and Communication Conference (GPECOM), Cappadocia, Turkey, 12–15 June 2019; pp. 19–23.
133. Mekki, K.; Bajic, E.; Chaxel, F.; Meyer, F. A comparative study of LPWAN technologies for large-scale IoT deployment. *ICT Express* **2019**, *5*, 1–7. [CrossRef]

MDPI
St. Alban-Anlage 66
4052 Basel
Switzerland
Tel. +41 61 683 77 34
Fax +41 61 302 89 18
www.mdpi.com

Energies Editorial Office
E-mail: energies@mdpi.com
www.mdpi.com/journal/energies